Applied Probability
Control
Economics
Information and Communication
Modeling and Identification
Numerical Techniques
Optimization

Applications of Mathematics

14

Edited by A. V. Balakrishnan

N. V. Krylov

Controlled Diffusion Processes

Translated by A. B. Aries

Springer-Verlag
New York Heidelberg Berlin

N. V. Krylov
Department of Mechanics and Mathematics
Moscow State University
Moscow 17234
USSR

Editor
A. V. Balakrishnan
Systems Science Department
University of California
Los Angeles, California 90024
USA

AMS Classification: 60H10, 60H20, 60J60, 93E20

Library of Congress Cataloging in Publication Data
Krylov, Nikolai Vladimirovich.
Controlled diffusion processes.
(Applications of mathematics; v. 14)
Translation of Upravliaemye protsessy diffuzionnogo tipa.
Bibliography: p.
1. Control theory. 2. Diffusion processes.
I. Balakrishnan, A. V. II. Title.
QA402.3.K74413 629.8′312 79-26631

Exclusively authorized English translation of the original Russian edition *Uproavlyaemye Protsessy Diffuzionnogo Tipa* published by Nauka, Moscow, 1977.

Printed in the United States of America.

9 8 7 6 5 4 3 2 1

ISBN 0-387-90461-1 Springer-Verlag New York
ISBN 3-540-90461-1 Springer-Verlag Berlin Heidelberg

Preface

Stochastic control theory is a relatively young branch of mathematics. The beginning of its intensive development falls in the late 1950s and early 1960s. During that period an extensive literature appeared on optimal stochastic control using the quadratic performance criterion (see references in Wonham [76]). At the same time, Girsanov [25] and Howard [26] made the first steps in constructing a general theory, based on Bellman's technique of dynamic programming, developed by him somewhat earlier [4].

Two types of engineering problems engendered two different parts of stochastic control theory. Problems of the first type are associated with multistep decision making in discrete time, and are treated in the theory of discrete stochastic dynamic programming. For more on this theory, we note in addition to the work of Howard and Bellman, mentioned above, the books by Derman [8], Mine and Osaki [55], and Dynkin and Yushkevich [12].

Another class of engineering problems which encouraged the development of the theory of stochastic control involves time continuous control of a dynamic system in the presence of random noise. The case where the system is described by a differential equation and the noise is modeled as a time continuous random process is the core of the optimal control theory of diffusion processes. This book deals with this latter theory.

The mathematical theory of the evolution of a system usually begins with a differential equation of the form

$$\dot{x}_t = f(t,x_t)$$

with respect to the vector of parameters x of such a system. If the function $f(t,x)$ can be measured or completely defined, no stochastic theory is needed. However, it is needed if $f(t,x)$ varies randomly in time or if the errors of measuring this vector cannot be neglected. In this case $f(t,x)$ is, as a rule, representable as $b(t,x) + \sigma(t,x)\dot{\xi}_t$ where b is a vector, σ is a matrix, and ξ_t is a random vector process. Then

$$\dot{x}_t = b(t,x_t) + \sigma(t,x_t)\dot{\xi}_t. \tag{1}$$

It is convenient to write the equation in the integral form

$$x_t = x_0 + \int_0^t b(t,x_s)\,ds + \int_0^t \sigma(s,x_s)\,d\xi_s, \tag{2}$$

where x_0 is the vector of the initial state of the system. We explain why Eq. (2) is preferable to Eq. (1). Usually, one tries to choose the vector of parameters x_t of the system in such a way that the knowledge of them at time t enables one to predict the probabilistic behavior of the system after time t with the same certainty (or uncertainty) to the same extent as would knowledge of the entire prior trajectory x_s ($s \leq t$). Such a choice of parameters is convenient because the vector x_t contains all the essential information about the system. It turns out that if x_t has this property, it can be proved under rather general conditions that the process ξ_t in (2) can be taken to be a Brownian motion process or, in other words, a Wiener process $\mathbf{w}_t$. The derivative of ξ_t is then the so-called "white noise," but, strictly speaking, ξ_t unfortunately cannot be defined and, in addition, Eq. (1) has no immediate meaning. However, Eq. (2) does make sense, if the second integral in (2) is defined as an Ito stochastic integral.

It is common to say that the process x_t satisfying Eq. (2) is a diffusion process. If, in addition, the coefficients b, σ of Eq. (2) depend also on some control parameters, we have a "controlled diffusion process."

The main subject matter of the book having been outlined, we now indicate how some parts of optimal control theory are related to the contents of the book.

Formally, the theory of deterministic control systems can be viewed as a special case of the theory of stochastic control. However, it has its own unique characteristics, different from those of stochastic control, and is not considered here. We mention only a few books in the enormous literature on the theory of deterministic control systems: Pontryagin, Boltyansky, Gamkrelidze, and Mishchenko [60] and Krassovsky and Subbotin [27].

A considerable number of works on controlled diffusion processes deal with control problems of linear systems of type (2) with a quadratic performance criterion. Besides Wonham [76] mentioned above, we can also mention Astrom [2] and Bucy and Joseph [7] as well as the literature cited in those books. We note that the control of such systems necessitates the construction of the so-called Kalman–Bucy filters. For the problems of the application of filtering theory to control it is appropriate to mention Lipster and Shiryayev [51].

Since the theory of linear control systems with quadratic performance index is represented well in the literature, we shall not discuss it here.

Control techniques often involve rules for stopping the process. A general and rather sophisticated theory of optimal stopping rules for Markov chains and Markov processes, developed by many authors, is described by Shiryayev [69]. In our book, problems of optimal stopping also receive considerable attention. We consider such problems for controlled processes with the help of the method of randomized stopping. It must be admitted, however, that our theory is rather crude compared to the general theory presented in [69] because of the fact that in the special case of controlled diffusion processes, imposing on the system only simply verifiable (and therefore crude) restrictions, we attempt to obtain strong assertions on the validity of the Bellman equation for the payoff function.

Concluding the first part of the Preface, we emphasize that in general the main aim of the book is to prove the validity of the Bellman differential equations for payoff functions, as well as to develop (with the aid of such equations) rules for constructing control strategies which are close to optimal for controlled diffusion processes.

A few remarks on the structure of the book may be helpful. The literature cited so far is not directly relevant to our discussion. References to the literature of more direct relevance to the subject of the book are given in the course of the presentation of the material, and also in the notes at the end of each chapter.

We have discussed only the main features of the subject of our investigation. For more detail, we recommend Section 1, of Chapter 1, as well as the introductions to Chapters 1–6.

The text of the book includes theorems, lemmas, and definitions, numeration of which is carried out throughout according to a single system in each section. Thus, the invoking of Theorem 3.1.5 means the invoking of the assertions numbered 5 in Section 1 in Chapter 3. In Chapter 3, Theorem 3.1.5 is referred to as Theorem 1.5, and in Section 1, simply as Theorem 5. The formulas are numbered in a similar way.

The initial constants appearing in the assumptions are, as a rule, denoted by K_i, δ_i. The constants in the assertions and in the proofs are denoted by the letter N with or without numerical subscripts. In the latter case it is assumed that in each new formula this constant is generally speaking unique to the formula and is to be distinguished from the previous constants. If we write $N = N(K_i, \delta_i, \ldots)$, this means that N depends only on what is inside the parentheses. The discussion of the material in each section is carried out under the same assumptions listed at the start of the section. Occasionally, in order to avoid the cumbersome formulation of lemmas and theorems, additional assumptions are given prior to the lemmas and theorems rather than in them.

Reading the book requires familiarity with the fundamentals of stochastic integral theory. Some material on this theory is presented in Appendix 1. The Bellman equations which we shall investigate are related to nonlinear partial differential equations. We note in this connection that we do not assume the reader to be familiar with the results related to differential equation theory.

In conclusion, I wish to express my deep gratitude to A. N. Shiryayev and all participants of the seminar at the Department of Control Probability of the Interdepartmental Laboratory of Statistical Methods of the Moscow State University for their assistance in our work in this book, and for their useful criticism of the manuscript.

N. V. Krylov

Contents

Notation

E_d denotes a Euclidean space of dimension d with a fixed orthonormal basis, x^i the ith coordinate of the point $x \in E_d$ $(i = 1,2,\ldots,d)$, $xy = (x,y)$ the scalar product of vectors x, $y \in E_d$, and $x^2 = xx$ the square of the length of x, $|x| = \sqrt{x^2}$.

$\sigma = (\sigma^{ij})$ denotes a matrix with elements σ^{ij}, σ^* the transpose of the matrix σ, σy the vector equal to the product of the matrix σ and the vector y, $x\sigma y = (x,\sigma y)$; $\operatorname{tr} a$ denotes the trace of the square matrix a, $\det a$ the determinant of the matrix a, and

$$\|\sigma\| = (\operatorname{tr} \sigma\sigma^*)^{1/2} = \left[\sum_{i,j} |\sigma^{ij}|^2\right]^{1/2},$$

where $\|\sigma\|$ is said to be the matrix norm of σ. $v_{x^i} = \partial v/\partial x^i$, $\operatorname{grad}_x v$ is the vector with the coordinates v_{x^i}, and $v_{x^i x^j} = \partial^2 v/\partial x^i \,\partial x^j$.

If σ is a matrix of dimension $d \times d_1$, and b is a d-dimensional vector, then

$$L^{\sigma,b}v(x) = \sum_{i,j=1} a^{ij} v_{x^i x^j}(x) + \sum_{i=1}^{d} b^i v_{x^i}(x),$$

where the matrix $(a^{ij}) = \frac{1}{2}\sigma\sigma^*$.

δ^{ij} denotes the Kronecker delta, $\chi_{x \in \Gamma} = \chi_\Gamma(x)$ is the indicator of a set Γ, that is, the function equal to unity on Γ and equal to zero outside Γ.

$x_{[0,T]}$ is the graph of the function x_t given on $[0,T]$.

$$\tau \wedge t = \min(\tau,t), \qquad \tau \vee t = \max(\tau,t),$$
$$t^+ = t_+ = \tfrac{1}{2}(|t| + t), \qquad t^- = t_- = \tfrac{1}{2}(|t| - t).$$

$\equiv$ means equal by definition.

$(\Omega,\mathscr{F},\mathsf{P})$ is a probability space; Ω denotes a set whose points are denoted by the ω with indices or without, $\mathscr{F}$ is a σ-algebra of the subset Ω, and P is a probability measure on $\mathscr{F}$.

$\mathsf{M}\xi$ denotes the mathematical expectation of a random variable ξ.

(A-a.s.) means almost surely on the set A.

(A-a.e.) means almost everywhere on the set A.

Introduction to the Theory of Controlled Diffusion Processes 1

The objective of Chapter 1 is to make the reader familiar with the general concepts, methods, and problems of the theory of controlled random diffusion processes. In Sections 1 and 2 we formulate the basic problems and indicate methods of solution. We omit rigorous proofs, our arguments being of a purely heuristic nature. In our opinion, the simplicity of the ideas at the basic level justifies our confidence in our ability to solve seemingly complex control problems. Many of the assertions in Sections 1 and 2 are proved rigorously further on in the book under appropriate conditions. In Section 3 we first explain heuristically how the notions of Section 2 can be applied in order to obtain estimates of various kinds, later making rigorous calculations for a specific example.

Beginning with Section 4 and throughout the sequel we try to adhere strictly to accepted norms of mathematical rigor.

Section 4 deals with the special case of 1-dimensional controlled processes, and Section 5 deals with the theory of optimal stopping rules for 1-dimensional controlled processes. In these two sections we distinguish a class of problems for which the main conclusions made in Sections 1 and 2 hold true.

In order to comprehend the material of Chapter 1, the reader ought to be familiar with some results related to the theory of stochastic integrals, the material, for instance, in Sections 1–6 of [23]. For the convenience of the reader, we summarize the fundamentals of stochastic integral theory in Appendix 1.

1. The Statement of Problems—Bellman's Principle—Bellman's Equation

We consider in an Euclidean space E_d (d is the dimension of the space) a random process x_t subject to the dynamic equations

$$x_t = x + \int_0^t b(\alpha_s, x_s)\, ds + \int_0^t \sigma(\alpha_s, x_s)\, d\mathbf{w}_s, \tag{1}$$

where $\sigma(\alpha, y)$, $b(\alpha, y)$ are given functions of $y \in E_d$ and of a control parameter α, x is the initial value of the process x_t, $\mathbf{w}_t$ is a d_1-dimensional Wiener process, and d_1 an integer. Naturally, $b(\alpha, y)$ is a d-dimensional vector: $b(\alpha, y) = (b^1(\alpha, y), \ldots, b^d(\alpha, y))$; $\sigma(\alpha, y)$ is a matrix of dimension $d \times d_1$, $\sigma(\alpha, y) = (\sigma^{ij}(\alpha, y))$.

We denote by A the set of admissible controls, i.e., values of the parameter α. Choosing appropriately the random process α_s with values in A we can obtain various solutions of Eq. (1). We can thereby "control" the process x_t considered. This gives rise to the question as to whether there exists a solution of Eq. (1) for the process $\{\alpha_s\}$ chosen, and if this is the case, whether it is unique, that is, whether the process $\{x_s\}$ can be defined uniquely after $\{\alpha_s\}$ has been chosen. We put off considering these questions and finding answers to them until later.

From the practical viewpoint, it is reasonable to consider that the values of the control process α_s at the time s are to be chosen on the basis of observations of the controlled process $\{x_t\}$ before time s. In other words, α_s has to be a function of the trajectory $x_{[0,s]} = \{(t, x_t) : 0 \le t \le s\} : \alpha_s = \alpha_s(x_{[0,s]})$.

Suppose that a cost functional is given for evaluating the control performance. Suppose also that on each trajectory of the process x_t from the time t to the time $t + \Delta t$ the "cost" is $f^{\alpha_t}(x_t)\, \Delta t + o(\Delta t)$, where $f^\alpha(y)$ is the given function.

Then the total loss for the control used is given by

$$\rho^\alpha = \int_0^\infty f^{\alpha_t}(x_t)\, dt,$$

for each individual trajectory of x_t. Corresponding to the strategy $\alpha = \{\alpha_s(x_{[0,s]})\}$, the "mean" loss for the process x_t with initial point x is given by

$$v^\alpha(x) = \mathsf{M} \int_0^\infty f^{\alpha_t}(x_t)\, dt.$$

This gives rise to the problem of finding a strategy $\alpha^0 = \{\alpha_s^0(x_{[0,s]})\}$ such that (for fixed x)

$$v^{\alpha^0}(x) = v(x) \equiv \inf_\alpha v^\alpha(x). \tag{2}$$

In the case where there exists no strategy α^0 (the lower bound in (2) is not attained), we may wish to construct for each $\varepsilon > 0$ a strategy $\alpha^\varepsilon = \{\alpha_s^\varepsilon(x_{[0,s]})\}$ such that $v^{\alpha^\varepsilon}(x) \le v(x) + \varepsilon$. The strategy α^ε is said to be ε-optimal for the (initial) point x; the strategy α^0 is said to be optimal for the point x. The function $v(x)$ is called a "performance function," determining which strategy will be of interest to us.

Setting aside temporarily the questions about the convergence of integrals which define ρ^α and $v^\alpha(x)$, we show how to solve the problem of finding $v(x)$ and α^ε using Bellman's principle.

Bellman's principle states that

$$v(x) = \inf_\alpha \mathsf{M}\left[\int_0^t f^{\alpha_s}(x_s)\,ds + v(x_t)\right] \tag{3}$$

for each $t \geq 0$.

To make the relation (3) clear we imagine that an interval of time t has elapsed from the initial instant of time. For the interval t the loss is given by

$$\int_0^t f^{\alpha_s}(x_s)\,ds. \tag{4}$$

The trajectory of the process has reached a point, say y, at the instant of time t. What is to be done after the time t to minimize the total loss? Since the quantity (4) has already been lost, we should find out how to minimize the loss occurring after the instant of time t. We note that increments of a Wiener process after the time t, together with the point y, define completely the behavior of a trajectory of x_s for $s \geq t$. The increments of $\mathbf{w}_s$ after the time t do not depend on the "past" prior to the time t, and they behave in the same way as the pertinent increments do after the initial instant of time. Furthermore, coefficients of (1) do not depend explicitly on time. Hence obtaining as a trajectory after the time t, say a function $\{y_s : s \geq t\}$, is equivalent to that of obtaining a function $\{y_{s+t} : s \geq 0\}$ as a trajectory after the initial instant of time (if we start from the point y). We note also that the loss $f^\alpha(x)$ does not depend on time explicitly. Therefore, we can solve the problem of minimizing the loss after the time t, assuming that the trajectory starts from a point y at the initial instant of time. It can readily be seen that the mean loss after the time t, under the condition $x_t = y$, cannot be smaller than $v(y)$ and can be made arbitrarily close to $v(y)$. Therefore, if we proceed after the time t in the optimal way, the mean loss during the entire control operation will be given by

$$\mathsf{M}\left[\int_0^t f^{\alpha_s}(x_s)\,ds + v(x_t)\right]. \tag{5}$$

In general, the quantity (5) is smaller than $v^\alpha(x)$. Even if there exists no optimal control, nevertheless we can get arbitrarily close to (5) by changing α_s for $s \geq t$. Therefore the lower bounds of (5) and $v^\alpha(x)$ with respect to all strategies coincide, as stated in (3).

Further, assume that v is a sufficiently smooth function. Applying Ito's formula to $v(x_t)$, we have[1]

$$v(x) = \mathsf{M}v(x_t) - \mathsf{M}\int_0^t L^{\alpha_s}(x_s)v(x_s)\,ds,$$

[1] Recall that $x_0 = x$.

where

$$L^\alpha(x) \equiv \sum_{i,j=1}^{d} a^{ij}(\alpha,x)\frac{\partial^2}{\partial x^i\,\partial x^j} + \sum_{i=1}^{d} b^i(\alpha,x)\frac{\partial}{\partial x^i},$$

$$a^{ij}(\alpha,x) = \frac{1}{2}\sum_{k=1}^{d_1} \sigma^{ik}(\alpha,x)\sigma^{jk}(\alpha,x).$$

Therefore, it follows from Bellman's principle that

$$0 = \inf_\alpha \left\{ \mathsf{M}\left[\int_0^t f^{\alpha_s}(x_s)\,ds + v(x_t)\right] - v(x)\right\}$$
$$= \inf_\alpha \mathsf{M} \int_0^t [f^{\alpha_s}(x_s) + L^{\alpha_s}v(x_t)]\,ds,$$

where we divide all the expressions by t and also we let t tend to zero, obtaining thereby the equation

$$\inf_{\alpha\in A}[L^\alpha(x)v(x) + f^\alpha(x)] = 0. \tag{6}$$

Equation (6) is known as Bellman's differential equation for the optimal control problem under consideration.

We started from $v^\alpha(x)$ and arrived at $v(x)$ and, then, at Eq. (6). We can do it in the backward direction, namely, we can show that if a function w satisfies Bellman's equation, the function w coincides with v. In doing so we can also see how to find optimal and ε-optimal strategies with the aid of Bellman's equation.

For the function w

$$\inf_{\alpha\in A}[L^\alpha w + f^\alpha] = 0. \tag{7}$$

Therefore, $-L^\alpha w \le f^\alpha$; using Ito's formula we obtain

$$w(x) = \mathsf{M}w(x_t) - \mathsf{M}\int_0^t L^{\alpha_s}(x_s)w(x_s)\,ds \le \mathsf{M}w(x_t) + \mathsf{M}\int_0^t f^{\alpha_s}(x_s)\,ds.$$

We pass to the limit in the last inequality as $t \to \infty$, assuming that for the function $w(x)$ and the process x_t for any strategy α as $t \to \infty$

$$\mathsf{M}w(x_t) \to 0, \qquad \mathsf{M}\int_0^t f^{\alpha_s}(x_s)\,ds \to \mathsf{M}\int_0^\infty f^{\alpha_s}(x_s)\,ds.$$

This yields[2]

$$w(x) \le \mathsf{M}\int_0^\infty f^{\alpha_s}(x_s)\,ds = v^\alpha(x), \qquad w(x) \le \inf v^\alpha(x) = v(x).$$

We show, in turn, that $w(x) \ge v(x)$. We assume that for each x the lower bound in (7) can be attained for some $\alpha = \alpha^0(x)$. We assume further that there

[2] We note that a sufficient condition for the inequality $w \le v$ to be satisfied is that the left-hand side of (7) be nonnegative.

exists a solution of the equation

$$x_t^0 = x + \int_0^t \sigma(\alpha^0(x_s^0), x_s^0)\, d\mathbf{w}_s + \int_0^t b(\alpha^0(x_s^0), x_s^0)\, ds.$$

Since $-L^{\alpha^0(x)}(x)w(x) = f^{\alpha^0(x)}(x)$, by Ito's formula

$$w(x) = \mathsf{M}w(x_t^0) - \mathsf{M}\int_c^t L^{\alpha^0(x_s^0)}(x_s^0)w(x_s^0)\, ds$$

$$= \mathsf{M}w(x_t^0) + \mathsf{M}\int_0^t f^{\alpha^0(x_s^0)}(x_s^0)\, ds,$$

from which it follows for a strategy $\alpha^0 = \{\alpha^0(x_s)\}$ as $t \to \infty$ that

$$w(x) = v^{\alpha^0}(x) \geq v(x).$$

Therefore, $w(x) = v(x)$ and α^0 is the optimal strategy (for any point x).

In the case where the lower bound in (7) cannot be attained, in order to prove the inequality $w(x) \geq v(x)$ we take a function $g(x) > 0$ such that for an arbitrary x and a strategy α

$$\mathsf{M}\int_0^\infty g(x_t)\, dt \leq 1.$$

For $\varepsilon > 0$ we find the function $\alpha^\varepsilon(x)$ from the condition

$$L^{\alpha^\varepsilon(x)}(x)w(x) + f^{\alpha^\varepsilon(x)}(x) \leq \varepsilon g(x)$$

and consider the strategy $\alpha^\varepsilon = \{\alpha^\varepsilon(x_t)\}$. Let x_t^ε be a process corresponding to the strategy α^ε and starting from a point x. By Ito's formula

$$w(x) = \mathsf{M}w(x_t^\varepsilon) - \mathsf{M}\int_0^t L^{\alpha^\varepsilon(x_s^\varepsilon)}(x_s^\varepsilon)w(x_s^\varepsilon)\, ds$$

$$\geq \mathsf{M}w(x_t^\varepsilon) + \mathsf{M}\int_0^t f^{\alpha^\varepsilon(x_s^\varepsilon)}(x_s^\varepsilon)\, ds - \varepsilon.$$

Therefore, $w(x) \geq v^{\alpha^\varepsilon}(x) - \varepsilon \geq v(x) - \varepsilon$, i.e., we have again that $v(x) = w(x)$ and the strategy α^ε is ε-optimal (for any point x).

In fact, Bellman's equation provides a technique for finding the performance function $v(x)$ as well as optimal and ε-optimal strategies.

We note that the ε-optimal strategies constructed above determine the choice of a control at time t on the basis of the instantaneous value of x_t rather than the entire segment of the trajectory $x_{[0,t]}$. In other words, these strategies are characterized by the fact that the control at a point y is always the same, namely, it is equal to $\alpha^\varepsilon(y)$, regardless in what way and at which instant of time the trajectory has arrived at the point y.

Intuitive reasoning suggests that we could have restricted ourselves to the aforementioned strategies from the very beginning. Indeed, the knowledge of how the trajectory has arrived at the point y cannot help us, by any means, to influence the "future" behavior of the trajectory x_t because increments of the process $\mathbf{w}_t$, which determine this behavior, do not depend on the "past." Furthermore, the "cost" we have to pay after the trajectory has arrived at the point y is not a function of the preceding segment of the trajectory. If it is therefore advantageous, for any reason, to use a control at least once after the trajectory has reached the point y, it will be advantageous, for the same reasons, to use this control each time when the trajectory reaches y.

A strategy of the form $\{\alpha(x_t)\}$ is said to be Markov since its corresponding process is Markov; the behavior of the latter after the instant of time t depends only on the position at the time t and does not depend on the "prehistory."

Therefore, one need seek optimal and ε-optimal strategies only among Markov strategies. However, it turns out that in order to prove the validity of our heuristic arguments justifying Bellman's equation, we need to consider all possible strategies. For example, in explaining Bellman's principle it was important that the controls applied after an instant of time t did not depend on the preceding ones. It becomes possible to apply various controls at a point y before time t as well as after time t.

It turns out that it is sometimes convenient to broaden the notion of a strategy. Taking a strategy $\alpha = \alpha_t(x_{[0,t]})$ and solving Eq. (1), we obtain a process x_t dependent on the trajectory $\mathbf{w}_t$:$x_t = x_t(\mathbf{w}_{[0,t]})$. Putting this solution into the expression for α_t, we write α_t as $\beta_t(\mathbf{w}_{[0,t]})$. It is clearly desirable to include all processes $\beta_t = \beta_t(\mathbf{w}_{[0,t]})$ with values in A in the set of the strategies considered, and to determine the resulting controlled process we solve in fact the equation

$$x_t = x + \int_0^t \sigma^{\beta_s(\mathbf{w}_{[0,s]})}(x_s)\,d\mathbf{w}_s + \int_0^t b^{\beta_s(\mathbf{w}_{[0,s]})}(x_s)\,ds.$$

It can be seen that admitting strategies of the form $\beta_t(\mathbf{w}_{[0,t]})$ is equivalent to being allowed to choose a control on the basis of observations of the process $\mathbf{w}_t$, the observations of the process $\mathbf{w}_t$ providing us simultaneously with all the data about the process x_t. It can also be seen that including these strategies in the set of all admissible strategies we leave intact the preceding arguments concerning Bellman's equation as well as Markov strategies. In particular, the inclusion of new strategies does not decrease the performance function, and can, thus, be approximated with the aid of Markov strategies.

We make now a few remarks on the structure of Bellman's equation which may simplify the notation in some cases.

Equation (6) can be written in a more expanded form as

$$\inf_{\alpha \in A}\left[\sum_{i,j=1}^{d} a^{ij}(\alpha,x)v_{x^i x^j}(x) + \sum_{i=1}^{d} b^i(\alpha,x)v_{x^i}(x) + f^\alpha(x)\right] = 0, \tag{8}$$

where x assumes values from that space in which the controlled process (1) assumes its values, and $b^i(\alpha,x)$ is the velocity of the deterministic component of the motion of the ith component of the process, with the position of the process at a point x and with a control α applied. The matrix $a(\alpha,x) = (a^{ij}(\alpha,x)) = \frac{1}{2}\sigma(\alpha,x)\sigma^*(\alpha,x)$ is symmetric; i.e., $a^* = \frac{1}{2}(\sigma\sigma^*)^* = \frac{1}{2}\sigma\sigma^* = a$ and nonnegative definite, i.e., $(a\lambda,\lambda) = \frac{1}{2}(\sigma^*\lambda,\sigma^*\lambda) = \frac{1}{2}|\sigma^*\lambda|^2 \geq 0$ and, in addition, characterizes the diffusion component of the process. Furthermore, as can easilv be seen,

$$a^{dd} = \frac{1}{2} \sum_{k=1}^{d_1} (\sigma^{dk})^2;$$

therefore $a^{dd}(\alpha,x) = 0$ for each x, α if and only if $\sigma^{dk}(\alpha,x) = 0$ for all $k = 1, \ldots, d_1$, x, α, i.e., if the last coordinate of the process has no diffusion component. In this case $a^{id}(\alpha,x) = 0$, $a^{di}(\alpha,x) = 0$; the first summand in (8) becomes

$$\sum_{i,j=1}^{d-1} a^{ij}(\alpha,x) v_{x^i x^j}(x).$$

1. Exercise

Let $\sigma_1(\alpha,x)$, $\sigma_2(\alpha,x)$ be two square matrices of dimension $d \times d$. To construct a new matrix $\sigma_3(\alpha,x)$ having d rows and $2d$ columns, suppose that the first d columns of the matrix $\sigma_3(\alpha,x)$ form a matrix $\sigma_1(\alpha,x)$ and that the last d columns form a matrix $\sigma_2(\alpha,x)$. We denote by $a_i(\alpha,x)$ the matrices corresponding to $\sigma_i(\alpha,x)\sigma_i^*(\alpha,x)$. Prove that

$$a_3(\alpha,x) = a_1(\alpha,x) + a_2(\alpha,x).$$

2. Examples of the Bellman Equations—The Normed Bellman Equation

The examples given in Section 2 are intended to show that in spite of the rather specialized form of the controlled system (1.1) and the performance functional

$$\rho^\alpha = \int_0^\infty f^{\alpha_t}(x_t)\, dt,$$

many stochastic control problems can be reduced to the problem examined in Section 1.

Suppose that we need to maximize $\mathsf{M}\rho^\alpha$ instead of minimizing it. Since

$$-\sup_\alpha \mathsf{M} \int_0^\infty f^{\alpha_t}(x_t)\, dt = \inf_\alpha \mathsf{M} \int_0^\infty [-f^{\alpha_t}(x_t)]\, dt,$$

the payoff function

$$v_1(x) \equiv \sup_\alpha \mathsf{M} \int_0^\infty f^{\alpha_t}(x_t)\, dt$$

satisfies the following Bellman equation:

$$0 = \inf_{\alpha \in A} [L^\alpha(-v_1) + (-f^\alpha)] = -\sup_{\alpha \in A} [L^\alpha v_1 + f^\alpha],$$

which yields

$$\sup_{\alpha \in A} [L^\alpha v_1 + f^\alpha] = 0. \tag{1}$$

We note that in the minimization problem we derived Bellman's equation, (1.6), which contained inf; in the maximization problem (1.6) contains sup.

In some cases, in order to ensure the existence of the functional ρ^α, i.e., the convergence of the corresponding integral, one need introduce "discounting." For example, we consider the problem of finding

$$v_2(x) \equiv \sup_\alpha \mathsf{M} \int_0^\infty e^{-t} f^{\alpha_t}(x_t)\, dt, \tag{2}$$

where x_t is a solution of Eq. (1.1). If $f^\alpha(x)$ is a bounded function, the integral in (2) exists. The multiplicative factor e^{-t} is said to be a discounting multiplicative factor and can be interpreted as the probability that the trajectory of the process does not vanish before the instant of time t, and in fact we obtain the payoff $f^{\alpha_t}(x_t)\,dt$ during the interval of time from t up to $t + dt$.

We show how to reduce the last problem to the previous one. For $x \in E_d$, $y \in (-\infty, \infty)$, we put $f^\alpha(x, y) = e^{-y} f^\alpha(x)$ and consider in $E_d \times E_1$ a control process whose first d coordinates move according to Eq. (1.1) and whose last coordinate is subject to the following "equation":

$$y_t = y + t = y + \int_0^t 1\, ds.$$

Let

$$v_2(x, y) = \sup_\alpha \mathsf{M} \int_0^\infty f^{\alpha_t}(x_t, y_t)\, dt.$$

It is seen that $v_2(x, y) = e^{-y} v_2(x)$ and if our conclusions about the Bellman equation are valid, then

$$\sup_{\alpha \in A} \left[\sum_{i,j=1}^d a_{ij}(\alpha, x) v_{2x^i x^j}(x, y) + \sum_{i=1}^d b^i(\alpha, x) v_{2x^i}(x, y) + v_{2y}(x, y) + f^\alpha(x, y) \right] = 0.$$

Putting $v_2(x, y) = e^{-y} v_2(x)$, $f^\alpha(x, y) = e^{-y} f^\alpha(x)$ into the last equality and canceling out by e^{-y}, we find

$$\sup_{\alpha \in A} \left[\sum_{i,j=1}^d a_{ij}(\alpha, x) v_{2x^i x^j}(x) + \sum_{i=1}^d b^i(\alpha, x) v_{2x^i}(x) - v_2(x) + f^\alpha(x) \right] = 0. \tag{3}$$

By the same token, for

$$v_3(x) = \sup_\alpha \mathsf{M} \int_0^\infty f^{\alpha_t}(x_t) \exp\left[-\int_0^t c^{\alpha_s}(x_s)\, ds \right] dt,$$

where $c^\alpha(x)$ is the given function of (α, x), we arrive at the equation

$$\sup_{\alpha \in A} [L^\alpha v_3 - c^\alpha v_3 + f^\alpha] = 0, \tag{4}$$

with the help of the equation

$$y_t = y + \int_0^t c^{\alpha_s}(x_s)\, ds$$

for the last coordinate y_t.

We see that introducing the "discounting" factor $\exp[-\int_0^t c^{\alpha_s}(x_s)\,ds]$ results in the appearance in Eq. (4) of the expression $-c^\alpha v_3$, which was absent in (1).

1. Exercise

Let $b_1(\alpha,x)$ be a d_1-dimensional vector. Introducing the additional coordinate

$$y_t = y + \int_0^t b_1(\alpha_s,x_s)\,d\mathbf{w}_s - \frac{1}{2}\int_0^t |b_1(\alpha_s,x_s)|^2\,ds,$$

explain why

$$\sup_\alpha \mathsf{M}\int_0^\infty f^{\alpha_t}(x_t)\exp\left\{\int_0^t b_1(\alpha_s,x_s)\,d\mathbf{w}_s - \frac{1}{2}\int_0^t |b_1(\alpha_s,x_s)|^2\,ds\right\}dt$$

satisfies the equation

$$\sup_\alpha\left[L^\alpha u + \sum_{i=1}^d (\sigma b_1)^i u_{x^i} + f^\alpha\right] = 0.$$

2. Exercise

Using Ito's formula, show that if a function $\alpha(x)$ furnishes the upper bound in (1) (or in (3) or in (4)) for each x, the strategy $\{\alpha(x_t)\}$ is optimal in the corresponding problem.

An important class of optimal control is the problems of optimal stopping, in which one needs to choose, in addition to the strategy α, a random stopping time τ such that the mean of the functional

$$\int_0^\tau f^{\alpha_t}(x_t)\,dt + g(x_\tau) \tag{5}$$

is maximized.

Surely, having made the decision to stop at time τ, we stop observing the process after this time τ. Hence we have to make the decision whether to stop the process at time t or not, on the basis of observation of the process only up to the time t. In other words, we shall treat Markov times as stopping times. On the set where $\tau = \infty$, we assume, as usual, that $g(x_\tau) = 0$, so that if stopping does not occur, we obtain

$$\int_0^\infty f^{\alpha_t}(x_t)\,dt.$$

It turns out that the problem of optimal stopping can be reduced to the problem mentioned above via the technique of randomized stopping. We will illustrate this by an example of stopping a Wiener process (in (1.1) $d_1 = d$, σ is a unit matrix, $b = 0$). Furthermore, we assume that $f^\alpha(x)$ does not depend on α: $f^\alpha(x) = f(x)$.

Defining the nonnegative process $r_t = r(\mathbf{w}_{[0,t]})$, we prescribe a rule for randomized stopping of the process $\mathbf{w}_t$. Let the trajectory $\mathbf{w}_t$ stop with probability $r_t\,\Delta t + o(\Delta t)$ during the interval of time from t up to $t + \Delta t$, under the condition that it does not stop before the time indicated. Stopping $\mathbf{w}_t$ at the instant of time t we have (compare with (5))

$$\rho_t = \int_0^t f(x + \mathbf{w}_s)\,ds + g(x + \mathbf{w}_t).$$

Then, as is easily seen, the probability that stopping does not occur on an individual trajectory before the time t is equal to $\exp(-\int_0^t r_s\,ds)$ (in particular, $\exp(-\int_0^\infty r_s\,ds)$ for $t = \infty$). Therefore, the probability that stopping does in fact, occur in the interval $(t, t + \Delta t)$, is given by

$$r_t \exp\left(-\int_0^t r_s\,ds\right)\Delta t + o(\Delta t).$$

Hence the expected payoff on an individual trajectory is given by

$$\int_0^\infty \rho_t r_t \exp\left[-\int_0^t r_s\,ds\right]dt + \int_0^\infty f(x + \mathbf{w}_s)\,ds \exp\left[-\int_0^\infty r_s\,ds\right]$$

$$= \int_0^\infty g(x + \mathbf{w}_t) r_t \exp\left[-\int_0^t r_s\,ds\right]dt$$

$$- \int_0^\infty \left(\int_0^t f(x + \mathbf{w}_s)\,ds\right) d \exp\left[-\int_0^t r_s\,ds\right]$$

$$+ \int_0^\infty f(x + \mathbf{w}_s)\,ds \exp\left[-\int_0^\infty r_s\,ds\right].$$

Integrating the last expression by parts we have

$$\int_0^\infty [f(x + \mathbf{w}_t) + r_t g(x + \mathbf{w}_t)] \exp\left\{-\int_0^t r_s\,ds\right\}dt.$$

Therefore, using the above technique of randomized stopping, we obtain in the mean with the aid of the process r_t

$$\mathsf{M} \int_0^\infty f^{r_t}(x + \mathbf{w}_t) \exp\left[-\int_0^t c^{r_s}\,ds\right]dt,$$

where $f^r = f + rg$, $c^r = r$.

It can be readily seen that if it is really advantageous to stop the trajectory $x + w_t$ at time t with nonzero intensity of r_t, then one can as well stop the process at time t with probability one. Therefore, "randomized" stopping cannot give a better result than "nonrandomized" stopping. On the other hand, an instantaneous stopping rule can be approximated in a reasonable sense by means of randomized stopping, if we increase the stopping intensity of r_t after the time τ. Hence

$$v_4(x) \equiv \sup_\tau \mathsf{M}\left[\int_0^\tau f(x + \mathbf{w}_t)\,dt + g(x + \mathbf{w}_\tau)\right]$$

$$= \sup_r \mathsf{M} \int_0^\infty f^{r_t}(x + \mathbf{w}_t) \exp\left[-\int_0^t c^{r_s}\,ds\right]dt.$$

Similarly, in the general case

$$
\begin{aligned}
v_5(x) &\equiv \sup_{\alpha,\tau} \mathsf{M}\left[\int_0^\tau f^{\alpha_t}(x_t)\,dt + g(x_\tau)\right] \\
&= \sup_{\alpha,r} \mathsf{M}\int_0^\infty f^{\alpha_t,r_t}(x_t)\exp\left[-\int_0^t c^{\alpha_s,r_s}\,ds\right]dt,
\end{aligned}
\tag{6}
$$

where $f^{\alpha,r} = f^\alpha + rg$, $c^{\alpha,r} = r$. If we regard the pair (α,r) as one parameter, we can easily notice the similarity between functions v_5 and v_3. Hence, we write for v_5 a Bellman equation which is similar to (4). We note, however, that v_5 does not satisfy a Bellman equation in many cases.

The point is that the functions $f^{\alpha,r}$ and $c^{\alpha,r}$ in (6) are not bounded as functions of one of the control parameters r. With this in mind, let us go back to the derivation of the Bellman equation (1.6).

Taking the last limit before (1.6), as $t \to \infty$, we assumed that

$$\frac{1}{t}\int_0^t [f^{\alpha_s}(x_s) + L^{\alpha_s}v(x_s)]\,ds$$

is close to $f^{\alpha_0}(x) + L^{\alpha_0}v(x)$ in some sense uniformly with respect to strategies α. In order this to be the case, it is necessary at least that the process x_s does not move far away from the initial point x in a short interval of time. We therefore assume that the coefficients $\sigma(\alpha,x)$, $b(\alpha,x)$ are bounded, and furthermore the function $f^\alpha(x)$ is bounded as well. As we saw above, the scheme which involves "discounting" reduces to a scheme without "discounting" if we introduce an additional coordinate y_t and the equation

$$y_t = y + \int_0^t c^{\alpha_s}(x_s)\,ds.$$

For the scheme with "discounting" we have therefore the requirement that $c^\alpha(x)$ be bounded.

Therefore, if we wish to consider the controlled process

$$x_t = x + \int_0^t \sigma(\alpha_s,x_s)\,d\mathbf{w}_s + \int_0^t b(\alpha_s,x_s)\,ds$$

with the payoff function v_3 and unbounded σ, b, c, f, we have to apply, in general, methods other than those given in Section 1.

One of the methods to be applied is based on random change of time and enables us to go from the unbounded σ, b, c, f to the bounded σ, b, c, f. We take a positive function $m(\alpha,x)$ for which the expressions $\sigma_1 \equiv \sqrt{m}\,\sigma$, $b_1 \equiv mb$, $c_1 \equiv mc$, $f_1 \equiv mf$ are bounded functions of (α,x). Let

$$\psi_t^\alpha = \int_0^t m^{-1}(\alpha_s,x_s)\,ds$$

and suppose that because of some special features of a controlled process, we can consider only those strategies α for which $\psi_t^\alpha < \infty$ for each t, $\psi_\infty^\alpha = \infty$. Then the function τ_t^α, the inverse of the function ψ_t^α, will be defined on $[0,\infty)$ and $\tau_\infty^\alpha = \infty$.

Replacing the variable $t = \tau_s^\alpha$ and assuming that $\beta_s = \alpha_{\tau_s^\alpha}$, $z_s = x_{\tau_s^\alpha}$, we have

$$\int_0^\infty \exp\left\{-\int_0^t c^{\alpha_s}(x_s)\,ds\right\} f^{\alpha_t}(x_t)\,dt = \int_0^\infty \exp\left\{-\int_0^s c_1^{\beta_u}(z_u)\,du\right\} f_1^{\beta_s}(z_s)\,ds,$$

$$z_s = x + \int_0^s \sigma_1(\beta_u,z_u)\,d\xi_u + \int_0^s b_1(\beta_u,z_u)\,du,$$

where the process ξ_u is given by the formula

$$\xi_u = \int_0^{\tau_u^\alpha} m^{-1/2}(\alpha_p,x_p)\,d\mathbf{w}_p.$$

By [23, Chapter 1, §3, Theorem 3], the process ξ_u is Wiener. Therefore, it is quite likely that

$$v_3(x) = \sup_\beta \mathsf{M} \int_0^\infty f_1^{\beta_s}(z_s) \exp\left[-\int_0^s c_1^{\beta_u}(z_u)\,du\right] ds$$

and v_3 satisfies the equation

$$\sup_{\beta \in A}\left[\sum_{i,j=1}^d a_1^{ij}(\beta,x)v_{3x^ix^j} + \sum_{i=1}^d b_1^i(\beta,x)v_{3x^i} - c_1^\beta(x)v_3 + f_1^\beta(x)\right] = 0,$$

where $a_1 = \frac{1}{2}\sigma_1\sigma^* = ma$. In other words,

$$\sup_{\alpha \in A} m(\alpha,x)\left[\sum_{i,j=1}^d a^{ij}(\alpha,x)v_{3x^ix^j} + \sum_{i=1}^d b^i(\alpha,x)v_{3x^i} - c^\alpha(x)v_3 + f^\alpha(x)\right] = 0, \quad (7)$$

which is known as "normed" Bellman equation, and which differs from the Bellman equation (4) by the presence of the normalizing multiplier $m(\alpha,x)$. We have deduced Eq. (7) to take care of the unboundedness of the functions σ, b, c, f. If these functions σ, b, c, f are bounded, we can take as the normalizing multiplier $m(\alpha,x)$ a function identically equal to unity. Then $\sqrt{m}\sigma$, mb, mc, mf are bounded, and (7) is equivalent to (4).

In order to see what the normed Bellman equation ensures for v_5, we assume that the functions $f^\alpha(x)$ and $g(x)$ in (6) are bounded. Let $m(\alpha,r,x) = (1 + r)^{-1}$. As can readily be seen, the functions $\sqrt{m(\alpha,r,x)}\sigma(\alpha,x)$, $m(\alpha,r,x)b(\alpha,x)$, $m(\alpha,r,x)c^{\alpha,r}(x)$, $m(\alpha,r,x)f^{\alpha,r}(x)$ are bounded, and in conjunction with (6), (7)

$$\sup_{\alpha \in A, r \ge 0} \frac{1}{1+r}\left[\sum_{i,j=1}^d a^{ij}(\alpha,x)v_{5x^ix^j} + \sum_{i=1}^d b^i(\alpha,x)v_{5x^i} - rv_5 + f^\alpha(x) + rg(x)\right] = 0.$$

Assuming $\varepsilon = 1/(1 + r)$, we obtain

$$\sup_{\varepsilon \in [0,1]} \sup_{\alpha \in A} [\varepsilon(L^\alpha v_5 + f^\alpha) + (1 - \varepsilon)(g - v_5)] = 0,$$

$$\sup_{\varepsilon \in [0,1]}\left[\varepsilon \sup_{\alpha \in A}(L^\alpha v_5 + f^\alpha) + (1 - \varepsilon)(g - v_5)\right] = 0. \quad (8)$$

Noting that the expression in square brackets is a linear function of ε, we can easily prove that (8) is equivalent to the combination of the following conditions:

$$\sup_{\alpha\in A}(L^\alpha v_5 + f^\alpha) \le 0, \qquad g - v_5 \le 0,$$

$$\sup_{\alpha\in A}(L^\alpha v_5 + f^\alpha) = 0 \quad \text{for } g - v_5 < 0.$$

We have thus obtained three relations for finding the payoff function v_5 in the optimal stopping problem. We can easily write these relations as one equality if we compute the upper bound with respect to ε in (8):

$$g - v_5 + \sup_\alpha [L^\alpha v_5 + f^\alpha + v_5 - g]_+ = 0.$$

The last equation is said to be Bellman's equation for the problem of optimal stopping of a controlled process.

In the examples given above, the controlled process (1.1) was considered in the entire space E_d. In some cases, however, we regard as essential only the values of the process (1.1) before an instant of time τ_D of first departure of a trajectory of (1.1) from a domain D. For example, let $\mathsf{M}\tau_D$ be either the largest value or the smallest one. The problem of finding the minimum for

$$\mathsf{M}\left[\int_0^{\tau_D} f^{\alpha_t}(x_t)\,dt + g(x_{\tau_D})\right]$$

is a more general problem, which can be reduced to the original problem in the following way. We change $\sigma(\alpha,x)$, $b(\alpha,x)$ so that they remain the same in the domain D and are equal to zero outside D. We write the new $\sigma(\alpha,x)$, $b(\alpha,x)$ thus obtained as $\tilde{\sigma}(\alpha,x)$, $\tilde{b}(\alpha,x)$. Furthermore, we assume $\tilde{f}^\alpha(x) = f^\alpha(x)$ for $x \in D$, $\tilde{f}^\alpha(x) = g(x)$ for $x \notin D$, $c(x) = 0$ for $x \in D$, $c(x) = 1$ for $x \notin D$. Then the process, once having entered the boundary ∂D of the domain D, will remain permanently at a point x_{τ_D}. It is seen that

$$v_6(x) \equiv \inf_\alpha \mathsf{M}\left[\int_0^{\tau_D} f^{\alpha_t}(x_t)\,dt + g(x_{\tau_D})\right]$$

$$= \inf_\alpha \mathsf{M}\int_0^\infty \tilde{f}^{\alpha_t}(\tilde{x}_t)\exp\left[-\int_0^t c(\tilde{x}_s)\,ds\right]dt,$$

where $\tilde{x}_t$ is the solution of (1.1) with the modified $\sigma(\alpha,x)$, $b(\alpha,x)$.

The Bellman equation for $v_6(x)$ yields

$$\inf_{\alpha\in A}\left[\sum_{i,j=1}^d \tilde{a}^{ij}(\alpha,x)v_{6x^ix^j}(x) + \sum_{i=1}^d \tilde{b}^i(\alpha,x)v_{6x^i}(x) - c(x)v_6(x) + \tilde{f}^\alpha(x)\right] = 0,$$

where $(\tilde{a}^{ij}(\alpha,x)) = \frac{1}{2}\tilde{\sigma}(\alpha,x)\tilde{\sigma}^*(\alpha,x)$. In particular,

$$\inf_{\alpha\in A}[L^\alpha v_6 + f^\alpha] = 0 \quad \text{in } D, \qquad -v_6 + g = 0 \quad \text{on } \partial D.[3] \tag{9}$$

We see that the problem of control in the domain D with "discounting" $c^\alpha(x)$, as well as the problem of optimal stopping before first exit from D, can be investigated in the same way.

[3] $v_6 = g$ even everywhere outside D.

Finally, another class of optimal control problems, which is included in the scheme considered, is that involving time varying stochastic equations.

Rather frequently, time appears in the coefficients of Eq. (1.1) in the explicit manner. For example, let the controlled process x_t start at time r at a point x, and let this process satisfy the equation

$$x_t = x + \int_r^t b(\alpha_s,x_s,s)\,ds + \int_r^t \sigma(\alpha_s,x_s,s)\,d\mathbf{w}_s, \qquad t \geq r.$$

We assume that we need to minimize

$$v^\alpha(x,r) = \mathsf{M} \int_r^\infty f^{\alpha_t}(x_t,t)\,dt.$$

Let $v_7(x,r) = \inf_\alpha v^\alpha(x,r)$. We take the direct product $E_d \times E_1$ and consider E_1 as a time axis. The process (x_t,t) is at a point (x,r) at an initial instant of *time of control*, and this process (x_t,t) will move to a point $(x_{r+u}, r+u)$ with unit velocity along the time axis during a *control interval* u.

Let $y_u = (x_{r+u}, r+u)$, $x_{r+u} = y_u^{(1)}$, $r + u = y_u^{(2)}$, $\beta_u = \alpha_{r+u}$; then

$$y_u^{(1)} = y^{(1)} + \int_0^u b(\beta_s,y_s^{(1)},y_s^{(2)})\,ds + \int_0^u \sigma(\beta_s,y_s^{(1)},y_s^{(2)})\,d\mathbf{w}_{s+r},$$

$$y_u^{(2)} = y^{(2)} + \int_0^u 1\,ds,$$

$$v^\alpha(y^{(1)},y^{(2)}) = \mathsf{M} \int_0^\infty f^{\beta_t}(y_t^{(1)},y_t^{(2)})\,dt,$$

where $y^{(1)} = x$, $y^{(2)} = r$, which, in turn, yields

$$v_7(y^{(1)},y^{(3)}) = \inf_\beta \mathsf{M} \int_0^\infty f^{\beta_t}(y_t^{(1)},y_t^{(2)})\,dt$$

and

$$\inf_{\beta \in A} \left[\sum_{i,j=1}^d a^{ij}(\beta,y) v_{xy^{(1)i}y^{(1)j}} + \sum_{i=1}^d b^i(\beta,y) v_{xy^{(1)i}} + v_{xy^{(2)}} + f^\beta \right] = 0$$

Therefore,

$$\inf_{\alpha \in A} \left[\sum_{i,j=1}^d a^{ij}(\alpha,x,r) v_{7x^ix^j}(x,r) + \sum_{i=1}^d b^i(\alpha,x,r) v_{7x^i}(x,r) + v_{7r}(x,r) + f^\alpha(x,r) \right] = 0,$$

or, in other words,

$$\inf_{\alpha \in A} \left[L^\alpha(x,r) v_7(x,r) + \frac{\partial v_7(x,r)}{\partial r} + f^\alpha(x,r) \right] = 0.$$

The problems with different performance functionals can be considered in a similar way for a controlled nonstationary process. Thus, for example, the problem of finding

$$v_8(x,r) \equiv \inf_{\alpha} \mathsf{M}\left[\int_r^T f^{\alpha_t}(x_t,t)\,dt + g(x_T)\right]$$

is equivalent to that of minimizing for the process y_u.

$$\mathsf{M}\left[\int_0^{\tau_D} f^{\beta_t}(y_t)\,dt + \tilde{g}(y_{\tau_D})\right].$$

where $\tilde{g}(y) = g(y^{(1)})$ and τ_D is the instant of time of first departure of y_t from a strip $\{y : y^{(2)} < T\}$. Therefore, for $r < T$

$$\inf_{\alpha \in A}\left[L^{\alpha}(x,r)v_8(x,r) + \frac{\partial v_8(x,r)}{\partial r} + f^{\alpha}(x,r)\right] = 0, \qquad v_8(x,T) = g(x).$$

3. Exercise

Let the lower bound in the last equation be attained for each (x,r) if $\alpha = \alpha(x,r)$. Find a nonstationary Markov strategy using the formula $\alpha^0 = \{\alpha(x_t,t)\}$. Show that α^0 is an optimal strategy.

4. Exercise

Let S_1 be a sphere with radius equal to unity in E_d, $A = \{\alpha = (\alpha^{(1)},\alpha^{(2)}) : \alpha^{(1)} \in [0,1], \alpha^{(2)} \in S_1\}$, $\sigma(\alpha,x) = \sqrt{\alpha^{(1)}}\sigma(x)$, where $\sigma(x)$ is some square matrix of dimension $d \times d$, $b(\alpha,x) = \alpha^{(2)}(1 - \alpha^{(1)}), f^{\alpha}(x) = \alpha^{(1)}(f(x) + 1) - 1$; $f(x)$ is a fixed function. Let τ_D be the first exit time of trajectories of the solution of equations

$$x_t = x + \int_0^t \sqrt{\alpha_s^{(1)}}\sigma(x_s)\,d\mathbf{w}_s + \int_0^t \alpha_s^{(2)}(1 - \alpha_s^{(1)})\,ds$$

from the domain D,

$$v(x) = \sup_{\alpha} \mathsf{M}\int_0^{\tau_D} [\alpha_t^{(1)}(f(x_t) + 1) - 1]\,dt,$$

$$a(x) = (a^{ij}(x)) = \frac{1}{2}\sigma(x)\sigma^*(x), \qquad L_0 = \sum_{i,j=1}^{d} a^{ij}(x)\frac{\partial^2}{\partial x^i\,\partial x^j}.$$

Show that Bellman's equation is equivalent for $v(x)$ to the following relations:

$$L_0 v + f \le 0,$$
$$|\mathrm{grad}\, v| - 1,$$
$$(|\mathrm{grad}\, v| - 1)(L_0 v + f) = 0$$

on D, $v = 0$ on ∂D.

5. Exercise

Let $A = [0,\infty)$. We consider the one-dimensional process

$$x_t = x + \int_0^t \sqrt{2\alpha_s}\,d\mathbf{w}_s$$

and put

$$v(x) = \sup_{\alpha} \mathsf{M} \int_0^\infty f(x_t) \exp\left[-\int_0^t \alpha_s \, ds - t \right] dt,$$

where f is given negative bounded function. Write the Bellman equation for v and prove that there is no solution to this equation. Prove also that $v = 0$ and that the function v satisfies the normed Bellman equation.

3. Application of Optimal Control Theory—Techniques for Obtaining Some Estimates

In some problems in diffusion process theory one needs to estimate from above or from below expressions of the form

$$I = \mathsf{M}\left[\int_0^\tau \exp\left\{ -\int_0^t c_s \, ds \right\} f(z_t) \, dt + \exp\left\{ -\int_0^\tau c_t \, dt \right\} g(z_\tau) \right].$$

We shall show in this section how such estimates can be obtained with the aid of optimal control techniques. The technique we apply in this section has been used for finding most estimates given in our book.

Let a process z_t be represented as

$$z_t = z + \int_0^t \sigma_s \, d\mathbf{w}_s + \int_0^t b_s \, ds,$$

where τ is the time of first departure of z_t from the domain D, $f(z)$, $g(z)$ are the given functions, and c_s is a random process.

Assume that we have imbedded the process z_t into a family of controlled processes. In other words, we consider that the process z_t can be obtained as a solution of the equation

$$x_t = x + \int_0^t \sigma(\alpha_s, x_s) \, d\mathbf{w}_s + \int_0^t b(\alpha_s, x_s) \, ds \tag{1}$$

for $x = z$ and a strategy α chosen, say, for $\bar{\alpha} = \{\bar{\alpha}_s(x_{[0,s]})\}$. Also, assume that we can find the functions $c^\alpha(x)$, $f^\alpha(x)$ so that $c^{\bar{\alpha}_s}(z_s) = c_s$, $f^{\bar{\alpha}_s}(z_s) = f(z_s)$. Then, $I = v^{\bar{\alpha}}(z) \geq v(z)$, where

$$v^\alpha(x) = \mathsf{M}\left[\int_0^{\tau_D} f^{\alpha_t}(x_t) \exp\left\{ -\int_0^t c^{\alpha_s}(x_s) \, ds \right\} dt + g(x_{\tau_D}) \exp\left\{ -\int_0^{\tau_D} c^{\alpha_s}(x_s) \, ds \right\} \right],$$

$$v(x) = \inf_{\alpha} v^\alpha(x),$$

x_t is a solution of Eq. (1), and τ_D is the first exit time of solutions of (1) from D.

As in Section 1.2, we can conjecture that the function $v(x)$ will satisfy the corresponding Bellman equation:

$$\inf_{\alpha \in A} [L^\alpha v + f^\alpha] = 0 \quad \text{in } D, \qquad -v + g = 0 \text{ on } \partial D, \tag{2}$$

where

$$L^\alpha = \sum_{i,j=1}^{d} a^{ij}(\alpha,x)\frac{\partial^a}{\partial x^i\,\partial x^j} + \sum_{i=1}^{d} b^i(\alpha,x)\frac{\partial}{\partial x^i} - c^\alpha(x),$$

$$a(\alpha,x) = (a^{ij}(\alpha,x)) = \frac{1}{2}\sigma(\alpha,x)\sigma^*(\alpha,x).$$

To estimate I from below it remains to find an explicit solution to the boundary value problem (2). If we write this solution as $w(x)$, then $w(z) \le v^{\bar{\alpha}}(z) = I$. Note that we do not need to prove here the equality $w = v$. Hence (compare this with the arguments concerning Eq. (1.7)) in order to estimate I from below it suffices to solve a less difficult problem,

$$\inf_{\alpha\in A}[L^\alpha w + f^\alpha] \ge 0 \quad \text{in } D, \qquad -w + g \ge 0 \quad \text{on } \partial D. \tag{3}$$

The feasibility of an explicit solution of Eq. (2) depends, to a great extent, on the form of the solution, that is, it depends on the fact that the process z_t is adequately incorporated into the system of processes (1). Hence a controlled process is to be introduced so that $v(x)$ depends on some known function of d coordinates, for example, on $|x|$ rather than on all the d coordinates. In this case (2) can be reduced to an ordinary differential equation.

We make a few additional remarks on the inequality $w(z) \le I$. The constant $w(z)$ can be found via some other arguments unrelated to (2) and (3). It is, however, convenient to prove that $w(z) \le I$, using (2) (or (3)), that is, using the function $w(x)$ rather than the value of $w(x)$ at z.

As we have seen in Section 1.1, for proving the inequality $w(z) \le v^{\bar{\alpha}}(z)$ we need to apply Ito's formula to a solution of Bellman's equation for a problem modified appropriately in which the performance functional is given by

$$\int_0^\infty \tilde{f}^{\alpha_t}(\tilde{x}_t)\,dt.$$

In the present case Ito's formula is to be applied to

$$w(x_t)\exp\left(-\int_0^t c^{\bar{\alpha}_s}(x_s)\,ds\right).$$

where x_t is the solution of (1) for $x = z$, $\alpha_s = \bar{\alpha}_s$. Note that the latter expression is equal to $w(z_t)\exp(-\int_0^t c_s\,ds)$.

By Ito's formula,

$$w(z) = \mathsf{M}\left[\int_0^{t\wedge\tau}[c_s w(z_s) - L^{\sigma_s,b_s}w(z_s)]\exp\left\{-\int_0^s c_r\,dr\right\}ds\right.$$
$$\left. + w(z_{t\wedge\tau})\exp\left\{-\int_0^{t\wedge\tau} c_r\,dr\right\}\right].$$

in which by virtue of (2) (or (3))

$$c_s w(z_s) - L^{\sigma_s,b_s}w(z_s) = -L^{\bar{\alpha}_s}(x_s)w(x_s) \le f^{\bar{\alpha}_s}(x_s) = f(z_s).$$

In short,

$$c_s w(z_s) - L^{\sigma_s,b_s} w(z_s) \le f(z_s), \tag{4}$$

whence

$$w(z) \le \mathsf{M}\left[\int_0^{t\wedge\tau} f(z_s)\exp\left\{-\int_0^s c_r\,dr\right\}ds + w(z_{t\wedge\tau})\exp\left\{-\int_0^{t\wedge\tau} c_r\,dr\right\}\right]. \tag{5}$$

Since $w(z_\tau) \le g(z_\tau)$, we obtain $w(z) \le I$, letting $t = \infty$.

It should be noted that in the above considerations we used (2) (or (3)) in obtaining (4). In other words, $w(s) \le I$ for any function w for which (4) is satisfied and $w(z_\tau) \le g(z_\tau)$. Such functions w are said to be stochastic Lyapunov functions (see [45]). Therefore, a method based on Bellman's equation can also be used for finding stochastic Lyapunov functions.

The above considerations imply as well that, having obtained an explicit expression for a function w such that (4) is satisfied, we do not need to use the Bellman equation, a performance function, or any other notions from an optimal control theory in order to justify the inequality $w(z) \le I$.

However, it is easier, as can be seen later, to find an explicit form for the function w, if we express this function w as a performance function and apply the Bellman equation. Furthermore, the esetimate $w(z) \le v^\alpha(z)$ is exact (unimprovable) in the class of processes (2) in the case $w = v$.

We shall illustrate this with an example.

Consider the process $z_t = z + \int_0^t \sigma_s\,d\mathbf{w}_s$, where σ is a matrix of dimension $d \times d$ and $\mathbf{w}_t$ is a d-dimensional Wiener process. Let $\varepsilon < |z| < R$. We estimate from above the probability of the event that this process reaches the closure of a sphere $S_\varepsilon = \{x : |x| < \varepsilon\}$ before it leaves a sphere S_R. In other words, we need to estimate from above

$$\mathsf{M} g(z_\tau),$$

where $g = 1$ on $\partial S_\varepsilon = \{x : |x| = \varepsilon\}$, $g = 0$ on ∂S_R, and τ is the time of first exit of z_t from $D = D(\varepsilon) = S_R \backslash (S_\varepsilon \cup \partial S_\varepsilon)$.

Assume that σ_s is nondegenerate and bounded. Moreover, for all $\lambda \in E_d$, $s \ge 0$ and all ω, let

$$\mu|\lambda|^2 \le \tfrac{1}{2}(\sigma_s\sigma_s^*\lambda,\lambda) \le \nu|\lambda|^2.$$

where μ, ν are constants larger than zero.

We take as A the set of all matrices α of dimension $d \times d$ such that for all $\lambda \in E_d$

$$\mu|\lambda|^2 \le (\tfrac{1}{2}\alpha\alpha^*\lambda,\lambda) \le \nu|\lambda|^2. \tag{6}$$

For all $\alpha \in A$ we put $\sigma(\alpha,x) = \alpha$, $b(\alpha,x) = 0$, $c^\alpha(x) = 0$, $f^\alpha(x) = 0$. We consider the controlled process

$$x_t = x + \int_0^t \sigma(\alpha_s,x_s)\,d\mathbf{w}_s = x + \int_0^t \alpha_s\,d\mathbf{w}_s.$$

It is seen that the process z_t belongs to the process given above and that $\mathsf{M}g(z_\tau) \le v(z)$, where

$$v(x) = \sup_\alpha \mathsf{M}g(x_{\tau_D}). \tag{7}$$

Bellman's equation yields for $v(x)$

$$\sup_{\alpha \in A} \sum_{i,j=1}^{d} a^{ij}(\alpha,x) v_{x^i x^j}(x) = 0 \quad \text{in } D, \qquad v = g \quad \text{on } \partial D. \tag{8}$$

We note further that because the problem is spherically symmetric the function v depends only on $|x|: v(x) = u(|x|)$, and $a(\alpha,x)$ does not depend on x, $a(\alpha,x) = a(\alpha)$. We obtain from (8) the relation

$$\sup_{\alpha \in A} \left\{ \frac{1}{|x|^2} xa(\alpha)xu''(|x|) + \frac{1}{|x|}\left[\operatorname{tr} a(\alpha) - \frac{1}{|x|^2} xa(\alpha)x \right] u'(|x|) \right\} = 0 \tag{9}$$

for $u(|x|)$.

We fix x, assume $r = |x|$, and take an orthogonal matrix T such that $x = |x| Te_1$, where $e_1 = (1,0,\dots,0)$. Then, (9) becomes

$$\sup_{\alpha \in A} \left\{ e_1 \tilde{a}(\alpha) e_1 u''(r) + \frac{1}{r}[\operatorname{tr} a(\alpha) - e_1 \tilde{a}(\alpha) e_1] u'(r) \right\} = 0,$$

where $\tilde{a}(\alpha) = T^* a(\alpha) T = a(T^*\alpha)$. We note that $\operatorname{tr} a(\alpha) = \operatorname{tr} \tilde{a}(\alpha)$, $e_1 \tilde{a}(\alpha) e_1 = \tilde{a}^{11}(\alpha)$, and that the matrices $T^*\alpha$ run through all A when α runs through A. Hence

$$\sup_{\alpha \in A} \left[a^{11}(\alpha) u''(r) + \frac{1}{r} \sum_{i=2}^{d} a^{ii}(\alpha) u'(r) \right] = 0, \tag{10}$$

(10) being equivalent to (9) for $r = |x|$.

Equation (10) is an ordinary differential equation of order 2. Let us solve Eq. (10) with respect to a second derivative. We have

$$\sup_{\alpha \in A} a^{11}(\alpha) \left[u''(r) + \sum_{i=2}^{d} \frac{a^{ii}(\alpha)}{a^{11}(\alpha)} \frac{1}{r} u'(r) \right] = 0. \tag{11}$$

Note that $a(\alpha) = \frac{1}{2}\alpha\alpha^*$ and that the inequality $a^{11}(\alpha) \ge \mu$ follows from (6) for $\lambda^1 = 1, \lambda^2 = \cdots = \lambda^d = D$. Hence the first factor in (11) cannot approximate zero; therefore

$$\begin{gathered} \sup_{\alpha \in A} \left(u''(r) + \sum_{i=2}^{d} \frac{a^{ii}(\alpha)}{a^{11}(\alpha)} \frac{1}{r} u'(r) \right) = 0, \\ u''(r) + \frac{1}{r} \sup_{\alpha \in A} u'(r) \sum_{i=2}^{d} \frac{a^{ii}(\alpha)}{a^{11}(\alpha)} = 0. \end{gathered} \tag{12}$$

It is easily seen that $u(r)$ decreases as r increases, that is, $u'(r) \le 0$, which fact follows from a version of Bellman's principle. To show this, let $\varepsilon < r < |x|$,

$D_1 = S_R \backslash (S_r \cup \partial S_r)$; then

$$u(|x|) = \sup_\alpha \mathsf{M}v(x_{\tau_{D_1}}) = u(r) \sup_\alpha \mathsf{P}\{|x_{\tau_{D_1}}| = r\} \le u(r).$$

Therefore, (12) yields

$$u''(r) + \frac{1}{r} u'(r) \inf_{\alpha \in A} \sum_{i=2}^{d} \frac{a^{ii}(\alpha)}{a^{11}(\alpha)} = 0.$$

The lower bound given above can be easily computed ($a^{ii}(\alpha)$ are changeable independently for different i in an interval $[\mu,\nu]$), and, finally, we obtain

$$u''(r) + \frac{d-1}{r}\frac{\mu}{\nu} u'(r) = 0, \qquad r \in [\varepsilon,R], \qquad u(\varepsilon) = 1, \qquad u(R) = 0,$$

whence

$$u(r) = \frac{-r^\gamma + R^\gamma}{-\varepsilon^\gamma + R^\gamma}, \qquad v(x) = \frac{R^\gamma - |x|^\gamma}{R^\gamma - \varepsilon^\gamma}, \tag{13}$$

where $\gamma = 1 - (d-1)(\mu/\nu)$.[4]

The considerations we used in deriving (13) are of heuristic nature. Hence it remains to prove (13). Keeping in mind the problem stated above, we prove first that

$$\mathsf{P}\{|z_\tau| = \varepsilon\} = \mathsf{M}g(z_\tau) \le \frac{R^\gamma - |z|^\gamma}{R^\gamma - \varepsilon^\gamma}. \tag{14}$$

Note that $u(r)$ and $v(x)$ defined in (13) are infinitely differentiable functions of their arguments for $r > 0$, $x \ne 0$, respectively. Furthermore, $u'(r) < 0$. Hence u actually satisfies (12), (11), (10), and (9). Therefore v satisfies (8). The above implies, because $v(x)$ is smooth, that for any t

$$\mathsf{M}v(z_{t \wedge \tau}) \le v(z)$$

(compare with (4) and (5)). Letting t go to infinity and taking advantage of Fatou's lemma and the fact that v is nonnegative, we arrive at (14).

We have thus solved our main problem. However, it is desirable to know to what extent the estimate (14) is exact and whether it can be strengthened. In other words, we wish to prove that $v(x)$ in (13) is in fact a payoff function.

Using Bellman's equation we find the optimal control $\alpha(x)$. For $x^1 = r$, $x^2 = \cdots = x^d = 0$, the upper bound in (8) can be attained for the same α as that in (9) and (10). As our investigation demonstrates, this upper bound in (10) can be attained by a diagonal matrix $a(\alpha)$ such that $a^{ii}(\alpha) = \mu$, $i \ge 2$, $a^{11}(\alpha) = \nu$. The eigenvectors of this matrix, except for the first vector, are orthogonal to the x^1 axis, with eigenvalues equal to μ; the first eigenvector is along the x^1 axis, with eigenvalue ν.

Because of the spherical symmetry of the problem, the upper bound in (8) can be attained for a different x on a matrix a_x, with one eigenvector being

[4] We assume that $\gamma \ne 0$.

parallel to x and corresponding to the eigenvalue ν, and the remaining eigenvalues being equal to μ.

Therefore, the matrix a_x is characterized by the fact that $a_x x = \nu x$, $a_x y = \mu y$ if $(x,y) = 0$. Since any vector

$$y = x(x,y)\frac{1}{|x|^2} + \left(y - \frac{(x,y)}{|x|^2}x\right)$$

and

$$\left(y - \frac{(x,y)}{|x|^2}x, x\right) = 0,$$

we have

$$a_x y = \nu x \frac{(x,y)}{|x|^2} + \mu\left(y - x\frac{(x,y)}{|x|^2}\right) = (\nu - \mu)x\frac{(x,y)}{|x|^2} + \mu y.$$

Therefore,

$$a_x^{ij} = \mu\delta^{ij} + (\nu - \mu)\frac{x^i x^j}{|x|^2}$$

Assuming $\alpha(x) = \sqrt{2a_x}$, we obtain the function $\alpha(x)$ for which $a(\alpha(x),x)$ $(=a_x)$ yields the upper bound in (8).

We can easily determine $\alpha(x)$ by noting that the eigenvectors of $\alpha(x)$ are the same as those of a_x, and that the eigenvalues are equal to $\sqrt{2\nu}$ and $\sqrt{2\mu}$. We have

$$\alpha^{ij}(x) = \sqrt{2\mu}\delta^{ij} + (\sqrt{2\nu} - \sqrt{2\mu})\frac{x^i x^j}{|x|^2}.$$

The function $\alpha(x)$ is smooth everywhere except for $x = 0$. Therefore, if we take a strategy $\alpha^0 = \{\alpha_t(x_t)\}$, the equation

$$x_t = x + \int_0^t \sigma(\alpha(x_s),x_s)\,d\mathbf{w}_s = x + \int_0^t \alpha(x_s)\,d\mathbf{w}_s \tag{15}$$

has a solution uniquely defined up to the time of first entry at the zero point.

Applying Ito's formula to the above solution, we have

$$v(x) = \mathsf{M}v(x_{t\wedge\tau_D}),$$

whence, as $t \to \infty$,

$$v(x) = \mathsf{M}v(x_{\tau_D}) + \lim_{t\to\infty} \mathsf{M}v(x_t)\chi_{\tau_D > t} = v^{\alpha^0}(x) + \lim_{t\to\infty} \mathsf{M}v(x_t)\chi_{\tau_D > t}.$$

In order to prove that $v(x)$ from (13) is in fact a payoff function, α^0 is an optimal strategy, and the estimate (14) is therefore exact, it sufficies to prove that the last summand is equal to zero. Since v is a bounded function, we need only to show that τ_D is finite with probability 1. Let $g_1(x) = -x^2 + R^2$. By Ito's formula we have

$$g_1(x) = \mathsf{M}[g_1(x_{t\wedge\tau_D}) + 2(\nu + (d-1)\mu)(t\wedge\tau_D)] \geq 2(\nu + (d-1)\mu)\mathsf{M}(t\wedge\tau_D),$$

which implies, by Fatou's lemma, that $\mathsf{M}\tau_D 2(\nu + (d-1)\mu) \le g_1(x)$ and $\mathsf{M}\tau_D < \infty$.

We make a few more remarks on the payoff function $v(x)$ and the optimal process (15). Let $\gamma > 0$, i.e., $(d-1)\mu < \nu$, and let $D(0) = S_1 \setminus \{0\}$. From the equality $v^{\alpha^0}(x) = v(x)$ we obtain

$$\mathsf{P}\{x_{\tau_{D(0)}} = 0\} = \lim_{\varepsilon \to 0} \mathsf{P}\{|x_{\tau_{D(\varepsilon)}}| = \varepsilon\} = 1 - \left|\frac{x}{R}\right|^{\nu},$$

from which it follows that the process (15) reaches with nonzero probability the zero point before it reaches ∂S_R. Furthermore, this probability tends to 1 when the initial point of the process tends to zero. We emphasize the fact that the process x_t is nondegenerate and has no drift.

4. One-Dimensional Controlled Processes

We shall prove in Sections 4 and 5 that a payoff function is twice continuously differentiable and satisfies Bellman's equation if a one-dimensional controlled process is nondegenerate for no strategy at all, and if, in addition, some assumptions of a technical nature can be satisfied. Furthermore, we justify in these sections the rule for finding ε-optimal strategies using Bellman's equation.

We wish to explain the relationship between the theory presented in Section 1.4 and the theory of multidimensional controlled processes which will be discussed in the subsequent chapters. One-dimensional controlled processes constitute a particular case of multidimensional controlled processes. Hence a general theory can provide a considerable amount of information about a payoff function. Bellman's equation, and ε-optimal strategies. However, taking advantage of the specific nature of one-dimensional processes, we can prove stronger assertions in some cases. At the same time, we should point out that the results given below do not include everything that follows from the general theory.

Let A be a (nonempty) convex subset of some Euclidean space, and let $\sigma(\alpha,x)$, $b(\alpha,x)$, $c^\alpha(x)$, $f^\alpha(x)$ be real functions given for $\alpha \in A$, $x \in (-\infty,\infty)$. Assume that $c^\alpha(x) \ge 0$; $\sigma(\alpha,x)$, $b(\alpha,x)$, $c^\alpha(x)$, and $f^\alpha(x)$ are bounded and satisfy a Lipschitz condition with respect to (α,x), i.e., there exists a constant K such that for all α, $\beta \in A$, x, $y \in E_1$

$$|a(\alpha,x)| + |b(\alpha,x)| + |c^\alpha(x)| + |f^\alpha(x)| \le K,$$

$$|\sigma(\alpha,x) - \sigma(\beta,y)| + |b(\alpha,x) - b(\beta,y)| + |c^\alpha(x) - c^\beta(y)| + |f^\alpha(x) - f^\beta(y)| \\ \le K(|x-y| + |\alpha - \beta|),$$

where, as usual, $a(\alpha,x) = \frac{1}{2}[\sigma(\alpha,x)]^2$.

Furthermore, we assume that the controlled processes are uniformly nondegenerate, that is, for some constant $\delta > 0$ for all $\alpha \in A$, $x \in E_1$

$$a(\alpha,x) \geq \delta.$$

Let a Wiener process $(\mathbf{w}_t,\mathscr{F}_t)$ be given on some complete probability space $(\Omega,\mathscr{F},\mathsf{P})$, and let σ-algebras of $\mathscr{F}_t$ be complete with respect to measure P.

1. Definition. By a strategy we shall mean a random process $\alpha_t(\omega)$ with values in A, which is progressively measurable with respect to the system of σ-algebras of $\{\mathscr{F}_t\}$. We denote by $\mathfrak{A}$ the set of all strategies.

To each strategy $\alpha \in \mathfrak{A}$ and a point x we set into correspondence a solution $x_t^{\alpha,x}$ of the equation

$$x_t = x + \int_0^t \sigma(\alpha_s,x_s)\, d\mathbf{w}_s + \int_0^t b(\alpha_s,x_s)\, ds. \tag{1}$$

By Ito's theorem the solution of Eq. (1) exists and is unique. We fix numbers $r_1 < r_2$ and a function $g(x)$ given for $x = r_1$, $x = r_2$.

We denote by $\tau^{\alpha,x}$ the first exit time of $x_t^{\alpha,x}$ from (r_1,r_2) and we set

$$v^\alpha(x) = \mathsf{M}\left\{\int_0^{\tau^{\alpha,x}} f^{\alpha_t}(x_t^{\alpha,x}) \exp\left[-\int_0^t c^{\alpha_s}(x_s^{\alpha,x})\, ds\right] dt \right.$$
$$\left. + \, g(x_{\tau^{\alpha,x}}^{\alpha,x}) \exp\left[-\int_0^{\tau^{\alpha,x}} c^{\alpha_s}(x_s^{\alpha,x})\, ds\right]\right\},$$

$$v(x) = \sup_{\alpha \in \mathfrak{A}} v^\alpha(x).$$

We shall frequently need to write mathematical expectations of the expressions which include repeatedly the indices α and x, where α is a strategy, and x is a point of the interval $[r_1,r_2]$. We agree to write these indices as subindices or superindices or the expectation sign. For example, we write $\mathsf{M}_x^\alpha \int_0^\tau f^{\alpha_t}(x_t)\, dt$ instead of $\mathsf{M} \int_0^{\tau^{\alpha,x}} f^{\alpha_t}(x_t^{\alpha,x})\, dt$, etc. In addition, it is convenient to introduce the following notation:

$$\varphi_t^{\alpha,x} = \int_0^t c^{\alpha_s}(x_s^{\alpha,x})\, ds.$$

The definition of $v^\alpha(x)$ thus becomes the following:

$$v^\alpha(x) = \mathsf{M}_x^\alpha\left[\int_0^\tau f^{\alpha_t}(x_t)e^{-\varphi_t}\, dt + g(x_\tau)e^{-\varphi_\tau}\right].$$

The definition given above of the strategy enables us to use the information about the behavior of the process $\mathbf{w}_t$ in controlling the solution of Eq. (1). From the practical point of view, this situation is not natural. Hence we shall consider some other control techniques.

Let $C[0,\infty)$ be the space of continuous real functions x_t, $t \in [0,\infty)$; let $\mathscr{N}_t$ be the smallest σ-algebra of subsets of $C[0,\infty)$, which contains all sets of the form $\{x_{[0,\infty)}: x_s \leq a\}$, $s \leq t$, $a \in (-\infty,\infty)$.

2. Definition. A function $\alpha_t(x_{[0,\infty)}) = \alpha_t(x_{[0,t]})$ with values in A, given for $t \in [0,\infty)$, $x_{[0,\infty)} \in C[0,\infty)$, is said to be a natural strategy admissible at the point $x \in [r_1,r_2]$ if this function is progressively measurable with respect to $\mathcal{N}_t$ and if there exists at least one solution of the stochastic equation

$$x_t = x + \int_0^t \sigma(\alpha_s(x_{[0,s]}),x_s)\,d\mathbf{w}_s + \int_0^t b(\alpha_s(x_{[0,s]}),x_s)\,ds \tag{2}$$

which is $\mathscr{F}_t$-measurable for each t.

We denote by $\mathfrak{A}_E(x)$ the set of all natural strategies admissible at a point x. To each strategy $\alpha \in \mathfrak{A}_E(x)$ we set into correspondence one (fixed) solution $x_t^{\alpha,x}$ of Eq. (2).

3. Definition. A natural strategy $\alpha_t(x_{[0,t]})$ is said to be (stationary) Markov if $\alpha_t(x_{[0,t]}) = \alpha(x_t)$ for some function $\alpha(x)$. We denote by $\mathfrak{A}_M(x)$ the set of all Markov strategies admissible at a point x.

Note that to each natural strategy $\alpha_t(x_{[0,t]})$ admissible at x, we can set into correspondence a strategy $\beta \in \mathfrak{A}$ such that $x_t^{\alpha,x} = x_t^{\beta,x}$.

In fact, let us take a solution $x_t(\omega) = x_t^{\alpha,x}(\omega)$ of Eq. (2), and let us assume that $\beta_t(\omega) = \alpha_t(x_{[0,t]}(\omega))$. It is seen that $\{\beta_t\}$ is a strategy and that the equation $dx_t = \sigma(\beta_t,x_t)\,d\mathbf{w}_t + b(\beta_t,x_t)\,dt$ with the initial $x_0 = x$ given can be satisfied for $x_t = x_t^{\alpha,x}$. By the uniqueness theorem this equation has no other solutions; therefore, $x_t^{\beta,x} = x_t^{\alpha,x}$. It is clear that the inclusions $\mathfrak{A}_M(x) \subset \mathfrak{A}_E(x) \subset \mathfrak{A}$ have precise meaning.

In order to show that $\mathfrak{A}_M(x) \neq \varnothing$, we take a function $\alpha(x)$ with values in A such that $|\alpha(x) - \alpha(y)| \leq N|x - y|$ for all x, y for some constant N. Since the composition of functions satisfying a Lipschitz condition satisfies also a Lipschitz condition, there exists a solution of the equation

$$x_t = x + \int_0^t \sigma(\alpha(x_s),x_s)\,d\mathbf{w}_s + \int_0^t b(\alpha(x_s),x_s)\,ds,$$

therefore, $\{\alpha(x_t)\} \in \mathfrak{A}_M(x)$.

In the same way as we introduced the function $v(x)$ for computing the upper bounds on the basis of $\mathfrak{A}_E(x)$, $\mathfrak{A}_M(x)$, we introduce here the functions $v_{(E)}(x)$, $v_{(M)}(x)$. It is seen that

$$v_{(M)}(x) \leq v_{(E)}(x) \leq v(x).$$

4. Definition. Let $\varepsilon \geq 0$. A strategy $\alpha \in \mathfrak{A}$ is said to be ε-optimal for a point x if $v(x) \leq v^{\alpha}(x) + \varepsilon$. A 0-optimal strategy is said to be optimal.

Our objective is to prove the following theorem.

5. Theorem. *$v_{(M)}(x) = v_{(E)}(x) = v(x)$ for $x \in [r_1,r_2]$, $v(r_1) = g(r_1)$, $v(r_2) = g(r_2)$, $v(x)$ and its derivatives up to and including the second order are continuous on*

$[r_1,r_2]$,[5] *and* $v''(x)$ *satisfies a Lipschitz condition on* $[r_1,r_2]$. *For all* $x \in [r_1,r_2]$,

$$\sup_{\alpha \in A} [a(\alpha,x)v''(x) + b(\alpha,x)v'(x) - c^\alpha(x)v(x) + f^\alpha(x)] = 0. \tag{3}$$

Furthermore, v is the unique solution of (3) *in the class of functions which are twice continuously differentiable on* $[r_1,r_2]$ *and equal to g at the end points of this interval.*

In order to prove the theorem we need four lemmas and additional notation.

Let

$$F(x,y,p,r) \equiv \sup_{\alpha \in A} [a(\alpha,x)r + b(\alpha,x)p - c^\alpha(x)y + f^\alpha(x)],$$

$$F_1(x,y,p) \equiv \sup_{\alpha \in A} \left[\frac{b(\alpha,x)}{a(\alpha,x)} p - \frac{c^\alpha(x)}{a(\alpha,x)} y + \frac{f^\alpha(x)}{a(\alpha,x)}\right],$$

$$L^\alpha u(x) \equiv L^\alpha(x)u(x) \equiv a(\alpha,x)u''(x) + b(\alpha,x)u'(x) - c^\alpha(x)u(x),$$

$$F[u] \equiv F[u](x) \equiv \sup_{\alpha \in A} [L^\alpha u(x) + f^\alpha(x)] = F(x,u,u',u''),$$

$$\|u\|_B \equiv \|u\|_{B[r_1,r_2]} \equiv \sup_{[r_1,r_2]} |u(x)|,$$

$$\|u\|_{\mathscr{L}_1} \equiv \|u\|_{\mathscr{L}_1[r_1,r_2]} \equiv \int_{r_1}^{r_2} |u(x)|\, dx.$$

6. Lemma. *Let* $a(x)$, $b(x)$, $c(x)$, $f(x)$ *be continuous functions on* $[r_1,r_2]$ *and let* $a \geq \delta$, $c \geq 0$, $|a| + |b| + |c| \leq K$ *on this interval. Then there exists a unique function* $u(x)$ *which is twice continuously differentiable on* $[r_1,r_2]$, *is equal to g at the end points of the interval* $[r_1,r_2]$, *and is such that for all* $x \in [r_1,r_2]$

$$a(x)u''(x) + b(x)u'(x) - c(x)u(x) + f(x) = 0. \tag{4}$$

Furthermore,

$$\|u''\|_B + \|u'\|_B + \|u\|_B \leq N_1(\|f\|_B + 1) \tag{5}$$

and, for $g(r_1) = g(r_2) = 0$,

$$\|u\|_B \leq N_2 \|f\|_{\mathscr{L}_1}, \tag{6}$$

where N_1 *depends only on* r_1, r_2, δ, K, $g(r_1)$, $g(r_2)$; N_2 *depends only on* r_1, r_2, δ, K.

PROOF. Assertions of this kind are well known from the theory of differential equations (see, for example, [46]). Hence we shall just sketch the proof.

First by considering instead of the function u the function $u - \psi$ in which ψ is linear on $[r_1,r_2]$ and $\psi(r_i) = g(r_i)$, we convince ourselves that it suffices to prove the lemma for $g = 0$. Further, we can define a function $y(x)$ explicitly

[5] By definition, $v'(r_1)$ ($v''(r_1)$) is assumed to be equal to the limit $v'(x)$ ($v''(x)$) as $x \downarrow r_1$. We define $v'(r_2)$, $v''(r_2)$ in a similar way.

such that the replacement of the unknown function with the aid of the formula $u(x) = \tilde{u}(y(x))$ turns Eq. (4) into

$$a_1(y)\tilde{u}''(y) - c_1(y)\tilde{u}(y) + f_1(y) = 0, \qquad y \in [0,1].$$

Dividing both sides of the last equation by a_1, we arrive at the equation

$$\tilde{u}''(y) = c_2(y)\tilde{u}(y) - f_2(y), \qquad y \in [0,1] \tag{7}$$

satisfying the boundary condition $\tilde{u}(0) = \tilde{u}(1) = 0$.

Note that $c_2(y) \geq 0$ in (7). Analogous properties of the solution (4) can readily be derived from the properties of (7). Therefore, it suffices to prove our lemma for Eq. (7).

Let $g_0(x,y) = (x \wedge y)(1 - x \vee y)$ and for $\lambda > 0$ let

$$g_\lambda(x,y) = \frac{1}{\sqrt{\lambda}\,\mathrm{sh}\sqrt{\lambda}}\,\mathrm{sh}\sqrt{\lambda}(x \wedge y)\,\mathrm{sh}\sqrt{\lambda}(1 - x \vee y).$$

Elementary computation indicates that the combination of the relations $\tilde{u}'' - \lambda\tilde{u} = -h$, $\tilde{u}(0) = \tilde{u}(1) = 0$ is equivalent to the following:

$$\tilde{u}(x) = \int_0^1 g_\lambda(x,y)h(y)\,dy.$$

In addition, (7) is equivalent to $\tilde{u}'' - \lambda\tilde{u} = (c_2 - \lambda)\tilde{u} - f_2$ and therefore to the equation

$$\tilde{u}(x) = \int_0^1 g_\lambda(x,y)f_2(y)\,dy + \int_0^1 g_\lambda(x,y)(\lambda - c_2(y))\tilde{u}(y)\,dy \equiv T_\lambda\tilde{u}. \tag{8}$$

Let $\lambda = \|c_2\|_{B[0,1]}$; then

$$\|T_\lambda u_1 - T_\lambda u_2\|_{B[0,1]} \leq \|u_1 - u_2\|_{B[0,1]}\lambda \max_x \int_0^1 g_\lambda(x,y)\,dy = \varepsilon\|u_1 - u_2\|_{B[0,1]}.$$

As can easily be verified,

$$\varepsilon = 1 - \frac{1}{\mathrm{ch}\,\frac{1}{2}\sqrt{\lambda}} < 1.$$

Consequently, T_λ is a contraction operator; Eqs. (8), (7), and (4) have unique solutions satisfying zero boundary conditions. We deduce the estimates (5) and (6) for the solution of (7). It follows from (8) that

$$\|\tilde{u}\|_{B[0,1]} \leq \max_x \max_y g_\lambda(x,y)\|f_2\|_{\mathscr{L}_1[0,1]} + \varepsilon\|\tilde{u}\|_{B[0,1]},$$

from which we find $\|\tilde{u}\|_{B[0,1]} \leq N\|f_2\|_{\mathscr{L}_1[0,1]}$. Further, from (7) we obtain an estimate for $\|\tilde{u}''\|_{B[0,1]}$. Finally, we obtain the estimate $\tilde{u}'$ using the representation $\tilde{u}'(x) = \int_{x_0}^x \tilde{u}''(y)\,dy$, where x_0 is a point in $[0,1]$ at which $\tilde{u}' = 0$. We have thus proved the lemma. □

7. Lemma. *There exists a constant N depending only on r_1, r_2, δ, K, and such that $\mathsf{M}_x^\alpha\tau \leq N$ for each $\alpha \in \mathfrak{A}$, $x \in [r_1,r_2]$. In particular, $v^\alpha(x)$ and $v(x)$ are finite functions.*

PROOF. We can consider without loss of generality that $r_1 = -r_2$. Let

$$w(x) = \frac{\delta}{K^2}\left(e^{(K/\delta)r_2} - e^{(K/\delta)|x|}\right) - \frac{r_2 - |x|}{K}.$$

Regardless of the fact that $w(x)$ is the difference between two nondifferentiable functions, we can easily verify that $w(x)$ is twice continuously differentiable and that for each $a \geq \delta$, $b \in [-K,K]$, $x \in [r_1,r_2]$

$$aw''(x) + bw'(x) \leq -1.$$

In addition, $w \geq 0$ on $[r_1,r_2]$, $w(r_i) = 0$. By Ito's formula, for each $\alpha \in \mathfrak{A}$, $x \in [r_1,r_2]$, $t \geq 0$

$$\mathsf{M}_x^\alpha w(x_{\tau\wedge t}) = w(x) + \mathsf{M}_x^\alpha \int_0^{\tau\wedge t} \left[a(\alpha_s,x_s)w''(x_s) + b(\alpha_s,x_s)w'(x_s)\right] ds,$$

from which we conclude, using properties of the function w, that $w(x) \geq \mathsf{M}_x^\alpha(\tau \wedge t)$ and, as $t \to \infty$, $w(x) \geq \mathsf{M}_x^\alpha \tau$. □

8. Lemma. *Let the function $\alpha(x)$ satisfy a Lipschitz condition. We define a Markov strategy $\alpha_t(x_{[0,t]})$ by $\alpha_t(x_{[0,t]}) = \alpha(x_t)$. If $f(x)$ is continuous on $[r_1,r_2]$, the function*

$$u(x) \equiv \mathsf{M}_x^\alpha\left\{\int_0^\tau f(x_t)e^{-\varphi_t}\,dt + g(x_\tau)e^{-\varphi_\tau}\right\}$$

is twice continuously differentiable on $[r_1,r_2]$ and is the unique solution of the equation

$$L^{\alpha(x)}(x)u(x) + f(x) = 0, \qquad x \in [r_1,r_2], \tag{9}$$

in the class of twice continuously differentiable functions which are equal to g at the end points of $[r_1,r_2]$. In particular, $L^{\alpha(x)}(x)v^\alpha(x) + f^\alpha(x) = 0$. Moreover, if some function $w(x)$ has two continuous derivatives on $[r_1,r_2]$, $w(r_i) = g(r_i)$ and

$$L^{\alpha(x)}(x)w(x) + f^{\alpha(x)}(x) + h(x) \geq 0, \qquad x \in [r_1,r_2], \tag{10}$$

we have $w \leq v^\alpha + N\|h\|_{\mathscr{L}_1}$, where N is the constant from (6).

PROOF. By Lemma 6, Eq. (9) satisfying the boundary conditions $u(r_i) = g(r_i)$ has a smooth solution. Writing this solution as u_1 and applying Ito's formula to the expression $u_1(x_t^{\alpha,x})e^{-\varphi_t^{\alpha,x}}$ we easily find

$$u_1(x) = \mathsf{M}_x^\alpha\left\{\int_0^{\tau\wedge t} f(x_s)e^{-\varphi_s}\,ds + u_1(x_{\tau\wedge t})e^{-\varphi_{\tau\wedge t}}\right\},$$

from which the equality $u_1(x) = u(x)$ follows as $t \to \infty$ because $\tau^{\alpha,x}$, $\mathsf{M}_x^\alpha\tau$ are finite, f, u_1 are bounded, and c^α is nonnegative. We have thus proved the first assertion. In order to prove the second assertion, we set $L^\alpha w + f^\alpha = -h_1$ and note that the function h_1 is continuous and $L^\alpha[w - v^\alpha] = -h_1$. Then, according to the first assertion, we have

$$w(x) - v^\alpha(x) = \mathsf{M}_x^\alpha \int_0^\tau h_1(x_t)e^{-\varphi_t}\,dt \le \mathsf{M}_x^\alpha \int_0^\tau h_1^+(x_t)e^{-\varphi_t}\,dt \equiv u_2(x).$$

It remains to estimate $u_2(x)$. h_1^+ as well as h_1 is a continuous function. Therefore, $u_2(x)$ satisfies the equation $L^\alpha u_2 = -h_1^+$, and, in addition, $u_2(r_i) = 0$. By Lemma 6 we have $u_2 \le N\|h_1^+\|_{\mathscr{L}_1}$. Since $h_1 \le h$, $h_1^+ \le h^+$, $\|h_1^+\|_{\mathscr{L}_1} \le \|h\|_{\mathscr{L}_1}$, $u_2 \le N\|h\|_{\mathscr{L}_1}$, thus proving the lemma. □

9. Lemma. *Let $u(x)$, $u_1(x)$, $u_2(x)$ be bounded Borel functions on $[r_1,r_2]$, $\varepsilon > 0$. There exists a function $\alpha(x)$ with values in A, such that $a(\alpha(x),x)u_2(x) + b(\alpha(x),x)u_1(x) - c^{\alpha(x)}(x)u(x) + f^{\alpha(x)}(x) + \varepsilon \ge F(x,u(x),u_1(x),u_2(x))$ for all $x \in [r_1,r_2]$. Furthermore, there exists $\alpha(x)$ satisfying the Lipschitz condition and also there exists a real-valued nonnegative function $h(x)$ such that $\|h\|_{\mathscr{L}_1} \le \varepsilon$ and for all $x \in [r_1,r_2]$*

$$\begin{aligned} a(\alpha(x),x)u_2(x) + b(\alpha(x),x)u_1(x) - c^{\alpha(x)}(x)u(x) + f^{\alpha(x)}(x) + h(x) \\ = F(x,u(x),u_1(x),u_2(x)). \end{aligned}$$

PROOF. We fix some countable set $\{\alpha(i)\}$ everywhere dense in A. Because a, b, c, f are continuous in the argument α

$$\begin{aligned} F(x,u(x),u_1(x),u_2(x)) = \sup_i \big[a(\alpha(i),x)u_2(x) \\ + b(\alpha(i),x)u_1(x) - c^{\alpha(i)}(x)u(x) + f^{\alpha(i)}(x)\big]. \end{aligned}$$

We conclude from the above that for each $x \in [r_1,r_2]$ there is an i such that

$$\begin{aligned} a(\alpha(i),x)u_2(x) + b(\alpha(i),x)u_1(x) - c^{\alpha(i)}(x)u(x) + f^{\alpha(i)}(x) + \varepsilon \\ \ge F(x,u(x),u_1(x),u_2(x)). \end{aligned}$$

Next, we denote by $i(x)$ the smallest value of i for which the last given inequality can be satisfied. It is seen that the (measurable) function $\alpha(x) \equiv \alpha(i(x))$ yields the function stated in the first assertion.

In order to prove the second assertion of the lemma, we extend the function $i(x)$ outside $[r_1,r_2]$ assuming $i(x) = 1$ for $x \notin [r_1,r_2]$. Let

$$\alpha_{t,n}(x) = \mathsf{M}\alpha(n \wedge i(x + \mathbf{w}_t)) = \frac{1}{\sqrt{2\pi t}} \int_{-\infty}^{\infty} \alpha(n \wedge i(y))e^{-(y-x)^2/2t}\,dt.$$

It is easily seen that $\alpha_{t,n}(x)$ is an infinitely differentiable function. Furthermore, $\alpha_{t,n}(x) \in A$ due to the convexity property of the set A. Note that, as is well known (see, for example, [10]), for each measurable bounded function $\gamma(x)$ as $t \downarrow 0$ the function $\mathsf{M}\gamma(x + \mathbf{w}_t) \to \gamma(x)$ (a.s.) Hence $\alpha_{t,n}(x) \to \alpha(n \wedge i(x))$ (a.s.) as $t \downarrow 0$, and, clearly, $\alpha(n \wedge i(x)) \to \alpha(i(x))$ as $n \to \infty$.

Defining

$$\begin{aligned} h^{t,n}(x) = F(x,u(x),u_1(x),u_2(x)) - \big[a(\alpha_{t,n}(x),x)u_2(x) \\ + b(\alpha_{t,n}(x),x)u_1(x) - c^{\alpha_{t,n}(x)}(x)u(x) + f^{\alpha_{t,n}(x)}(x)\big], \end{aligned}$$

we obtain

$$\lim_{n\to\infty} \lim_{t\downarrow 0} h^{t,n}(x) \le \varepsilon \qquad \text{(a.s.)},$$
$$h^{t,n}(x) \ge 0.$$

Further, since u, u_1, u_2 are bounded, the class of functions $h^{t,n}$ is class equibounded. Therefore,

$$\lim_{n\to\infty} \lim_{t\downarrow 0} \|h^{t,n}\|_{\mathscr{L}_1} \le \varepsilon(r_2 - r_1),$$

and t, n can be chosen so that $\|h^{t,n}\|_{\mathscr{L}_1} \le 2\varepsilon(r_2 - r_1)$. The lemma has thus been proved. □

10. Proof of Theorem 5. We use the so-called method of successive approximation in a space of strategies. This method, known as the Bellman–Howard method, enables us to find ε-optimal strategies and approximate values of a payoff function without solving nonlinear differential equations.

We take as $\alpha_0(x)$ any function with values in A satisfying the Lipschitz condition. We define the Markov strategy α_0 using the formula $\alpha_t(x_{[0,t]}) = \alpha_0(x_t)$. Let $v_0(x) = v^{\alpha_0}(x)$. If $\alpha_0, \alpha_1, \ldots, \alpha_n$, $v_0(x)$, $v_1(x), \ldots, v_n(x)$ have been constructed, we choose a function $\alpha_{n+1}(x)$ such that it satisfies the Lipschitz condition and

$$L^{\alpha_{n+1}} v_n + f^{\alpha_{n+1}} + h_{n+1} = F[v_n], \tag{11}$$

where h_{n+1} is a function with a small norm: $\|h_{n+1}\|_{\mathscr{L}_1[r_1,r_2]} \le 1/(n+1)(n+2)$. Assume that $v_{n+1}(x) = v^{\alpha_{n+1}}(x)$, where the strategy $\alpha_{n+1} \in \mathfrak{A}_M(x)$ can be found with the aid of the function α_{n+1}.

We prove that the sequence $\{v_n(x)\}$ has a limit and that this limit satisfies Eq. (3). We shall also prove that the limit of v_n coincides with v.

First, we investigate the behavior of v_n, v_n' as $n \to \infty$. Applying Lemma 8, we obtain $F[v_n] \ge 0$ since $L^{\alpha_n} v_n + f^{\alpha_n} = 0$. We conclude from (11) that

$$v_n \le v_{n+1} + N \frac{1}{(n+1)(n+2)}.$$

Therefore

$$v_n - N \frac{1}{n+1} \le v_{n+1} - N \frac{1}{n+2},$$

i.e., the sequence of functions

$$u_n = v_n - N \frac{1}{n+1}$$

is monotone increasing.

Furthermore, by Lemma 7 the totality of functions v_n as well as u_n is bounded, hence $\lim_{n\to\infty} u_n$ exists. It is seen that v_n has a limit as well. For

$x \in [r_1,r_2]$ let

$$\tilde{v}(x) = \lim_{n\to\infty} v_n(x).$$

By Lemma 6, it follows from the equality $L^{\alpha_n}v_n + f^{\alpha_n} = 0$ that

$$\|v_n\|_B + \|v_n'\|_B + \|v_n''\|_B \le N,$$

where N does not depend on n. By the Lagrange theorem, $|v_n(x) - v_n(y)| \le N|x - y|$. Therefore, the functions v_n as well as u_n are equicontinuous and uniformly bounded. By the Arzela theorem, some subsequence of the functions u_n converges to the limit uniformly in x. Since the function u_n increases as n increases, the entire sequence of u_n converges to the limit uniformly in x. It then follows that v_n converges to $\tilde{v}$ uniformly in x. In particular. $\tilde{v}(x)$ is continuous on $[r_1,r_2]$.

Further, using the Lagrange theorem, we derive from the uniform estimate $\|v_n''\|$ that $|v_n'(x) - v_n'(y)| \quad N|x - y|$, where N does not depend on n. By the Arzelá theorem, the sequence $\{v_n'\}$ is compact in the sense of uniform convergence on $[r_1,r_2]$. Assume that $\{v_{n_k}'\}$ is a uniformly convergent subsequence and that $\tilde{v}^1$ is the limit of this subsequence. Taking the limit in the equality

$$v_{n_k}(x) = v_{n_k}(r_1) = \int_{r_1}^{x} v_{n_k}'(s)\,ds,$$

we have

$$\tilde{v}(x) - \tilde{v}(r_1) = \int_{r_1}^{x} \tilde{v}^1(s)\,ds.$$

Therefore, $\tilde{v}^1 = \tilde{v}'$, $\{v_n'\}$ has a single limit point, and $v_n' \to \tilde{v}'$ uniformly on $[r_1,r_2]$. In addition, $|\tilde{v}'(x) - \tilde{v}'(y)| \le N|x - y|$.

Let us consider $\tilde{v}''$ and $F[\tilde{v}]$. We use Eq. (11). As was noted above, $F[v_n] \ge 0$ and $f^{\alpha_{n+1}} + L^{\alpha_{n+1}}v_{n+1} = 0$. Therefore, (11) yields

$$L^{\alpha_{n+1}}(v_n - v_{n+1}) + h_{n+1} = F[v_n] \ge 0.$$

Dividing the last inequality by $a(\alpha_{n+1}(x),x)$, we easily find

$$v_n'' - v_{n+1}'' + \delta^{-1}K(|v_n' - v_{n+1}'| + |v_n - v_{n+1}|) + \delta^{-1}h_{n+1} \ge K^{-1}F[v_n] \ge 0. \quad (12)$$

Also, we note that

$$F(x,y,p,r) = \sup_{\alpha\in A} a(\alpha,x)\left\{r + \left[\frac{b(\alpha,x)}{a(\alpha,x)}p - \frac{c^\alpha(x)}{a(\alpha,x)}y + \frac{f^\alpha(x)}{a(\alpha,x)}\right]\right\}.$$

From this representation of F it follows that the equation $F = 0$ is equivalent to $r + F_1 = 0$, and that $0 \le r + F_1 \le \delta^{-1}\varepsilon$ if $0 \le F \le \varepsilon$. Therefore, (12) yields

$$\begin{aligned}\delta^{-1}K[v_n'' - v_{n+1}'' + \delta^{-1}K(|v_n' - v_{n+1}'| + |v_n - v_{n+1}|) + \delta^{-1}h_{n+1}]\\ \ge (v_n'' + F_1(x,v_n,v_n') \ge 0.\end{aligned}$$

Integrating over x and letting $n \to \infty$,

$$\tilde{v}'(x) - \tilde{v}'(r_1) + \lim_{n\to\infty} \int_{r_1}^{x} F_1(s,v_n(s),v_n'(s))\, ds = 0 \tag{13}$$

by virtue of the proven properties of v_n, v_n'.

Next we exploit a property of the function $F_1(x,y,p)$. Since the magnitude of the difference between the upper bounds does not exceed the upper bound of the magnitude of this difference, we have

$$|F_1(x,y_1,p_1) - F_1(x,y_2,p_2)| \le \sup_{\alpha \in A} \left| \frac{b(\alpha,x)}{a(\alpha,x)}(p_1 - p_2) - \frac{c^{\alpha}(x)}{a(\alpha,x)}(y_1 - y_2) \right|$$
$$\le \frac{K}{\delta}(|p_1 - p_2| + |y_1 - y_2|).$$

In particular, $|F_1(s,v_n,v_n') - F_1(s,\tilde{v},\tilde{v}')| \le (K/\delta)(|v_n' - \tilde{v}'| + |v_n - \tilde{v}|)$, and (13) yields

$$\tilde{v}'(x) - \tilde{v}'(r_1) + \int_{r_1}^{x} F_1(s,\tilde{v}(s),\tilde{v}'(s))\, ds = 0. \tag{14}$$

Further, applying the elementary inequality

$$\left| \frac{b_1}{a_1} - \frac{b_2}{a_2} \right| \le \frac{|b_1 - b_2|}{|a_1|} + |b_2| \frac{|a_1 - a_2|}{|a_1 a_2|}$$

and the property of the upper bound noted, we find

$$|F_1(x_1,y,p) - F_1(x_2,y,p)| \le \frac{K}{\delta}\left(1 + \frac{K}{\delta}\right)(1 + |y| + |p|)|x_1 - x_2|,$$

from which we derive

$$\begin{aligned}
&|F_1(x_1,\tilde{v}(x_1),\tilde{v}'(x_1)) - F_1(x_2,\tilde{v}(x_2),\tilde{v}'(x_2)|\\
&\quad \le |F_1(x_1,\tilde{v}(x_1),\tilde{v}'(x_1)) - F_1(x_2,\tilde{v}(x_1),\tilde{v}'(x_1))|\\
&\qquad + |F_1(x_2,\tilde{v}(x_1),\tilde{v}'(x_1)) - F_1(x_2,\tilde{v}(x_2),\tilde{v}'(x_2))|\\
&\quad \le N|x_1 - x_2| + \frac{K}{\delta}(|\tilde{v}(x_1) - \tilde{v}(x_2)| + |\tilde{v}'(x_1) - \tilde{v}'(x_2)|)\\
&\quad \le N|x_1 - x_2|.
\end{aligned}$$

In short, $F_1(x,\tilde{v}(x),\tilde{v}'(x))$ satisfies a Lipschitz condition. Differentiating (14), we have first that $\tilde{v}''(x) + F_1(x,\tilde{v}(x),\tilde{v}'(x)) = 0$, and $F[\tilde{v}] = 0$; second, it follows from the equality $\tilde{v}'' = -F_1(x,\tilde{v},\tilde{v}')$ that $\tilde{v}''$ satisfies a Lipschitz condition.

In order to complete the proof of the theorem, we need only to show that for each twice continuously differentiable solution $u(x)$ of the equation $F[u] = 0$ satisfying the boundary conditions $u(r_1) = g(r_1)$, $u(r_2) = g(r_2)$, the equalities $u = v_{(M)} = v_{(E)} = v$ hold.

First of all, if $\varepsilon > 0$ and u is a function having the properties given above, by Lemma 9 there are $\alpha(x)$ and $h(x)$ such that $L^{\alpha(x)}u(x) + f^{\alpha(x)}(x) + h(x) = 0$

and $\|h\|_{\mathscr{L}_1} \le \varepsilon$. This implies, by Lemma 8, that $u \le v^\alpha + N\varepsilon$ and, therefore, $u \le v_{(M)} \le v_{(E)} \le v$.

On the other hand, the equality $F[u] = 0$ for each $\alpha \in \mathfrak{A}$, $x \in [r_1, r_2]$ for $t \le \tau$ yields

$$L^{\alpha_t} u(x_t^{\alpha,x}) + f^{\alpha_t}(x_t^{\alpha,x}) \le 0.$$

We apply Ito's formula to the expression $u(x_t^{\alpha,x})e^{-\varphi_t^{\alpha,x}}$; then

$$\begin{aligned} u(x) &= \mathsf{M}_x^\alpha \left\{ -\int_0^{\tau \wedge t} L^{\alpha_s} u(x_s) e^{-\varphi_s}\, ds + u(x_{\tau \wedge t}) e^{-\varphi_{\tau \wedge t}} \right\} \\ &\ge \mathsf{M}_x^\alpha \left\{ \int_0^{\tau \wedge t} f^{\alpha_s}(x_s) e^{-\varphi_s}\, ds + u(x_{\tau \wedge t}) e^{-\varphi_{\tau \wedge t}} \right\} \end{aligned} \tag{15}$$

which gives the inequality $u \ge v^\alpha$ as $t \to \infty$, due to the finiteness of $\tau^{\alpha,x}$, $\mathsf{M}_x^\alpha \tau$ (Lemma 7), the boundedness of f, u, the equality $u(x_\tau^{\alpha,x}) = g(x_\tau^{\alpha,x})$, and the nonnegativeness of c^α. Hence, $u \ge v$ and $u \ge v_{(E)} \ge v_{(M)} \ge u$. The theorem has thus been proved.

11. Remark. Since $\tilde{v} = v$, $v = \lim_{n\to\infty} v^{\alpha_n}$. Furthermore, the proof of the theorem provides us with a technique for finding ε-optimal strategies and the approximate value of the payoff function v. Using this technique, one has to know how to solve equations of the form $L^{\alpha(x)}u(x) + f^{\alpha(x)}(x) = 0$ and how to find $\beta(x)$ such that $L^{\beta(x)}u(x) + f^{\beta(x)}(x) \ge F[u]x - \varepsilon$. The equality $v = \lim_{n\to\infty} v^{\alpha_n}$ enables us to estimate v'',v',v as follows. By Lemmas 6 and 8,

$$\|v_n\|_B + \|v_n'\|_B + \|v_n''\|_B \le N_1\left(\sup_{\alpha,x} |f^\alpha(x)| + 1\right) \equiv N_2,$$

where N_1 depends only on the maximum magnitudes of $a(\alpha,x)$, $b(\alpha,x)$, $c^\alpha(x)$ $g(r_1)$, $g(r_2)$, r_1, r_2, δ. Therefore, by the Lagrange theorem, for all x, $y \in [r_1, r_2]$

$$|v_n(x)||x - y| + |v_n(x) - v_n(y)| + |v_n'(x) - v_n'(y)| \le N_2|x - y|.$$

Letting $n \to \infty$, we obtain a similar inequality for the function v. We divide the both sides of the inequality thus obtained by $|x - y|$. Letting y go to x, we then have that for all $x \in [r_1, r_2]$ the sum $|v(x)| + |v'(x)| + |v''(x)|$ does not exceed N_2. Therefore,

$$\|v\|_B + \|v'\|_B + \|v''\|_B \le N_1\left(\sup_{\alpha,x} |f^\alpha(x)| + 1\right).$$

12. Remark. (The smooth pasting condition.) At each point $x \in (r_1, r_2)$,

$$v(x - 0) = v(x + 0), \qquad v'(x - 0) = v'(x + 0),$$
$$v''(x - 0) = v''(x + 0),$$

which fact together with the boundary conditions $v(r_1) = g(r_1)$, $v(r_2) = g(r_2)$ helps us to find x_0, c_1, c_2, d_1, d_2; if, for example, it has been proved that on some interval $[r_1,x_0]$ the function v is representable as $v_1(x,c_1,c_2)$, and on $[x_0,r_2]$ as $v_2(x,d_1,d_2)$, with v_1 and v_2 being known.

13. Exercise

Let $A = [-1,1]$,

$$x_t^\alpha = x + \mathbf{w}_t + \int_0^t \alpha_s\,ds.$$

Prove that the third derivative of the function $v(x) = \sup_{\alpha\in\mathfrak{A}} \mathsf{M}\tau^{\alpha,x}$ is discontinuous at a point $(r_1 + r_2)/2$.

14. Exercise

Using the inequality $v_n'' + F_1(x,v_n,v_n') \geq 0$, prove that $v'' = \underline{\lim}_{n\to\infty} v_n''$ (a.s.).

We shall make a few more remarks on Theorem 5 proved above. We have first proved the existence of a solution of the equation $F[u] = 0$. Second, we have proved the fact that u is equivalent to v. We have thus obtained a theorem on the uniqueness of the solution of the equation $F[u] = 0$ satisfying the boundary condition $u(r_i) = g(r_i)$. One should keep in mind that in the theory of differential equations the existence and uniqueness theorems are proved for a wider class of equations than that of equations of the type (3) (see [3, 33, 43, 46, 47]).

The result of Exercise 13 shows that a payoff function need not have three continuous derivatives even if σ, b, c, f is analytic in (α,x). We note here that if, for example, the function $F_1(x,y,p)$ has 10 continuous derivatives in (x,y,p), v has 12 continuous derivatives. We can easily deduce this by induction from the fact that the equation $F[v] = 0$ is equivalent to the equation $v'' + F_1(x,v,v') = 0$.

The next theorem follows from Remark 11 and the uniform convergence of v^{α_n} to v.

15. Theorem. *For each $\varepsilon > 0$ there exists a function $\alpha(x)$ satisfying a Lipschitz condition and such that the Markov strategy $\alpha_t(x_{[0,t]}) = \alpha(x_t)$ is ε-optimal for all x.*

If the payoff function v has been found, ε-optimal Markov strategies can easily be found with the aid of Lemmas 8 and 9. In fact, using Lemma 9, one can find a function $\alpha(x)$ satisfying a Lipschitz condition and such that

$$L^{\alpha(x)}(x)v(x) + f^{\alpha(x)}(x) + h(x) = 0.$$

In this case $\|h\|_{\mathscr{L}_1} \leq \varepsilon$ and, by Lemma 8, $v(x) \leq v^\alpha(x) + N\varepsilon$.

The problem with optimal strategies is much more complicated. To show that it is true we assume for $\varepsilon \geq 0$ that

$$A^{\varepsilon}(x) = \{\alpha \in A : L^{\alpha}(x)v(x) + f^{\alpha}(x) \geq -\varepsilon\}.$$

It is seen that for some x the set $A^0(x)$ can be empty.

The next theorem is given without proof.

16. Theorem. (a) *If a strategy α_t is optimal for a point $z \in (r_1,r_2)$, the random vector $\alpha_t \in A^0(x_t^{\alpha,z})$ almost surely on a set $\{\tau^{\alpha,z} > t\}$ for almost all t.*

(b) *A Markov strategy $\alpha_t(x_{[0,t]}) = \alpha(x_t)$ admissible for a point $z \in (r_1,r_2)$, is optimal for z if and only if $\alpha(x) \in A^0(x)$ for almost all $x \in (r_1,r_2)$.*

It follows from our theorem that the requirement for optimality imposes a very strict limitation on a strategy. The reader who has solved Exercise 13 can easily understand that the sets $A^0(x)$ $(x \neq 0)$ are empty if $A = (-1,1)$, $x_t^{\alpha} = x + \mathbf{w}_t + \int_0^t \alpha_s\,ds$, $v(x) = \sup_{\alpha} \mathsf{M}_x^{\alpha}\tau$. Consequently, there is no optimal strategy in this case. The case where $A = [-1,1]$ is more interesting (as in Exercise 13). Here $A^0(x) = \{1\}$ for $x \in [r_1,(r_1 + r_2)/2)$; $A^0((r_1 + r_2)/2) = [-1,1]$, $A^0(x) = \{-1\}$ for $x \in ((r_1 + r_2)/2, r_2]$. In this case the function $\alpha(x)$ determining an optimal strategy is to satisfy (at least for almost all x) the following conditions: $\alpha(x) = +1$ for $x \in [r_1,(r_1 + r_2)/2]$; $\alpha((r_1 + r_2)/2) \in [-1,1]$, $\alpha(x) = -1$ for $x \in ((r_1 + r_2)/2, r_2]$. There arises the question of admissibility of the strategy $\alpha(x_t)$ with the function $\alpha(x)$, that is, the question of solvability of the equation

$$x_t = x + \mathbf{w}_t + \int_0^t \alpha(x_s)\,ds$$

with a discountinuous drift coefficient. That the last equation is solvable can easily be proved with the aid of an appropriate transformation $y_t = f(x_t)$ which reduces the initial equation to $y_t = f(x) + \int_0^t \sigma(y_s)\,d\mathbf{w}_s$, where $\sigma(y)$ satisfies a Lipschitz condition. Hence, there exists an optimal strategy in Exercise 13. Equations with coefficients which do not satisfy a Lipschitz condition have not been studied adequately (see, however, [75, 78] and Section 11.6).

In Section 1.1, we used Bellman's principle to deduce the Bellman equation. We now prove this principle.

17. Theorem (Bellman's Principle). *For all $x \in [r_1,r_2]$, $\alpha \in \mathfrak{A}$, let a Markov time $\gamma^{\alpha,x} \leq \tau^{\alpha,x}$ be given. Then, for each function $u(x)$ twice continuously differentiable on the interval $[r_1,r_2]$, and such that $F(u) = 0$, we have on $[r_1,r_2]$ the equality*

$$u(x) = \sup_{\alpha \in \mathfrak{A}} \mathsf{M}_x^{\alpha}\left\{\int_0^{\gamma} f^{\alpha_t}(x_t)e^{-\varphi_t}\,dt + u(x_{\gamma})e^{-\varphi_{\gamma}}\right\}, \tag{16}$$

which, in particular, holds for $u = v$.

PROOF. We denote by $\tilde{u}(x)$ the right side of (16). Taking in (15) instead of $\tau \wedge t$ the expression $\gamma \wedge t$, we easily find that $u(x) \geq \tilde{u}(x)$. On the other hand, for $\varepsilon > 0$, a smooth function $\alpha(x)$ and a function $h(x)$ such that $\|h\|_{\mathscr{L}_1} \leq \varepsilon$,

$$L^{\alpha(x)}(x)u(x) + f^{\alpha(x)}(x) + h(x) = 0,$$

we define the Markov strategy $\alpha_t(x_{[0,t]}) = \alpha(x_t)$. By Ito's formula

$$u(x) = \mathsf{M}_x^\alpha \left\{ \int_0^\gamma [f^{\alpha(x_t)}(x_t) + h(x_t)]e^{-\varphi_t}\, dt + u(x_\gamma)e^{-\varphi_\gamma} \right\},$$

which yields

$$u(x) \leq \tilde{u}(x) + \mathsf{M}_x^\alpha \int_0^\gamma h(x_t)e^{-\varphi_t}\, dt \leq \tilde{u}(x) + \mathsf{M}_x^\alpha \int_0^\tau h(x_t)e^{-\varphi_t}\, dt.$$

We have used the fact that $h \geq 0$ since $0 = F[u] \geq L^{\alpha(x)}u(x) + f^{\alpha(x)}(x) = -h(x)$. By Lemma 8, the last expectation as a function of x satisfies the equation

$$L^{\alpha(x)}u_1(x) + h(x) = 0.$$

Therefore, by Lemma 6, the above expectation does not exceed $N\|h\|_{\mathscr{L}_1} \leq N\varepsilon$. Finally, $u(x) \leq \tilde{u} + N\varepsilon$ for each $\varepsilon > 0$. Hence $u \leq \tilde{u}$, which, together with the converse inequality proved before, yields the equality $u = \tilde{u}$, thus proving the theorem. □

18. Exercise

Prove that Eq. (16) will hold if we require only that $f^\alpha(x)$ be measurable in x, continuous in α, and bounded with respect to (α, x).

19. Exercise

For $h \in \mathscr{L}_1[r_1, r_2]$ let

$$u(x) = \sup_{\alpha \in \mathfrak{A}} \mathsf{M}_x^\alpha \int_0^\tau h(x_t)\, dt.$$

Prove that $|u(x)| \leq N\|h\|_{\mathscr{L}_1}$, where N does not depend on h, x.

5. Optimal Stopping of a One-Dimensional Controlled Process

We consider the control scheme given in Section 1.4. To this end, we take the same set A and the functions $\sigma(\alpha,x)$, $b(\alpha,x)$, $c^\alpha(x)$, $f^\alpha(x)$ satisfying the conditions given in Section 1.4. For simplicity of notation, we assume that $c^\alpha(x) \equiv 0$. In contrast to what we did in Section 1.4, we assume here that the function $g(x)$ is given on the entire interval $[r_1, r_2]$ and that $g(x)$ is twice continuously differentiable on this interval. As in Section 1.4, we denote by $x_t^{\alpha,x}$ a solution of Eq. (4.1), and assume that $\tau = \tau^{\alpha,x}$ is the time of first departure of $x_t^{\alpha,x}$ from $[r_1,r_2]$.

For a Markov time γ we set

$$v^{\alpha,\gamma}(x) = \mathsf{M}_x^\alpha\left\{\int_0^{\gamma\wedge\tau} f^{\alpha_t}(x_t)\,dt + g(x_{\gamma\wedge\tau})\right\}$$

and introduce a payoff function in the optimal stopping problem defined by

$$w(x) = \sup_{\alpha\in\mathfrak{A},\,\gamma} v^{\alpha,\gamma}(x).$$

In this section, we deal with the problem of finding a strategy $\alpha \in \mathfrak{A}$ and a Markov time γ, such that $v^{\alpha,\gamma}(x) \geq w(x) - \varepsilon$.

1. Definition. Let $\varepsilon \geq 0$. A Markov (with respect to $\{\mathscr{F}_t\}$) time $\gamma = \gamma^\alpha$ is said to be ε-optimal for a point x if

$$\sup_{\alpha\in\mathfrak{A}} v^{\alpha,\gamma^\alpha}(x) \geq w(x) - \varepsilon.$$

A 0-optimal Markov time is said to be an optimal Markov time.

We shall investigate the optimal stopping problem using the method of randomized stopping. We denote by $\mathfrak{B}_n$ the set of pairs (α,r), where $\alpha \in \mathfrak{A}$, $r = r_t$, is a nonnegative progressively measurable (with respect to $\{\mathscr{F}_t\}$) process such that $r_t(\omega) \leq n$ for all (t,ω). Let $\mathfrak{B} = \bigcup_n \mathfrak{B}_n$.

For $\alpha \in A$, $r \geq 0$, let $f^{\alpha,r}(x) = f^\alpha(x) + rg(x)$ and for $(\alpha,r) \in \mathfrak{B}$ let

$$\tilde{v}^{\alpha,r}(x) = \mathsf{M}\left\{\int_0^{\tau^{\alpha,x}} f^{\alpha_t,r_t}(x_t^{\alpha,x})\exp\left(-\int_0^t r_s\,ds\right)dt + g(x_{\tau^{\alpha,x}}^{\alpha,x})\exp\left(-\int_0^{\tau^{\alpha,x}} r_s\,ds\right)\right\},$$

$$\tilde{v}_n(x) = \sup_{(\alpha,r)\in\mathfrak{B}_n} \tilde{v}^{\alpha,r}(x), \qquad \tilde{v}(x) = \lim_{n\to\infty}\tilde{v}_n(x) = \sup_{(\alpha,r)\in\mathfrak{B}} \tilde{v}^{\alpha,r}(x).$$

The main properties of functions $\tilde{v}_n(x)$ and the relationship of these functions with $w(x)$ will be proved in the following lemma, whose first assertion justifies, in addition, the application of the method of randomized stopping.

2. Lemma. (a) $w(x) = \tilde{v}(x)$ *on* $[r_1,r_2]$.

(b) $|w(x) - \tilde{v}_n(x)| \leq (1/n)N$ *for all* $x \in [r_1,r_2]$, *where* N *depends only on* K *and the function* g.

(c) $\tilde{v}_n(x)$ *are twice continuously differentiable on* $[r_1,r_2]$, $\tilde{v}_n''(x)$ *satisfies a Lipschitz condition*, $\tilde{v}_n(r_i) = g(r_i)$, *and*

$$F[\tilde{v}_n] + n(g - \tilde{v}_n)_+ = 0 \quad on\ [r_1,r_2]. \tag{1}$$

(d) $\|\tilde{v}_n''\|_{B[r_1,r_2]} + \|\tilde{v}_n'\|_{B[r_1,r_2]} + \|\tilde{v}_n\|_{B[r_1,r_2]} \leq N$, *where* N *does not depend on* n.

PROOF. The function $\tilde{v}_n(x)$ is representable as the payoff function given in Section 1.4 if we take $B_n = A \times [0,n]$ instead of the set A, and if we assume that $\sigma(\beta,x) = \sigma(\alpha,x)$, $b(\beta,x) = b(\alpha,x)$, $c^\beta(x) = r$, $f^\beta(x) = f^\alpha(x) + rg(x)$ for $\beta = (\alpha,r) \in B_n$. β becomes a control parameter, and $\mathfrak{B}_n$ replaces a set of strategies $\mathfrak{A}$. Hence, Theorem 4.5 immediately implies the assertion on smoothness of $\tilde{v}_n(x)$ and the fact that $\tilde{v}_n(x)$ satisfies the corresponding Bellman equation. This equation is the following:

$$0 = \sup_{\alpha \in A, r \in [0,n]} [a(\alpha,x)\tilde{v}_n''(x) + b(\alpha,x)\tilde{v}_n'(x) - r\tilde{v}_n(x) + rg(x) + f^\alpha(x)]$$

$$\sup_{\alpha \in A} [L^\alpha \tilde{v}_n(x) + f^\alpha(x)] + \sup_{r \in [0,n]} r[g(x) - \tilde{v}_n(x)] = F[\tilde{v}_n] + n(g - \tilde{v}_n)_+,$$

which proves (c).

In order to prove (b), we write (1) as

$$\sup_{\alpha \in A} [a(\alpha,x)\tilde{v}_n''(x) + b(\alpha,x)\tilde{v}_n'(x) + f_n^\alpha(x)] = 0,$$

where $f_n^\alpha = f^\alpha + n(g - \tilde{v}_n)_+$.

From this, using Theorem 4.17, we have for all Markov times $\gamma = \gamma^{\alpha,x}$ that

$$\tilde{v}_n(x) = \sup_{\alpha \in \mathfrak{A}} \mathsf{M}_x^\alpha \left[\int_0^{\gamma \wedge \tau} f_n^{\alpha_t}(x_t)\, dt + \tilde{v}_n(x_{\gamma \wedge \tau}) \right] \tag{2}$$

where $\tilde{v}_n \geq g - (g - \tilde{v}_n)_+ \equiv g_n$ and $f_n^\alpha \geq f^\alpha$; therefore,

$$\tilde{v}_n(x) \geq \sup_{\alpha \in \mathfrak{A}, \gamma} \mathsf{M}_x^\alpha \left[\int_0^{\gamma \wedge \tau} f^{\alpha_t}(x_t)\, dt + g_n(x_{\gamma \wedge \tau}) \right]. \tag{3}$$

On the other hand, if we take in (2) $\gamma = \gamma_0 = \gamma_0^{\alpha,x} = \inf\{t : g(x_t^{\alpha,x}) \geq \tilde{v}_n(x_t^{\alpha,x})\}$, then for $0 \leq t \leq \gamma_0^{\alpha,x}$ we have $f_n(x_t^{\alpha,x}) = f(x_t^{\alpha,x})$ and $\tilde{v}_n(x_{\tau \wedge \gamma_0}^{\alpha,x}) = g_n(x_{\tau \wedge \gamma_0}^{\alpha,x})$. Therefore,

$$\tilde{v}_n(x) = \sup_{\alpha \in \mathfrak{A}} \mathsf{M}_x^\alpha \left[\int_0^{\gamma_0 \wedge \tau} f^{\alpha_t}(x_t)\, dt + g_n(x_{\gamma_0 \wedge \tau}) \right].$$

Comparing the last equality with (3), we obtain

$$\tilde{v}_n(x) = \sup_{\alpha \in \mathfrak{A}, \gamma} \mathsf{M}_x^\alpha \left[\int_0^{\gamma \wedge \tau} f^{\alpha_t}(x_t)\, dt + g_n(x_{\gamma \wedge \tau}) \right],$$

which makes a crucial point in our proof. It turns out that if g is replaced with $g_n = g - (g - \tilde{v}_n)_+$, $\tilde{v}_n(x)$ will become a payoff function in the optimal stopping problem.

Further, using the inequality connecting the magnitude of the difference between the upper bounds and the upper bound of the magnitude of the differences, we find

$$|w(x) - \tilde{v}_n(x)| \leq \sup_{\alpha \in \mathfrak{A}, \gamma} \mathsf{M}_x^\alpha |g - g_n|(x_{\gamma \wedge \tau}).$$

Therefore,

$$|w(x) - \tilde{v}_n(x)| \qquad \sup_{x \in [r_1, r_2]} |(g(x) - \tilde{v}_n(x))_+|. \tag{4}$$

In order to estimate $(g - \tilde{v}_n)_+$, we write (1) as follows:

$$\sup_{\alpha \in A} \left[a(\alpha,x)(\tilde{v}_n(x) - g(x))'' + b(\alpha,x)(\tilde{v}_n(x) - g(x))' - n(\tilde{v}_n(x) - g(x)) + \tilde{f}_n^\alpha(x)\right] = 0,$$

where $\tilde{f}_n^\alpha = f^\alpha + n(g - \tilde{v}_n)_+ + n(\tilde{v}_n - g) + L^\alpha g$, which yields, by Theorem 4.17,

$$\tilde{v}_n(x) - g(x) = \sup_{\alpha \in \mathfrak{A}} \mathsf{M}_x^\alpha \int_0^\tau e^{-nt} \tilde{f}_n^{\alpha_t}(x_t)\, dt.$$

Note that $\tilde{f}_n^\alpha \geq f^\alpha + L^\alpha g \geq -(f^\alpha + L^\alpha g)_-$; hence

$$\tilde{v}_n(x) - g(x) \geq -\sup_{\alpha \in \mathfrak{A}} \mathsf{M} \int_0^\infty e^{-nt}\, dt \sup_{\alpha,x} (f^\alpha + L^\alpha g)_- = -\frac{1}{n} N,$$

$$g - \tilde{v}_n \leq \frac{1}{n} N, \qquad (g - \tilde{v}_n)_+ \leq \frac{1}{n} N, \tag{5}$$

which, together with (4), completes the proof of (b). (a) follows obviously from (b). Next, we prove (d).

We have from (2) that

$$\tilde{v}_n(x) = \sup_{\alpha \in \mathfrak{A}} \mathsf{M}_x^\alpha \left[\int_0^\tau \tilde{f}_n^{\alpha_t}(x_t)\, dt + g(x_\tau)\right].$$

It can easily be verified that the function $\tilde{f}_n^\alpha(x)$ satisfies a Lipschitz condition. According to Remark 4.11,

$$\|\tilde{v}_n\|_B + \|\tilde{v}_n'\|_B + \|\tilde{v}_n''\|_B \leq N\left(\sup_{\alpha,x} |\tilde{f}_n^\alpha(x)| + 1\right),$$

where N depends only on the maximum magnitudes of $a(\alpha,x)$, $b(\alpha,x)$, $g(r_i)$, r_1, r_2, δ. It remains only to note that $|\tilde{f}_n^\alpha| \leq |f^\alpha| + n(g - \tilde{v}_n)_+$, and that the both terms in the right side can be estimated with the aid of constants not depending on n. We have thus proved the lemma. □

We can easily derive the desired properties of the function $w(x)$ from the lemma proved. It follows from (d) and the fact that $\tilde{v}_n$ converges to w in the same way as in 4.10 that w, w' are continuous on $[r_1,r_2]$, $\tilde{v}_n' \to w'$ uniformly on $[r_1,r_2]$, $|w'(x) - w'(y)| \leq N|x - y|$, which implies that $w'(x)$ is absolutely continuous, $w''(x)$ exists almost everywhere on $[r_1,r_2]$, and $w''(x)| \leq N$ on the set on which w'' exists. From now on we shall denote by $w''(x)$ the function defined everywhere on $[r_1,r_2]$, equal to the second derivative of $w(x)$ at those points at which this derivative exists, and equal to zero at the remaining points.

It is seen that $w \geq g$ on $[r_1,r_2]$, $w(r_i) = g(r_i)$. Furthermore, it follows from Lemma 2c that $F[\tilde{v}_n] \leq 0$; therefore $\tilde{v}_n'' + F_1(x,\tilde{v}_n,\tilde{v}_n') \leq 0$. Integrating this inequality and passing to the limit, we find for $x > y$ that

$$w'(x) - w'(y) + \int_y^x F_1(s,w(s),w'(s))\, ds \leq 0.$$

We divide the both sides of the last inequality by $x - y$ and take the limit as $y \uparrow x$. Then, $w'' + F_1(x,w,w') \le 0$ (a.s.), i.e., $F[w] \le 0$ (a.s.) on $[r_1,r_2]$.

Further, let $\Gamma = \{x : w(x) = g(x)\}$. Γ is a closed nonempty ($r_i \in \Gamma$) subset of the interval $[r_1,r_2]$. Let $[\rho_1,\rho_2]$ be a subinterval not intersecting Γ. Then $w(x) > g(x)$ for $x \in [\rho_1,\rho_3]$. Since $\tilde{v}_n \to w$ uniformly on $[r_1,r_2]$, the inequality $\tilde{v}_n(x) > g(x)$ will be satisfied for $x \in [\rho_1,\rho_2]$, beginning from some n. Therefore, $(g - \tilde{v}_n)_+ = 0$, and by Lemma 2c $F[\tilde{v}_n] = 0$ on $[\rho_1,\rho_2]$. Hence $\tilde{v}_n'' + F_1(x,\tilde{v}_n,\tilde{v}_n') = 0$ on $[\rho_1,\rho_2]$ for sufficiently large n, which leads us, as in 4.10, to the assertion that $w''(x)$ is continuous on $[\rho_1,\rho_2]$, $w'' + F_1(x,w,w') = 0$ on $[\rho_1,\rho_2]$.

The above implies, in turn, two facts. First, $w'' = -F_1(x,w,w')$ on any subinterval not intersecting Γ; therefore w'' outside Γ satisfies a Lipschitz condition. Second, $F[w] = 0$ outside Γ.

Finally, noting that it follows from $w - g \ge 0$ and $w - g = 0$ on Γ that $w' - g' = 0$ on $\Gamma \cap (r_1,r_2)$, we have the following theorem.

3. Theorem. (a) *w together with its derivative is continuous on $[r_1,r_2]$, w' is absolutely continuous, and w'' is bounded on $[r_1,r_2]$. The function w'' satisfies a Lipschitz condition outside the set $\Gamma = \{x \in [r_1,r_2] : w(x) = g(x)\}$.*

(b) *$w \ge g$, $w(r_i) = g(r_i)$, $F[w] \le 0$ (a.s.) and $F[w] = 0$ on $[r_1,r_2]\backslash\Gamma$.*

(c) *$w' = g'$ on the set $\Gamma \cap (r_1,r_2)$.*

Next, we investigate ε-optimal strategies and optimal stopping times.

4. Theorem. (a) *For $\alpha \in \mathfrak{A}$, $x \in [r_1,r_2]$ we denote by $\gamma_0 = \gamma^{\alpha,x}$ the time of first entry of a process $x_t^{\alpha,x}$ into the set Γ; then, γ_0 is an optimal stopping time.*

(b) *We define for $\varepsilon > 0$ a function $\alpha(x)$ satisfying a Lipschitz condition. In addition, we define a numerical function $h(x)$ such that $\|h\|_{\mathscr{L}_1} \le \varepsilon$ and*

$$L^{\alpha(x)}w(x) + f^{\alpha(x)}(x) + h(x) \ge F[w](x), \qquad x \in [r_1,r_2] \tag{6}$$

(see Lemma 4.9). Also, we define a Markov strategy α using the formula $\alpha_t(x_{[0,t]}) = \alpha(x_t)$. Then, α is an $N\varepsilon$-optimal strategy for any point x and, moreover,

$$w(x) \le \mathsf{M}_x^{\alpha}\left[\int_0^{\gamma_0} f^{\alpha_t}(x_t)\,dt + g(x_{\gamma_0})\right] + N\varepsilon, \tag{7}$$

N not depending on ε and x.

Proof. It is easily seen that only (7) is to be proved, and that the remaining assertions of our theorem follow from (7). It is obvious that (7) is to be proved only for $x \notin \Gamma$.

Let (ρ_1,ρ_2) be a subinterval not intersecting Γ. The function w is twice continuously differentiable and w'' satisfies a Lipschitz condition on (ρ_1,ρ_2). Hence, the limit $w''(\rho_1 + 0)$ and the limit $w''(\rho_2 - 0)$ exist.

Furthermore, it follows from (6) that on (ρ_1,ρ_2)

$$L^{\alpha(x)}w(x) + f^{\alpha(x)}(x) + h(x) \ge 0.$$

We apply Lemma 4.8 to the function w and the strategy α taking an interval $[\rho_1,\rho_2]$ as an initial one. Then

$$w(x) \leq \mathsf{M}_x^\alpha\left[\int_0^{\gamma_0} f^{\alpha_t}(x_t)\,dt + g(x_{\gamma_0})\right] + N\|h\|_{\mathscr{L}_1[\rho_1,\rho_2]}.$$

A simple analysis of the deduction of (4.6) shows that the constant N can be taken to be the same for all ρ_1, $\rho_2 \in [r_1,r_2]$. It shows also that

$$\|h\|_{\mathscr{L}_1[\rho_1,\rho_2]} \leq \|h\|_{\mathscr{L}_1[r_1,r_2]} \leq \varepsilon,$$

thus proving the theorem. □

5. Exercise

Let $\gamma^{\alpha,x} = \gamma_\varepsilon^{\alpha,x}$ be the first exit time of the process $x_t^{\alpha,x}$ from $\{x: w(x) > g(x) + \varepsilon\}$; then $\gamma^{\alpha,x}$ is an ε-optimal stopping time.

6. Exercise

Let $w(x) > g(x)$ on (ρ_1,ρ_2) and let $w(\rho_i) \leq g(\rho_i) + \varepsilon$. Find the strategy α_ε which is ε-optimal for a point $x_0 \in (\rho_1,\rho_2)$ in the maximization problem

$$\mathsf{M}_x^\alpha\left[\int_0^{\tau_1} f^{\alpha_t}(x_t)\,dt + g(x_{\tau_1})\right],$$

where τ_1 is the first exit time from (ρ_1,ρ_2) (see Theorem 4.15). Then

$$v(x_0) \leq \mathsf{M}\left[\int_0^{\tau_1} f^{\alpha_t}(x_t^{\alpha,x_0})\,dt + g(x_{\tau_1}^{\alpha,x_0})\right] + 2\varepsilon,$$

that is, τ_1 is a 2ε-optimal stopping time and α is a 2ε-optimal strategy in the primary problem for the point x_0.

We shall explain how the preceding results can be used for finding ε-optimal strategies and ε-optimal stopping times. First, we find n such that $|\tilde{v}_n - w| \leq \varepsilon/4$ (see Lemma 2b). The function $\tilde{v}_n$ is a solution of the equation

$$\sup_{\alpha\in A,\, r\in[0,n]} [a(\alpha,x)\tilde{v}_n''(x) + b(\alpha,x)\tilde{v}_n'(x) - r\tilde{v}_n(x) + rg(x) + f^\alpha(x)] = 0;$$

hence $\tilde{v}_n$ can be found via the method of successive approximation in the space of strategies (see 4.10). Let $\tilde{v}_n^m \to \tilde{v}_n$ as $m \to \infty$. We take m such that $|\tilde{v}_n(x) - \tilde{v}_n^m(x)| \leq \varepsilon/4$ for $x \in [r_1,r_2]$. Let

$$G = \left\{x \in (r_1,r_2): \tilde{v}_n^m(x) > g(x) + \frac{\varepsilon}{2}\right\}.$$

It can be easily verified that $w(x) \quad g(x) + \varepsilon$ on $[r_1,r_2]\backslash G$, and if $(\rho_1,\rho_2) \subset G$, then $w > g$ on (ρ_1,ρ_2). Therefore, the ε-optimal strategy for the points of $[r_1,r_2]\backslash G$ consists in instantaneous stopping. For the points of any interval $(\rho_1,\rho_2) \subset G$ the first exit time from G is an ε-optimal stopping time (Exercise 5); we can find ε-optimal strategies using Exercise 6.

In some cases it is difficult to do this. However, sometimes it is possible to define a function $u(x)$ explicitly such that $u(x)$ seems to coincide with w. In such cases the following theorem can be useful.

7. Theorem (On Uniqueness). *Let the function $u(x)$ together with its first derivative be defined and continuous on $[r_1,r_2]$. Assume that $u'(x)$ is absolutely continuous on $[r_1,r_2]$. Finally, let $u \geq g$, $u(r_i) = g(r_i)$, $F[u] \leq 0$ (a.s.). $F(u) = 0$ (a.s.) on the set $\{x \in [r_1,r_2]:u(x) > g(x)\}$. Then $u(x) = w(x)$.*

PROOF. We first prove that $u \leq w$. Let $\Gamma = \{x:u(x) = g(x)\}$. Since $g \leq w$, it suffices to establish the inequality $u \leq w$ on any subinterval (ρ_1,ρ_2) not intersecting Γ.

Let the sequences ρ_i^n be such that $\rho_i^n \neq \rho_i$, $\rho_1^n \downarrow \rho_1$, $\rho_2^n \uparrow \rho_2$. We immediately note that by the Lagrange theorem,

$$|g(\rho_i^n) - u(\rho_i^n)| \leq |u(\rho_i) - u(\rho_i^n)| + |g(\rho_i) - g(\rho_i^n)|$$
$$\leq (\|u'\|_B + \|g'\|_B)|\rho_i - \rho_i^n| \to 0$$

as $n \to \infty$.

Further, on (ρ_1,ρ_2) we have $F[u] = 0$ (a.s.), which yields $u'' + F_1(x,u,u') = 0$, $u'' = -F_1(x,u,u')$ (a.s.). The expression $F_1(x,u,u')$ is continuous in x, therefore u'' coincides with a continuous function almost everywhere on (ρ_1,ρ_2). This readily implies that the function u'' itself is continuous on (ρ_1,ρ_2).

Next, we apply Theorem 4.5 to the function u on the interval $[\rho_1^n,\rho_2^n]$. Denoting by $\gamma^n = \gamma^{n,\alpha,x}$ the first exit time of $x_t^{\alpha,x}$ from $[\rho_1^n,\rho_2^n]$, and noting that $u(x_{\gamma^n}^{\alpha,x})$ is equal to $u(\rho_1^n)$ or $u(\rho_2^n)$, we have

$$u(x) = \sup_{\alpha \in \mathfrak{A}} \mathsf{M}_x^\alpha \left[\int_0^{\gamma^n} f^{\alpha_t}(x_t)\,dt + u(x_{\gamma^n}) \right]$$
$$\leq \sup_{\alpha \in \mathfrak{A}} \mathsf{M}_x^\alpha \left[\int_0^{\gamma^n} f^{\alpha_t}(x_t)\,dt + g(x_{\gamma^n}) \right] + |g(\rho_1^n) - u(\rho_1^n)| + |g(\rho_2^n) - u(\rho_2^n)|$$
$$\leq w(x) + |g(\rho_1^n) - u(\rho_1^n)| + |g(\rho_2^n) - u(\rho_2^n)|$$

which implies, as $n \to \infty$, that $u(x) \leq w(x)$. Further, we prove the converse inequality.

We denote by $u''(x)$ a Borel function equal almost everywhere to a derivative of $u'(x)$. We can take a function $u''(x)$ such that the inequality $F[u] \leq 0$ is satisfied at all points $[r_1,r_2]$. In fact, by assumption this inequality can be satisfied almost everywhere. We can redefine $u''(x)$ at the points at which $F[u](x) > 0$ for a randomly selected u'', if we note that, in conjunction with the obvious inequality $F(x,y,p,r) \leq \delta r + K(|p| + |y| + 1)$ for $r \leq 0$, for any x, y, p one can choose $r \leq 0$ so that $F(x,y,p,r) \leq 0$. In this case, $L^\alpha u + f^\alpha \leq F[u] \leq 0$ everywhere on $[r_1,r_2]$ for each $\alpha \in A$.

Since $u'(x)$ is absolutely continuous, $\int_{r_1}^{r_2}|u''(x)|dx < \infty$. By Theorem 2.10.1, the fact that this condition is satisfied is sufficient for Ito's formula to be applied to the expression $u(x_t^{\alpha,x})$. Using Ito's formula and the inequalities $u \geq g$, $L^\alpha u + f^\alpha \leq 0$ for each $\alpha \in \mathfrak{A}$ and for each Markov time γ, we conclude that

$$u(x) = \mathsf{M}\left[-\int_0^{\gamma \wedge \tau} L^{\alpha_t}u(x_t^{\alpha,x})\,dt + u(x_{\gamma\wedge\tau}^{\alpha,x})\right]$$

$$\leq \mathsf{M}\left[\int_0^{\gamma \wedge \tau} f^{\alpha_t}(x_t^{\alpha,x})\,dt + g(x_{\gamma\wedge\tau}^{\alpha,x})\right] = v^{\alpha,\gamma}(x).$$

Therefore, $u(x) \geq \sup_{\alpha,\gamma} v^{\alpha,\gamma}(x) = w(x)$. thus proving the theorem. □

The arguments in the above proof can prove as well the following theorem.

8. Theorem. *Let a function $u(x)$, together with its first derivative, be definite and continuous on $[r_1,r_2]$. Assume that $u'(x)$ is absolutely continuous on $[r_1,r_2]$. If $u \geq g$ and $F[u] \leq 0$ (a.s.), then $u(x) \geq w(x)$ on $[r_1,r_2]$.*

In other words, w is the smallest function among those satisfying the inequalities $u \geq g$, $F[u] \leq 0$ on $[r_1,r_2]$.

9. Exercise

We take the function $c^\alpha(x)$ from Section 1.4 and redefine $v^{\alpha,\gamma}(x)$, letting

$$v^{\alpha,\gamma}(x) = \mathsf{M}_x^\alpha\left\{\int_0^{\gamma\wedge\tau} f^{\alpha_t}(x_t)\exp\left[-\int_0^t c^{\alpha_s}(x_s)\,ds\right]dt + g(x_{\gamma\wedge\tau})\exp\left[-\int_0^{\gamma\wedge\tau} c^{\alpha_s}(x_s)\,ds\right]\right\}.$$

We encourage the reader to develop the actual arguments required in this section.

Notes

Section 1. Girsanov in [25] was apparently the first who justified the application of Bellman's equation to some control problems, relying to a great extent upon differential equations theory. Using the same theory, Fleming in [13–16] made further steps in the development of optimal control theory; also, see Fleming and Rishel [17]. Speaking of the relationship between differential equations theory and optimal control theory, it is appropriate to draw the reader's attention to [20, 21, 30, 31, 35, 66]. The control variables depending upon the entire "past" for processes with continuous time is first discussed by Fleming in [15].

Section 2. The normed Bellman equation was first introduced in [37]. The method of randomized stopping is developed in [29–31, 37]. The optimal stopping problem of a Markov (uncontrolled) process is discussed by Shiryayev in [69]. It is essential to note that the equations from optimal stopping theory are, in many cases, equivalent to the equations from the theory of differential (variational) inequalities; see Lions [49, 50], Lewy and Stampacchia [48], and Tobias [73, 74]. The comparison of Exercise

4 with Example 3.7 in Chapter 1 in [50] shows the relationship of other kinds between the optimal control theory of diffusion processes and the variational inequalities theory.

Section 3. Other examples illustrating the application of optimal control theory for obtaining estimates which play an essential role in this theory may be found in [39,40].

Section 4. In a sense, our discussion consists of carrying the results and methods of Fleming in [13] over to the one-dimensional case. In the multivariate ($d \geq 3$) case, using the methods mentioned above, it is possible to consider *only* optimal control problems in which the control parameter is not included in the diffusion coefficients; see [13–17, 21]. The reason for this is that it is impossible (in view of a well-known example due to N. N. Uraltseva) to prove a suitable analog of Lemma 6 for $d \geq 3$. At the same time it is possible to work out a theory rather similar to that described in this section for the plane ($d = 2$), and, to allow the control parameter to be included in the coefficients of diffusion as well as drift; see [30, 31, 35, 66].

The method used in proving Theorem 5 differs, in fact, from the Bellman–Howard method to the extent that in the latter method, α_{n+1} is determined by the condition $F[v_n] = L^{\alpha_{n+1}}v_n + f^{\alpha_{n+1}}$. P. Mosolov drew the author's attention to the fact that the Bellman–Howard method follows the Newton–Kantorowich method for solving nonlinear function equations (see [3]). We note that the Bellman–Howard method applied to functional equations led to the quasilinearization method (see Bellman and Kalaba [5]). For one-dimensional control problems, see Mandl [52, 53], Prokhorov [64], Arkin, Kolemayev, and Shiryayev [1], and Safonov [65].

Section 5. The methods developed in this section have been borrowed from [29, 31]. Some hints as to how to find the set Γ can be found in Section 3.4 as well as in [56]. For the solution of equations arising from the sequential analysis, see also Shiryayev [69].

Auxiliary Propositions 2

1. Notation and Definitions

In addition to the notation given on pages xi and xii we shall use the following:

T is a nonnegative number, and the interval $[0,T]$ is interpreted as an interval of time; the points on this interval are, as a rule, denoted by t, s.

D denotes an open set in Euclidean space, $\bar{D}$ the closure of D, and ∂D the boundary of D.

Q denotes an open set in E_{d+1}; the points of Q are expressed as (t,x) where $t \in E_1$, $x \in E_d$. $\partial' Q$ denotes the parabolic boundary of Q (see Section 4.5).

$$S_R = \{x \in E_d : |x| < R\}, \qquad C_{T,R} = (0,T) \times S_R, \qquad C_R = C_{\infty,R},$$
$$H_T = (0,T) \times E_d.$$

If v is a countably additive set function, then $|v|$ is the variation of v, $v_+ = \frac{1}{2}(|v| + v)$ is the positive part of v, and $v_- = \frac{1}{2}(|v| - v)$ is the negative part of v.

If Γ denotes a measurable set in Euclidean space, meas Γ is the Lebesgue measure of this set.

For $p \geq 1$ $\mathscr{L}_p(\Gamma)$ denotes a set of real-valued Borel functions $f(x)$ on Γ such that

$$\|f\|_{p,\Gamma} \equiv \left(\int_\Gamma |f(x)|^p \, dx \right)^{1/p} < \infty.$$

In the cases where the middle expression is equal to infinity, we continue to denote it by $\|f\|_{p,\Gamma}$ as before. In general, we admit infinite values for various integrals (and mathematical expectations) of measurable functions. These values are considered to be defined if either the positive part or the negative part of the function has a finite integral. In this case the integral is assumed to be equal to $+\infty$ ($-\infty$) if the integral of the positive (negative) part of the function is infinite.

For any (possibly, nonmeasurable) function $f(x)$ on Γ we define an exterior norm in $\mathscr{L}_p(\Gamma)$, using the formula

$$]\!|f|\!|_{p,\Gamma} = \inf \|h\|_{p,\Gamma},$$

where the lower bound is taken over the set of all Borel functions $h(x)$ on Γ such that $|f| \le h$ on Γ. We shall use the fact that the exterior norm satisfies the triangle inequality: $]\!|f_1 + f_2|\!|_{p,\Gamma} \le]\!|f_1|\!|_{p,\Gamma} +]\!|f_2|\!|_{p,\Gamma}$. Also, we shall use the fact that if $]\!|f_n|\!|_{p,\Gamma} \to 0$ as $n \to \infty$, there is a subsequence $\{n'\}$ for which $f_{n'}(x) \to 0$ as $n' \to \infty$ (Γ-a.s.).

$B(\Gamma)$ denotes the set of bounded Borel functions on Γ with the norm

$$\|f\|_{B(\Gamma)} = \sup_{x \in \Gamma} |f(x)|.$$

$C(\Gamma)$ denotes the set of continuous (possibly, unbounded) functions on Γ.

f *is a smooth function* means that f is infinitely differentiable. We say that f has compact support in a region D if it vanishes outside some compact subset of D.

$C_0^\infty(D)$ denotes the set of all smooth functions with compact support in the region D.

$$f_{(y)}(t,x) \equiv \frac{1}{|y|} \sum_i y^i f_{x^i}(t,x) \quad \text{if } |y| \ne 0; \qquad f_{(0)}(t,x) \equiv 0;$$

$$f_{(y_1)(y_2)}(t,x) \equiv \frac{1}{|y_1| \cdot |y_2|} \sum_{i,j} y_1^i y_2^j f_{x^i x^j}(t,x) \quad \text{if } |y_1| \cdot |y_2| \ne 0;$$

$$f_{(y_1)(y_2)}(t,x) \equiv 0 \quad \text{if } |y_1| \cdot |y_2| = 0.$$

We introduce $f_{(y_1)\cdots(y_n)}$. These elements are derivatives of $f(t,x)$ along spacial directions. The time derivative is always expressed as $(\partial/\partial t) f(t,x)$.

$C^2(\bar{D})$ denotes the set of functions $u(x)$ twice continuously differentiable in $\bar{D}$ (i.e., twice continuously differentiable in D and such that $u(x)$ as well as all first and second derivatives of $u(x)$ have extensions continuous in $\bar{D}$).

$C^{1,2}(\bar{Q})$ denotes the set of functions $u(t,x)$ twice continuously differentiable in x and once continuously differentiable in t in $\bar{Q}$.

Let D be a bounded region in E_d, and let $u(x)$ be a function in $\bar{D}$. We write $u \in W^2(D)$ if there exists a sequence of functions $u^n \in C^2(\bar{D})$ such that

$$\|u - u^n\|_{B(\bar{D})} \to 0, \qquad \|u^n - u^m\|_{W^2(D)} \to 0 \tag{1}$$

as $n, m \to \infty$, where

$$\|f\|_{W^2(D)} \equiv \sum_{i,j=1}^d \|f_{x^i x^j}\|_{d,D} + \sum_{i=1}^d \|f_{x^i}\|_{d,D} + \|f\|_{B(\bar{D})}.$$

Under the first condition of (1) and due to the continuity property of u^n, the functions in $W^2(D)$ are continuous in $\bar{D}$. The second condition in (1) implies that the sequences $u^n_{x^i}$, $u^n_{x^ix^j}$ are fundamental in $\mathscr{L}_d(D)$. Hence there exist (Borel) functions u_i, $u_{ij} \in \mathscr{L}_d(D)$, to which $u^n_{x^i}$, $u^n_{x^ix^j}$ converge in $\mathscr{L}_d(D)$. These sequences $u^n_{x^i}$, $u^n_{x^ix^j}$ converge weakly as well to the functions given above. In particular, assuming $\varphi \in C_0^\infty(D)$, and integrating by parts, we obtain

$$\int_D \varphi u^n_{x^i}\,dx = -\int_D \varphi_{x^i} u^n\,dx,$$

Letting $n \to \infty$, we obtain

$$\int_D \varphi u_i\,dx = -\int_D \varphi_{x^i} u\,dx. \tag{2}$$

1. Definition. Let $D \subset E_d$, let v and h be Borel functions locally summable in D, and let $l_1, \ldots, l_n \in E_d$. The function h is said to be a generalized derivative (in the region D) of the function v of order n in the $l_1, \ldots, l_n$ directions and this function h is denoted by $v_{(l_1)\cdots(l_n)}$ if for each $\varphi \in C_0^\infty(D)$

$$\int_D \varphi(x)h(x)\,dx = (-1)^n \int_D v(x)\varphi_{(l_1)\cdots(l_n)}\,dx.$$

In the case where the l_i direction coincides with the direction of the r_ith coordinate vector, the above function is expressed in terms of $v_{x^{r_1}\cdots x^{r_n}} = v_{(l_1)\cdots(l_n)}$.

The properties of a generalized derivative are well known (see [57, 71, 72]. We shall list below only those properties which we use frequently, without proving them. Note first that a generalized derivative can be defined uniquely almost everywhere.

Equation (2) shows that $u_i = u_{x^i}$ in the sense of Definition 1. Similarly, $u_{ij} = u_{x^ix^j}$. Therefore, the functions $u \in W^2(D)$ have generalized derivatives up to and including the second order. Furthermore, these derivatives belong to $\mathscr{L}_d(D)$. We assume that the values of first and second derivatives of each function $u \in W^2(D)$ are fixed at each point. By construction, for the sequence u^n entering (1),

$$\|u^n_{x^i} - u_{x^i}\|_{d,D} \to 0, \qquad \|u^n_{x^ix^j} - u_{x^ix^j}\|_{d,D} \to 0.$$

The set of functions $W^2(D)$ introduced resembles the well-known Sobolev space $W_d^2(D)$ (see [46, 71, 72]). If the boundary of the region D is sufficiently regular, for example, it is once continuously differentiable; Sobolev's theorem on imbedding (see [46, 47]) shows that, in fact, $W^2(D) = W_d^2(D)$. In this case $u \in W^2(D)$ if and only if u is continuous in $\bar{D}$, has generalized derivatives up to and including the second order, and, furthermore, these derivatives are summable in D to the power d.

It is seen that if the function u is once continuously differentiable in D, its ordinary first derivatives coincide with its first generalized derivatives (almost everywhere). It turns out (a corollory of Fubini's theorem) that, for example, a generalized derivative u_{x^1} exists in the region D if for almost all $(x_0^2, \ldots, x_0^d)$ the function $u(x^1, x_0^2, \ldots, x_0^d)$ is absolutely continuous in x^1 on $\{x^1:(x^1, x_0^2, \ldots, x_0^d) \in D\}$ and its usual derivative with respect to x^1 is locally summable in D. The converse is also true. However, we ought to replace then the function u by a function equivalent with respect to Lebesgue measure. It is well known that if for almost all $(x_0^{i+1}, \ldots, x_0^d)$ the function $u(x^1, \ldots, x^i, x_0^{i+1}, \ldots, x_0^d)$ has a generalized derivative on $\{(x^1, \ldots, x^i):(x^1, \ldots, x^i, x_0^{i+1}, \ldots, x_0^d) \in D\}$ and, in addition, this derivative is locally summable in D, u will have a generalized derivative in D.

Using the notion of weak convergence, we can easily prove that if the functions φ, v^n $(n = 0,1,2, \ldots)$ are uniformly bounded in D, $v^n \to v^0$ (D-a.s.), for some $l_1, \ldots, l_k$ for $n \geq 1$ the generalized derivatives $v^n_{(l_1)\cdots(l_k)}$ exist, and $|v^n_{(l_1)\cdots(l_k)}| \leq \varphi$ (D-a.s.), the generalized derivative $v^0_{(l_1)\cdots(l_k)}$ also exists, $|v^0_{(l_1)\cdots(l_k)}| \leq \varphi$ (D-a.s.), and

$$v^n_{(l_1)\cdots(l_k)} \to v^0_{(l_1)\cdots(l_k)}$$

weakly in $\mathscr{L}_2$ in any bounded subset of the region D.

In many cases, one needs to "mollify" functions to be smooth. We shall do this in a standard manner. Let $\zeta(x)$, $\zeta_1(t)$, $\zeta(t,x) \equiv \zeta_1(t)\zeta(x)$ be nonnegative, infinitely differentiable functions of the arguments $x \in E_d$, $t \in E_1$, equal to zero for $|x| > 1$, $|t| > 1$ and such that

$$\int_{E_d} \zeta(x)\, dx = 1, \qquad \int_{-\infty}^{\infty} dt \int_{E_d} \zeta(t,x)\, dx = 1.$$

For $\varepsilon \neq 0$ and the functions $u(x)$, $u(t,x)$ locally summable in E_d, $E_1 \times E_d$, let

$$u^{(\varepsilon)}(x) = \varepsilon^{-d}\zeta\left(\frac{x}{\varepsilon}\right) * u(x) \quad \text{(convolution with respect to } x\text{)},$$

$$u^{(0,\varepsilon)}(t,x) = \varepsilon^{-d}\zeta\left(\frac{x}{\varepsilon}\right) * u(t,x) \quad \text{(convolution with respect to } x\text{)},$$

$$u^{(\varepsilon)}(t,x) = \varepsilon^{-(d+1)}\zeta\left(\frac{t}{\varepsilon},\frac{x}{\varepsilon}\right) * u(t,x) \quad \text{(convolution with respect to } (t,x)\text{)}.$$

The functions $u^{(\varepsilon)}(x)$, $u^{(0,\varepsilon)}(t,x)$, $u^{(\varepsilon)}(t,x)$ are said to be mean functions of the functions $u(x)$, $u(t,x)$. It is a well-known fact (see [10, 71]) that $u^{(\varepsilon)} \to u$ as $\varepsilon \to 0$:

a. at each Lebesgue point of the function u, therefore almost everywhere;
b. at each continuity point of the function u; uniformly in each bounded region, if u is continuous;
c. in the norm $\mathscr{L}_p(D)$ if $u \in \mathscr{L}_p(D)$ and in computing the convolution of $u^{(\varepsilon)}$ the function u is assumed to be equal to zero outside D.

Furthermore, $u^{(\varepsilon)}$ is infinitely differentiable. If a generalized derivative $u_{(l)}$ exists in E_d, then $[u_{(l)}]^{(\varepsilon)} = [u^{(\varepsilon)}]_{(l)}$. Finally, for $p \geq 1$

$$\|u^{(\varepsilon)}\|_{p,E_d} \leq \|u\|_{p,E_d}, \qquad \|u^{(\varepsilon)}\|_{B(E_d)} \leq \|u\|_{B(E_d)}.$$

Considering the functions $u^{(\varepsilon)}$, we prove that the generalized derivative u_{x^1} of the function $u(x)$ continuous in D does not exceed a constant N_1 almost everywhere if and only if the function $u(x)$ satisfies in D the Lipschitz condition with respect to x^1 having this constant, that is, if for any points $x_1, x_2 \in D$ such that an interval with the end points x_1, x_2 lies in D and $x_1^i = x_2^i$ $(i = 2, \ldots, d)$, the inequality $|u(x_1) - u(x_2)| \le N_1|x_1 - x_2|$ can be satisfied. It turns out that if a bounded function σ has a bounded generalized derivative, σ^2 has as well a generalized derivative, and one can use usual formulas to find this generalized derivative.

In addition to the space $W^2(D)$ we need spaces $\bar{W}^2(D)$, $W^{1,2}(Q)$, and $\bar{W}^{1,2}(Q)$, which are introduced for bounded regions D, Q in a way similar to the way $W^2(D)$ was, starting from sets of functions $C^2(\bar{D})$, $C^{1,2}(\bar{Q})$, and $C^{1,2}(\bar{Q})$, respectively, and using the norms

$$\|f\|_{\bar{W}^2(D)} = \|f\|_{\bar{W}^2(D)} + \sum_{i=1}^{d} \|f_{x^i}\|_{2d,D},$$

$$\|f\|_{W^{1,2}(Q)} = \left\|\frac{\partial}{\partial t} f\right\|_{d+1,Q} + \sum_{i,j=1}^{d} \|f_{x^i x^j}\|_{d+1,Q}$$

$$+ \sum_{i=1}^{d} \|f_{x^i}\|_{d+1,Q} + \|f\|_{B(\bar{Q})},$$

$$\|f\|_{\bar{W}^{1,2}(Q)} = \|f\|_{\bar{W}^{1,2}(Q)} + \sum_{i=1}^{d} \|f_{x^i}\|_{2(d+1),Q}.$$

For proving existence of generalized derivatives of a payoff function another notion proves to be useful.

2. Definition. Let a function $u(x)$ be given, and let it be locally summable in a region D. Let $v(\Gamma)$ be a function of a set Γ which is definite, σ-additive, and finite on the σ-algebra of Borel subsets of each bounded region $D' \subset \bar{D}' \subset D$. We say that the set function v on D is a generalized derivative of the function u in the $l_1, \ldots, l_k$ directions, and we write

$$v(dx) = u_{(l_1)\ldots(l_k)}(x)\,(dx), \tag{3}$$

if for each function $\varphi \in C_0^\infty(D)$,

$$\int_D u\varphi_{(l_1)\ldots(l_k)}\,dx = (-1)^k \int_D \varphi v(dx). \tag{4}$$

The generalized derivative $(\partial/\partial t)u(t,x)(dt\,dx)$ for the function $u(t,x)$ locally summable in the region Q can be found in a similar way.

The definitions given above immediately imply the following properties. It is easily seen that there exists only one function $v(dx)$ satisfying (4) for all $\varphi \in C_0^\infty(D)$. If the function $u_{(l_1)\cdots(l_k)}(x)$ exists, which is a generalized derivative of u in the $l_1, \ldots, l_k$ directions in the sense of Definition 1, assuming that $v(dx) = u_{(l_1)\cdots(l_k)}(x)\,dx$, we obtain in an obvious manner a set function

v, being the generalized derivative of u in the $l_1, \ldots, l_k$ directions in the sense of Definition 2.

Conversely, if the set function v in Definition 2 is absolutely continuous with respect to Lebesgue measure, its Radon–Nikodym derivative will satisfy Definition 1 in conjunction with (4). Therefore, this Radon–Nikodym derivative is the generalized derivative $u_{(l_1)\cdots(l_k)}(x)$. This fact justifies the notation of (3). In the case where the direction l_i coincides with the direction of the r_ith coordinate vector, we shall write

$$u_{(l_1)\cdots(l_k)}(x)(dx) = u_{x^{r_1}\cdots x^{r_k}}(x)(dx).$$

Using the uniqueness property of a generalized derivative, we easily prove that if the derivatives $u_{(l_1)\cdots(l_k)}(x)(dx)$ for some k exist for all $l_1, \ldots, l_k$, then

$$u_{(l_1)\cdots(l_k)}(x)(dx) = \frac{1}{|l_1|\cdots|l_k|} \sum_{r_1,\ldots,r_k} u_{x^{r_1}\cdots x^{r_k}}(x)(dx) l_1^{r_1}\cdots l_k^{r_k}$$

for $|l_1|\cdots|l_k| \neq 0$. Further, if the derivatives $u_{(l)(l)}(x)(dx)$ exist for all l, all the derivatives $u_{(l_1)(l_2)}(x)(dx)$ exist as well. In this case, if $|l_1|\cdot|l_2| \neq 0$, then

$$u_{(l_1)(l_2)}(x)(dx) = \frac{1}{4|l_1|\cdot|l_2|}[(l_1 + l_2)^2 u_{(l_1+l_2)(l_1+l_2)}(x)(dx) - (l_1 - l_2)^2 u_{(l_1-l_2)(l_1-l_2)}(x)(dx)].$$

In fact, using Definition 2 we easily prove that the right side of this formula satisfies Definition 2 for $k = 2$.

Theorem V of [67, Chapter 1, §1] constitutes the main tool enabling us to prove the existence of $u_{(l_1)\cdots(l_k)}(x)(dx)$. In accord with this theorem from [67], the nonnegative generalized function is a measure. Regarding

$$\int u\varphi_{(l_1)\cdots(l_k)}(dx) - (-1)^k \int \varphi v(dx) \tag{5}$$

as a generalized function, we have the following.

3. Lemma. *Let $u(x)$, $v(\Gamma)$ be the same as those in the first two propositions of Definition 2. For each nonnegative $\varphi \in C_0^\infty(D)$ let the expression (5) be nonnegative. Then there exists a generalized derivative $u_{(l_1)\ldots(l_k)}$ in the sense of Definition 2. In this case, inside D*

$$(-1)^k u_{(l_1)\cdots(l_k)}(x)(dx) \geq (-1)^k v(dx),$$

that is, for all bounded Borel $\Gamma \subset \bar{\Gamma} \subset D$

$$(-1)^k u_{(l_1)\cdots(l_k)}(\Gamma) \geq (-1)^k v(\Gamma).$$

To conclude the discussion in this section we summarize more or less conventional agreements and notation.

$(\mathbf{w}_t, \mathscr{F}_t)$ is a Wiener process (see Appendix 1).

$\mathscr{F}_\tau$ is the σ-algebra consisting of all those sets A for which the set $A \cap \{\tau \leq t\} \in \mathscr{F}_t$ for all t.

$\mathfrak{M}(t)$ denotes the set of all Markov (with respect to $\{\mathscr{F}_t\}$) times τ not exceeding t (see Appendix 1).

$C([0,T],E_d)$ denotes a Banach space of continuous functions on $[0,T]$ with range in E_d, $\mathscr{N}_t$ the smallest σ-algebra of the subsets of $C([0,T],E_d)$ which contains all sets of the form

$$\{x_{[0,T]} \in C([0,T],E_d): x_s \in \Gamma\},$$

where $s \leq t$, Γ denotes a Borel subset of E_d.

l.i.m. reads the mean square limit.

ess sup reads the essential upper bound (with respect to the measure which is implied).

$$\inf \varnothing = \infty, \qquad f(x_\tau) \equiv f(x_\tau)\chi_{\tau<\infty}.$$

When we speak about measurable functions (sets), we mean, as a rule, Borel functions (sets). The words "nonnegative," "nonpositive," "it does not increase," "it does not decrease," mean the same as the words "positive," "negative," "it decreases," "it increases," respectively.

Finally,

$$\Delta = \sum_{i=1}^{d} \frac{\partial^2}{\partial (x^i)^2}$$

denotes the Laplace operator. The operators L^α, $F[u]$, $F_1[u]$, used in Chapters 4–6 are defined in the introductory section in Chapter 4.

2. Estimates of the Distribution of a Stochastic Integral in a Bounded Region

Let A be a set of pairs (σ,b), where σ is a matrix of dimension $d \times d_1$ and b is a d-dimensional vector. We assume that a random process $(\sigma_t,b_t) \in A$ for all (ω,t), and that the process

$$x_t = x_0 + \int_0^t \sigma_s \, d\mathbf{w}_s + \int_0^t b_s \, ds$$

is defined.

We shall see further that in stochastic control, estimates of the form

$$\mathsf{M} \int_0^{\tau_D} |f(t,x_t)| \, dt \leq N \|f\|_{p,Q} \tag{1}$$

play an essential role, in (1) f is an arbitrary Borel function, τ_D is the first exit time of x_t from the region D, and $Q = (0,\infty) \times D$. A crucial fact here is that the constant N does not depend on a specified process (σ_t,b_t), but is given instead by the set A. In this section, our objective is to deduce a few versions of the estimate (1).

We assume that D is a bounded region in E_d, x_0 is a fixed point of D, an integer $d_1 \geq d$, $(\mathbf{w}_t, \mathscr{F}_t)$ is a d_1-dimensional Wiener process, $\sigma_t(\omega)$ is a matrix of dimension $d \times d_1$, $b_t(\omega)$ is a d-dimensional vector, and $c_t(\omega)$, $r_t(\omega)$ are nonnegative numbers. Assume in addition that σ_t, b_t, c_t, r_t are progressively measurable with respect to $\{\mathscr{F}_t\}$ and that they are bounded functions of (t,ω). Let $a_t = \frac{1}{2}\sigma_t\sigma_t^*$.

Next, let p be a fixed number, $p \geq d$, and let

$$y_{s,t} = \int_s^t r_u\,du, \qquad \varphi_{s,t} = \int_s^t c_u\,du, \qquad \psi_t = c_t^{1-[(d+1)/(p+1)]}(r_t \det a_t)^{1/(p+1)}.$$

One should keep in mind that for $p = d$ the expression $c_t^{(p-d)/(p+1)}$ is equal to unity even if $c_t = 0$; therefore $\psi_t = (r_t \det a_t)^{1/(d+1)}$ for $p = d$.

1. Definition. A nonnegative function $F(c,a)$ defined on the set of all nonnegative numbers c as well as nonnegative definite symmetric matrices a of dimension $d \times d$ is said to be regular if for each $\varepsilon > 0$ there is a constant $k(\varepsilon)$ such that for all c, a and unit vectors λ

$$F(c,a) \leq \varepsilon \operatorname{tr} a + k(\varepsilon)[c + (a\lambda,\lambda)].$$

2. Theorem. *Assume that $|b_t| \leq F(c_t,a_t)$ for all (t,ω) for some regular function $F(c,a)$. There exist constants N_1,N_2 depending only on d, the function $F(c,a)$ and the diameter of the region D, and such that for all $s \geq 0$, Borel $f(t,x)$ and $g(x)$, on a set $\{\tau_D \geq s\}$, almost surely*

$$\mathsf{M}\left\{\int_s^{\tau_D} e^{-\varphi_{s,t}}\psi_t|f(y_{s,t},x_t)|\,dt \,\Big|\, \mathscr{F}_s\right\} \leq N_1\|f\|_{p+1,Q}, \tag{2}$$

$$\mathsf{M}\left\{\int_s^{\tau_D} e^{-\varphi_{s,t}}c_t^{1-(d/p)}(\det a_t)^{1/p}|g(x_t)|\,dt \,\Big|\, \mathscr{F}_s\right\} \leq N_s\|g\|_{p,D}. \tag{3}$$

Before proving our theorem, we discuss the assertions of the theorem and give examples of regular functions. Note that the left sides of the inequalities (2) and (3) make sense because of the measurability requirements.

It is seen that the function $F(c,a) \equiv c$ is regular. Next, in conjunction with Young's inequality,

$$xy = \left(\frac{x}{\varepsilon}\right)(\varepsilon y) \leq \frac{\varepsilon^q y^q}{q} + \frac{x^p}{\varepsilon^p p},$$

if $x,y \geq 0$, $p^{-1} + q^{-1} = 1$. Hence for $\alpha \in (0,1)$, $\varepsilon \in (0,1)$

$$c^\alpha(\operatorname{tr} a)^{1-\alpha} \leq \varepsilon(1-\alpha)\operatorname{tr} a + \alpha\varepsilon^{1-(1/\alpha)}c \leq \varepsilon \operatorname{tr} a + \varepsilon^{1-(1/\alpha)}c.$$

Therefore, $c^\alpha(\operatorname{tr} a)^{1-\alpha}$ is a regular function for $\alpha \in (0,1)$.

We show that the function $(\det a)^{1/d}$ not depending on c is regular. Let $\mu_1 \le \mu_2 \le \cdots \le \mu_d$ be eigenvalues of a matrix a. We know that $\mu_1 \le (a\lambda,\lambda)$ if $|\lambda| = 1$. Further, $\det a = \mu_1\mu_2 \cdots \mu_d$, $\operatorname{tr} a = \mu_1 + \mu_2 + \cdots + \mu_d$. From this, in conjunction with the Young's inequality, we have

$$(\det a)^{1/d} = \mu_1^{1/d}(\mu_2 \cdots \mu_d)^{1/d} \le \frac{1}{d}\varepsilon^{-(d-1)}\mu_1 + \frac{d-1}{d}(\varepsilon\mu_2 \cdots \varepsilon\mu_d)^{1/(d-1)}$$

$$\le \frac{1}{d}\varepsilon^{-(d-1)}(a\lambda,\lambda) + \frac{1}{d}\varepsilon(\mu_2 + \cdots + \mu_d)$$

$$\le (\varepsilon \operatorname{tr} a + \varepsilon^{-(d-1)}(a\lambda,\lambda).$$

Using the regular functions given above, we can construct many other regular functions, noting that a linear combination with positive coefficients of regular functions is a regular function.

The function $\operatorname{tr} a$ is the limit of regular functions $c^{\alpha}(\operatorname{tr} a)^{1-\alpha}$ as $\alpha \downarrow 0$. However, for $d \ge 2$ the function $\operatorname{tr} a$ is not regular. To prove this, we suggest the reader should consider

3. Exercise

For $p = d$, $c_t \equiv 0$, $s = 0$, $g \equiv 1$ it follows from (3) that

$$\mathsf{M} \int_0^{\tau_D} (\det a_t)^{1/d}\, dt \le N_2 (\operatorname{meas} D)^{1/d}. \tag{4}$$

From the statement of Theorem 2 we take $D = S_R$, $F(c,a) = K \operatorname{tr} a$, with $K > R^{-1}$.

It is required to prove that for $d \ge 2$ there exists no constant N_2 depending only on d, K, R, for which (4) can be satisfied.

This exercise illustrates the fact that the requirement $|b_t| \le F(c_t,a_t)$, where F is a regular function, is essential. In contrast to this requirement, we can weaken considerably the assumption about boundedness of σ, b, c, r. For example, considering instead of the process x_t, $y_{r,t}$ the processes

$$\bar{x}_t = x_0 + \int_0^t \chi_{u<\tau_D}\sigma_u\, d\mathbf{w}_u + \int_0^t \chi_{u<\tau_D} b_u\, du, \qquad \bar{y}_{s,t} = \int_s^t \chi_{u<\tau_D} r_u\, du,$$

where τ_D is the time of first departure of x_t from D, and noting that $x_t = \bar{x}_t$, $y_{s,t} = \bar{y}_{s,t}$ for $t < \tau_D$, we immediately establish the assertion of Theorem 2 in the case where $\chi_{t<\tau_D}\sigma_t$, $\chi_{t<\tau_D} b_t$, $\chi_{t<\tau_D} c_t$, $\chi_{t<\tau_D} r_t$ are bounded functions of (t,ω).

We think that the case where $s = 0$, $r_t \equiv 1$, $p = d$ is the most important particular case of Theorem 2. It is easily seen, in fact, that the proof of our theorem follows generally from the particular case indicated. The formal proof is rather difficult, however. It should be noted that according to our approach to the proof of the theorem, assuming $s \ne 0$, $r_t \not\equiv 1$ makes the proving of estimates for $s = 0$, $r_t \equiv 1$ essentially easier. In the future, it will be convenient to use the following weakened version of the assertions of Theorem 2.

4. Theorem. *Let τ be a Markov time (with respect to $\{\mathscr{F}_t\}$), not exceeding τ_D. Also, let there exist constants $K, \delta > 0$ such that for all $t < \tau(\omega)$, $\lambda \in E_d$*

$$|b_t(\omega)| \le K, \qquad \sum_{i,j=1}^{d} a_t^{ij}(\omega)\lambda^i\lambda^j \ge \delta|\lambda|^2.$$

Then there exists a constant N depending only on d, K, δ, and the diameter of the region D such that for all $s \ge 0$ and Borel $f(t,x)$ and $g(x)$ on the set $\{s \le \tau\}$, almost surely

$$\mathsf{M}\left\{\int_s^\tau |f(t,x_t)|\,dt \,\Big|\, \mathscr{F}_s\right\} \le N\|f\|_{d+1,Q},$$

$$\mathsf{M}\left\{\int_s^\tau |g(x_t)|\,dt \,\Big|\, \mathscr{F}_s\right\} \le N\|g\|_{d,D}.$$

This theorem follows immediately from Theorem 2 for $r_t \equiv 1$, $c_t \equiv 0$, $p = d$. In fact, we have

$$x_{t\wedge\tau} = x_0 + \int_0^t \chi_{u<\tau}\sigma_u\,d\mathbf{w}_u + \int_0^t \chi_{u<\tau}b_u\,du,$$

$$\mathsf{M}\left\{\int_s^\tau |f(t,x_t)|\,dt \,\Big|\, \mathscr{F}_s\right\} \le \delta^{-d/(d+1)}\mathsf{M}\left\{\int_s^{\tau_D} e^{-\varphi_{s,t}}\chi_{t<\tau}\psi_t|f(s + y_{s,t},x_t)|\,dt \,\Big|\, \mathscr{F}_s\right\}$$

since $e^{-\varphi_{s,t}}\psi_t = (\det a_t)^{1/(d+1)}$ and $\det a_t$, which is equal to the product of eigenvalues of the matrix a_t for $t \le \tau$, is not smaller than δ^d. Furthermore, $|\chi_{t<\tau}b_t| \le K\delta^{-1}(\det \chi_{t<\tau}a_t)^{1/d}$, the function $F(c,a) = K\delta^{-1}(\det a)^{1/d}$ is regular and, in addition, $\{s \le \tau_D\} \supset \{s \le \tau\}$.

Next, in order to prove Theorem 2, we need three lemmas.

5. Lemma. *Let $|b_t| \le F(c_t,a_t)$ for all (t,ω) for some regular function $F(c,a)$. There exists a constant N depending only on the function $F(c,a)$ and the diameter of the region D such that on the set $\{\tau_D \ge s\}$ almost surely*

$$\mathsf{M}\left\{\int_s^{\tau_D} e^{-\varphi_{s,t}}|b_t|\,dt \,\Big|\, \mathscr{F}_s\right\} \le \mathsf{M}\left\{\int_s^{\tau_D} e^{-\varphi_{s,t}}F(c_t,a_t)\,dt \,\Big|\, \mathscr{F}_s\right\} \le N.$$

PROOF. We can assume without loss of generality that $x_0 = 0$. We denote by R the diameter of the region D and set $u(x) = \beta - \operatorname{ch}\alpha|x|$ for $\alpha > 0$, $\beta > \operatorname{ch}(\alpha R)$. We note that $u(x)$ is twice continuously differentiable and $u(x) \ge 0$ for $x \in D$. Applying Ito's formula to $e^{-\varphi_{s,t}}u(x_t)$, we have for $t \ge s$ on the set $\{\tau_D \ge s\}$ that

$$u(x_s) = \mathsf{M}\left\{\int_s^{t\wedge\tau_D} e^{-\varphi_{s,r}}[c_r u(x_r) - L^{\sigma_r,b_r}u(x_r)]\,dr + e^{-\varphi_{s,t\wedge\tau_D}}u(x_{t\wedge\tau_D}) \,\Big|\, \mathscr{F}_s\right\}$$

$$\ge \mathsf{M}\left\{\int_s^{t\wedge\tau_D} e^{-\varphi_{s,r}}[c_r u(x_r) - L^{\sigma_r,b_r}u(x_r)]\,dr \,\Big|\, \mathscr{F}_s\right\} \qquad \text{(a.s.)}$$

Assume that for all $x \in D$, $r \geq 0$

$$c_r u(x) - L^{\sigma_r,b_r}u(x) \geq F(c_r,a_r). \tag{5}$$

Then

$$\beta \geq u(x_s) \geq \mathsf{M}\left\{\int_s^{t\wedge\tau_u} e^{-\varphi_{s,r}}F(c_r,a_r)\,dr \,\Big|\, \mathscr{F}_s\right\},$$

which proves the assertion of the lemma as $t \to \infty$, with the aid of Fatou's lemma.

Therefore, it remains only to choose constants α, β such that (5) is satisfied, assuming obviously that $x \neq 0$. For simplicity of notation, we shall not write the subscript r in c_r, σ_r, a_r, b_r. In addition, let $\lambda = x/|x|$, $\rho = |x|$. A simple computation shows that

$$\begin{aligned} I &\equiv (1 + \alpha\,\mathrm{sh}\,\alpha|x|)^{-1}[cu(x) - L^{\sigma,b}u(x) - F(c,a)] \\ &= (1 + \alpha\,\mathrm{sh}\,\alpha\rho)^{-1}\{c(\beta - \mathrm{ch}\,\alpha\rho) + \alpha\,\mathrm{sh}\,\alpha\rho(b,\lambda) \\ &\quad + \alpha^2\,\mathrm{ch}\,\alpha\rho(a\lambda,\lambda) + \frac{\alpha}{\rho}\,\mathrm{sh}\,\alpha\rho[\mathrm{tr}\,a - (a\lambda,\lambda)] - F(c,a)\} \\ &\geq c\frac{\beta - \mathrm{ch}\,\alpha\rho}{1 + \alpha\,\mathrm{sh}\,\alpha\beta} + (a\lambda,\lambda)\frac{\alpha^2\,\mathrm{ch}\,\alpha\rho}{1 + \alpha\,\mathrm{sh}\,\alpha\rho} \\ &\quad + \frac{\alpha\,\mathrm{sh}\,\alpha\rho}{1 + \alpha\,\mathrm{sh}\,\alpha\rho}\frac{1}{\rho}[\mathrm{tr}\,a - (a\lambda,\lambda)] - F(c,a). \end{aligned} \tag{6}$$

We note that $\mathrm{ch}\,\alpha\rho \geq 1, \mathrm{ch}\,\alpha\rho \geq \mathrm{sh}\,\alpha\rho$, $\alpha\,\mathrm{sh}\,\alpha\rho \geq \alpha^2\rho$ and for $x \in D$ the number $\rho \leq R$. Hence

$$\frac{\beta - \mathrm{ch}\,\alpha\rho}{1 + \alpha\,\mathrm{sh}\,\alpha\rho} \geq \frac{\beta - \mathrm{ch}\,\alpha R}{1 + \alpha\,\mathrm{sh}\,\alpha R}, \qquad \frac{\alpha^2\,\mathrm{ch}\,\alpha\rho}{1 + \alpha\,\mathrm{sh}\,\alpha\rho} \geq \frac{\alpha^2\,\mathrm{ch}\,\alpha\rho}{\mathrm{ch}\,\alpha\rho + \alpha\,\mathrm{ch}\,\alpha\rho} = \frac{\alpha^2}{1 + \alpha},$$

$$\frac{1}{\rho}\frac{\alpha\,\mathrm{sh}\,\alpha\rho}{1 + \alpha\,\mathrm{sh}\,\alpha\rho} \geq \frac{1}{\rho}\frac{\alpha^2\rho}{1 + \alpha^2\rho} \geq \frac{\alpha^2}{1 + \alpha^2 R}.$$

Therefore, it follows from (6) that

$$I \geq c\frac{\beta - \mathrm{ch}\,\alpha R}{1 + \alpha\,\mathrm{sh}\,\alpha R} + (a\lambda,\lambda)\frac{\alpha^2}{1 + \alpha} + \frac{\alpha^2}{1 + \alpha^2 R}[\mathrm{tr}\,a - (a\lambda,\lambda)] - F(c,a).$$

We recall that $F(c,a)$ is a regular function. Also, we fix some $\varepsilon < 1/R$ and choose α large enough that $\alpha^2/(1 + \alpha^2 R) > \varepsilon$, $\alpha^2/(1 + \alpha) \geq k(\varepsilon) + \varepsilon$. Next, we take a number β so large that

$$\frac{\beta - \mathrm{ch}\,\alpha R}{1 + \alpha\,\mathrm{sh}\,\alpha R} \geq k(\varepsilon).$$

Then $I \geq k(\varepsilon)[c + (a\lambda,\lambda)] + \varepsilon\,\mathrm{tr}\,a - F(c,a) \geq 0$, thus proving the lemma. □

6. Corollary. *Let $G(c,a)$ be a regular function. There exists a constant N depending only on $F(c,a)$, $G(c,a)$ and the diameter of the region D such that*

$$\mathsf{M}\left\{\int_s^{\tau_D} e^{-\varphi_{s,t}}G(c_t,a_t)\,dt\,\Big|\,\mathscr{F}_s\right\} \le N \qquad (\{\tau_D \ge s\}\text{-a.s.}).$$

In fact, let $F_1(c,a) = F(c,a) + G(c,a)$. Then $|b_t| \le F_1(c_t,a_t)$, $G(c_t,a_t) \le F_1(c_t,a_t)$, and the assertion of our lemma is proved for $F_1(c,a)$.

7. Lemma. *Let $R > 0$, $h(t,x) \ge 0$, $h \in \mathscr{L}_{d+1}(C_R)$, $h(t,x) = 0$ for $t \le 0$, $h(t,x) = 0$ for $|x| \ge R$. Then on $(-\infty,\infty) \times E_d$ there exists a bounded function $z(t,x) \le 0$ equal to zero for $t < 0$ and such that for all sufficiently small $\varepsilon > 0$ and non-negative definite symmetric matrices $a = (a^{ij})$ on a cylinder C_R.*

$$N(d)(\det a)^{1/(d+1)}h^{(\varepsilon)} \le -\frac{\partial}{\partial t}z^{(\varepsilon)} + \sum_{i,j=1}^{d} a^{ij}z^{(\varepsilon)}_{x^ix^j},$$

where $N(d) > 0$. Furthermore, if the vector b and the number c are such that $|b| \le (R/2)c$, then on the same set $\sum_{i=1}^d b^iz^{(\varepsilon)}_{x^i} \ge cz^{(\varepsilon)}$, if ε is sufficiently small. Finally, for all $t \ge 0$, $x \in E_d$

$$|z(t,x)|^{d+1} \le N(d,R)\int_{S_R}\int_0^t h^{d+1}(s,y)\,ds\,dy$$

This lemma is proved in [42] by geometric arguments.

8. Lemma. *Let $|b_t| \le F(c_t,a_t)$ for all (t,ω) for a regular function $F(c,a)$. There exists a constant N depending only on d, $F(c,a)$, and the diameter of D, and such that for all $s \ge 0$, $f(t,x)$ on a set $\{\tau_D \ge s\}$, almost surely*

$$\mathsf{M}\left\{\int_s^{\tau_D}\exp\left\{-\int_s^t c_u\,du\right\}(r_t\det a_t)^{1/(d+1)}\left|f\left(\int_s^t r_u\,du,x_t\right)\right|dt\,\Big|\,\mathscr{F}_s\right\} \le N\|f\|_{d+1,Q}.$$

In other words, the inequality (2) holds for $p = d$.

PROOF. Let us use the notation introduced above:

$$\varphi_{s,t} = \int_s^t c_u\,du, \qquad \psi_t = (r_t\det a_t)^{1/(d+1)}, \qquad y_{s,t} = \int_s^t r_u\,du.$$

We denote by R the diameter of D and we consider without loss of generality that $x_0 = 0$. In this case $D \subset S_R$. Also, we assume that τ_R is the first exit time of x_t from S_R. It is seen that $\tau_R \ge \tau_D$.

Suppose that we have proved the inequality

$$\mathsf{M}\left\{\int_s^{\tau_R} e^{-\varphi_{s,t}}\psi_t|f(y_{s,t},x_t)|\,dt\,\Big|\,\mathscr{F}_s\right\} \le N\|f\|_{d+1,C_R} \tag{7}$$

($\{\tau_R \ge s\}$-a.s.) for arbitrary s, f, where $N = N(d,F,R)$. Furthermore, taking in (7) the function f equal to zero for $x \notin D$, we obtain

$$\mathsf{M}\left\{\int_s^{\tau_D} e^{-\varphi_{s,t}}\psi_t|f(y_{s,t},x_t)|\,dt\,|\mathscr{F}_s\right\} \le \mathsf{M}\left\{\int_s^{\tau_R} e^{-\varphi_{s,t}}\psi_t|f(y_{s,t},x_t)|\,dt\,|\mathscr{F}_s\right\}$$
$$\le N\|f\|_{d+1,C_R} = N\|f\|_{d+1,Q}$$

($\{\tau_R \ge s\}$-a.s.) and, *a fortiori*, ($\{\tau_D \ge s\}$-a.s.).

It suffices therefore to prove (7). Usual reasoning (using, for example, the results given in [54, Chapter 1, §2]) shows that it suffices to prove (7) only for bounded continuous nonnegative $f(t,x)$. Noting in addition that by Fatou's lemma, for such a function

$$\mathsf{M}\left\{\int_s^{\tau_R} e^{-\varphi_{s,t}}\psi_t f(y_{s,t},x_t)\,dt\,|\mathscr{F}_s\right\} \le \varliminf_{\varepsilon\downarrow 0} \mathsf{M}\left\{\int_s^{\tau_R} e^{-\varphi_{s,t}}[(r_t+\varepsilon)\det a_t]^{1/(d+1)}\right.$$
$$\left.\times f\left(\int_s^t (r_u+\varepsilon)\,du, x_t\right)dt\,|\mathscr{F}_s\right\} \qquad \text{(a.s.)},$$

we conclude that it is enough to consider the case where $r_t(\omega) > 0$ for all (t,ω).

We fix $T > 0$ and assume that $h(y,x) = f(T-y,x)$ for $0 < y < T$, $x \in S_R$, and $h = 0$ in all the remaining cases. Using Lemma 7, we find an appropriate function z. Let $\tau = \tau_{T,R}$ be the first exit time of a process $(y_{s,t},x_t)$ considered for $t \ge s$ from a set $[0,T) \times S_R$.

We apply Ito's formula to the expression $e^{-\varphi_{s,t}}z^{(\varepsilon)}(T-y_{s,t},x_t)$ for $\varepsilon > 0$, $t \ge s$. Then

$$-z^{(\varepsilon)}(T,x_s) = \mathsf{M}\left\{\int_s^{t\wedge\tau}\left[-r_u\frac{\partial}{\partial t}z^{(\varepsilon)}(T-y_{s,u},x_u)\right.\right.$$
$$\left.-c_u z^{(\varepsilon)}(T-y_{s,u},x_u) + L^{\sigma_u,b_u}z^{(\varepsilon)}(T-y_{s,u},x_u)\right]e^{-\varphi_{s,t}}\,du$$
$$\left.-e^{-\varphi_{s,t\wedge\tau}}z^{(\varepsilon)}(T-y_{s,t\wedge\tau},x_{t\wedge\tau})|\mathscr{F}_s\right\} \qquad \text{(a.s.)}.$$

Using the properties of $z^{(\varepsilon)}$ for small $\varepsilon > 0$, we find

$$-r_u\frac{\partial}{\partial t}z^{(\varepsilon)} + L^{\sigma_u,b_u}z^{(\varepsilon)} = r_u\left[-\frac{\partial}{\partial t}z^{(\varepsilon)} + \sum_{i,j=1}^d \frac{1}{r_u}a_u^{ij}z^{(\varepsilon)}_{x^ix^j}\right] + \sum_{i=1}^d b_u^i z^{(\varepsilon)}_{x^i}$$
$$\ge N(d)\psi_u h^{(\varepsilon)} + \frac{2}{R}|b_u|z^{(\varepsilon)}.$$

Furthermore, $z^{(\varepsilon)} \le 0$. Hence

$$-z^{(\varepsilon)}(T,x_s) \ge \mathsf{M}\left\{\int_s^{t\wedge\tau} e^{-\varphi_{s,u}}\left[N(d)\psi_u h^{(\varepsilon)}(T-y_{s.u,}x_u)\right.\right.$$
$$\left.\left.+\frac{2}{R}|b_u|z^{(\varepsilon)}(T-y_{s,u},x_u)\right]du\,|\mathscr{F}_s\right\} \qquad \text{(a.s.)},$$

in which we carry the term containing $z^{(\varepsilon)}$ from the right side to the left side. Also, we use the estimate $|z^{(\varepsilon)}| \leq \sup_{t,x} |z| \leq N\|h\|_{d+1,C_R} \leq N\|f\|_{d+1,C_R}$:

$$N(d,R)\|f\|_{d+1,C_R}\left(1 + \mathsf{M}\left\{\int_s^{t\wedge\tau} e^{-\varphi_{s,u}}|b_u|\,du \,\Big|\, \mathscr{F}_s\right\}\right)$$

$$\geq \mathsf{M}\left\{\int_s^{t\wedge\tau} e^{-\varphi_{s,u}}\psi_u h^{(\varepsilon)}(T - y_{s,u}, x_u)\,du \,\Big|\, \mathscr{F}_s\right\} \qquad \text{(a.s.)},$$

where $y_{s,u} \in (0,T)$ for $u \in (s,\tau)$ by virtue of the condition $r_t > 0$, and in addition, $x_u \in S_R$; hence the function h is continuous at a point $(T - y_{s,u}, x_u)$ and $h(T - y_{s,u}, x_u) = f(y_{s,u}, x_u)$. Letting ε to zero in the last inequality, we obtain, using Fatou's lemma,

$$N(d,R)\|f\|_{d+1,C_R}\left(1 + \mathsf{M}\left\{\int_s^{t\wedge\tau} e^{-\varphi_{s,u}}|b_u|\,du \,\Big|\, \mathscr{F}_s\right\}\right)$$

$$\geq \mathsf{M}\left\{\int_s^{t\wedge\tau} e^{-\varphi_{s,u}}\psi_u f(y_{s,u}, x_u)\,du \,\Big|\, \mathscr{F}_s\right\} \qquad \text{(a.s.)}.$$

Further, on the set $\{\tau_R \geq s\}$ it is seen that $\tau \leq \tau_R$. Therefore, by Lemma 5,

$$\chi_{\tau_R \geq s}\mathsf{M}\left\{\int_s^{t\wedge\tau} e^{-\varphi_{s,u}}|b_u|\,du \,\Big|\, \mathscr{F}_s\right\}$$

$$= \mathsf{M}\left\{\chi_{\tau_R \geq s}\int_s^{t\wedge\tau} e^{-\varphi_{s,u}}|b_u|\,du \,\Big|\, \mathscr{F}_s\right\}$$

$$\leq \chi_{\tau_R \geq s}\mathsf{M}\left\{\int_s^{\tau_R} e^{-\varphi_{s,u}}|b_u|\,du \,\Big|\, \mathscr{F}_s\right\} \leq N(F,R) \qquad \text{(a.s.)}.$$

Finally, on the set $\{\tau_R \geq s\}$ for all $T > 0$, $t > s$, we obtain

$$\mathsf{M}\left\{\int_s^{t\wedge\tau_{T,R}} e^{-\varphi_{s,u}}\psi_u f(y_{s,u}, x_u)\,du \,\Big|\, \mathscr{F}_s\right\} \leq N(d,F,R)\|f\|_{d+1,C_R} \qquad \text{(a.s.)}.$$

It remains only to let first $t \to \infty$, second $T \to \infty$, and then to use Fatou's lemma as well as the fact that obviously $\tau_{T,R} \to \tau_R$ as $T \to \infty$ on the set $\{\tau_R \geq s\}$. We have thus proved the lemma. □

9. Proof of Theorem 2. We note first that it suffices to prove Theorem 2 only for $p = d$. In fact, for $p > d$ in accord with Hölder's inequality, for example,

$$\mathsf{M}\left\{\int_s^{\tau_D} e^{-\varphi_{s,t}} c_t^{1-(d/p)}(\det a_t)^{1/p}|g(x_t)|\,dt \,\Big|\, \mathscr{F}_s\right\}$$

$$\leq \left(\mathsf{M}\left\{\int_s^{\tau_D} e^{-\varphi_{s,t}}(\det a_t)^{1/d}|g(x_t)|^{p/d}\,dt \,\Big|\, \mathscr{F}_s\right\}\right)^{p/d}\left(\mathsf{M}\left\{\int_s^{\tau_D} e^{-\varphi_{s,t}} c_t\,dt \,\Big|\, \mathscr{F}_s\right\}\right)^{1-(d/p)}$$

In this case, $\int_s^{\tau_D} e^{-\varphi_{s,t}} c_t\,dt = 1 - e^{-\varphi_{s,\tau_D}} \le 1$, and if we have proved the theorem for $p = d$, the first factor does not exceed

$$[N(d,F,D)\|g^{p/d}\|_{d,D}]^{d/p} = N^{d/p}(d,F,D)\|g\|_{p,D} \le (N(d,F,D) + 1)\|g\|_{p,D}.$$

The inequality (2) was proved for $p = d$ in Lemma 8. Therefore, it suffices to prove that

$$\mathsf{M}\left\{\int_s^{\tau_D} e^{-\varphi_{s,t}}(\det a_t)^{1/d}|g(x_t)|\,dt \,\Big|\, \mathscr{F}_s\right\} \le N(d,F,D)\|g\|_{d,D}$$

($\{\tau_D \le s\}$-a.s.) for all g. We can consider without loss of generality that g is a nonnegative bounded function. In this case, since $(\det a)^{1/d}$ is a regular function

$$v = \sup_{s\ge 0} \operatorname{ess\,sup}_{\omega} \chi_{\tau_D \ge s}\mathsf{M}\left\{\int_s^{\tau_D} (\det a_t)^{1/d} g(x_t) e^{-\varphi_{s,t}}\,dt \,\Big|\, \mathscr{F}_s\right\}$$

is finite by Corollary 6. If $v = 0$, we have nothing to prove. Hence we assume that $v > 0$.

Using Fubini's theorem or integrating by parts, we obtain for any numbers $t_1 < t_2$ and nonnegative functions $h(t)$, $r(t)$ that

$$\int_{t_1}^{t_2} h(t)\,dt = \int_{t_1}^{t_2} h(t)\exp\left\{-\int_{t_1}^{t} r(u)\,du\right\}dt + \int_{t_1}^{t_2} \exp\left\{-\int_{t_1}^{t} r(u)\,du\right\} r(t)\left(\int_t^{t_2} h(u)\,du\right)dt.$$

From this for $s \ge 0$, $A \in \mathscr{F}_s$, $r_t = (1/v)(g(x_t)(\det a_t)^{1/d}$, $h_t = (\det a_t)^{1/d} g(x_t)$, we find

$$\begin{aligned}
\mathsf{M}\chi_{A,\tau_D\ge s}&\int_s^{\tau_D} h_t e^{-\varphi_{s,t}}\,dt \\
&= \mathsf{M}\chi_{A,\tau_D\ge s}\int_s^{\tau_D} h_t \exp\left\{-\varphi_{s,t} - \int_s^t r_u\,du\right\}dt \\
&\quad + \mathsf{M}\chi_{A,\tau_D\ge s}\int_s^{\tau_D} \exp\left\{-\int_s^t r_u\,du\right\} r_t\left(\int_t^{\tau_D} h_u e^{-\varphi_{s,u}}\,du\right)dt,
\end{aligned}$$

where the last term is equal to

$$\begin{aligned}
&\int_s^\infty \left[\mathsf{M}\chi_{A,\tau_D\ge s}\exp\left\{-\int_s^t r_u\,du - \varphi_{s,t}\right\} r_t \chi_{\tau_D\ge t}\left(\int_t^{\tau_D} h_u e^{-\varphi_{t,u}}\,du\right)\right]dt \\
&= \int_s^\infty \left[\mathsf{M}\chi_{A,\tau_D\ge s}\exp\left\{-\int_s^t r_u\,du - \varphi_{s,t}\right\} r_t \chi_{\tau_D\ge t}\mathsf{M}\left\{\int_t^{\tau_D} h_u e^{-\varphi_{t,u}}\,du \,\Big|\, \mathscr{F}_t\right\}\right]dt \\
&\le \int_s^\infty \mathsf{M}\chi_{A,\tau_D\ge s}\exp\left\{-\int_s^t r_u\,du - \varphi_{s,t}\right\}\chi_{\tau_D\ge t} r_t v\,dt \\
&= \mathsf{M}\chi_{A,\tau_D\ge s}\int_s^{\tau_D} h_t \exp\left\{-\int_s^t r_u\,du - \varphi_{s,t}\right\}dt.
\end{aligned}$$

Therefore,

$$\mathsf{M}\chi_{A,\tau_D\geq s}\int_s^{\tau_D} h_t e^{-\varphi_{s,t}}\,dt$$

$$\leq 2\mathsf{M}\chi_{A,\tau_D\geq s}\int_s^{\tau_D} h_t \exp\left\{-\int_s^t r_u\,du-\varphi_{s,t}\right\}dt$$

$$=2v^{1/(d+1)}\mathsf{M}\chi_{A,\tau_D\geq s}\int_s^{\tau_D}(r_t\det a_t)^{1/(d+1)}f\left(\int_s^t r_u\,du,x_t\right)e^{-\varphi_{s,t}}\,dt,$$

where $f(t,x)=e^{-t}g^{d/(d+1)}(x)$. Consequently, by Lemma 8,

$$\mathsf{M}\chi_{A,\tau_D\geq s}\int_s^{\tau_D} h_t e^{-\varphi_{s,t}}\,dt\leq N\|f\|_{d+1,Q}v^{1/(d+1)}\mathsf{P}\{A,\tau_D\geq s\}$$

$$\leq N\|g\|_{d,D}^{d/(d+1)}v^{1/(d+1)}\mathsf{P}\{A,\tau_D\geq s\},$$

where the constants N (which differ from one another) depend only on d, the function $F(c,a)$, and the diameter of D. The last inequality is equivalent to the fact that $\{\tau_D\geq s\}$-a.s.)

$$\mathsf{M}\left\{\int_s^{\tau_D}(\det a_t)^{1/d}g(x_t)e^{-\varphi_{s,t}}\,dt\,\Big|\,\mathscr{F}_s\right\}=\mathsf{M}\left\{\int_s^{\tau_D}h_t e^{-\varphi_{s,t}}\,dt\,\Big|\,\mathscr{F}_s\right\}$$

$$\leq N\|g\|_{d,D}^{d/(d+1)}v^{1/(d+1)}$$

From this, taking the upper bounds, we find

$$v\leq N\|g\|_{d,D}^{d/(d+1)}v^{1/(d+1)},\qquad v^{d(d+1)}\leq N\|g\|_{d,D}^{d/(d+1)}$$

and $v\leq N\|g\|_{d,D}$, thus completing the proof of Theorem 2. □

10. Remark. Let $\delta>0$. The function $F(c,a)$ is said to be δ-regular if for *some* $\varepsilon\in(0,\delta)$ there is a constant $k(\varepsilon)$ such that for all c, a, and unit vectors λ

$$F(c,a)\leq\varepsilon\operatorname{tr}a+k(\varepsilon)[c+(a\lambda,\lambda)].$$

In the sense of the above definition, the function which is δ-regular for all $\delta>0$, is a regular function.

Repeating almost word for word the proofs of Lemmas 5 and 8 and the proof of Theorem 2, we convince ourselves that if the region D belongs to a circle of radius R, $|b_t|\leq F(c_t,a_t)$ for all (t,ω) and if $F(c,a)$ is an R^{-1}-regular function, there exist constants N_1,N_2 depending only on d, $F(c,a)$, and R such that the inequalities (2) and (3) are satisfied.

11. Exercise

Let $d\geq 2$, $D=S_R$, $\varepsilon>0$. Give an example illustrating the $(R^{-1}+\varepsilon)$-regular function $F(c,a)$ for which the assertions of Theorem 2 do not hold. (*Hint*: See Exercise 3.)

12. Exercise

Let $z^{(\varepsilon)}$ be the function from Lemma 7. Prove that for sufficiently small ε the function $z^{(\varepsilon)}(t,x)$ decreases in t and is convex downward with respect to x on the cylinder C_R.

3. Estimates of the Distribution of a Stochastic Integral in the Whole Space

In this section[1] we shall estimate expressions of the form $\mathsf{M}\int_0^\infty |f(t,x_t)|\,dt$ using the $\mathscr{L}_p$-norm of f, that is, we extend the estimates from Section 2.2 to the case $D = E_d$.

We use in this section the assumptions and notation introduced at the beginning of Section 2.2. Furthermore, let

$$\varphi_t = \varphi_{0,t} = \int_0^t c_u\,du, \qquad y_t = y_{0,t} = \int_0^t r_u\,du.$$

Throughout this section we shall have two numbers $K_1,K_2 > 0$ fixed and assume permanently that

$$|b_t(\omega)| \le K_1 c_t(\omega), \qquad \operatorname{tr} a_t(\omega) \le K_2 c_t(\omega)$$

for all (t,ω). Note immediately that under this condition $|b_t|$ does not exceed the regular function $F(c_t,a_t) \equiv K_1 c_t$.

First we prove a version of Theorem 2.2.

1. Lemma. *Let $R > 0$, let τ be a Markov time with respect to $\{\mathscr{F}_t\}$, and let $\tau_R = \inf\{t \ge \tau : |x_t - x_\tau| \ge R\}$.[2] Then there exists a constant $N = N(d,K_1,R)$ such that for any Borel $f(t,x)$*

$$\mathsf{M}\left\{\int_\tau^{\tau_R} e^{-\varphi_t}\psi_t|f(y_t,x_t)|\,dt\,\Big|\,\mathscr{F}_\tau\right\}$$

$$\le e^{-\varphi_\tau}\chi_{\tau<\infty}N\left(\int_{y_\tau}^\infty \int |f(t,x)|^{p+1}\,dx\,dt\right)^{1/(p+1)} \qquad \text{(a.s.).}$$

PROOF. First, let τ be nonrandom finite. For $t \ge 0$ we set $\mathscr{F}'_t = \mathscr{F}_{\tau+t}$, $\mathbf{w}'_t = \mathbf{w}_{\tau+t} - \mathbf{w}_\tau$,

$$\sigma'_t = \sigma_{\tau+t}, \qquad b'_t = b_{\tau+t}, \qquad c'_t = c_{\tau+t}, \qquad r'_t = r_{\tau+t},$$

$$\psi'_t = \psi_{\tau+t}, \qquad x'_t = \int_0^t \sigma'_u\,d\mathbf{w}'_u + \int_0^t b'_u\,du = x_{\tau+t} - x_\tau,$$

$$y'_t = \int_0^t r'_u\,du, \qquad \varphi'_t = \int_0^t c'_u\,du;$$

τ' is the first exit time of the process x'_t from S_R. It is then seen that

$$e^{\varphi_\tau}\mathsf{M}\left\{\int_\tau^{\tau_R} e^{-\varphi_t}\psi_t|f(y_t,x_t)|\,dt\,\Big|\,\mathscr{F}_\tau\right\}$$

$$= \mathsf{M}\left\{\int_0^{\tau'} e^{-\varphi'_t}\psi'_t|f(y'_t + y_\tau,\, x'_t + x_\tau)|\,dt\,\Big|\,\mathscr{F}'_0\right\} \qquad \text{(a.s.).}$$

[1] Also, see Theorem 4.1.8.

[2] $\inf \phi = \infty$.

Furthermore, $(w'_t, \mathscr{F}'_t)$ is a Wiener process. In addition, by Theorem 2.2

$$\mathsf{M}\left\{\int_0^{\tau'} e^{-\varphi'_t}\psi'_t|f(y'_t + y, x'_t + x)|\,dt\,|\mathscr{F}'_0\right\}$$

$$\leq N\left(\int_0^\infty \int |f(t+y,z)|^{p+1}\,dz\,dt\right)^{1/(p+1)} \qquad \text{(a.s.)}$$

for any $x \in E_d$, $y \geq 0$. In order to prove our lemma for the constant τ, it remains to replace y, x by the $\mathscr{F}'_0$-measurable variables y_τ, x_τ in the last inequality. To do as indicated, we let $\kappa_n(t) = (k+1)/2^n$ for $t \in (k/2^n, (k+1)/2^n]$, $\kappa_n(x) = \kappa_n(x^1, \ldots, x^d) = (\kappa_n(x), \ldots, \kappa_n(x^d))$. Note that $\kappa_n(t) \downarrow t$ for all $t \in (-\infty, \infty)$, $\kappa_n(x) \to x$ for all $x \in E_d$.

From the very start, we can consider without loss of generality that f is a continuous nonnegative function. We denote by Γ_n^1, Γ_n^d the sets of values of the functions $\kappa_n(t)$, $\kappa_n(x)$ respectively. Using Fatou's lemma, we obtain for the function f mentioned,

$$\mathsf{M}\left\{\int_0^{\tau'} e^{-\varphi'_t}\psi'_t f(y'_t + y_\tau, x'_t + x_\tau)\,dt\,|\mathscr{F}'_0\right\}$$

$$\leq \varliminf_{n\to\infty} \mathsf{M}\left\{\int_0^{\tau'} e^{-\varphi'_t}\psi'_t f(y'_t + \kappa_n(y_\tau), x'_t + \kappa_n(x_\tau))\,dt\,|\mathscr{F}'_0\right\}$$

$$= \varliminf_{n\to\infty} \sum_{y\in\Gamma_n^1}\sum_{x\in\Gamma_n^d} \mathsf{M}\left\{\int_0^{\tau'} e^{-\varphi'_t}\psi'_t f(y'_t + y_\tau, x'_t + x)\,dt\,|\mathscr{F}'_0\right\}\chi_{\kappa_n(y_\tau)=y,\kappa_n(x_\tau)=x}$$

$$\leq N \varliminf_{n\to\infty}\left(\int_{\kappa_n(y_\tau)} \int f^{p+1}(t,x)\,dx\,dt\right)^{1/(p+1)}$$

$$\leq N\left(\int_{y_\tau}^\infty \int f^{p+1}(t,x)\,dx\,dt\right)^{1/(p+1)}$$

Further, we prove the lemma in the general case. Taking $A \in \mathscr{F}_\tau$ and setting $\tau^n = \kappa_n(\tau)$,

$$\tau_R^n = \inf\{t \geq \tau^n : |x_t - x_{\tau^n}| \geq R\},$$

we can easily see that

$$\tau^n \downarrow \tau, \qquad \varliminf_{n\to\infty} \tau_R^n \geq \tau_R,$$

$$\chi_{\tau<\infty} \varliminf_{n\to\infty} \int_{\tau^n}^{\tau_R^n} e^{-\varphi_t}\psi_t|f(y_t, x_t)|\,dt \geq \int_\tau^{\tau_R} e^{-\varphi_t}\psi_t|f(y_t, x_t)|\,dt,$$

and that for $s \in \Gamma_n^1$ the set

$$\{A, \tau^n = s\} = \left\{A, \tau \in \left(s - \frac{1}{2^n}, s\right]\right\} \in \mathscr{F}_s.$$

Therefore, in accord with what has been proved,

$$\mathsf{M}\chi_A \int_\tau^{\tau_R} e^{-\varphi_t}\psi_t|f(y_t,x_t)|\,dt$$

$$\leq \varliminf_{n\to\infty} \mathsf{M}\chi_{A,\tau<\infty} \int_{\tau^n}^{\tau_R^n} e^{-\varphi_t}\psi_t|f(y_t,x_t)|\,dt$$

$$= \varliminf_{n\to\infty} \sum_{s\in\Gamma_n^1} \mathsf{M}\chi_{A,\tau^n=s} \int_s^{\tau_R^n} e^{-\varphi_t}\psi_t|f(y_t,x_t)|\,dt$$

$$= \varliminf_{n\to\infty} \sum_{s\in\Gamma_n^1} \mathsf{M}\chi_{A,\tau^n=s}\mathsf{M}\left\{\int_s^{\tau_R^n} e^{-\varphi_t}\psi_t|f(y_t,x_t)|\,dt\,\Big|\,\mathscr{F}_s\right\}$$

$$\leq N \varliminf_{n\to\infty} \mathsf{M}\chi_{A,\tau^n<\infty} e^{-\varphi_{\tau^n}}\left(\int_{y_{\tau^n}}^{\infty}\int |f(t,x)|^{p+1}\,dx\,dt\right)^{1/(p+1)}$$

$$= N\mathsf{M}\chi_{A,\tau<\infty} e^{-\varphi_\tau}\left(\int_{y_\tau}^{\infty}\int |f(t,x)|^{p+1}\,dx\,dt\right)^{1/(p+1)},$$

thus completing the proof of our lemma. □

2. Lemma. *As in the preceding lemma, we introduce R, τ, τ_R. Also, we denote by $\mu = \mu(\lambda)$ the positive root of the equation $\lambda - \mu K_1 - \mu^2 K_2 = 0$ for $\lambda > 0$. Then*

$$\mathsf{M}\{\chi_{\tau_R<\infty} e^{-\lambda\varphi_{\tau_R}}|\mathscr{F}_\tau\} \leq \frac{1}{\operatorname{ch}\mu R} e^{-\lambda\varphi_\tau}\chi_{\tau<\infty} \qquad \text{(a.s.)}.$$

PROOF. Let $\pi(x) = \operatorname{ch}\mu|x|$. Simple computations show that

$$\lambda c_t\pi(x) - L^{\sigma_t,b_t}\pi(x) = \lambda c_t \operatorname{ch}\mu|x| - \mu \operatorname{sh}\mu|x|\left(b_t,\frac{x}{|x|}\right)$$

$$-\mu^2 \operatorname{ch}\mu|x|\frac{(a_t x,x)}{|x|^2} - \mu^2\frac{1}{\mu|x|}\operatorname{sh}\mu|x|\left[\operatorname{tr} a_t - \frac{(a_t x,x)}{|x|^2}\right].$$

Taking advantage of the fact that $\operatorname{sh}\mu|x| \leq \operatorname{ch}\mu|x|$, $\operatorname{sh}\mu|x| \leq \mu|x|\operatorname{ch}\mu|x|$, we obviously obtain

$$\lambda c_t\pi(x) - L^{\sigma_t,b_t}\pi(x) \geq c_t \operatorname{ch}\mu|x|(\lambda - \mu K_1 - \mu^2 K_2) = 0.$$

Further, using Ito's formula applied to $e^{-\lambda\varphi_t}\pi(x_t + x)$, we have from the last inequality that

$$e^{-\lambda\varphi_{t\wedge\tau}}\pi(x_{t\wedge\tau} + x) = \mathsf{M}\left\{\int_{t\wedge\tau}^{t\wedge\tau_R} e^{-\lambda\varphi_u}[\lambda c_u\pi(x_u + x)\right.$$

$$\left.- L^{\sigma_u,b_u}\pi(x_u + x)]\,du + e^{-\lambda\varphi_{t\wedge\tau_R}}\pi(x_{t\wedge\tau_R} + x)\,\Big|\,\mathscr{F}_{t\wedge\tau}\right\}$$

$$\geq \mathsf{M}\{e^{-\lambda\varphi_{\tau_R}}\pi(x_{t\wedge\tau_R} + x)|\mathscr{F}_{t\wedge\tau}\}.$$

Using the continuity property of $\pi(x)$, we replace x with a variable $(-x_{t\wedge\tau})$ in the last inequality. Then

$$e^{-\lambda\varphi_{t\wedge\tau}} \geq \mathsf{M}\{e^{-\lambda\varphi_{\tau_R}}\pi(x_{t\wedge\tau_R} - x_{t\wedge\tau})|\mathscr{F}_{t\wedge\tau}\},$$

which yields for $A \in \mathscr{F}_\tau$

$$\begin{aligned}\mathsf{M}\chi_{A,\tau<\infty}e^{-\lambda\varphi_\tau} &= \lim_{t\to\infty} \mathsf{M}\chi_{A,\tau\leq t}e^{-\lambda\varphi_{t\wedge\tau}} \\ &\geq \varliminf_{t\to\infty} \mathsf{M}\chi_{A,\tau\leq t}e^{-\lambda\varphi_{\tau_R}}\pi(x_{t\wedge\tau_R} - x_{t\wedge\tau}) \\ &\geq \mathsf{M}\chi_A e^{-\lambda\varphi_{\tau_R}}\pi(x_{\tau_R} - x_\tau)\chi_{\tau_R<\infty} = \operatorname{ch}\mu R \mathsf{M}\chi_A e^{-\lambda\varphi_{\tau_R}}\chi_{\tau_R<\infty}. \quad \square\end{aligned}$$

We have proved the lemma; further, we shall prove the main theorem of Section 2.3.

3. Theorem. *There exist constants $N_i = N_i(d,K_1,K_2)$ $(i = 1,2)$ such that for all Markov times τ and Borel functions $f(t,x)$, $g(x)$*

$$\mathsf{M}\left\{\int_\tau^\infty e^{-\varphi_t}\psi_t|f(y_t,x_t)|\,dt\,|\mathscr{F}_\tau\right\}$$

$$\leq N_1 e^{-\varphi_\tau}\left(\int_{y_\tau}^\infty \int |f(t,x)|^{p+1}\,dx\,dt\right)^{1/(p+1)} \qquad \text{(a.s.)},$$

$$\mathsf{M}\left\{\int_\tau^\infty e^{-\varphi_t}c_t^{1-(d/p)}(\det a_t)^{1/p}|g(x_t)|dt|\mathscr{F}_\tau\right\} \leq N_2 e^{-\varphi_\tau}\|g\|_{p,H_d} \qquad \text{(a.s.)}.$$

PROOF. We regard f, g as nonnegative bounded functions and in addition, we introduce the Markov times recursively as follows:

$$\tau^0 = \tau,$$
$$\tau^{n+1} = \inf\{t \geq \tau^n : |x_t - x_{\tau^n}| \geq 1\}.$$

Note that by Lemma 2,

$$\begin{aligned}\mathsf{M}\{\chi_{\tau^{n+1}<\infty}e^{-\varphi_{\tau^{n+1}}}|\mathscr{F}_\tau\} &= \mathsf{M}\{\mathsf{M}\{\chi_{\tau^{n+1}<\infty}e^{-\varphi_{\tau^{n+1}}}|\mathscr{F}_{\tau^n}\}|\mathscr{F}_\tau\} \\ &\leq \frac{1}{\operatorname{ch}\mu}\mathsf{M}\{\chi_{\tau^n<\infty}e^{-\varphi_{\tau^n}}|\mathscr{F}_\tau\} \\ &\leq \left(\frac{1}{\operatorname{ch}\mu}\right)^{n+1} e^{-\varphi_\tau}\chi_{\tau<\infty} \qquad \text{(a.s.)},\end{aligned}$$

where μ is the positive root of the equation $1 - \mu K_1 - \mu^2 K_2 = 0$.

It is seen that τ^n increase as n increases; the variables

$$\chi_{\tau^n<\infty}e^{-\varphi_{\tau^n}}$$

decrease as n increases. The estimate given shows that as $n \to \infty$

$$\mathsf{M}\chi_{\tau^n<\infty}e^{-\varphi_{\tau^n}} \to 0, \qquad \chi_{\tau^n<\infty}e^{-\varphi_{\tau^n}} \to 0 \qquad \text{(a.s.)}.$$

Due to the boundedness of the function $c_t(\omega)$ we immediately have that $\tau^n \to \infty$ (a.s.) as $n \to \infty$.

Therefore, using Lemma 1, we obtain (a.s.)

$$\mathsf{M}\left\{\int_\tau^\infty e^{-\varphi_t}\psi_t f(y_t,x_t)\,dt \,\Big|\, \mathscr{F}_\tau\right\}$$

$$= \sum_{n=0}^{\infty} \mathsf{M}\left\{\int_{\tau^n}^{\tau^{n+1}} \,\Big|\, \mathscr{F}_\tau\right\}$$

$$= \sum_{n=0}^{\infty} \mathsf{M}\left\{\mathsf{M}\left\{\int_{\tau^n}^{\tau^{n+1}} e^{-\varphi_t}\psi_t f(y_t,x_t)\,dt \,\Big|\, \mathscr{F}_{\tau_n}\right\}\Big|\,\mathscr{F}_\tau\right\}$$

$$\le N(d,K_1)\sum_{n=0}^{\infty} \mathsf{M}\left\{e^{-\varphi_{\tau^n}}\chi_{\tau^n<\infty}\left(\int_{y_{\tau^n}}^\infty \int f^{p+1}(t,x)\,dx\,dt\right)^{1/(p+1)}\Bigg|\,\mathscr{F}_t\right\}$$

$$\le N\left(\int_{y_\tau}^\infty \int f^{p+1}(t,x)\,dx\,dt\right)^{1/(p+1)} \sum_{n=0}^{\infty} \mathsf{M}\{e^{-\varphi_{\tau^n}}\chi_{\tau^n<\infty}\,|\,\mathscr{F}_\tau\}$$

$$\le N\frac{\operatorname{ch}\mu}{\operatorname{ch}\mu - 1}e^{-\varphi_\tau}\left(\int_{y_\tau}^\infty \int f^{p+1}(t,x)\,dx\,dt\right)^{1/(p+1)}.$$

Having proved the first assertion of the theorem, we proceed to proving the second.

To this end, we use the same technique as in 2.9. The function g is bounded and

$$c_t^{1-(d/p)}(\det a_t)^{1/p} \le c_t^{1-(d/p)}(\operatorname{tr} a_t)^{d/p} \le K_2^{d/p}c_t.$$

Hence

$$\int^\infty e^{-\varphi_{\tau,t}}c_t^{1-(d/p)}(\det a_t)^{1/p}g(x_t)\,dt \le N\int_\tau^\infty e^{-\varphi_{\tau,t}}c_t\,dt = N(1-e^{-\varphi_{\tau,\infty}}) \le N,$$

and the number

$$v = \sup_\tau \operatorname{ess\,sup}_\omega \mathsf{M}\left\{\int_\tau^\infty e^{-\varphi_{\tau,t}}\,c_t^{1-(d/p)}(\det a_t)^{1/p}g(x_t)\,dt \,\Big|\, \mathscr{F}_\tau\right.$$

is finite. We assume that $v > 0$, and that $r_t = (1/v)c_t^{1-(d/p)}(\det a_t)^{1/p}g(x_t)$, $h_t = c_t^{1-(d/p)}(\det a_t)^{1/p}g(x_t)$. Using Fubini's theorem, we obtain

$$\mathsf{M}\left\{\int_\tau^\infty e^{-\varphi_{\tau,t}}h_t\,dt \,\Big|\, \mathscr{F}_\tau\right\}$$

$$= \mathsf{M}\left\{\int_\tau^\infty h_t \exp\left\{-\varphi_{\tau,t} - \int_\tau^t r_u\,du\right\}dt \,\Big|\, \mathscr{F}_\tau\right\}$$

$$+ \mathsf{M}\left\{\int_\tau^\infty r_t \exp\left\{-\varphi_{\tau,t} - \int_\tau^t r_u\,du\right\}\left(\int_t^\infty h_u e^{-\varphi_{t,u}}\,du\right)dt \,\Big|\, \mathscr{F}_\tau\right\} \qquad \text{(a.s.)},$$

from which it follows, as in 2.9, that

$$\mathsf{M}\left\{\int_\tau^\infty e^{-\varphi_{\tau,t}}h_t\,dt \,\Big|\, \mathscr{F}_\tau\right\} \le 2\mathsf{M}\left\{\int_\tau^\infty h_t \exp\left\{-\varphi_{\tau,t} - \int_\tau^t r_u\,du\right\}dt \,\Big|\, \mathscr{F}_\tau\right\}.$$

Noting that the last expression equals zero on a set $\{\tau = \infty\}$ we transform it into

$$\exp\left\{\int_0^\tau r_u\,du + \varphi_\tau\right\}\chi_{\tau<\infty}2v^{1/(p+1)}$$

$$\times \mathsf{M}\left\{\int_\tau^\infty e^{-\varphi_t}c_t^{(p-d)/(p+1)}(r_t \det a_t)^{1/(p+1)}f\left(\int_0^t r_u du, x_t\right)dt \,\Big|\, \mathscr{F}_\tau\right\},$$

where $f(t,x) = e^{-t}g^{p/(p+1)}(x)$. Therefore, according to the first assertion of the theorem

$$\mathsf{M}\left\{\int_\tau^\infty e^{-\varphi_{\tau,t}}h_t\,dt \,\Big|\, \mathscr{F}_\tau\right\} \le N_1 v^{1/(p+1)}e^{y_\tau}\chi_{\tau<\infty}\left(\int_{y_\tau}^\infty \int e^{-(p+1)t}g^p(x)\,dx\,dt\right)^{1/(p+1)}$$

$$\le N_1 v^{1/(p+1)}\|g\|_{p,E_d}^{p/(p+1)}\left(\frac{1}{p+1}\right)^{1/(p+1)} \qquad \text{(a.s.)}.$$

Consequently,

$$v \le N_1 v^{1/(p+1)}\|g\|_{p,E_d}^{p/(p+1)}, \qquad v \le N_1^{1+(1/p)}\|g\|_{p,E_d} \le (1 + N_1^2)\|g\|_{p,E_d},$$

which is equivalent to the second assertion, thus completing the proof of our theorem. □

We give one essential particular case of the theorem proved above.

4. Theorem. *Let* K_3, $K_4 < \infty$, $\lambda > 0$, $\delta > 0$, $s \ge 0$, *for all* $t \ge s$, $\omega \in \Omega$, $\xi \in E_d$

$$|b_t(\omega)| \le K_3, \qquad \sum_{i=1}^d a_t^{ii}(\omega) \le K_4, \qquad \sum_{i,j=1}^d a_t^{ij}(\omega)\xi^i\xi^j \ge \delta|\xi|^2. \tag{1}$$

There exist constants $N_i = N_i(d,p,\lambda,\delta,K_3,K_4)$ $(i = 1,2)$ *such that for all Borel functions* $f(t,x)$, $g(x)$

$$\mathsf{M}\int_s^\infty e^{-\lambda t}|f(t,x_t)|\,dt \le N_1\|f\|_{p+1,H_\infty}, \tag{2}$$

$$\mathsf{M}\int_s^\infty e^{-\lambda t}|g(x_t)|\,dt \le N_2\|g\|_{p,E_d}. \tag{3}$$

This theorem follows from the preceding theorem. In fact, for example, let $r_t = 1$, $c_t = \lambda$ for $t \ge s$, $K_1 = K_3/\lambda$, $K_2 = K_4/\lambda$. Then $|b_t| \le K_1 c_t$, $\operatorname{tr} a_t \le K_2 c_t$ for $t \ge s$. For $t < s$, let us take c_t such that the above inequalities still hold, noting that $(\det a_t)^{1/(p+1)} \ge \delta^{d/(p+1)}$. Therefore

$$\lambda^{(p-d)/(p+1)}\delta^{d/(p+1)}\mathsf{M}\int_s^\infty e^{-\lambda t}|f(t,x_t)|\,dt$$

$$\le e^{-\lambda s}\mathsf{M}\int_s^\infty \exp\left\{-\int_s^t c_u\,du\right\}c_t^{(p-d)/(p+1)}(r_t \det a_t)^{1/(p+1)}|f(y_t,x_t)|\,dt$$

$$\le \mathsf{M}\int_s^\infty e^{-\varphi_{s,t}}\psi_t|f(y_t,x_t)|\,dt.$$

5. Exercise

We replace the third inequality in (1) so that $\det a_t \geq \delta$, and we preserve the first two inequalities. Using the self-scaling property of a Wiener process, and also, using the fact that in (3) $g(x)$ can be replaced by $g(cx)$, prove

$$\mathsf{M}\int_0^\infty e^{-\lambda t}|g(x_t)|\,dt \leq \frac{1}{\sqrt{\lambda}}\delta^{-1/d}K_4^{1/2}N\left(d,\frac{K_2}{\sqrt{\lambda K_4}}\right)\|g\|_{d,E_d},$$

where $N(d,K_3)$ is a finite function nondecreasing with respect to K_3.

4. Limit Behavior of Some Functions

Theorems 6 and 7 are most crucial for the discussion in this section. We shall use them in Chapter 4 in deducing the Bellman equation. However, we use only Corollary 8 in the case of uniform nondegenerate controlled processes. In this regard, we note that the assertion of Corollary 8 follows obviously from intuitive considerations since the lower bound with respect to $\alpha \in \mathfrak{B}(s,x)$ which appears in the assertion of Corollary 8 is the lower bound with respect to a set of uniform nondegenerate diffusion processes with bounded coefficients (see Definition of $\mathfrak{B}(s,x)$ prior to Theorem 5).

We fix the integer d. Also, let the number $p \geq d$ and the numbers $K_1 > 0$, $K_2 > 0$, $K_3 > 0$. We denote by α an arbitrary set of the form

$$(\Omega,\mathscr{F},\mathsf{P},d_1,\mathbf{w}_t,\mathscr{F}_t,\sigma_t,b_t,c_t,r_t), \tag{1}$$

where $(\Omega,\mathscr{F},\mathsf{P})$ is a probability space, the integer $d_1 \geq d$, $(\mathbf{w}_t,\mathscr{F}_t)$ is a d_1-dimensional Wiener process on $(\Omega,\mathscr{F},\mathsf{P})$, $\sigma_t = \sigma_t(\omega)$ is a matrix of dimension $d \times d_1$, $b_t = b_t(\omega)$ is a d-dimensional vector, $c_t = c_t(\omega)$, $r_t = r_t(\omega)$ are nonnegative numbers, and σ_t, b_t, c_t, r_t are progressively measurable with respect to $\{\mathscr{F}_t\}$ and are bounded functions of (t,ω) for $t \geq 0$, $\omega \in \Omega$. In the case where the set (1) is written as α, we write $\Omega = \Omega^\alpha$, $\mathscr{F} = \mathscr{F}^\alpha$, etc.

Denote by $\mathfrak{A}(K_1,K_2,K_3)$ the set of all sets α satisfying the conditions

$$|b_t^\alpha| \leq K_1 c_t^\alpha, \qquad \operatorname{tr}\tfrac{1}{2}\sigma_t^\alpha|\sigma_t^\alpha|^* \leq K_2 c_t^\alpha, \qquad r_t^\alpha \leq K_3 c_t^\alpha$$

for all (t,ω). For $x \in E_d$, $\alpha \in \mathfrak{A}(K_1,K_2,K_3)$, let

$$x_t^{\alpha,x} = x + \int_0^t \sigma_u^\alpha\,d\mathbf{w}_u^\alpha + \int_0^t b_u^\alpha\,du,$$

$$y_t^{\alpha,s} = s + \int_0^t r_u^\alpha\,du, \qquad \varphi_t^\alpha = \int_0^t c_u^\alpha\,du, \qquad a_t^\alpha = \tfrac{1}{2}\sigma_t^\alpha[\sigma_t^\alpha]^*,$$

$$\psi_t^\alpha = (c_t^\alpha)^{(p-d)/(p+1)}(r_t^\alpha \det a_t^\alpha)^{1/(p+1)}.$$

As usual, for $p = d$, $\psi_t^\alpha = (r_t^\alpha \det a_t^\alpha)^{1/(d+1)}$.

For the Borel function $f(t,y)$, $s \in (-\infty,\infty)$, $x \in E_d$ let

$$v(s,x) = v(f,s,x) = v(K_1,K_2,K_3,f,s,x)$$
$$= \sup_{\alpha \in \mathfrak{A}(K_1,K_2,K_3)} \mathsf{M}^\alpha \int_0^\infty e^{-\varphi_t^\alpha} \psi_t^\alpha f(y_t^{\alpha,s}, x_t^{\alpha,x})\, dt,$$

where M^α denotes the integration over Ω^α with respect to a measure P^α. In addition to the elements mentioned, we shall use the elements given prior to Theorem 5.

1. Theorem. *Let $f \in \mathscr{L}_{p+1}(E_{d+1})$. Then $v(s,x)$ is a continuous function of (s,x) on E_{d+1}, and, furthermore,*

$$|v(s,x)| \leq N(d,K_1,K_2)\left(\int_s^\infty \int |f(t,y)|^{p+1}\, dy\, dt\right)^{1/(p+1)}$$

Proof. Since $|b_t^\alpha| \leq K_1 c_t^\alpha$, $\operatorname{tr} a_t^\alpha \leq K_2 c_t^\alpha$, the estimate of v follows from Theorem 3.3. In this case, we can take $N(d,K_1,K_2) = N_1(d,K_1,K_2)$, where N_1 is the constant given in Theorem 3.3.

Further, we note that for any families of numbers h_1^α, h_2^α,

$$\left|\sup_\alpha h_1^\alpha - \sup_\alpha h_2^\alpha\right| \leq \sup_\alpha |h_1^\alpha - h_2^\alpha|.$$

Hence

$$|v(s_1,x_1) - v(s_2,x_2)| \leq \sup_\alpha \mathsf{M}^\alpha \int_0^\infty e^{-\varphi_t^\alpha} \psi_t^\alpha |f(y_t^{\alpha,s_1}, x_t^{\alpha,x_1}) - f(y_t^{\alpha,s_2}, x_t^{\alpha,x_2})|\, dt.$$

If $f(t,x)$ is a smooth function of (t,x), with compact support, then

$$|f(y_t^{\alpha,s_1}, x_t^{\alpha,x_1}) - f(y_t^{\alpha,s_2}, x_t^{\alpha,x_2})|$$
$$\leq \sup_{(t,x)}\left(|\operatorname{grad}_x f(t,x)| + \left|\frac{\partial f(t,x)}{\partial t}\right|\right)(|y_t^{\alpha,s_1} - y_t^{\alpha,s_2}| + |x_t^{\alpha,x_1} - x_t^{\alpha,x_2}|)$$
$$= N(|s_1 - s_2| + |x_1 - x_2|).$$

Morever,

$$\psi_t^\alpha \leq (c_t^\alpha)^{(p-d)/(p+1)}(K_2 c_t^\alpha t d^{-d}(\operatorname{tr} a^\alpha)^d)^{1/(p+1)} \leq K_3^{1/(p+1)} K_2^{d/(p+1)} c_t^\alpha.$$

Therefore

$$\int_0^\infty e^{-\varphi_t^\alpha} \psi_t^\alpha\, dt \leq K_3^{1/(p+1)} K_2^{d/(p+1)} \int_0^\infty e^{-\varphi_t^\alpha}\, d\varphi_t^\alpha \leq K^{1/(p+1)} K_2^{d/(p+1)}.$$

Consequently, we have $|v(s_1,x_1) - v(s_2,x_2)| \leq N(|s_1 - s_2| + |x_1 - x_2|)$ for $f(t,x)$, with v being a continuous function.

If f is an arbitrary function in $\mathscr{L}_{p+1}(E_{d+1})$, we take a sequence of smooth functions f_n with compact support so that $\|f - f_n\|_{p+1,E_{d+1}} \to 0$. Using the property of the magnitude of the difference between the upper bounds, which we used before, we obtain

$$|v(f,s,x) - v(f_n,s,x)| \leq v(|f - f_n|,s,x) \leq N\|f - f_n\|_{p+1,E_{d+1}}.$$

This implies that the continuous functions $v(f_n,s,x)$ converge to $v(f,s,x)$ uniformly in E_{d+1}. Therefore, $v(f,s,x)$ is continuous, thus proving the theorem. □

The continuity property of $v(s,x)$ implies the measurability of this function. For investigating the integrability property of $v(s,x)$ we need the following lemma.

2. Lemma. *Let $R > 0$, let $\tau_R^{\alpha,x}$ be the time of first entry of a process $x_t^{\alpha,x}$ into a set $\bar{S}_R$, let γ^α be a random variable on Ω^α, $\gamma^\alpha \geq \tau_R^{\alpha,x}$ and let ε be the positive root of the equation $K_2\varepsilon^2 + K_1\varepsilon - 1 = 0$. Then, for all t_1, s*

$$\mathsf{M}^\alpha \chi_{\gamma^\alpha < \infty} e^{-\varphi_{\gamma^\alpha}^\alpha} \leq e^{\varepsilon R - \varepsilon|x|},$$

$$\mathsf{M}^\alpha \chi_{t_1 \leq y_{\gamma^\alpha}^{\alpha,s}, \gamma^\alpha < \infty} e^{-\varphi_{\gamma^\alpha}^\alpha} \leq \exp\left\{\frac{\varepsilon}{2}R - \frac{\varepsilon}{2}|x| - \frac{1}{2K_3}(t_1 - s)\right\}.$$

PROOF. We fix α, x. For the sake of simplicity we do not write the superscripts α, x. In addition, we write $y_{\gamma^\alpha}^{\alpha,s} = s + y_{\gamma^\alpha}^{\alpha,0}$ as $s + y_\gamma$.

The first assertion of the lemma is obvious for $|x| \leq R$; therefore we assume that $|x| > R$. In accord with Ito's formula applied to $e^{-\varphi_t - \varepsilon|x_t|}$ we obtain

$$e^{-\varepsilon R}\mathsf{M}e^{-\varphi_\gamma}\chi_{\gamma \leq t} \leq e^{-\varepsilon R}\mathsf{M}e^{-\varphi_{\tau_R}}\chi_{\tau_R \leq t}$$

$$\leq \mathsf{M}e^{-\varphi_{t \wedge \tau_R} - \varepsilon|x_{t \wedge \tau_R}|} = e^{-\varepsilon|x|} + \mathsf{M}\int_0^{t \wedge \tau_R} e^{-\varphi_s - \varepsilon|x_s|} I_s(x_s)\, ds,$$

where

$$I_s(x) = \varepsilon^2\frac{(a_s x,x)}{|x|^2} - \varepsilon\frac{1}{|x|}\left[\operatorname{tr} a_s - \frac{(a_s x,x)}{|x|^2}\right] - \varepsilon b_s \frac{x}{|x|} - c_s$$

$$\leq \varepsilon^2\frac{(a_s x,x)}{|x|^2} + \varepsilon K_1 c_s - c_s \leq \varepsilon^2 K_2 c_s + \varepsilon K_1 c_s - c_s = 0.$$

Hence $e^{-\varepsilon R}\mathsf{M}e^{-\varphi_\gamma}\chi_{\gamma \leq t} \leq e^{-\varepsilon|x|}$. Using Fatou's lemma, as $t \to \infty$, we arrive at the former inequality.

In order to prove the latter inequality, we note that under the assumption $r_t \leq K_3 c_t$ we have on the set $\{t_1 \leq y_\gamma + s\}$ that $t_1 - s \leq K_3\varphi_\gamma$, from which it follows that

$$-\varphi_\gamma \leq -K_3^{-1}(t_1 - s),$$

$$\mathsf{M}\chi_{t_1 \leq y_\gamma + s,\, \gamma < \infty} e^{-\varphi_\gamma} \leq e^{-K_3^{-1}(t_1 - s_\gamma)}.$$

Furthermore,

$$\mathsf{M}\chi_{t_1 \leq y_\gamma + s,\, \gamma < \infty} e^{-\varphi_\gamma} \leq \mathsf{M}\chi_{\gamma < \infty} e^{-\varphi_\gamma} \leq e^{\varepsilon R - \varepsilon|x|}.$$

Having multiplied the extreme terms in the last two inequalities we establish the second assertion of the lemma, thus completing the proof of the lemma. □

3. Theorem. *There exists a finite function* $N(d,K_1)$ *increasing with respect* K_1 *and such that for all* $f \in \mathscr{L}_{p+1}(E_{d+1})$

$$\|v(f,\cdot,\cdot)\|_{p+1,E_{d+1}} \le K_3^{1/(p+1)}K_2^{d/(p+1)}N\left(d,\frac{K_1}{\sqrt{K_2}}\right)\|f\|_{p+1,E_{d+1}}.$$

PROOF. Suppose that we have proved the theorem under the condition that $K_2 = K_3 = 1$. In order to prove the theorem under the same assumption in the general case, we use arguments which replace implicitly the application of the self-scaling property of a Wiener process (see Exercise 3.5).

If $\alpha \in \mathfrak{A} = \mathfrak{A}(K_1,K_2,K_3)$, let

$$\alpha' = \left(\Omega^\alpha,\mathscr{F}^\alpha,\mathrm{P}^\alpha,d_1^\alpha,\mathrm{w}_t^\alpha,\mathscr{F}_t^\alpha,\frac{1}{\sqrt{K_2}}\sigma_t^\alpha,\frac{1}{\sqrt{K_2}}b_t^\alpha,c_t^\alpha,\frac{1}{K_3}r_t^\alpha\right).$$

It is seen that $\alpha' \in \mathfrak{A} = \mathfrak{A}(K_1/\sqrt{K_2},1,1)$. It is also seen that α' runs through the entire set $\mathfrak{A}(K_1/\sqrt{K_2},1,1)$ when α runs through the entire set $\mathfrak{A}(K_1,K_2,K_3)$.

Further, for $f \in \mathscr{L}_{p+1}(E_{d+1})$ let $f'(t,x) = f(K_3t,\sqrt{K_2}x)$. We have

$$\begin{aligned}
&v(K_1,K_2,K_3,f,s,x)\\
&\quad= \sup_{\alpha\in\mathfrak{A}} \mathrm{M}^\alpha \int_0^\infty e^{-\varphi_t^\alpha}\psi_t^\alpha f(y_t^{\alpha,s},x_t^{\alpha,x})\,dt\\
&\quad= K_3^{1/(p+1)}K_2^{d/(p+1)} \sup_{\alpha'\in\mathfrak{A}'} \mathrm{M}^{\alpha'} \int_0^\infty e^{-\varphi_t^{\alpha'}}\psi_t^{\alpha'} f(s + K_3 y_t^{\alpha',0}, x + \sqrt{K_2}x_t^{\alpha',0})\,dt\\
&\quad= K_3^{1/(p+1)}K_2^{d/(p+1)} \sup_{\alpha'\in\mathfrak{A}'} \mathrm{M}^{\alpha'} \int_0^\infty e^{-\varphi_t^{\alpha'}}\psi_t^{\alpha'} f'\left(\frac{s}{K_3} + y_t^{\alpha',0}, \frac{x}{\sqrt{K_2}} + x_t^{\alpha',0}\right)dt\\
&\quad= K_3^{1/(p+1)}K_2^{d/(p+1)}v\left(\frac{K_1}{\sqrt{K_2}},1,1,f',\frac{s}{K_3},\frac{x}{\sqrt{K_2}}\right).
\end{aligned}$$

Therefore, if we have proved our theorem for $K_2 = K_3 = 1$, then

$$\begin{aligned}
&\|v(K_1,K_2,K_3,f,\cdot,\cdot)\|_{p+1,E_{d+1}}^{p+1}\\
&\quad= K_3K_2^d \int_{-\infty}^\infty \int \left|v\left(\frac{K_1}{\sqrt{K_2}},1,1,f',\frac{s}{K_3},\frac{x}{\sqrt{K_2}}\right)\right|^{p+1} dx\,ds\\
&\quad= K_3^2K_2^{(3/2)d} \int_{-\infty}^\infty \int \left|v\left(\frac{K_1}{\sqrt{K_2}},1,1,f',s,x\right)\right|^{p+1} dx\,ds\\
&\quad\le K_3^2K_2^{(3/2)d}N^{p+1}\left(d,\frac{K_1}{\sqrt{K_2}}\right)\|f'\|_{p+1,E_{d+1}}^{p+1}\\
&\quad= K_3K_2^dN^{p+1}\left(d,\frac{K_1}{\sqrt{K_2}}\right)\|f\|_{p+1,E_{d+1}}^{p+1}.
\end{aligned}$$

Therefore it suffices to prove this theorem only for $K_2 = K_3 = 1$. We use in our proof in this case the expression

$$I^\alpha(s,x) \equiv \mathsf{M}^\alpha \int_0^\infty e^{-\varphi_t^\alpha}\psi_t^\alpha f(y_t^{\alpha,s},x_t^{\alpha,x})\,dt$$

representable as the "sum" of terms each of which incorporates the change which occurs while the process $(y_t^{\alpha,s},x_t^{\alpha,s})$ moves across the region associated with the given term.

We assume without loss of generality that $f \geq 0$.

Let R be such that the volume of S_R is equal to unity. We denote by $w(t,x)$ the indicator of a set $\bar{C}_{1,R}$. Let $f_{(t_1,x_t)}(t,x) = w(t_1 - t, x_1 - x)f(t,x)$. It is seen that

$$f(t,x) = \int_{-\infty}^{\infty} \int f_{(t_1,x_1)}(t,x)\,dx_1\,dt_1,$$

$$I^\alpha(s,x) = \int_{-\infty}^{\infty} dt_1 \int dx_1\, \mathsf{M}^\alpha \int_0^\infty e^{-\varphi_t^\alpha}\psi_t^\alpha f_{(t_1,x_1)}(y_t^{\alpha,s},x_t^{\alpha,x})\,dt.$$

In order to estimate the last expectation for fixed t_1, x_1, we note that $f_{(t_1,x_1)}(t,x)$ can be nonzero only for $0 \leq t_1 - t \leq 1$, $|x_1 - x| \leq R$. Hence, if γ^α is the time of first entry of the process $(t_1 - y_t^{\alpha,s}\, x_1 - x_t^{\alpha,x})$ into the set $\bar{C}_{1,R}$, then

$$\mathsf{M}^\alpha \int_0^\infty e^{-\varphi_t^\alpha}\psi_t^\alpha f_{(t_1,x_1)}(y_t^{\alpha,s},x_t^{\alpha,x})\,dt = \mathsf{M}^\alpha \chi_{\gamma^\alpha<\infty} \int_{\gamma^\alpha}^\infty e^{-\varphi_t^\alpha}\psi_t^\alpha f_{(t_1,x_1)}(y_t^{\alpha,s},x_t^{\alpha,x})\,dt.$$

Furthermore, on the set $\{\gamma^\alpha < \infty\}$

$$0 \leq t_1 - y_{\gamma^\alpha}^{\alpha,s} \leq 1 \qquad \text{and} \qquad R \geq |x_1 - x_{\gamma^\alpha}^{\alpha,x}| = |x_{\gamma^\alpha}^{\alpha,x-x_1}|.$$

The last inequality in the preceding lemma implies the inequality $\gamma^\alpha \geq \tau_R^{\alpha,x-x_1}$. By Theorem 3.3 and the preceding lemma we obtain

$$\begin{aligned}
&\mathsf{M}^\alpha \int_0^\infty e^{-\varphi_t^\alpha}\psi_t^\alpha f_{(t_1,x_1)}(y_t^{\alpha,s},x_t^{\alpha,x})\,dt \\
&= \mathsf{M}^\alpha \chi_{t_1-1\leq y_{\gamma^\alpha}^{\alpha,s},\,\gamma^\alpha<\infty} \times \mathsf{M}^\alpha\left\{\int_{\gamma^\alpha}^\infty e^{-\varphi_t^\alpha}\psi_t^\alpha f_{(t_1,x_1)}(y_t^{\alpha,s},x_t^{\alpha,x})\,dt \,\middle|\, \mathscr{F}_{\gamma^\alpha}^\alpha\right\} \\
&\leq N_1 \|f_{(t_1x_1)}\|_{p+1,E_{d+1}} \mathsf{M}^\alpha \chi_{t_1-1\leq y_{\gamma^\alpha}^{\alpha,s},\,\gamma^\alpha<\infty} e^{-\varphi_{\gamma^\alpha}^\alpha} \\
&\leq N_1 \|f_{(t_1,x_1)}\|_{p+1,E_{d+1}} \exp\left\{\frac{\varepsilon}{2}R - \frac{\varepsilon}{2}|x - x_1| - \frac{1}{2}(t_1 - s - 1)\right\},
\end{aligned}$$

where $N_1 = N_1(d,K_1,1)$ is the constant given in Theorem 3.3. Also, we note that for $t_1 < s$ the first expression in the above computations is equal to zero since $t_1 - y_t^{\alpha,s} \leq t_1 - s < 0$ and $\gamma^\alpha = \infty$. Hence

$$I^\alpha(s,x) \leq N_1 \int_{-\infty}^\infty dt_1 \int dx_1 \|f_{(t_1,x_1)}\|_{p+1,E_{d+1}} \pi(s - t_1, x - x_1),$$

where $\pi(t,x) = \exp[(\varepsilon/2)R - (\varepsilon/2)|x| + \frac{1}{2}(t + 1)]$ for $t \leq 0$, $\pi(t,x) = 0$ for $t > 0$. Therefore, since $v = \sup_\alpha I^\alpha$,

$$v(s,x) \leq N_1 \int_{-\infty}^\infty \int \|f_{(t_1,x_1)}\|_{p+1,E_{d+1}} \pi(s - t_1, x - x_1)\,dx_1\,dt_1.$$

In the right side of the last expression there is a convolution (with respect to (t_1,x_1)) of the two functions $\|f_{(t_1,x_1)}\|_{p+1,E_{d+1}}$ and $\pi(t_1,x_1)$. It is a well-known fact that the norm of the convolution in $\mathscr{L}_p$ does not exceed the product of the norm of one function in $\mathscr{L}_p$ and the norm of the other function in $\mathscr{L}_1$. Using this fact, we conclude that

$$\|v\|_{p+1,E_{d+1}} \le N_1 \int_{-\infty}^{\infty}\int \pi(t,x)\,dx\,dt \big\| \|f_{(t_1,x_1)}\|_{p+1,E_{d+1}}\big\|_{p+1,E_{d+1}}$$

$$= N(d,K_1)\left[\int_{-\infty}^{\infty}\int dt_1\,dx_1 \right.$$

$$\left.\times\left(\int_{-\infty}^{\infty}\int w(t_1-t,x_1-x)f^{p+1}(t,x)\,dx\,dt\right)\right]^{1/(p+1)}$$

$$= N(d,K_1)\|f\|_{p+1,E_{d+1}}.$$

To complete the proof of the theorem it remains only to show that the last constant $N(d,K_1)$ can be regarded as an increasing function of K_1.

Let

$$\tilde{N}(d,K_1) = \sup \|v(K_1,1,1,|f|,\cdot,\cdot)\|_{p+1,E_{d+1}}\|f\|_{p+1,E_{d+1}}^{-1},$$

where the upper bound is taken over all $f \in \mathscr{L}_{p+1}(E_{d+1})$ such that $\|f\|_{p+1,E_{d+1}} > 0$. According to what has been proved above, $\tilde{N}(d,K_1) < \infty$. In addition, the sets $\mathfrak{A}$ increase with respect to K_1. Hence v, $\tilde{N}(d,K_1)$ increase with respect to K_1. Finally, it is seen that $|v(f,s,x)| \le v(|f|,s,x)$ and

$$\|v(K_1,1,1,f,\cdot,\cdot)\|_{p+1,E_{d+1}} \le \tilde{N}(d,K_1)\|f\|_{p+1,E_{d+1}}.$$

The theorem has been proved. □

We extend the assertions of Theorem 1 and 3 to the case where the function $f(t,x)$ does not depend on t. However, we do not consider here the process r_t, as we did in the previous sections. Let

$$v(x) = v(g,x) = v(K_1,K_2,g,x)$$

$$= \sup_{\alpha\in\mathfrak{A}(K_1,K_2,0)} \mathsf{M}^\alpha \int_0^\infty e^{-\varphi_t^\alpha}(c_t^\alpha)^{(p-d)/p}(\det a_t^\alpha)^{1/p} g(x_t^{\alpha,x})\,dt.$$

4. Theorem. (a) *Let $g \in \mathscr{L}_p(E_d)$; then $v(x)$ is a continuous function,*

$$|v(x)| \le N(d,K_1,K_2)\|g\|_{p,E_d}.$$

(b) *There exists a finite function $N(d,K_1)$ increasing with respect to K_1 and such that for all $g \in \mathscr{L}_p(E_d)$*

$$\|v(g,\cdot)\|_{p,E_d} \le K_2^{d/p} N\left(d,\frac{K_1}{\sqrt{K_2}}\right)\|g\|_{p,E_d}.$$

This theorem can be proved in the same way as Theorems 1 and 3.

We proceed now to consider the main results of the present section. Let numbers $K > 0$, $\delta > 0$ be fixed, and let each point $(t,x) \in E_{d+1}$ $(x \in E_d)$ be associated with some nonempty set $\mathfrak{B}(t,x)$ (respectively, $\mathfrak{B}(x)$) consisting of sets α of type (1). Let $\mathfrak{B}$ be the union of all sets $\mathfrak{B}(t,x)$, $\mathfrak{B}(x)$. We assume that a function $c_u^\alpha(\omega)$ is bounded on $\mathfrak{B} \times [0,\infty) \times \bigcup_\alpha \Omega^\alpha$ and that for all $\alpha \in \mathfrak{B}$, $u \in [0,\infty)$, $\omega \in \Omega^\alpha$, $y \in E_d$

$$|b_u^\alpha| \leq K, \qquad \operatorname{tr} \sigma_u^\alpha[\sigma_u^\alpha]^* \leq K, \qquad r_u^\alpha = 1,$$
$$|[\sigma_u^\alpha]^* y| \geq \delta |y|. \tag{2}$$

It is useful to note that (2) can be rewritten as

$$(a_t^\alpha y, y) = \sum_{i,j=1}^{d} (a_t^\alpha)^{ij} y^i y^j \geq \frac{1}{2}\delta^2 |y|^2,$$

since $(a_t^\alpha y, y) = \frac{1}{2}(\sigma_t^\alpha(\sigma_t^\alpha)^* y, y) = \frac{1}{2}|(\sigma_t^\alpha)^* y|^2$.

5. Theorem. (a) *Let $\lambda \geq \lambda_0 > 0$, let $Q \subset E_{d+1}$, let Q be an open set, and let $f \in \mathscr{L}_{p+1}(Q)$,*

$$\tau^\alpha = \tau^{\alpha,s,x} = \inf\{t \geq 0 : (t+s, x_t^{\alpha,x}) \notin Q\},$$

$$= \sup_{\alpha \in \mathfrak{B}(s,x)} \mathsf{M}^\alpha \int_0^{\tau^\alpha} e^{-\varphi_t^\alpha - \lambda t} f(t+s, x_t^{\alpha,x})\, dt.$$

Then

$$\lambda]\|z^\lambda\|_{p+1,Q} \leq N(d,K,\delta,\lambda_0)\|f\|_{p+1,Q}.$$

(b) *Let $\lambda \geq \lambda_0 > 0$, let $D \subset E_d$, let D be an open set, and let $g \in \mathscr{L}_p(D)$,*

$$\tau^\alpha = \tau^{\alpha,x} = \inf\{t \geq 0 : x_t^{\alpha,x} \notin D\},$$

$$z^\lambda(x) = \sup_{\alpha \in \mathfrak{B}(x)} \mathsf{M}^\alpha \int_0^{\tau^\alpha} e^{-\varphi_t^\alpha - \lambda t} g(x_t^{\alpha,x})\, dt.$$

Then

$$\lambda]\|z^\lambda\|_{p,D} \leq N(d,K,\delta,\lambda_0)\|g\|_{p,D}.$$

Proof. Since all eigenvalues of the matrix a_t^α are greater than $\frac{1}{2}\delta^2$, $\det a_t^\alpha \geq 2^\alpha \delta^{2d}$. From this, assuming that $\tilde{f} = |f|\chi_Q$, $\tilde{c}_t^\alpha = c_t^\alpha + \lambda$, $\tilde{\varphi}_t^\alpha = \varphi_t^\alpha + \lambda t$ and noting that $\tilde{c}_t^\alpha \geq \lambda$, we find

$$|z^\lambda(s,x)| \leq N(\delta)\lambda^{(d-p)/(p+1)} \sup_{\alpha \in \mathfrak{B}(s,x)} \mathsf{M}^\alpha \int_0^\infty e^{-\tilde{\varphi}_t^\alpha}(\tilde{c}_t^\alpha)^{(p-d)/(p+1)}$$
$$\times (r_t^\alpha \det a_t^\alpha)^{1/(p+1)} \tilde{f}(y_t^{\alpha,s}, x_t^{\alpha,s})\, dt.$$

It is seen that

$$|b_t^\alpha| \leq \frac{K}{\lambda}\tilde{c}_t^\alpha, \qquad \operatorname{tr} a_t^\alpha \leq \frac{K}{\lambda}\tilde{c}_t^\alpha, \qquad \text{and} \qquad r_t^\alpha \leq \frac{1}{\lambda}\tilde{c}_t^\alpha.$$

Therefore,

$$|z^\lambda(s,x)| \leq v\left(\frac{K}{\lambda}, \frac{K}{\lambda}, \frac{1}{\lambda}, \tilde{f}, s, x\right) N(\delta)\lambda^{(d-p)/(p+1)},$$

which implies, by Theorem 3, that

$$]\|z^\lambda\|_{p+1,Q} \le N(\delta)\lambda^{(d-p)/(p+1)}\|v\|_{p+1,E_{d+1}}$$
$$\le N(\delta)\lambda^{-1}K^{d/(p+1)}N\left(d,\sqrt{\frac{K}{\lambda}}\right)\|\tilde{f}\|_{p+1,E_{d+1}},$$

thus proving assertion (a) of our theorem since

$$\|\tilde{f}\|_{p+1,E_{d+1}} = \|f\|_{p+1,Q},\quad K^{d/(p+1)} \le 1 + K,\quad N\left(d,\sqrt{\frac{K}{\lambda}}\right) \le N\left(d,\sqrt{\frac{K}{\lambda_0}}\right).$$

Proceeding in the same way, we can prove assertion (b) with the aid of Theorem 4. The theorem is proved. □

6. Theorem. (a) *Suppose that Q is a region in E_{d+1}, $f_1(t,x)$ is a bounded Borel function, $f \in \mathscr{L}_{p+1}(Q)$, $\lambda > 0$,*

$$\tau^\alpha = \tau^{\alpha,s,x} = \inf\{t \ge 0 : (s + t, x_t^{\alpha,x}) \notin Q\},$$

$$z^\lambda(s,x) = z^\lambda(f,s,x)$$
$$= \sup_{\alpha\in\mathfrak{B}(s,x)} \mathsf{M}^\alpha\left[\int_0^{\tau^\alpha} e^{-\varphi_t^\alpha - \lambda t} f(s + t, x_t^{\alpha,x})\,dt + e^{-\varphi_{\tau^\alpha}^\alpha - \lambda\tau^\alpha} f_1(s + \tau^\alpha, x_{\tau^\alpha}^{\alpha,x})\right].$$

Then, there exists a sequence $\lambda_n \to \infty$ such that $\lambda_n z^{\lambda_n}(s,x) \to f(s,x)$ (Q-a.s.).

(b) *Suppose that D is a region in E_d, $g_1(x)$ is a bounded Borel function, $g \in \mathscr{L}_p(D)$, $\lambda > 0$,*

$$\tau^\alpha = \tau^{\alpha,x} = \inf\{t \ge 0 : x_t^{\alpha,x} \notin D\},$$

$$z^\lambda(x) = z^\lambda(g,x)$$
$$= \sup_{\alpha\in\mathfrak{B}(x)} \mathsf{M}^\alpha\left[\int_0^{\tau^\alpha} e^{-\varphi_t^\alpha - \lambda t} g(x_t^{\alpha,x})\,dt + e^{-\varphi_{\tau^\alpha}^\alpha - \lambda\tau^\alpha} g_1(x_{\tau^\alpha}^{\alpha,x})\right].$$

Then, there exists a sequence $\lambda_n \to \infty$ such that $\lambda_n z^{\lambda_n}(x) \to g(x)$($D$-a.s.).

7. Theorem. (a) *We introduce another element in Theorem 6a. Suppose that Q' is a bounded region $Q' \subset \bar{Q}' \subset Q$. Then $]\|\lambda z^\lambda - f\|_{p+1,Q'} \to 0$ as $\lambda \to \infty$. If $f_1 \equiv 0$, we can take $Q' = Q$.*

(b) *Suppose that in Theorem 6b D' is a bounded region, $D' \subset \bar{D}' \subset D$; then $]\|\lambda z^\lambda - g\|_{p,D'} \to 0$ as $\lambda \to \infty$. If $g_1 \equiv 0$, we can take $D' = D$.*

PROOF OF THEOREMS 6 AND 7. It was noted in Section 2.1 that the property of convergence with respect to an exterior norm implies the existence of a subsequence convergent almost everywhere. Using this fact, we can easily see that only Theorem 7 is to be proved.

PROOF OF THEOREM 7a. First, let $f_1 \equiv 0$. We take a sequence of functions $f^n \in C_0^\infty(Q)$ such that $\|f^n - f\|_{p+1,Q} \to 0$. It is seen that

$$|\lambda z^\lambda(f,s,x) - f(s,x)| \le \lambda|z^\lambda(f,s,x) - z^\lambda(f^n,s,x)|$$
$$+ |\lambda z^\lambda(f^n,s,x) - f^n(s,x)| + |f^n(s,x) - f(s,x)|,$$

from which, noting that $|z^\lambda(f,s,x) - z^\lambda(f^n,s,x)| \le z^\lambda(|f - f^n|,s,x)$, *we obtain*, in accord with Theorem 5a,

$$\varlimsup_{\lambda\to\infty}]\|\lambda z^\lambda(f,\cdot,\cdot) - f(\cdot,\cdot)\|_{p+1,Q}$$
$$\le N(d,K,\delta,1)\|f - f^n\|_{p+1,Q} + \varlimsup_{\lambda\to\infty}]\|\lambda z^\lambda(f^n,\cdot,\cdot)$$
$$- f^n(\cdot,\cdot)\|_{p+1,Q} + \|f^n - f\|_{p+1,Q}$$

In the last inequality the left side does not depend on n; the first and third terms in the right side can be made arbitrarily small by choosing an appropriate n. In order to make sure that the left side of the last inequality is equal to zero, we need only to show that for each n

$$\varlimsup_{\lambda\to\infty}]\|\lambda z^\lambda(f^n,\cdot,\cdot) - f^n(\cdot,\cdot)\|_{p+1,Q} = 0.$$

In short, it suffices to prove assertion (a) for $f_1 \equiv 0$ in the case $f \in C_0^\infty(Q)$.

In conjunction with Ito's formula applied to $f(s + t, x_t^{\alpha,x})l^{-\varphi_t^\alpha - \lambda t}$ for each $\alpha \in \mathfrak{B}(s,x)$, $t \ge 0$ we have

$$f(s,x) = \mathsf{M}^\alpha\Bigg\{\int_0^{t\wedge\tau^\alpha} e^{-\varphi_r^\alpha - \lambda r}[\lambda f(s + r, x_r^{\alpha,x}) - L_r^\alpha f(s + r, x_r^{\alpha,x})]\,dr$$
$$+ f(s + t\wedge\tau^\alpha, x_{t\wedge\tau^\alpha}^{\alpha,x})e^{-\varphi_{t\wedge\tau^\alpha}^\alpha - \lambda t\wedge\tau^\alpha}\Bigg\}, \tag{3}$$

where

$$L_r^\alpha f(t,x) \equiv \frac{\partial}{\partial t}f(t,x) + \sum_{i,j-1}^d (a_r^\alpha)^{ij} f_{x^i x^j}(t,x) + \sum_{i-1}^d (b_r^\alpha)^i f_{x^i}(t,x) - c_r^\alpha f(t,x).$$

Since a_r^α, b_r^α, c_r^α are bounded, $|L_r^\alpha f(t,x)|$ does not exceed the expression

$$N\left[\left|\frac{\partial}{\partial t}f(t,x)\right| + \sum_{i,j=1}^d |f_{x^i x^j}(t,x)| + \sum_{i=1}^d |f_{x^i}(t,x)| + |f(t,x)|\right]$$

Denoting the last expression by $h(t,x)$, we note that $h(t,x)$ is a bounded finite function; in particular, $h \in \mathscr{L}_{p+1}(Q)$.

Using the Lebesgue bounded convergence theorem, we pass to the limit in (3) as $t \to \infty$. Thus we have

$$f(s,x) = \lambda\mathsf{M}^\alpha \int_0^{\tau^\alpha} e^{-\varphi_t^\alpha - \lambda t} f(s + t, x_t^{\alpha,x})\,dt - \mathsf{M}^\alpha \int_0^{\tau^\alpha} e^{-\varphi_t^\alpha - \lambda t} L_t^\alpha f(s + t, x_t^{\alpha,x})\,dt,$$

which immediately yields

$$|\lambda z^\lambda(f,s,x) - f(s,x)| = \left|\sup_{\alpha\in\mathfrak{B}(s,x)} \mathsf{M}^\alpha \int_0^{\tau^\alpha} e^{-\varphi_t^\alpha - \lambda t} L_t^\alpha f(s + t, x_t^{\alpha,x})\,dt\right|$$
$$\le \sup_{\alpha\in\mathfrak{B}(s,x)} \mathsf{M}^\alpha \int_0^{\tau^\alpha} e^{-\varphi_t^\alpha - \lambda t}|L_t^\alpha f(s + t, x_t^{\alpha,x})|\,dt \le z^\lambda(h,s,x).$$

In short, we have

$$|\lambda z^\lambda(fs,x) - f(s,x)| \le z^\lambda(h,s,x),$$

which, according to Theorem 5a yields

$$\varlimsup_{\lambda\to\infty}]|\lambda z^\lambda(f,\cdot,\cdot) - f(\cdot,\cdot)\|_{p+1,Q} \le \lim_{\lambda\to\infty}]|z^\lambda(h,\cdot,\cdot)\|_{p+1,Q}$$

$$\le N\|h\|_{p+1,Q} \lim_{\lambda\to\infty} \frac{1}{\lambda} = 0,$$

thus proving Theorem 7a for $f_1 \equiv 0$. In the general case

$$|\lambda z^\lambda(f,s,x) - f(s,x)| \le \lambda \sup_{\alpha\in\mathfrak{B}(s,x)} \mathsf{M}^\alpha e^{-\varphi_\tau^\alpha - \lambda\tau^\alpha}|f_1(s+\tau^\alpha, x_\tau^{\alpha,x})|$$

$$+ \left|\lambda \sup_{\alpha\in\mathfrak{B}(s,x)} \mathsf{M}^\alpha \int_0^{\tau^\alpha} e^{-\varphi_t^\alpha - \lambda t} f(s+t, x_t^{\alpha,x})\,dt - f(s,x)\right|,$$

where the exterior norm of the second term tends to zero; due to the boundedness of f_1 the first term does not exceed the product of a constant and the expression

$$\pi^\lambda(s,x) \equiv \lambda \sup_{\alpha\in\mathfrak{B}(s,x)} \mathsf{M}^\alpha e^{-\lambda\tau^{\alpha,s,x}}.$$

Therefore, in order to complete proving Theorem 7a, it remains only to show that $]|\pi^\lambda\|_{p+1,Q'} \to 0$ as $\lambda \to \infty$ for any bounded region Q' lying together with the closure in Q. To this end, it suffices in turn to prove that $\pi^\lambda(s,x) \to 0$ uniformly on Q'. In addition, each region Q' can be covered with a finite number of cylinders of the type $C_{r,R}(s,y) = \{(t,x): |y - x| < R, |t - s| < r\}$, so that $C_{2r,2R}(s,y) \subset Q$. It is seen that we need only to prove that $\pi^\lambda(t,x) \to 0$ uniformly on any cylinder of this type.

We fix a cylinder $C_{r,R}(s,y)$ such that $C_{2r,2R}(s,y) \subset Q$. Let $\tau_R^\alpha(x) = \inf\{t \ge 0: |x - x_t^{\alpha,x}| \ge R\}$. Finally, we denote by $\mu(\lambda)$ the positive root of the equation $\lambda - \mu K - \mu^2 K = 0$. Also, we note that for $(t,x) \in C_{r,R}(s,y)$ we have $\tau^{\alpha,t,x} \ge r \wedge \tau_R^\alpha(x)$. Hence

$$\lambda\mathsf{M}^\alpha e^{-\lambda\tau^{\alpha,t,x}} \le \lambda\mathsf{M}^\alpha e^{-\lambda r\wedge\tau_R^\alpha(x)} \le \lambda e^{-\lambda r} + \lambda\mathsf{M}^\alpha e^{-\lambda\tau_R^\alpha(x)}.$$

Furthermore, by Lemma 3.2[3] the inequality

$$\mathsf{M}^\alpha e^{-\lambda\tau_R^\alpha(x)} \le \operatorname{ch}^{-1}\mu R$$

holds true. Therefore, the function $\pi^\lambda(t,x)$ does not exceed $\lambda e^{-\lambda r} + \lambda(\operatorname{ch}\mu(\lambda)R)^{-1}$ on $C_{r,R}(s,y)$. Simple computations show that the last constant tends to zero as $\lambda \to \infty$. Therefore, $\pi^\lambda(t,x)$ tends uniformly to zero on $C_{r,R}(s,y)$. This completes the proof of Theorem 7a.

Theorem 7b can be proved in a similar way, which we suggest the reader should do as an exercise. We have thus proved Theorems 6 and 7. □

[3] In Lemma 3.2, one should take $\tau = 0$, $c_t \equiv 1$.

8. Corollary. *Let $f \in \mathscr{L}_{p+1}(Q)$, $f \geq 0$ (Q-a.s.) and for all $(s,x) \in Q$ let*

$$\inf_{\alpha \in \mathfrak{B}(s,x)} \mathsf{M}^{\alpha} \int_0^{\tau^{\alpha}} e^{-\varphi_t^{\alpha}} f(s+t, x_t^{\alpha,x})\,dt = 0. \tag{4}$$

Then $f = 0$ (Q-a.s.).

In fact, by Theorem 2.4 the equality (4) still holds if we change f on the set of measure zero. It is then seen that for $\lambda \geq 0$

$$\inf_{\alpha \in \mathfrak{B}(s,x)} \mathsf{M}^{\alpha} \int_0^{\tau^{\alpha}} e^{-\varphi_t^{\alpha} - \lambda t} f(s+t, x_t^{\alpha,x})\,dt = 0.$$

Furthermore, for $f_1 \equiv 0$

$$z^{\lambda}(-f,s,x) = -\inf_{\alpha \in \mathfrak{B}(s.x)} \mathsf{M}^{\alpha} \int_0^{\tau^{\alpha}} e^{-\varphi_t^{\alpha} - \lambda t} f(s+t, x_t^{\alpha,x})\,dt.$$

Therefore, $z^{\lambda} = 0$ in Q and $-f = \lim_{n\to\infty} \lambda_n z^{\lambda_n} = 0$ (Q-a.s.).

5. Solutions of Stochastic Integral Equations and Estimates of the Moments

In this section, we list some generalizations of the kind we need of well-known results on existence and uniqueness of solutions of stochastic equations. Also, we present estimates of the moments of the above solutions. The moments of these solutions are estimated when the condition for the growth of coefficients to be linear is satisfied. The theorem on existence and uniqueness is proved for the case where the coefficients satisfy the Lipschitz condition (condition $(\mathscr{L})$).

We fix two constants, $T > 0$, $K > 0$. Also, we adopt the following notation: $(\mathbf{w}_t, \mathscr{F}_t)$ is a d_1-dimensional Wiener process; x, y denote points of E_d; σ_t, $\sigma_t(x)$, $\tilde{\sigma}_t(x)$ are random matrices of dimension $d \times d_1$; $b_t(x)$, $\tilde{b}_t(x)$, ξ_t, $\tilde{\xi}_t$ are random d-dimensional vectors; r_t, h_t are nonnegative numbers. We assume all the processes to be given for $t \in [0,T]$, $x \in E_d$ and progressively measurable with respect to $\{\mathscr{F}_t\}$. If for all $t \in [0,T]$, ω, x, y

$$\|\sigma_t(x) - \sigma_t(y)\| \leq K|x-y|, \qquad |b_t(x) - b_t(y)| \leq K^2|x-y|,$$

we say that the condition $(\mathscr{L})$ is satisfied. If for all $t \in [0,T]$, ω, x

$$\|\sigma_t(x)\|^2 \leq 2r_t^2 + 2K^2|x|^2, \qquad |b_t(x)| \leq h_t + K^2|x|,$$

we say that the condition (R) is satisfied.

Note that we do not impose the condition $(\mathscr{L})$ and the condition (R) on $\tilde{\sigma}_t(x)$, $\tilde{b}_t(x)$. Furthermore, it is useful to have in mind that if the condition $(\mathscr{L})$ is satisfied, the condition (R) will be satisfied for $r_t = \|\sigma_t(0)\|$, $h_t = |b_t(0)|$

(with the same constant K) since, for example, $\|\sigma_t(x)\|^2 \le 2\|\sigma_t(0)\|^2 + 2\|\sigma_t(x) - \sigma_t(0)\|^2$.

As usual, by a solution of the stochastic equation

$$x_t = \xi_t + \int_0^t \sigma_s(x_s)\,dw_s + \int_0^t b_s(x_s)\,ds \tag{1}$$

we mean a progressively measuable (with respect to $\{\mathscr{F}_t\}$), process x_t for which the right side of (1) is defined[4] and, in addition, $x_t(\omega)$ coincides with the right side of (1) for some set Ω' of measure one for all $t \in [0,T]$, $\omega \in \Omega'$.

1. Lemma. *Let x_t be a solution of Eq.* (1) *for $\xi_t \equiv 0$. Then for $q \ge 1$*

$$\begin{aligned} d|x_t|^{2q} = {} & [2q|x_t|^{2q-2}x_t b_t(x_t) + q|x_t|^{2q-2}\|\sigma_t(x_t)\|^2 \\ & + 2q(q-1)|x_t|^{2q-4}|\sigma_t^*(x_t)x_t|^2]\,dt \\ & + 2q|x_t|^{2q-2}x_t\sigma_t(x_t)\,d\mathbf{w}_t \le q|x_t|^{2q-2}(2|x_t|\,|b_t(x_t)| \\ & + (2q-1)\|\sigma_t(x_t)\|^2)dt + 2q|x_t|^{2q-2}x_t\sigma_t(x_t)d\mathbf{w}_t. \end{aligned}$$

We prove this lemma by applying Ito's formula to the twice continuously differentiable function $|x|^{2q}$ and using the inequalities

$$x_t b_t \le |x_t|\,|b_t|, \qquad |\sigma_t^*(x_t)x_t|^2 \le \|\sigma_t(x_t)\|^2|x_t|^2.$$

2. Lemma. *Let the condition* (R) *be satisfied and let x_t be a solution of Eq.* (1) *for $\xi_t \equiv 0$. Then, for all $q \ge 1$, $\varepsilon > 0$, $t \in [0,T]$*

$$(\mathsf{M}|x_t|^{2q})^{1/q} \le \frac{1}{\varepsilon}\int_0^t e^{\lambda(t-s)}[\mathsf{M}h_s^{2q}]^{1/q}\,ds + 2(2q-1)\int_0^t e^{\lambda(t-s)}[\mathsf{M}r_s^{2q}]^{1/q}\,ds, \tag{2}$$

where $\lambda = 4qK^2 + \varepsilon \equiv \lambda_{q,\varepsilon}$. If the condition ($\mathscr{L}$) *is satisfied, one can take in* (2) $h_s = |b_s(0)|$, $r_s = \|\sigma_s(0)\|$.

PROOF. We fix $q \ge 1$, $\varepsilon > 0$, $t_0 \in [0,T]$. Also, denote by $\psi(t)$ the right side of (2). We prove (2) for $t = t_0$. We can obviously assume that $\psi(t_0) < \infty$. We make one more assumption which we will drop at the end of the proof. Assume that $x_t(\omega)$ is a bounded function of ω, t.

Using the preceding lemma and the condition (R), we obtain

$$\begin{aligned} d|x_t|^{2q} \le {} & [4q^2K^2|x_t|^{2q} + 2q|x_t|^{2q-1}h_t \\ & + 2q(2q-1)|x_t|^{2q-2}r_t^2]\,dt + 2q|x_t|^{2q-2}x_t\sigma_t(x_t)\,d\mathbf{w}_t. \end{aligned}$$

Next, we integrate the last inequality over t. In addition, we take the expectation from the both sides of this inequality. In this case, the expectation of the stochastic integral disappears because, due to the boundedness of

[4] Recall that the stochastic integral in (1) is defined and continuous in t for $t \le T$ if

$$\int_0^T \|\sigma_s(x_s)\|^2\,ds < \infty \quad \text{(a.s.)}.$$

$x_t(\omega)$, finiteness of $\psi(t_0)$, and, in addition, Hölder's inequality,

$$\begin{aligned}\mathsf{M}\int_0^{t_0}|x_t|^{4q-4}|\sigma_t^*(x_t)x_t|^2\,dt &\le N\mathsf{M}\int_0^{t_0}\|\sigma_t(x_t)\|^2\,dt\\ &\le N\mathsf{M}\int_0^{t_0}|x_t|^2\,dt + N\int_0^{t_0}e^{\lambda(t_0-t)}\mathsf{M}r_t^2\,dt\\ &\le N + N\int_0^{t_0}e^{\lambda(t_0-t)}[\mathsf{M}r_t^{2q}]^{1/q}\,dt < \infty.\end{aligned}$$

Furthermore, we use the following inequalities:

$$\begin{aligned}\mathsf{M}|x_t|^{2q-1}h_t &\le (\mathsf{M}|x_t|^{2q})^{1-(1/2q)}(\mathsf{M}h_t^{2q})^{1/2q}\\ &= (\mathsf{M}|x_t|^{2q})^{1/2}[(\mathsf{M}|x_t|^{2q})^{1-(1/q)}(\mathsf{M}h_t^{2q})^{1/q}]^{1/2}\\ &\le \frac{\beta}{2}\mathsf{M}|x_t|^{2q} + \frac{1}{2\beta}(\mathsf{M}|x_t|^{2q})^{1-(1/q)}(\mathsf{M}h_t^{2q})^{1/q},\\ \mathsf{M}|x_t|^{2q-2}r_t^2 &\le (\mathsf{M}|x_t|^{2q})^{1-(1/q)}(\mathsf{M}r_t^{2q})^{1/q}.\end{aligned}$$

Also, let

$$m(t) = \mathsf{M}|x_t|^{2q}, \qquad g_t = \frac{q}{\varepsilon}(\mathsf{M}h_t^{2q})^{1/q} + 2q(2q-1)(\mathsf{M}r_t^{2q})^{1/q}.$$

In accord with what has been said above for $t \le t_0$,

$$m(t) \le \int_0^t [\lambda q m(s) + g_s m^{1-(1/q)}(s)]\,ds. \tag{3}$$

Further, we apply a well-known method of transforming such inequalities. Let $\delta > 0$. We introduce an operator F_δ on nonnegative functions of one variable, on $[0,t_0]$, by defining

$$F_\delta u(t) = \int_0^t [\lambda q u(s) + g_s u^{1-(1/q)}(s)]\,ds + \delta, \qquad t \in [0,t_0].$$

It is easily seen that F_δ is a monotone operator, i.e., if $0 \le u^1(t) \le u^2(t)$ for all t, then $0 \le F_\delta u^1(t) \le F_\delta u^2(t)$ for all t. Furthermore, if all the nonnegative functions u_t^n are bounded and if they converge for each t, then $\lim_{n\to\infty} F_\delta u_n(t) = F_\delta \lim_{n\to\infty} u_n(t)$. Finally, for the function $v(t) = Ne^{\lambda qt}$ for all sufficiently large N and $\delta \le 1$ we have $F_\delta v(t) \le v(t)$ if $t \in [0,t_0]$. In fact,

$$F_\delta v(t) \le N(e^{\lambda qt} - 1) + N^{1-(1/q)}e^{\lambda qt}\int_0^t g_s e^{-\lambda s}\,ds + \delta \le Ne^{\lambda qt}$$

for

$$N^{-1/q}e^{\lambda q t_0}\int_0^{t_0} g_s e^{-\lambda s}\,ds + \delta N^{-1} \le 1.$$

It follows from (3) and the aforementioned properties, with N such that $m(t) \le v(t)$, that $m(t) \le F_\delta m(t) \le \cdots \le F_\delta^n m(t) \le v(t)$. Therefore, the limit $\lim_{n\to\infty} F_\delta^n m(t)$ exists. If we denote this limit by $v_\delta(t)$, then $m(t) \le v_\delta(t)$. Taking the limit in the equality $F_\delta^{n+1}m(t) = F_\delta(F_\delta^n m)(t)$, we conclude that $v_\delta = F_\delta v_\delta$. Therefore, for each $\delta \in (0,1)$ the function $m(t)$ does not exceed some nonnegative solution of the equation

$$v_\delta(t) = \int_0^t [\lambda q v_\delta(s) + g_s v_\delta^{1-(1/q)}(s)]\, ds + \delta.$$

We solve the last given equation, from which it follows that $v_\delta(t) \geq \delta$, $v_\delta(0) = \delta$, and

$$v_\delta'(t) = \lambda q v_\delta(t) + g v_\delta^{1-(1/q)}(t). \tag{4}$$

Equation (4), after we have multiplied it by $v_\delta^{(1/q)-1}$ (which is possible due to the inequality $v_\delta \geq \delta$) becomes a linear equation with respect to $v_\delta^{1/q}$. Having solved this equation, we find $v_\delta^{1/q}(t) = \delta^{1/q} + \psi(t)$.

Therefore, $m(t) \leq (\delta^{1/q} + \psi(t))^q$ for all $t \in [0,t_0]$, $\delta \in (0,1)$. We have proved the lemma for the bounded $x_t(\omega)$ as $\delta \to 0$.

In order to prove the lemma in the general case, we denote by τ_R the first exit time of x_t from S_R. Then $x_{t \wedge \tau_R}(\omega)$ is the bounded function of (ω,t), and, as is easily seen,

$$x_{t \wedge \tau_R} = \int_0^t \chi_{s<\tau_R} \sigma_s(x_{s \wedge \tau_R})\, d\mathbf{w}_s + \int_0^t \chi_{s<\tau_R} b_s(x_{s \wedge \tau_R})\, ds.$$

Therefore the process $x_{t \wedge \tau_R}$ satisfies the same equation as the process x_t does; however, $\sigma_s(x)$, $b_s(x)$ are to be replaced by $\chi_{s<\tau_R}\sigma_s(x)$, $\chi_{s<\tau_R} b_s(x)$, respectively. In accord with what has been proved above, $\mathsf{M}|x_{t \wedge \tau_R}|^{2q} \leq [\psi(t)]^q$. It remains only to allow $R \to \infty$, to use Fatou's lemma, and in addition, to take advantage of the fact that due to the continuity of x_t the time $\tau_R \to \infty$ as $R \to \infty$. We have thus proved our lemma. □

3. Corollary. *Let* $\int_0^t \|\sigma_s\|^2\, ds < \infty$ *with probability* 1, *and let* τ *be a Markov time with respect to* $\{\mathscr{F}_s\}$. *Then, for all* $q \geq 1$

$$\mathsf{M}\left|\int_0^{\tau \wedge t} \sigma_s\, d\mathbf{w}_s\right|^{2q} \leq 2^q(2q-1)^q \left\{\int_0^t [\mathsf{M}\|\sigma_s\|^{2q}\chi_{s<\tau}]^{1/q}\, ds\right\}^q$$

$$\leq 2^q(2q-1)^q t^{q-1} \mathsf{M} \int_0^{\tau \wedge t} \|\sigma_s\|^{2q}\, ds.$$

In fact, we have obtained the second inequality using Hölder's equality. The first inequality follows from the lemma, if we take $\sigma_s(x) = \sigma_s \chi_{s<\tau}$, $b_s(x) = 0$, write the assertion of the lemma with arbitrary K, ε, and, finally, assume that $K \downarrow 0$, $\varepsilon \downarrow 0$.

4. Exercise

In the proof of the lemma, show that the factor 2^q in Corollary 3 can be replaced by unity.

5. Corollary. *Let the condition* $(\mathscr{L})$ *be satisfied, let* x_t *be a solution of Eq.* (1), *and let* $\tilde{x}_t$ *be a solution of the equation*

$$\tilde{x}_t = \tilde{\xi}_t + \int_0^t \tilde{\sigma}_s(\tilde{x}_s)\, d\mathbf{w}_s + \int_0^t \tilde{b}_s(\tilde{x}_s)\, ds.$$

Then, for all $q \geq 1$, $t \in [0,T]$

$$\mathsf{M}|x_t - \tilde{x}_t|^{2q} \leq 4^q \mathsf{M}|\xi_t - \tilde{\xi}_t|^{2q} + N(q,K)t^{q-1}\mathsf{M}\int_0^t e^{\mu(t-s)}|\xi_s - \tilde{\xi}_s|^{2q}\,ds$$
$$+ N(q)t^{q-1}\mathsf{M}\int_0^t e^{\mu(t-s)}\{|b_s(\tilde{x}_s) - \tilde{b}_s(\tilde{x}_s)|^{2q}$$
$$+ \|\sigma_s(\tilde{x}_s) - \tilde{\sigma}_s(\tilde{x}_s)\|^{2q}\}\,ds,$$

where $\mu = 4q^2K^2 + q$.

PROOF. Let $y_t = (x_t - \tilde{x}_t) - (\xi_t - \tilde{\xi}_t)$. Then, as is easily seen,

$$y_t = \int_0^t [\sigma_s(y_s + \tilde{x}_s + \xi_s - \tilde{\xi}_s) - \tilde{\sigma}_s(\tilde{x}_s)]\,d\mathbf{w}_s + \int_0^t [b_s(y_s + \tilde{x}_s + \xi_s - \tilde{\xi}_s) - \tilde{b}_s(\tilde{x}_s)]\,ds,$$

in this case

$$[\sigma_s(x + \tilde{x}_s + \xi_s - \tilde{\xi}_s) - \tilde{\sigma}_s(\tilde{x}_s)], \qquad [b_s(x + \tilde{x}_s + \xi_s - \tilde{\xi}_s) - \tilde{b}_s(\tilde{x}_s)]$$

satisfy the condition ($\mathscr{L}$). From this, according to the lemma applied to the process y_t, we have

$$(\mathsf{M}|y_t|^{2q})^{1/q} \leq \int_0^t e^{(1/q)\mu(t-s)}[\mathsf{M}|b_s(\tilde{x}_s + \xi_s - \tilde{\xi}_s) - \tilde{b}_s(\tilde{x}_s)|^{2q}]^{1/q}\,ds$$
$$+ 2(2q-1)\int_0^t e^{(1/q)\mu(t-s)}[\mathsf{M}\|\sigma_s(\tilde{x}_s + \xi_s - \tilde{\xi}_s) - \tilde{\sigma}_s(\tilde{x}_s)\|^{2q}]^{1/q}\,ds.$$

We raise both sides of the last inequality to the qth power. We use Hölder's inequality as well as the fact that

$$|b_s(\tilde{x}_s + \xi_s - \tilde{\xi}_s) - \tilde{b}_s(\tilde{x}_s)| \leq |b_s(\tilde{x}_s + \xi_s - \tilde{\xi}_s) - b_s(\tilde{x}_s)| + |b_s(\tilde{x}_s) - \tilde{b}_s(\tilde{x}_s)|$$
$$\leq K^2|\xi_s - \tilde{\xi}_s| + |b_s(\tilde{x}_s) - \tilde{b}_s(\tilde{x}_s)|,$$
$$(a+b)^q \leq 2^{q-1}(a^q + b^q),$$

which yields

$$\mathsf{M}|y_t|^{2q} \leq 2^{q-1}t^{q-1}\mathsf{M}\int_0^t e^{\mu(t-s)}[2^{2q-1}K^{4q}|\xi_s - \tilde{\xi}_s|^{2q}$$
$$+ 2^{2q-1}|b_s(\tilde{x}_s) - \tilde{b}_s(\tilde{x}_s)|^{2q} + 2^q(2q-1)^q 2^{2q-1}K^{2q}|\xi_s - \tilde{\xi}_s|^{2q}$$
$$+ 2^q(2q-1)^q 2^{2q-1}\|\sigma_s(\tilde{x}_s) - \tilde{\sigma}_s(\tilde{x}_s)\|^{2q}]\,ds.$$

It remains to note that $|x_t - \tilde{x}_t| \leq |y_t| + |\xi_t - \tilde{\xi}_t|$, $|x_t - \tilde{x}_t|^{2q} \leq 2^{2q-1}|y_t|^{2q} + 2^{2q-1}|\xi_t - \tilde{\xi}_t|^{2q}$, thus proving Corollary 5. □

6. Corollary. *Let the condition (R) be satisfied, and let x_t be a solution of* (1). *Then there exists a constant $N = N(q,K)$ such that for all $q \geq 1$, $t \in [0,T]$*

$$\mathsf{M}|x_t|^{2q} \leq N\mathsf{M}|\xi_t|^{2q} + Nt^{q-1}\mathsf{M}\int_0^t [|\xi_s|^{2q} + h_s^{2q} + r_s^{2q}]e^{N(t-s)}\,ds.$$

In fact, the process $y_t \equiv x_t - \xi_t$ satisfies the equation

$$dy_t = \sigma(y_t + \xi_t)\,d\mathbf{w}_t + b(y_t + \xi_t)\,dt, \qquad y_0 = 0,$$

the coefficients of this equation satisfying the condition (R), however with different h_t, r_t, K. For example,

$$\|\sigma_t(x + \xi_t)\|^2 \leq 2r_t^2 + 2K^2\|x + \xi_t\|^2 \leq 2r_t^2 + 4K^2|\xi_t|^2 + 4K^2|x|.$$

Therefore, using this lemma we can estimate $\mathsf{M}|y_t|^{2q}$. Having done this, we need to use the fact that $|x_t|^{2q} \leq 2^{2q-1}|y_t| + 2^{2q-1}|\xi_t|$.

In our previous assertions we assumed that a solution of Eq. (1) existed and we also wrote the inequalities which may sometimes take the form $\infty \leq \infty$. Further, it is convenient to prove one of the versions of the classical Ito theorem on the existence of a solution of a stochastic equation. Since the proofs of these theorems are well known, we shall dwell here only on the most essential points.

7. Theorem. *Let the condition* ($\mathscr{L}$) *be satisfied and let*

$$\mathsf{M}\int_0^T [|\xi_t|^2 + |b_t(0)|^2 + \|\sigma_t(0)\|^2]\, dt < \infty.$$

Then for $t \leq T$ Eq. (1) *has a solution such that* $\mathsf{M}\int_0^T |x_t|^2\, dt < \infty$. *If x_t, y_t are two solutions of* (1), *then* $\mathsf{P}\{\sup_{t\in[0,T]}|x_t - y_t| > 0\} = 0$.

PROOF. Due to Corollary 5, $\mathsf{M}|x_t - y_t|^2 = 0$ for each t. Furthermore, the process $x_t - y_t$ can be represented as the sum of stochastic integrals and ordinary integrals. Hence the process $x_t - y_t$ is continuous almost surely. The equality $x_t = y_t$ (a.s.) for each t implies that $x_t = y_t$ for all t (a.s.), thus proving the last assertion of the theorem.

For proving the first assertion of the theorem we apply, as is usually done in similar cases, the method of successive approximation. We define the operator I using the formula

$$Ix_t = \int_0^t \sigma_s(x_s)\, d\mathbf{w}_s + \int_0^t b_s(x_s)\, ds. \tag{5}$$

This operator is defined on those processes x_t for which the right side of (5) makes sense, and, furthermore, this operator maps these processes into processes Ix_t whose values can be found with the aid of the formula (5).

Denote by V a space of progressively measurable processes x_t with values in E_d such that

$$\|x_t\| = \left(\mathsf{M}\int_0^T |x_t|^2\, dt\right)^{1/2} < \infty.$$

It can easily be shown that the operator I maps V into V. In addition, it can easily be deduced from the condition ($\mathscr{L}$) that

$$\mathsf{M}|Ix_t - Iy_t|^2 \leq \alpha\mathsf{M}\int_0^t |x_s - y_s|^2\, ds, \tag{6}$$

where $\alpha = 2K^2(1 + TK^2)$.

Let $x_t^{(0)} \equiv 0$, $x_t^{(n+1)} = \xi_t + Ix_t^{(n)}$ ($n = 0,1,2,\ldots$). It follows from (6) that

$$\mathsf{M}|x_t^{(n+1)} - x_t^{(n)}|^2 \leq \alpha\mathsf{M}\int_0^t |x_s^{(n)} - x_s^{(n-1)}|^2\, ds.$$

Iterating the last inequality, we find

$$\|x_t^{(n+1)} - x_t^{(n)}\|^2 \leq \frac{T^n \alpha^n}{n!} \|x_t^{(1)}\|^2. \tag{7}$$

Since a series of the numbers $(T\alpha)^{n/2}(n!)^{-n/2}$ converges, it follows from (7) that a series of functions $x_t^{(n+1)} - x_t^{(n)}$ converges in V. In other words, the functions $x_t^{(n+1)}$ converge in V, and furthermore, there exists a process $\tilde{x}_t \in V$ such that $\|x_t^{(n)} - \tilde{x}_t\| \to 0$ as $n \to \infty$.

Further, integrating (6), we obtain

$$\|Ix_t - Iy_t\| \leq \alpha T \|x_t - y_t\|. \tag{8}$$

In particular, the operator I is continuous in V. Passing to the limit in the equality $\|x_t^{(n+1)} - (\xi_t + Ix_t^{(n)})\| = 0$, we conclude $\|\tilde{x}_t - (\xi_t + I\tilde{x}_t)\| = 0$, from which and also from (8) it follows that $I\tilde{x}_t = I(\xi + I\tilde{x})_t$ for almost all t, ω. However, the both sides of this equality are continuous with respect to t for almost all ω. Hence they coincide for all t at once almost surely. Finally, taking $x_t = \xi_t + I\tilde{x}_t$, we have $x_t = \xi_t + I(\xi + I\tilde{x})_t = \xi_t + Ix_t$ for all t almost surely. Therefore, x_t is a solution of the primary equation, (1), thus completing the proof of the theorem. □

8. Exercise

Noting that $\sigma_s(x) = [\sigma_s(x) - \sigma_s(0)] + \sigma_s(0)$, prove that the assertions of the theorem still hold if $\mathsf{M} \int_0^T |\eta_t|^2 \, dt < \infty$, where

$$\eta_t = \xi_t + \int_0^t \sigma_s(0) \, d\mathbf{w}_s + \int_0^t b_s(0) \, ds.$$

We continue estimating the moments of solutions of a stochastic equation.

9. Theorem. *Suppose the condition ($\mathscr{L}$) is satisfied, x_t is a solution of Eq.* (1), *and $\tilde{x}_t$ is a solution of the equation*

$$\tilde{x}_t = \tilde{\xi}_t + \int_0^t \tilde{\sigma}_s(\tilde{x}_s) \, d\mathbf{w}_s + \int_0^t \tilde{b}_s(\tilde{x}_s) \, ds.$$

Then, if the process $\xi_t - \tilde{\xi}_t$ is separable, the process $x_t - \tilde{x}_t$ is also separable, and for all $q \geq 1$, $t \in [0,T]$

$$\mathsf{M} \sup_{s \leq t} |x_s - \tilde{x}_s|^{2q} \leq N e^{Nt} \mathsf{M} \sup_{s \leq t} |\xi_s - \tilde{\xi}_s|^{2q}$$
$$+ N t^{q-1} e^{Nt} \mathsf{M} \int_0^t [|b_s(\tilde{x}_s) - \tilde{b}_s(\tilde{x}_s)|^{2q} + \|\sigma_s(\tilde{x}_s) - \tilde{\sigma}_s(\tilde{x}_s)\|^{2q}] \, ds,$$

where $N = N(q,K)$.

PROOF. It is seen that $x_t - \tilde{x}_t$ is the sum of $\xi_t - \tilde{\xi}_t$, stochastic integrals and Lebesque integrals. Both types of integrals are continuous with respect to t. Hence, the separability property of $\xi_t - \tilde{\xi}_t$ implies that $x_t - \tilde{x}_t$ is separable, and in particular, the quantity $\sup_{s \leq t} |x_s - \tilde{x}_s|$ is measurable with respect to ω.

As was done in proving Corollary 5, the assertion of the theorem in the general case can easily be deduced from that in the case where $\xi_t = \tilde{\xi}_t = \tilde{x}_t = 0$, $\tilde{\sigma}_s(x) = 0$, $\tilde{b}_s(x) = 0$. It is required to prove in the latter case that

$$\mathsf{M} \sup_{s \le t} |x_s|^{2q} \le N t^{q-1} e^{Nt} \mathsf{M} \int_0^t [|b_s(0)|^{2q} + \|\sigma_s(0)\|^{2q}]\, ds. \tag{9}$$

Reasoning in the same way as in proving Lemma 2, we convince ourselves that it is possible to consider only the case with bounded functions $x_t(\omega)$ and to assume, in addition, that the right side of (9) is finite.

First, we prove that the process

$$\eta_t = |x_t| e^{K^2 t} + \int_0^t e^{K^2 s} |b_s(0)|\, ds$$

is a submartingale. We fix $\varepsilon > 0$ and introduce an auxiliary function of the real variable r using the formula $\varphi(r) = \sqrt{r^2 + \varepsilon^2}$. Note that $\varphi(|x|)$ is a smooth function on E_d. In conjunction with Ito's formula,

$$d[\varphi(|x_t|)e^{K^2 t}] = e^{K^2 t} \Bigg\{ K^2 \varphi(|x_t|) + \varphi'(|x_t|) \frac{b_t(x_t)x_t}{|x_t|}$$
$$+ \varphi''(|x_t|) \frac{1}{2} \frac{|\sigma_t^*(x_t)x_t|^2}{|x_t|^2} + \varphi'(|x_t|) \frac{1}{2} \frac{1}{|x_t|} \times$$
$$\left[\|\sigma_t(x_t)\|^2 - \frac{|\sigma_t^*(x_t)x_t|^2}{|x_t|^2} \right] \Bigg\} dt + e^{K^2 t} \varphi'(|x_t|) \frac{x_t \sigma_t(x_t)}{|x_t|} d\mathbf{w}_t.$$

Let us integrate the last expression over t from s_1 up to $s_2 \ge s_1$, and also, let us take the conditional expectation under the condition $\mathscr{F}_{s_1}$. In this case, the expectation of the stochastic integral disappears (see Proof of Lemma 2). In addition, we use the fact that since

$$b_t(x_t)x_t \ge -|b_t(x_t)|\,|x_t| \ge -K^2|x_t|^2 - |b_t(0)|\,|x_t|, \qquad 0 \le \varphi'(r) \le 1,\ |r| \le \varphi(r),$$

then

$$K^2 \varphi(|x_t|) + \varphi'(|x_t|) \frac{b_t(x_t)x_t}{|x_t|} \ge -|b_t(0)|.$$

Furthermore, $\varphi'' \ge 0$, $|x_t|^2 \|\sigma_t(x_t)\|^2 \ge |\sigma_t^*(x_t)x_t|^2$. Therefore,

$$\mathsf{M}\{\varphi(|x_{s_2}|)e^{K^2 s_2} | \mathscr{F}_{s_1}\} - \varphi(|x_{s_1}|)e^{K^2 s_1} \ge \mathsf{M}\left\{ \int_{s_1}^{s_2} e^{K^2 t} |b_t(0)|\, dt \,\Big|\, \mathscr{F}_{s_1} \right\}.$$

from which, letting ε go to zero, we obtain, using the theorem on bounded convergence, $\mathsf{M}\{\eta_{s_2} | \mathscr{F}_{s_1}\} \ge \eta_{s_1}$. Therefore η_t is a submartingale.

From well-known inequalities for submartingales (see Appendix 2) as well as Hölder's inequality we have

$$\mathsf{M} \sup_{s \le t} |x_s|^{2q} \le \mathsf{M} \sup_{s \le t} \eta_s^{2q} \le 4 \mathsf{M} \eta_t^{2q}$$
$$\le 4 \cdot 2^{2q-1} e^{2qK^2 t} \mathsf{M} |x_t|^{2q} + 4 \cdot 2^{2q-1} e^{2qK^2 t} t^{2q-1} \mathsf{M} \int_0^t |b_s(0)|^{2q}\, ds.$$

It remains only to use Lemma 2 or Corollary 6 for estimating $\mathsf{M}|x_t|^{2q}$, and, furthermore, to note that $t^a e^{bt} \le N(a,b)e^{2bt}$ for $a > 0$, $b > 0$, $t > 0$. The theorem is proved. □

10. Corollary. *Let the condition (R) be satisfied, and let x_t be a solution of Eq. (1). Then there exists a constant $N(q,K)$ such that for all $q \ge 1$, $t \in [0,T]$*

$$\mathsf{M} \sup_{s \le t} |x_s - \xi_s|^{2q} \le N t^{q-1} e^{Nt} \mathsf{M} \int_0^t [|\xi_s|^{2q} + h_s^{2q} + r_s^{2q}]\, ds.$$

If ξ_t is a separable process, then

$$\mathsf{M} \sup_{s \le t} |x_s|^{2q} \le N \mathsf{M} \sup_{s \le t} |\xi_t|^{2q} + N t^{q-1} e^{Nt} \mathsf{M} \int_0^t [|\xi_s|^{2q} + h_s^{2q} + r_s^{2q}]\, ds.$$

First, we note that the second inequality follows readily from the first expression. In order to prove the first inequality, we introduce the process $y_t = x_t - \xi_t$. It is seen that $dy_t = \sigma_t(y_t + \xi_t)\, d\mathbf{w}_t + b_t(y_t + \xi_t)\, dt$, $y_0 = 0$. In estimating y_t it suffices, as was done in proving Lemma 2, to consider only the case where $y_t(\omega)$ is a bounded function. Similarly to what we did in proving our theorem above, we use here the inequality $b_t(y_t + \xi_t)y_t \ge -K^2|y_t|^2 - (K^2|\xi_t| + h_t)|y_t|$, thus obtaining that the process

$$\eta_t = |y_t| e^{K^2 t} + \int_0^t e^{K^2 s}(K^2|\xi_s| + h_s)\, ds$$

is a submartingale.

From the above, using the inequalities for submartingales as well as Hölder's inequality, we find

$$\mathsf{M} \sup_{s \le t} |y_t|^{2q} \le \mathsf{M} \sup_{s \le t} \eta_s^{2q} \le 4\mathsf{M}\eta_t^{2q}$$
$$\le N e^{Nt} \mathsf{M}|y_t|^{2q} + N e^{Nt} t^{2q-1} \mathsf{M} \int_0^t (|\xi_s|^{2q} + h_s^{2q})\, ds.$$

For estimating $\mathsf{M}|y_t|^{2q}$, it remains to apply Lemma 2, noting that $\sigma_t(x + \xi_t)$, $b_t(x + \xi_t)$ satisfy the condition (R) in which we replace r_t^2, h_t, K by $r_t^2 + 2K^2|\xi_t|^2$, $h_t + K^2|\xi_t|$, $2K$, respectively.

11. Corollary. *Let $\int_0^t \|\sigma_s\|^2\, ds < \infty$ (a.s.). Then for all $q \ge 1$*

$$\mathsf{M} \sup_{s \le t} \left| \int_0^s \sigma\, d\mathbf{w}_s \right|^{2q} \le 2^{q+2}(2q-1)^q t^{q-1} \mathsf{M} \int_0^t \|\sigma_s\|^{2q}\, ds.$$

This corollary as well as Corollary 10 can be proved by arguments similar to those used to prove the theorem. Taking $\sigma_s(x) = \sigma_s$, $b_s(x) = 0$, we have the process $x_t = \int_0^t \sigma_s\, d\mathbf{w}_s$. The proof of the theorem for $K = 0$ shows that $|x_t|$ is a submartingale. Hence $\mathsf{M}\ \sup_{s \ge t} |x_s|^{2q} \le 4\mathsf{M}|x_t|^{2q}$, which can be estimated with the aid of Corollary 3.

12. Corollary. *Let there exist a constant K_1 such that $\|\sigma_t(x)\| + |b_t(x)| \le K_1(1 + |x|)$ for all t, ω, x. Let x_t be a solution of Eq. (1) for $\xi_t \equiv x_0$, where x_0 is a fixed point on E_d. There exists a constant $N(q,K_1)$ such that for all $q \ge 0$, $t \in [0,T]$*

$$\mathsf{M} \sup_{s \le t} |x_s - x_0|^q \le N t^{q/2} e^{Nt} (1 + |x_0|)^q,$$

$$\mathsf{M} \sup_{s \le t} |x_s|^q \le N e^{Nt} (1 + |x_0|)^q.$$

In fact, for $q \ge 2$ these estimates are particular cases of the estimates given in Corollary 10. To prove these inequalities for $q \in [0,2]$ we need only to take $\eta_1 = \sup_{s \le t} |x_s - x_0|(1 + |x_0|)^{-1}$, $\eta_2 = \sup_{s \le t} |x_s|(1 + |x_0|)^{-1}$ and, furthermore, use the fact that in conjunction with Hölder's inequality, $\mathsf{M}|\eta_t|^q \le (\mathsf{M}|\eta_t|^2)^{q/2}$.

13. Remark. The sequential approximations x_t^n defined in proving Theorem 7 have the property that

$$\lim_{n \to \infty} \mathsf{M} \sup_{t \le T} |x_t^n - x_t|^2 = 0,$$

where x_t is a solution of Eq. (1). Indeed,

$$x_t^{n+1} = \xi_t + I x_t^n, \qquad x_t = \xi_t + I \tilde{x}_t, \qquad x_t^{n+1} - x_t = I x_t^n - I \tilde{x}_t,$$

from which, using Corollary 11 and the Cauchy inequality, we obtain

$$\begin{aligned} \mathsf{M} \sup_{t \le T} |x_t^{n+1} - x_t|^2 &\le 2\mathsf{M} \sup_{t \le T} \left| \int_0^t [\sigma_s(x_s^n) - \sigma_s(\tilde{x}_s)] \, d\mathbf{w}_s \right|^2 \\ &\quad + 2\mathsf{M} \sup_{t \le T} \left| \int_0^t [b_s(x_s^n) - b_s(\tilde{x}_s)] \, ds \right|^2 \\ &\le N\mathsf{M} \int_0^T \|\sigma_s(x_s^n) - \sigma_s(\tilde{x}_s)\|^2 \, ds \\ &\quad + N\mathsf{M} \int_0^T |b_s(x_s^n) - b_s(\tilde{x}_s)|^2 \, ds \le N\mathsf{M} \int_0^T |x_s^n - \tilde{x}_s|^2 \, ds. \end{aligned}$$

As was seen in the proof of Theorem 7, the expression given above tends to zero as $n \to \infty$.

6. Existence of a Solution of a Stochastic Equation with Measurable Coefficients

In this section, using the estimates obtained in Sections 2.2–2.5 we prove that in a wide class of cases there exists a probability space and a Wiener process on this space such that a stochastic equation having measurable coefficients as well as this Wiener process is solvable. In other words, ac-

acording to conventional terminology, we construct here "weak" solutions of a stochastic equation. The main difference between "weak" solutions and usual ("strong") solutions consists in the fact that the latter can be constructed on any a priori given probability space on the basis of any given Wiener process.

Let $\sigma(t,x)$ be a matrix of dimension $d \times d$, and let $b(t,x)$ be a d-dimensional vector. We assume that $\sigma(t,x)$, $b(t,x)$ are given for $t \geq 0$, $x \in E_d$, and, in addition, are bounded and Borel measurable with respect to (t,x). Also, let the matrix $\sigma(t,x)$ be positive definite, and, moreover, let

$$(\sigma(t,x)\lambda,\lambda) \geq \delta|\lambda|^2$$

for some constant $\delta > 0$ for all (t,x), $\lambda \in E_d$.

1. Theorem. *Let $x \in E_d$. There exists a probability space, a Wiener process $(\mathbf{w}_t,\mathscr{F}_t)$ on this space, and a continuous process x_t which is progressively measurable with respect to $\{\mathscr{F}_t\}$, such that almost surely for all $t \geq 0$*

$$x_t = x + \int_0^t \sigma(s,x_s)\,d\mathbf{w}_s + \int_0^t b(s,x_s)\,ds.$$

For proving our theorem we need two assertions due to A. V. Skorokhod.

2. Lemma.[5] *Suppose that d_1-dimensional random processes ξ_t^n $(t \geq 0, n = 0, 1,2, \ldots)$ are defined on some probability space. Assume that for each $T \geq 0$, $\varepsilon > 0$*

$$\lim_{c\to\infty} \sup_n \sup_{t\leq T} \mathsf{P}\{|\xi_t^n| > c\} = 0,$$

$$\lim_{h\downarrow 0} \sup_n \sup_{\substack{t_1,t_2\leq T\\ |t_1-t_2|\leq h}} \mathsf{P}\{|\xi_{t_1}^n - \xi_{t_2}^n| > \varepsilon\} = 0.$$

Then, one can choose a sequence of numbers n', a probability space, and random processes $\tilde{\xi}_t$, $\tilde{\xi}_t^{n'}$ defined on this probability space such that all finite-dimensional distributions of $\tilde{\xi}_t^{n'}$ coincide with the pertinent finite-dimensional distributions of $\xi_t^{n'}$ and $\mathsf{P}\{|\tilde{\xi}_t^{n'} - \tilde{\xi}_t| > \varepsilon\} \to 0$ as $n' \to \infty$ for all $\varepsilon > 0$, $t \geq 0$.

3. Lemma.[6] *Suppose the assumptions of Lemma 2 are satisfied. Also, suppose that d_1-dimensional Wiener processes $(\mathbf{w}_t^n,\mathscr{F}_t^n)$ are defined on the aforegoing probability space. Assume that the functions $\xi_t^n(\omega)$ are bounded on $[0,\infty) \times \Omega$ uniformly in n and that the stochastic integrals $I_t^n = \int_0^t \xi_s^n\, d\mathbf{w}_s^n$ are defined. Finally, let $\xi_s^n \to \xi_s^0$, $\mathbf{w}_s^n \to \mathbf{w}_s^0$ in probability as $n \to \infty$ for each $s \geq 0$. Then $I_t^n \to I_t^0$ as $n \to \infty$ in probability for each $t \geq 0$.*

4. Proof of Theorem 1. We smooth out σ, b using the convolution. Let $\sigma_n(t,x) = \sigma^{(\varepsilon_n)}(t,x)$, $b_n(t,x) = b^{(\varepsilon_n)}(t,x)$[7] (see Section 2.1), where $\varepsilon_n \to 0$ as $n \to \infty$,

[5] See [70, Chapter 1, §6].

[6] See [70, Chapter 2, §6].

[7] In computing the convolution we assume that $\sigma^{ij}(t,x) = \delta\delta^{ij}$, $b^i(t,x) = 0$ for $t \leq 0$.

$\varepsilon_n \neq 0$. It is clear that σ_n, b_n are bounded, $\sigma_n \to \sigma$, $b_n \to b$ (a.s.) as $n \to \infty$,

$$(\sigma_n\lambda,\lambda) = (\sigma\lambda,\lambda)^{(\varepsilon_n)} \geq \delta|\lambda|^2$$

for all $\lambda \in E_d$, $n \geq 1$. Let $\sigma_0 = \sigma$, $b_0 = b$.

We take some d-dimensional Wiener process $(\mathbf{w}_t,\mathscr{F}_t)$. Furthermore, we consider for $n = 1, 2, \ldots$ solutions of the following stochastic equations: $dx_t^n = \sigma_n(t,x_t^n)\,d\mathbf{w}_t + b_n(t,x_t^n)\,dt$, $t \geq 0$, $x_0^n = x$. Note that the derivatives σ^n, b^n are bounded for each n. Hence the functions of σ^n, b^n satisfy the Lipschitz condition and the solutions of the aforegoing equations in fact exist.

According to Corollary 5.12, for each T

$$\sup_n \mathsf{M} \sup_{t \leq T} |x_t^n| < \infty.$$

Using Chebyshev's inequality we then obtain

$$\lim_{c \to \infty} \sup_n \sup_{t \leq T} \mathsf{P}\{|x_t^n| > c\} = 0.$$

Further, for $t_2 > t_1$

$$x_{t_2}^n - x_{t_1}^n = \int_{t_1}^{t_2} \sigma_n(s,x_s^n)\,d\mathbf{w}_s + \int_{t_1}^{t_2} b_n(s,x_s^n)\,ds,$$

from which, according to Corollary 5.3[8] for $t_2 - t_1 < 1$ we have

$$\begin{aligned}\mathsf{M}|x_{t_2}^n - x_{t_1}^n|^4 &\leq N\left\{\int_{t_1}^{t_2} [\mathsf{M}\|\sigma_n(s,x_s^n)\|^4]^{1/2}\,ds\right\}^2 + N\mathsf{M}\left|\int_{t_1}^{t_2} b_n(s,x_s^n)\,ds\right|^4 \\ &\leq N(t_2 - t_1)^2 + N(t_2 - t_1)^4 \leq N(t_2 - t_1)^2,\end{aligned}$$

where the constants N depend only on the upper bounds $\|\sigma\|$, $|b|$, and do not depend on n. In conjunction with Chebyshev's inequality

$$\lim_{h \downarrow 0} \sup_n \sup_{|t_1 - t_2| < h} \mathsf{P}\{|x_{t_2}^n - x_{t_1}^n| > \varepsilon\} = 0. \tag{1}$$

Using Lemma 2 we conclude that there exists a sequence of numbers n', a probability space, and random processes $(\tilde{x}_t^{n'};\tilde{\mathbf{w}}_t^{n'})$ on this probability space such that the finite-dimensional distributions of $(\tilde{x}_t^{n'},\tilde{\mathbf{w}}_t^{n'})$ coincide with the corresponding finite-dimensional distributions of the processes $(x_t^{n'};\mathbf{w}_t)$, and for all $t \geq 0$ the limit, say $(\tilde{x}_t^0;\tilde{\mathbf{w}}_t^0)$, exists in probability of the sequence $(\tilde{x}_t^{n'};\tilde{\mathbf{w}}_t^{n'})$ as $n' \to \infty$. For brevity of notation we assume that the sequence $\{n'\}$ coincides with $\{1,2,3,\ldots\}$.

The processes $(\tilde{x}_t^n;\tilde{\mathbf{w}}_t^n)$ can be regarded as separable processes for all $n \geq 0$. Since $\mathsf{M}|\tilde{x}_{t_2}^n - \tilde{x}_{t_1}^n|^4 = \mathsf{M}|x_{t_2}^n - x_{t_1}^n|^4 \leq N|t_2 - t_1|^2$ for $n > 0$, $|t_2 - t_1| \leq 1$ (by Fatou's lemma), the relationship between the extreme terms of this inequality holds for $n = 0$ as well. Then, by Kolmogorov's theorem $\tilde{x}_t^n$ is a continuous process for all $n \geq 0$. $\tilde{\mathbf{w}}_t^n$, being separable Wiener processes, are continuous as well.

[8] In Corollary 5.3 we need take $\tau = \infty$, $t = t_2$, $\sigma_s = \sigma_n(s.\,x_s^n)\chi_{s > t_1}$.

Further, we fix some $T > 0$. The processes $(x_t^n; \mathbf{w}_t)$ are measurable with respect to $\mathscr{F}_T$ for $t \le T$; the increments $\mathbf{w}_s$ after an instant of time T do not depend on $\mathscr{F}_T$. Therefore, the processes $(x_t^n; \mathbf{w}_t)$ $(t \le T)$ do not depend on the increments $\mathbf{w}_s$ after the instant of time T. Due to the coincidence of finite-dimensional distributions, the processes $(\tilde{x}_t^n; \tilde{\mathbf{w}}_t^n)$ $(t \le T)$ do not depend on the increments $\tilde{\mathbf{w}}_s^n$ after the time T for $n \ge 1$. This property obviously holds true for a limiting process as well, i.e., it holds for $n = 0$. This readily implies that for $n \ge 0$ the processes $\tilde{\mathbf{w}}_t^n$ are Wiener processes with respect to σ-algebras of $\mathscr{F}_t^{(n)}$, defined as the completion of $\sigma\{\tilde{x}_s^n, \tilde{\mathbf{w}}_s^n : s \le t\}$. Furthermore, for $n \ge 0$ and each $s \le t$ the variable $\tilde{x}_s^n$ is $\mathscr{F}_t^{(n)}$-measurable. Since $\tilde{x}_s^n$ is continuous with respect to s, $\tilde{x}_s^n$ is a progressively measurable process with respect to $\{\mathscr{F}_t^{(n)}\}$. These arguments show that the stochastic integrals given below make sense.

Let $\kappa_m(a) = 2^{-m}[2^m a]$, where $[a]$ is the largest integer $\le a$. Since $\sigma_n(t, \tilde{x}_t^n)$ for $n \ge 1$ are bounded functions of (ω, t), continuous with respect to t, and since $\kappa_m(t) \to t$ as $m \to \infty$, then

$$\lim_{m \to \infty} \mathsf{M} \int_0^T \|\sigma_n(t, \tilde{x}_t^n) - \sigma_n(\kappa_m(t), \tilde{x}_{\kappa_m(t)}^n)\|^2 \, dt \to 0$$

for $n \ge 1$ for each $T \ge 0$. Hence for each $t \ge 0$

$$\int_0^t \sigma_n(s, \tilde{x}_s^n) \, d\tilde{\mathbf{w}}_s^n = \operatorname*{l.i.m.}_{m \to \infty} \int_0^t \sigma_n(\kappa_m(s), x_{\kappa_m(s)}^n) \, d\tilde{\mathbf{w}}_s^n$$

$$= \operatorname*{l.i.m.}_{m \to \infty} \sum_{k2^{-m} \le t} \sigma_n(k2^{-m}, \tilde{x}_{k2^{-m}}^n)(\tilde{\mathbf{w}}_{(k+1)2^{-m}}^n - \tilde{\mathbf{w}}_{k2^{-m}}^n).$$

Writing similar relations for $\int_0^t \sigma_n(s, x_s^n) \, d\mathbf{w}_s$, $\int_0^t b_n(s, \tilde{x}_s^n) \, ds$, $\int_0^t b_n(s, x_s^n) \, ds$, and using the fact that the familiar finite-dimensional distributions coincide, we can easily prove that for all $n \ge 1$, $t \ge 0$

$$\mathsf{M} \left| \tilde{x}_t^n - x - \int_0^t \sigma_n(s, \tilde{x}_s^n) \, d\tilde{\mathbf{w}}_s^n - \int_0^t b_n(s, \tilde{x}_s^n) \, ds \right|^2 = 0.$$

In other words,

$$\tilde{x}_t^n = x + \int_0^t \sigma_n(s, \tilde{x}_s^n) \, d\tilde{\mathbf{w}}_s^n + \int_0^t b_n(s, \tilde{x}_s^n) \, ds \tag{2}$$

for each $t \ge 0$ almost surely. We have thus completed the first stage of proving Theorem 1. If we had so far the processes x_t^n, the convergence property of which we knew nothing about, we have now the convergent processes $\tilde{x}_t^n$. However, in contrast to x_t^n, the processes $\tilde{x}_t^n$ satisfy an equation containing a Wiener process which changes as n changes.

We take the limit in (2) as $n \to \infty$. For each $n_0 \ge 1$, we have

$$\int_0^t \sigma_n(s, \tilde{x}_s^n) \, d\tilde{\mathbf{w}}_s^n = \int_0^s \sigma_{n_0}(s, \tilde{x}_s^n) \, d\tilde{\mathbf{w}}_s^n + \int_0^t [\sigma_n - \sigma_{n_0}](s, \tilde{x}_s^n) \, d\tilde{\mathbf{w}}_s^n, \tag{3}$$

where $\sigma_{n_0}(s, x_s)$ satisfies the Lipschitz condition with respect to (s, x). Hence $\|\sigma_{n_0}(t_2, \tilde{x}_{t_2}^n) - \sigma_{n_0}(t_1, \tilde{x}_{t_1}^n)\| \le N(|t_2 - t_1| + |\tilde{x}_{t_2}^n - \tilde{x}_{t_1}^n|)$.

In addition, by virtue of (1)

$$\lim_{h\downarrow 0}\sup_{n}\sup_{|t_2-t_1|<h}\mathsf{P}\{\|\sigma_{n_0}(t_2,\tilde{x}^n_{t_2})-\sigma_{n_0}(t_1,\tilde{x}^n_{t_1})\|>\varepsilon\}=0.$$

From this it follows, according to Lemma 3, that the first term in (3) tends in probability to $\int_0^t \sigma_{n_0}(s,\tilde{x}^0_s)\,d\tilde{\mathbf{w}}^0_s$. Therefore, applying Chebyshev's inequality, we obtain

$$\begin{aligned}
&\lim_{n\to\infty}\mathsf{P}\left\{\left|\int_0^t \sigma_n(s,\tilde{x}^n_s)\,d\tilde{\mathbf{w}}^n_s-\int_0^t \sigma_0(s,\tilde{x}^0_s)\,d\tilde{\mathbf{w}}^0_s\right|>\varepsilon\right\}\\
&\quad\le \overline{\lim_{n\to\infty}}\,\mathsf{P}\left\{\left|\int_0^t \sigma_{n_0}(s,\tilde{x}^n_s)\,d\tilde{\mathbf{w}}^n_s-\int_0^t \sigma_{n_0}(s,\tilde{x}^0_s)\,d\tilde{\mathbf{w}}^0_s\right|>\frac{\varepsilon}{3}\right\}\\
&\qquad+\overline{\lim_{n\to\infty}}\,\mathsf{P}\left\{\left|\int_0^t [\sigma_n(s,\tilde{x}^n_s)-\sigma_{n_0}(s,\tilde{x}^n_s)]\,d\tilde{\mathbf{w}}^n_s\right|>\frac{\varepsilon}{3}\right\}\\
&\qquad+\mathsf{P}\left\{\left|\int_0^t [\sigma_{n_0}(s,\tilde{x}^0_s)-\sigma_0(s,\tilde{x}^0_s)]\,d\tilde{\mathbf{w}}^0_s\right|>\frac{\varepsilon}{3}\right\}\\
&\quad\le\frac{9}{\varepsilon^2}\left[\overline{\lim_{n\to\infty}}\,\mathsf{M}\int_0^t\|\sigma_n-\sigma_{n_0}\|^2(s,\tilde{x}^n_s)\,ds+\mathsf{M}\int_0^t\|\sigma_{n_0}-\sigma_0\|^2(s,\tilde{x}^0_s)\,ds\right].
\end{aligned}$$

We estimate the last expression. It is seen that

$$\mathsf{M}\int_0^t|f(s,\tilde{x}^n_s)|\,ds\le e^t\mathsf{M}\int_0^t e^{-s}|f(s,\tilde{x}^n_s)|\,ds\le e^t\mathsf{M}\int_0^\infty e^{-s}|f(s,\tilde{x}^n_s)|\,ds.$$

Therefore, by Theorem 3.4[9]

$$\mathsf{M}\int_0^T|f(s,\tilde{x}^n_s)|\,ds\le N\|f\|_{d+1,H_\infty}$$

for $n\ge 1$, where N does not depend on n. For $n=0$ the last inequality as well holds, which fact we can can easily prove for continuous f taking the limit as $n\to\infty$ and using Fatou's lemma. Furthermore, we can prove it for all Borel f applying the results obtained in [54, Chapter 1, §2]. Let $w(t,x)$ be a continuous function equal to zero for $t^2+|x|^2\ge 1$ and such that $w(0,0)=1$, $0\le w(t,x)\le 1$. Then, for $R>0$

$$\begin{aligned}
\overline{\lim_{n\to\infty}}\,\mathsf{M}\int_0^t\|\sigma_n-\sigma_{n_0}\|^2(s,\tilde{x}^n_s)\,ds&\le N\mathsf{M}\int_0^t\left[1-w\left(\frac{s}{R},\frac{\tilde{x}^0_s}{R}\right)\right]ds\\
&\quad+\overline{\lim_{n\to\infty}}\,\mathsf{M}\int_0^t w\left(\frac{s}{R},\frac{\tilde{x}^n_s}{R}\right)\\
&\quad\times\|\sigma_n-\sigma_{n_0}\|^2(s,\tilde{x}^n_s)\,ds\\
&\le N\mathsf{M}\int_0^t\left[1-w\left(\frac{s}{R},\frac{\tilde{x}^0_s}{R}\right)\right]ds\\
&\quad+N\|\|\sigma_0-\sigma_{n_0}\|^2\|_{d+1,C_{R,R}}.
\end{aligned}$$

[9] It should be noted that $\frac{1}{2}(\sigma_n\sigma_n^*\lambda)=\frac{1}{2}|\sigma_n^*\lambda|^2\ge\frac{1}{2}\delta^2|\lambda|^2$ since $\delta|\lambda|^2\le(\sigma_n\lambda,\lambda)=(\lambda,\sigma_n^*\lambda)\le|\lambda|\,|\sigma_n^*\lambda|$.

Estimating $\mathsf{M}\int_0^t \|\sigma_{n_0} - \sigma_0\|^2(r,\tilde{x}_r^0)\,dr$ in similar fashion, we find

$$\overline{\lim_{n\to\infty}}\,\mathsf{P}\left\{\left|\int_0^t \sigma_n(s,\tilde{x}_s^n)\,d\tilde{\mathbf{w}}_s^n - \int_0^t \sigma_0(s,\tilde{x}_s^0)\,d\tilde{\mathbf{w}}_s^0\right| > \varepsilon\right\}$$

$$\le \varepsilon^{-2}N\left\{\mathsf{M}\int_0^t\left[1 - w\left(\frac{s}{R},\frac{\tilde{x}_s^0}{R}\right)\right]ds + \|\|\sigma_0 - \sigma_{n_0}\|^2\|_{d+1,\,C_{R,R}}\right\}$$

for each $n_0 > 0$, $R > 0$. Finally, we note that the last expression tends to zero if we assume first that $n_0 \to \infty$, and next, that $R \to \infty$. Therefore,

$$\int_0^t \sigma_n(s,\tilde{x}_s^n)\,d\tilde{\mathbf{w}}_s^n \to \int_0^t \sigma_0(s,\tilde{x}_s^0)\,d\tilde{\mathbf{w}}_s^0$$

in probability. We have a similar situation for the second integral in (2). Therefore it follows from (2) that

$$\tilde{x}_t^0 = x + \int_0^t \sigma_0(s,\tilde{x}_s^0)\,d\tilde{\mathbf{w}}_s^0 + \int_0^t b_0(s,\tilde{x}_s^0)\,ds$$

for each $t \ge 0$ almost surely. It remains only to note that each side of the last equality is continuous with respect to t; hence both sides coincide on a set of complete probability. We have thus proved Theorem 1. □

7. Some Properties of a Random Process Depending on a Parameter

In investigating the smoothness property of a payoff function in optimal control problems it is convenient to use theorems on differentiability in the mean of random variables over some parameter. It turns out frequently that the random variable in question, say $J(p)$, depends on a parameter p in a complicated manner. For example, $J(p)$ can be given as a functional of trajectories of some process x_t^p which depends on p. In this section we prove the assertions about differentiability in the mean of such or other functionals of the process.

Three constants T, K, $m > 0$ will be fixed throughout the entire section.

1. Definition. Let a real random process $x_t(\omega)$ be defined for $t \in [0,T]$. We write $x_t \in \mathscr{L}$ if the process $x_t(\omega)$ is measurable with respect to (ω,t) and for all $q \ge 1$

$$\mathsf{M}\int_0^T |x_t|^q\,dt < \infty.$$

We write $x_t \in \mathscr{L}B$ if x_t is a separable process and for all $q \ge 1$

$$\mathsf{M}\sup_{t\le T}|x_t|^q < \infty.$$

The convergence property in the sets $\mathscr{L}$, $\mathscr{L}B$ can be defined in a natural way.

2. Definition. Let $x_t^0, x_t^1, \ldots, x_t^n, \ldots \in \mathscr{L}(\mathscr{L}B)$. We say that the $\mathscr{L}$-limit ($\mathscr{L}B$-limit) of the process x_t^n equals x_t^0, and we write $\mathscr{L}\text{-}\lim_{n\to\infty} x_t^n = x_t^0$ ($\mathscr{L}B\text{-}\lim_{n\to\infty} x_t^n = x_t^0$) if for all $q \geq 1$

$$\lim_{n\to\infty} \mathsf{M} \int_0^T |x_t^n - x_t^0|^q \, dt = 0 \qquad \left(\lim_{n\to\infty} \mathsf{M} \sup_{t\leq T} |x_t^n - x_t^0|^q = 0\right).$$

Having introduced the notions of the $\mathscr{L}$-limit ($\mathscr{L}B$-limit), it is clear what is meant by $\mathscr{L}$-continuity ($\mathscr{L}B$-continuity) of the process x_t^p with respect to the parameter p at a point p_0.

3. Definition. Suppose that $p_0 \in E_d$, unit vector $l \in E_d$, $y_t \in \mathscr{L}(\mathscr{L}B)$. Further, suppose that for each p from some neighborhood of the point p_0 a process $x_t^p \in \mathscr{L}(\mathscr{L}B)$ is given. We say that y_t is an $\mathscr{L}$-derivative ($\mathscr{L}B$-derivative) of x_t^p at the point p_0 along the l direction, and also, we write

$$y_t = \mathscr{L}\text{-}\frac{\partial}{\partial l} x_t^p\Big|_{p=p_0} \qquad \left(y_t = \mathscr{L}B\text{-}\frac{\partial}{\partial l} x_t^p\Big|_{p=p_0}\right),$$

if

$$y_t = \mathscr{L}\text{-}\lim_{r\to 0} \frac{1}{r}(x_t^{p_0+rl} - x_t^{p_0}) \qquad y_t = \mathscr{L}B\text{-}\lim_{r\to 0} \frac{1}{r}(x_t^{p_0+rl} - x_t^{p_0}) .$$

We say that the process x_t^p is once $\mathscr{L}$-differentiable ($\mathscr{L}B$-differentiable) at the point p_0 if this process x_t^p has $\mathscr{L}$-derivatives ($\mathscr{L}B$-derivatives) at the point p_0 along all l directions. The process x_t^p is said to be i times ($i \geq 2$) $\mathscr{L}$-differentiable ($\mathscr{L}$B-differentiable) at the point p_0 if this process x_t^p is once $\mathscr{L}$-differentiable ($\mathscr{L}B$-differentiable) in some neighborhood[10] of the point p_0 and, in addition, each (first) $\mathscr{L}$-derivative ($\mathscr{L}B$-derivative) of this process x_t^p is $i-1$ times $\mathscr{L}$-differentiable ($\mathscr{L}B$-differentiable) at the point p_0.

Definitions 1–3 have been given for numerical processes x_t only. They can be extended to vector processes and matrix processes x_t in the obvious way.

Further, as is commonly done in conventional analysis, we write $y_t^p = \mathscr{L}\text{-}(\partial/\partial l)x_t^p$ if $y_t^{p_0} = \mathscr{L}\text{-}(\partial/\partial l)x_t^p|_{p=p_0}$ for all p_0 considered, $\mathscr{L}\text{-}(\partial/\partial l_1\, \partial l_2)x_t^p \equiv \mathscr{L}\text{-}(\partial/\partial l_1)[\mathscr{L}\text{-}(\partial/\partial l_2)x_t^p]$, etc. We say that x_t^p is i times $\mathscr{L}$-continuously $\mathscr{L}$-differentiable if all $\mathscr{L}$-derivatives of x_t^p up to order i inclusive are $\mathscr{L}$-continuous. We shall not dwell in future on the explanation of such obvious facts.

We shall apply Definitions 1–3 to random variables as well as random processes, the former being regarded as time independent processes.

In order to grow familiar with the given definitions, we note a few simple properties these definitions imply. It is obvious that the notion of $\mathscr{L}$-continuity is equivalent to that of $\mathscr{L}B$-continuity for random variables. Furthermore, $|\mathsf{M}x^p - \mathsf{M}x^{p_0}| \leq \mathsf{M}|x^p - x^{p_0}|$. Hence the expectation of an $\mathscr{L}$-

[10] That is, at each point of this neighborhood.

continuous random variable is continuous. Since

$$\left|\frac{1}{r}(\mathsf{M}x^{p_0+rl} - \mathsf{M}x^{p_0}) - \mathsf{M}y\right| \le \mathsf{M}\left|\frac{1}{r}(x^{p_0+rl} - x^{p_0}) - y\right|,$$

the derivative of $\mathsf{M}x^p$ along the l direction at a point p_0 is equal to the expectation of the $\mathscr{L}$-derivative of x^p if the latter exists. Therefore, the sign of the first derivative is interchangeable with the sign of the expectation. Combining the properties listed in an appropriate way, we deduce that $(\partial/\partial l)\mathsf{M}x^p$ exists and it is continuous at the point p_0 if the variable x^p is $\mathscr{L}$-continuously $\mathscr{L}$-differentiable at the point p_0 along the l direction. A similar situation is observed for derivatives of higher orders.

Since for $\tau \le T$

$$\mathsf{M}|x_\tau^p - x_\tau^{p_0}|^q \le \mathsf{M}\sup_{t \le T}|x_t^p - x_t^{p_0}|^q,$$

x_τ^p is an $\mathscr{L}$-continuous variable if $\tau(\omega) \le T$ for all ω, x_t^p is an $\mathscr{L}B$-continuous process, and x_τ^p is a measurable function of ω. A similar inequality shows that for the same τ

$$\mathscr{L}\text{-}\frac{\partial}{\partial l}x_\tau^p = \left(\mathscr{L}B\text{-}\frac{\partial}{\partial l}x_t^p\right)\Bigg|_{t=\tau} \tag{1}$$

if x_t^p has an $\mathscr{L}B$-derivative along the l direction, and if x_τ^p and the right side of (1) are measurable functions of ω. These arguments allow us to derive the properties of $\mathscr{L}$-continuity and $\mathscr{L}$-differentiability of the random variable x_τ^p from the properties of $\mathscr{L}$B-continuity and $\mathscr{L}$B-differentiability of the process x_t^p. Furthermore, (1) shows that the order of the substitution of t for τ and the order of the computation of derivatives are interchangeable.

Suppose that the process x_t^p is continuous with respect to t and is $\mathscr{L}$B-continuous with respect to p at a point p_0. Also, suppose that $\tau(p)$ are random functions with values in $[0,T]$, continuous in probability at the point p_0. We assert that in this case $x_{\tau(p)}^{p_0}$, $x_{\tau(p)}^p$ are $\mathscr{L}$-continuous at the point p_0. In fact, the difference $|x_{\tau(p)}^{p_0} - x_{\tau(p_0)}^{p_0}|^q \to 0$ in probability as $p \to p_0$ and in addition, this difference is bounded by the summable quantity $2^{q-1}\sup_t|x_t^{p_0}|^q$. Therefore, the expectation of the difference indicated tends to zero, i.e., the variable $x_{\tau(p)}^{p_0}$ is $\mathscr{L}$-continuous. The $\mathscr{L}$-continuity of the second variable follows from the $\mathscr{L}$-continuity of the first variable and from the inequalities

$$\begin{aligned}\mathsf{M}|x_{\tau(p)}^p - x_{\tau(p_0)}^{p_0}|^q &\le 2^{q-1}\mathsf{M}|x_{\tau(p)}^p - x_{\tau(p)}^{p_0}|^q + 2^{q-1}\mathsf{M}|x_{\tau(p)}^{p_0} - x_{\tau(p_0)}^{p_0}|^q\\ &\le 2^{q-1}\mathsf{M}\sup_{t\in[0,T]}|x_t^p - x_t^{p_0}|^q + 2^{q-1}\mathsf{M}|x_{\tau(p)}^{p_0} - x_{\tau(p_0)}^{p_0}|^q.\end{aligned}$$

In conjunction with Hölder's inequality

$$\mathsf{M}\sup_{t\le T}\left|\int_0^t x_s^p\,ds - \int_0^t x_s^{p_0}\,ds\right|^q \le \mathsf{M}\left|\int_0^T |x_s^p - x_s^{p_0}|\,ds\right|^q \le T^{q-1}\mathsf{M}\int_0^T |x_s^p - x_s^{p_0}|^q\,ds.$$

Therefore, $\int_0^t x_s^p\,ds$ is an $\mathscr{L}B$-continuous process if the process x_t^p is $\mathscr{L}$-continuous. We prove in a similar way that this integral has an $\mathscr{L}B$-derivative along the l direction, which coincides with the integral of the $\mathscr{L}$-derivative of x_t^p along the l direction if the latter derivative exists. In other words, the derivative can be brought under the integral sign.

Combining the assertions given above in an appropriate way, we can obtain many necessary facts. They are, however, too simple to require formal proof.

It is useful to have in mind that if $\{\mathscr{F}_t\}$ is a family of σ-algebras in Ω and if the process x_t^p is k times $\mathscr{L}$-differentiable at the point p_0 and, in addition, progressively measurable with respect to $\{\mathscr{F}_t\}$, all the derivatives of the process x_t^p can be chosen to be progressively measurable with respect to $\{\mathscr{F}_t\}$. Keeping in mind that induction is possible in this situation, we prove the foregoing assertion only for $k = 1$. Let $y_t^p = \mathscr{L}\text{-}(\partial/\partial l)x_t^p$. Having fixed p, we find a sequence $r_n \to 0$ such that $(1/r_n)(x_t^{p+r_n l} - x_t^p) \to y_t^p$ almost everywhere $d\mathsf{P} \times dt$. Further, we take $\tilde{y}_t^p = \lim_{n\to\infty} (1/r_n)(x_t^{p+r_n l} x_t^p)$ for those ω, t for which this limit exists and $\tilde{y}_t^p = 0$ on the remaining set. It is seen that the process $\tilde{y}_t^p$ is progressively measurable. Also, it is seen that

$$\tilde{y}_t^p = \mathscr{L}\text{-}\frac{\partial}{\partial l} x_t^p$$

since $\tilde{y}_t^p = y_t^p$ ($d\mathsf{P} \times dt$-a.s.).

We shall take this remark into account each time we calculate $\mathscr{L}$-derivatives of a stochastic integral.

We have mentioned above that differentiation is interchangeable with the (standard) integration. Applying Corollary 5.11, we immediately obtain that if $(\mathbf{w}_t, \mathscr{F}_t)$ is a d_1-dimensional Wiener process, σ_t^p is a matrix of dimension $d_2 \times d_1$, which is progressively measurable with respect to $\{\mathscr{F}_t\}$ and is $\mathscr{L}$-continuous at a point p_0, the integral $\int_0^t \sigma_t^p\,d\mathbf{w}_t$ is $\mathscr{L}B$-continuous at the point p_0. If σ_t^p is $\mathscr{L}$-differentiable along the l direction at the point p_0, then for $p = p_0$

$$\mathscr{L}B\text{-}\frac{\partial}{\partial l}\int_0^t \sigma_s^p\,d\mathbf{w}_s = \int_0^t \left(\mathscr{L}\text{-}\frac{\partial}{\partial l}\sigma_s^p\right) d\mathbf{w}_s.$$

A similar assertion is valid in an obvious way for derivatives of higher orders.

4. Exercise

Prove that if the function x_t^p is continuous (continuously differentiable) with respect to p in the usual sense for all (t,ω) and, in addition, the function $\mathsf{M}\int_0^T |x_t^p|^q\,dt(\mathsf{M}\int_0^T |(\partial/\partial x)x_t^p|^q\,dt$ for each l, $|l| = 1$) is bounded in some region for each $q \geq 1$, the process x_t^p is $\mathscr{L}$-continuous ($\mathscr{L}$-differentiable and $\mathscr{L}\text{-}(\partial/\partial l)x_t^p = (\partial/\partial l)x_t^p$) in this region.

Further, we turn to investigating the continuity and differentiability properties of a composite function. To do this, we need three lemmas.

5. Lemma. *Suppose that for* $n = 1, 2, \ldots, t \in [0, T]$, $x \in E_d$, *the* d_1*-dimensional processes* x_t^n *measurable with respect to* (ω,t) *are defined, and, in addition, the variables* $h_t^n(x)$ *measurable with respect to* (ω,t,x) *are given. Assume that* $x_t^n \to 0$ *as* $n \to \infty$ *with respect to the measure* $d\mathsf{P} \times dt$, *and that the variable* $h_t^n(x)$ *is continuous in* x *for all* n, ω, t. *Furthermore, we assume that one of the following two conditions is satisfied:*

a. *for almost all* (ω,t)

$$\lim_{\delta \to 0} \varlimsup_{n \to \infty} w_t^n(\delta) = 0,$$

where $w_t^n(\delta) = \sup_{|x| \le \delta} |h_t^n(x)|$;

b. *for each* $\varepsilon > 0$

$$\lim_{\delta \to 0} \varlimsup_{n \to \infty} \int_0^T \mathsf{P}\{w_t^n(\delta) > \varepsilon\}\, dt = 0.$$

Then $|h_t^n(x_t^n)| \le w_t^n(|x_t^n|) \to 0$ *as* $n \to \infty$ *in measure* $d\mathsf{P} \times dt$.

Proof. We note that since $h_t^n(x)$ is continuous in x, $w_t^n(\delta)$ will be measurable with respect to (ω,t). Further, condition (b) follows from condition (a) since (a) implies that $w_t^n(\delta) \to 0$ as $n \to \infty$, $\delta \to 0$ almost everywhere; (b) implies the same although with respect to $d\mathsf{P} \times dt$.

Finally, for each $\varepsilon > 0$, $\delta > 0$

$$\varlimsup_{n \to \infty} \int_0^T \mathsf{P}\{w_t^n(|x_t^n|) > \varepsilon\}\, dt \le \varlimsup_{n \to \infty} \int_0^T \mathsf{P}\{|x_t^n| > \delta\}\, dt + \varlimsup_{n \to \infty} \int_0^T \mathsf{P}\{w_t^n(\delta) > \varepsilon\}\, dt,$$

where the first summand equals zero by assumption. Thus, letting $\delta \to 0$ and using (b), we have proved the lemma. □

6. Lemma. *Let* x_t^n *be* d_1*-dimensional processes measurable with respect to* (ω,t) $(n = 0,1,2,\ldots, t \in [0,T])$, *such that* $\mathscr{L}\text{-}\lim_{n \to \infty} x_t^n = x_t^0$. *Let* $f_t(x)$ *be random variables defined for* $t \in [0,T]$, $x \in E_d$, *measurable with respect to* (ω,t), *continuous in* x *for all* (ω,t), *and such that* $|f_t(x)| \le K(1 + |x|)^m$ *for all* ω, t, x. *Then* $\mathscr{L}\text{-}\lim_{n \to \infty} f_t(x_t^n) = f(x_t^0)$.

Proof. First we note that under the condition $|f_t(x)| \le K(1 + |x|)^m$ the processes $f_t(x_t^n) \in \mathscr{L}$ for all $n \ge 0$. Next, we write $f_t(x_t^n) - f_t(x_t^0)$ as $h_t(y_t^n)$, where $h_t(x) = f_t(x + x_t^0) - f_t(x_t^0)$, $y_t^n = x_t^n - x_t^0$. Since $\mathsf{M}\int_0^T |y_t^n|\, dt \to 0$, $y_t^n \to 0$ in measure $d\mathsf{P} \times dt$, from which we have, using Lemma 5 applied to y_t^n and $h_t(x)$, that $h_t(y_t^n) \to 0$ in measure $d\mathsf{P} \times dt$.

Since the function $|a|/(|a| + 1)$ is bounded and

$$g_t^n = |h_t(y_t^n)|[|h_t(y_t^n)| + 1]^{-1} \to 0$$

in measure $d\mathsf{P} \times dt$, then for each $q \geq 1$

$$\lim_{n\to\infty} \mathsf{M} \int_0^T |g_t^n|^{2q}\, dt = 0. \tag{2}$$

Moreover, in view of the estimate $|f_t(x)| \leq K(1 + |x|)^m$ and the fact that

$$\mathsf{M} \int_0^T |x_t^n - x_t^0|^{2qm}\, dt \to 0, \qquad \mathsf{M} \int_0^T |x_t^0|^{2qm}\, dt < \infty,$$

we have

$$\sup_n \mathsf{M} \int_0^T |x_t^n|^{2qm}\, dt < \infty,$$

$$\begin{aligned} \sup_n \mathsf{M} \int_0^T (1 + |h_t(y_t^n)|)^{2q}\, dt \leq \sup_n \mathsf{M} \int_0^T [1 + K(1 + |x_t^n|)^m \\ + K(1 + |x_t^0|)^m]^{2q}\, dt < \infty. \end{aligned} \tag{3}$$

Using the Cauchy inequality, we derive from (2) and (3) that

$$\begin{aligned} &\lim_{n\to\infty} \mathsf{M} \int_0^T |h_t(y_t^n)|^q\, dt \\ &\leq \lim_{n\to\infty} \left(\mathsf{M} \int_0^T |g_t^n|^{2q}\, dt\right)^{1/2} \left(\mathsf{M} \int_0^T (1 + |h_t(y_t^n)|)^{2q}\, dt\right)^{1/2} = 0 \end{aligned}$$

for each $q \geq 1$. The lemma is proved. □

We note a simple corollary of Lemma 6.

7. Corollary. *If for $n = 0, 1, 2, \ldots$ the one-dimensional processes x_t^n, y_t^n are defined and $\mathscr{L}$-$\lim_{n\to\infty} x_t^n = x_t^0$, $\mathscr{L}$-$\lim_{n\to\infty} y_t^n = y_t^0$, then*

$$\mathscr{L}\text{-}\lim_{n\to\infty} x_t^n y_t^n = x_t^0 y_t^0$$

Indeed, the two-dimensional process (x_t^n, y_t^n) has the $\mathscr{L}$-limit equal to (x_t^0, y_t^0). Furthermore, the function $f(x,y) \equiv xy$ satisfies the growth condition $|f(x,y)| \leq (1 + \sqrt{x^2 + y^2})^2$. Hence $\mathscr{L}$-$\lim_{n\to\infty} f(x_t^n, y_t^n) = f(x_t^0, y_t^0)$.

8. Lemma. *Suppose that the assumptions of Lemma 6 are satisfied. Also, suppose that for $n = 1, 2, \ldots, u \in [0,1]$ the d_1-dimensional random variables $x_t^n(u)$ are defined which are continuous in u, measurable with respect to (ω,t) and such that $|x_t^n(u) - x_t^0| \leq |x_t^n - x_t^0|$. Then*

$$\mathscr{L}\text{-}\lim_{n\to\infty} \int_0^1 f_t(x_t^n(u))\, du = f_t(x_t^0). \tag{4}$$

PROOF. In accord with Hölder's inequality, for $q \geq 1$

$$\begin{aligned} \left|\int_0^1 f_t(x_t^n(u))\, du - f_t(x_t^0)\right|^q &= \left|\int_0^1 [f_t(x_t^n(u)) - f_t(x_t^0)]\, du\right|^q \\ &\leq \int_0^1 \left|f_t(x_t^n(u)) - f_t(x_t^0)\right|^q du. \end{aligned}$$

It follows from the inequalities $|x_t^n(u) - x_t^0| \le |x_t^n - x_t^0|$, $|x_t^n(u)| \le |x_t^0| + |x_t^n - x_t^0|$ that $x_t^n(u) \in \mathscr{L}$ and $\mathscr{L}\text{-}\lim_{n\to\infty} x_t^n(u) = x_t^0$ for each $u \in [0,1]$. Therefore, by Lemma 6

$$I_n(u) \equiv \mathsf{M} \int_0^T |f_t(x_t^n(u)) - f_t(x_t^0)|^q \, dt \to 0.$$

Finally, by the inequalities

$$|f_t(x_t^n(u))| \le K(1 + |x_t^n(u)|)^m \le K(1 + |x_t^n - x_t^0| + |x_t^0|)^m$$

we have that the limiting expression in (4) belongs to $\mathscr{L}$, and the totality of variables $I_n(u)$ is bounded. By the Lebesgue theorem, as $n \to \infty$

$$\mathsf{M} \int_0^T \left| \int_0^1 f_t(x_t^n(u))\, du - f_t(x_t^0) \right|^q dt \le \int_0^1 I_n(u)\, du \to 0,$$

thus proving the lemma. □

Further, we prove a theorem on continuity and differentiability of a composite function.

9. Theorem. *Suppose for $x \in E_{d_1}$ and p in a neighborhood of a point $p_0 \in E_d$ the random processes $x_t^p = x_t^p(\omega)$, $f_t(x) = f_t(\omega,x)$ with values in E_d and E_1, respectively, are given for $t \in [0,T]$ and measurable with respect to (t,ω).*

(a) *For all t, ω let the function $f_t(x)$ be continuous in x, let $|f_t(x)| \le K(1 + |x|)^m$, and let the process x_t^p be $\mathscr{L}$-continuous at p_0. Then the process $f_t(x_t^p)$ is also $\mathscr{L}$-continuous at p_0.*

(b) *Suppose that for all t,ω the function $f_t(x)$ is i times continuously differentiable over x. Furthermore, suppose that for all t, ω the absolute values of the function $f_t(x)$ as well as those of its derivatives up to order i inclusively do not exceed $K(1 + |x|)^m$. Then, if the process x_t^p is i times ($\mathscr{L}$-continuously) $\mathscr{L}$-differentiable at the point p_0, the process $f_t(x_t^p)$ is i times ($\mathscr{L}$-continuously) $\mathscr{L}$-differentiable at the point p_0 as well. In addition, for the unit vector $l \in E_d$*

$$\mathscr{L}\text{-}\frac{\partial}{\partial l} f_t(x_t^p) = f_{t(y_t^p)}(x_t^p)|y_t^p|, \tag{5}$$

$$\mathscr{L}\text{-}\frac{\partial^2}{\partial l^2} f_t(x_t^p) = f_{t(z_t^p)}(x_t^p)|z_t^p| + f_{t(y_t^p)(y_t^p)}(x_t^p)|y_t^p|^2, \tag{6}$$

where

$$z_t^p = \mathscr{L}\text{-}\frac{\partial^2}{\partial l^2} x_t^p, \qquad y_t^p = \mathscr{L}\text{-}\frac{\partial}{\partial l} x_t^p,$$

for those i, p for which the existence of the left sides of (5) *and* (6) *has been established.*

PROOF. For proving (a) it suffices to take any sequence of points $p_n \to p_0$, to put $x_t^{(n)} = x_t^{p_n}$ and, finally, to make use of Lemma 6.

We shall prove (b) for $i = 1$. First we note that $f_t(x,y) \equiv f_{t(y)}(x)|y|$ is a continuous function of (x,y) and

$$|f_t(x,y)| = |f_{t(y)}(x)|\,|y| \le K(1 + |x|)^m|y| \le N(1 + \sqrt{|x|^2 + |y|^2})^{m+1}.$$

Further, we take the unit vector $l \in E_d$, a sequence of numbers $r_n \to 0$, and we put

$$x_t^{(n)}(u) = ux_t^{p_0 + r_n l} + (1 - u)x_t^{p_0}, \qquad y_t^{(n)} = \frac{1}{r_n}(x_t^{p_0 + r_n l} - x_t^{p_0}).$$

Using the Newton-Leibniz rule we have

$$\begin{aligned} \frac{1}{r_n}[f_t(x_t^{p_0 + r_n l}) - f_t(x_t^{p_0})] &= \frac{1}{r_n}\int_0^1 \frac{\partial}{\partial u} f_t(x_t^{(n)}(u))\,du \\ &= \int_0^1 f_t(x_t^{(n)}(u), y_t^{(n)})\,du, \end{aligned}$$

where $|x_t^{(n)}(u) - x_t^{p_0}|^2 + |y_t^{(n)} - y_t^{p_0}|^2 \le |x_t^{p_0 + r_n l} - x_t^{p_0}|^2 + |y_t^{(n)} - y_t^{p_0}|^2$ and where, by Lemma 8 applied to $x_t^{(n)}(u)$ and $y_t^{(n)}(u) \equiv y_t^{(n)}$,

$$\mathscr{L}\text{-}\lim_{n\to\infty} \int_0^1 f_t(x_t^{(n)}(u), y_t^{(n)})\,du = f(x_t^{p_0}, y_t^{p_0}).$$

Therefore,

$$\mathscr{L}\text{-}\lim_{r\to 0} \frac{1}{r}[f_t(x_t^{p_0 + rl}) - f_t(x_t^{p_0})] = f_t(x_t^{p_0}, y_t^{p_0}).$$

Finally, by (a), $f_t(x^{p_0}, y_t^{p_0})$ is $\mathscr{L}$-continuous with respect to p_0 if $x_t^{p_0}$ is $\mathscr{L}$-continuously $\mathscr{L}$-differentiable with respect to p_0. This proves assertion one in (b) for $i = 1$. At the same time we have proved Eq. (5), which we find convenient to write as follows:

$$\mathscr{L}\text{-}\frac{\partial}{\partial l} f_t(x_t^p) = f_t(x_t^p, y_t^p).$$

For proving (b) for all i we apply the method of induction. Assume that the first assertion in (b) is proved for $i \le j$ and for any processes $f_t(x)$, x_t^p satisfying condition (b). Let the pair $f_t(x)$, x_t^p satisfy the conditions of (b) for $i = j + 1$. We take a derivative $\mathscr{L}\text{-}(\partial/\partial l)f_t(x_t^p)$ and prove that this derivative is j times $\mathscr{L}$-differentiable at a point p_0. Let us write this derivative as $f_t(x_t^p, y_t^p)$. We note that the process (x_t^p, y_t^p) is j times $\mathscr{L}$-differentiable at the point p_0 by assumption, the function $f_t(x,y)$ is continuously differentiable j times with respect to the variables (x,y). Also, we note that the absolute values of the derivatives of the above function up to order j inclusively do not exceed $N(1 + \sqrt{|x|^2 + |y|^2})^{m+1}$. Therefore, by the induction assumption, $f_t(x_t^p, y_t^p)$ is j times $\mathscr{L}$-differentiable at the point p_0. Since l is a vector, $f_t(x_t^p)$ is, by definition, $j + 1$ times $\mathscr{L}$-differentiable at the point p_0.

In a similar way we can prove $\mathscr{L}$-continuity of $\mathscr{L}$-derivatives of $f_t(x_t^p)$ at the point p_0 if $\mathscr{L}$-derivatives of x_t^p are $\mathscr{L}$-continuous at the point p_0. Finally, in conjunction with (5)

$$\mathscr{L}\text{-}\frac{\partial}{\partial l} f_t(x_t^p, y_t^p) = f_{t(y_t^p, z_t^p)}(x_t^p, y_t^p)\sqrt{|y_t^p|^2 + |z_t^p|^2},$$

which, after simple transformations, yields (6). The theorem is proved. □

10. Remark. The theorem proved above can easily be used for proving the $\mathscr{L}$-continuity and $\mathscr{L}$-differentiability of various expressions which contain random processes. For example, arguing in the same way as in Corollary 7, we can prove that if x_t^p, y_t^p are real i times $\mathscr{L}$-differentiable processes, the product $x_t^p y_t^p$ is i times $\mathscr{L}$-differentiable as well. If the real nonnegative process x_t^p is i times $\mathscr{L}$-differentiable, the process $e^{-x_t^p}$ is i times $\mathscr{L}$-differentiable as well. In fact, notwithstanding that the function e^{-x} grows more rapidly than any polynomial as $x \to -\infty$, we consider the nonnegative process x_t^p, and moreover, we can take any smooth function $f(x)$ equal to zero for $x \leq -1$ and equal to e^{-x} for $x \geq 0$. In this situation the hypotheses of the theorem concerning $f(x)$ will be satisfied and $e^{-x_t^p} = f(x_t^p)$. Combining the foregoing arguments with the known properties of integrals of $\mathscr{L}$-continuous and $\mathscr{L}$-differentiable functions, we arrive at the following assertion.

11. Lemma. *Let the processes $x_t^p, f_t^1(x), f_t^2(x)$ satisfy the conditions of Theorem 9a (Theorem 9b), and, in addition, let $f_t^1(x) \geq 0$; then the process*

$$f_t^2(x_t^p)\exp\left\{-\int_0^t f_s^1(x_s^p)\,ds\right\}$$

is $\mathscr{L}$-continuous at the point p_0 (i times ($\mathscr{L}$-continuously) $\mathscr{L}$-differentiable at the point p_0).

Fixing $\tau \in [0,T]$, and regarding $\int_0^\tau f_s^1(x_s^p)\,ds$ as a time independent process, we conclude that the following lemma is valid.

12. Lemma. *Let the processes x_t^p, $f_t^1(x)$, $f_t^2(x)$ satisfy the hypotheses of Theorem 9a (Theorem 9b), and, in addition, let $f_t^1(x) \geq 0$. Let the random variable $\tau(\omega) \in [0,T]$ and let the random processes y_t^p, $f_t^2(x)$ be such that the processes $\tilde{x}_t \equiv y_\tau^p$, $\tilde{f}_t(x) \equiv f_\tau^2(x)$ satisfy the hypotheses of Theorem 9a (Theorem 9b). Then the random variable*

$$f_\tau^2(y_\tau^p)\exp\left\{-\int_0^\tau f^1(x_s^p)\,ds\right\}$$

is $\mathscr{L}$-continuous at the point p_0 (i times ($\mathscr{L}$-continuously) $\mathscr{L}$-differentiable at the point p_0).

13. Remark. Equation (5) shows that in computing an $\mathscr{L}$-derivative of a composite function the usual formulas familiar in analysis can be applied.

14. Exercise

Derive a formula for the derivative of a product, using (5). (*Hint*: Take a function $f(x,y) \equiv xy$.)

We have investigated the properties of the functions $f_t(x_t^n)$ in the case where $f_t(x)$ does not depend on n. We prove a few assertions for the case where $f_t(x)$ depends on the parameter n in an explicit manner.

15. Lemma. *Let $\xi(\omega)$ be a d_1-dimensional random vector. Further, let $h(x) = h(\omega,x)$, $w(R,\varepsilon) = w(\omega,R,\varepsilon)$ be measurable variables which are defined for $x \in E_{d_1}$, $R \geq 0$, $\varepsilon \geq 0$, $\omega \in \Omega$. Assume that $w(R,\varepsilon)$ increases with respect to R and to ε, $|h(x) - h(y)| \leq w(|x| \vee |y|, |x - y|)$ for all ω, x, y and $|h(x)| \leq K(1 + |x|)^m$ for all ω, x. Then, for all $R \geq 0$, $\varepsilon \in (0,1)$*

$$\mathsf{M}|h(\xi)| \leq K\mathsf{M}(1 + |\xi|)^m \chi_{|\xi| > R-1} + \mathsf{M}w(R,\varepsilon) + N(d_1)\varepsilon^{-d_1}\mathsf{M}\int_{|y| \leq R} |h(y)|\,dy.$$

PROOF. We fix $R \geq 0$, $\varepsilon \in (0,1)$, and also we take a d_1-dimensional vector η such that it does not depend on ξ, w and is uniformly distributed in the sphere $\{x \in E_{d_1} : |x| < \varepsilon\}$.

It is seen that

$$\mathsf{M}|h(\xi)| \leq \mathsf{M}|h(\xi)|\chi_{|\xi| > R-1} + \mathsf{M}|h(\xi) - h(\xi + \eta)|\chi_{|\xi| \leq R-1} + \mathsf{M}\,|h(\xi + \eta)|\chi_{\xi \leq R-1}.$$

The assertion of our lemma follows from the above expression as well as the assumptions of the lemma since $|\eta| < \varepsilon < 1$, for $|\xi| \leq R - 1$

$$|\xi + \eta| < R, \qquad |h(\xi) - h(\xi + \eta)| \leq w(R,\varepsilon),$$

and

$$\begin{aligned}\mathsf{M}|h(\xi + \eta)|\chi_{|\xi| \leq R-1} &= N(d_1)\varepsilon^{-d_1}\mathsf{M}\int \chi_{|x| < \varepsilon, |\xi| \leq R-1}|h(\xi + x)|\,dx \\ &= N(d_1)\varepsilon^{-d_1}\mathsf{M}\int \chi_{|y-\xi| < \varepsilon, |\xi| \leq R-1}|h(y)|\,dy \\ &\leq N(d_1)\varepsilon^{-d_1}\mathsf{M}\int_{|y| \leq R}|h(y)|\,dy.\end{aligned}$$

□

16. Lemma. *Suppose that for $x \in E_{d_1}$, $t \in [0,T]$, $n = 1, 2, 3, \ldots$, $R > 0$, $\varepsilon > 0$ d_1-dimensional processes x_t^n are defined which are measurable with respect to (ω,t). Furthermore, suppose that the variables $h_t^n(x)$ and $w_t^n(R,\varepsilon)$ increasing with respect to R and ε are defined, these variables being measurable with respect to (ω,t,x) and (ω,t), respectively. Assume that $w_t^n(|x| \vee |y|, |x - y|) \geq |h_t^n(x) - h_t^n(y)|$ for all ω, t, x, y,*

$$\lim_{R\to\infty} \overline{\lim_{n\to\infty}} \int_0^T \mathsf{P}\{|x_t^n| > R\}\,dt = 0, \tag{7}$$

and for each $R > 0$, $\delta > 0$

$$\lim_{\varepsilon \downarrow 0} \overline{\lim_{n \to \infty}} \int_0^T \mathsf{P}\{w_t^n(R,\varepsilon) > \delta\}\, dt = 0. \tag{8}$$

Finally, let $h_t^n(x) \to 0$ as $n \to \infty$ in measure $d\mathsf{P} \times dt$ for each $x \in E_d$.

Then, $h_t^n(x_t^n) \to 0$ as $n \to \infty$ in measure $d\mathsf{P} \times dt$.

We shall prove this lemma later. We derive from Lemma 16 (in the same way as we derived Lemma 6 from Lemma 5) the following theorem.

17. Theorem. *Let the hypotheses of Lemma 16 be satisfied. Furthermore, let $|h_t^n(x)| \le K(1 + |x|)^m$ for all n, ω, t, x and for all $q \ge 1$*

$$\sup_n \mathsf{M} \int_0^T |x_t^n|^q\, dt < \infty. \tag{9}$$

Then

$$\mathscr{L}\text{-}\lim_{n \to \infty} h_t^n(x_t^n) = 0.$$

18. Remark. By Chebyshev's inequality, (7) follows from (9). Using Chebyshev's inequality it can easily be proved that the condition (8) is satisfied if $w_t^n(R,\varepsilon)$ is nonrandom and

$$\lim_{\varepsilon \downarrow 0} \overline{\lim_{n \to \infty}} \int_0^T w_t^n(R,\varepsilon)\, dt = 0.$$

For $w_t^n(R,\varepsilon)$ it is convenient to take $K\varepsilon$ if $|h_t^n(x) - h_t^n(y)| \le K|x - y|$.

Proof of Lemma 16. Since the convergence of $h_t^n(x_t^n)$ to zero in measure is equivalent to the same convergence of $(2/\pi)\arctan h_t^n(x_t^n)$, and furthermore, since the latter variable is bounded and

$$|\arctan h_t^n(x) - \arctan h_t^n(y)| \le |h_t^n(x) - h_t^n(y)| \le w_t^n(|x| \vee |y|, |x - y|),$$

we can consider without loss of generality that $|h_t^n| \le 1$.

It is clear that in this case $2 \wedge w_t^n$ can be taken instead of w_t^n so that w_t^n will be assumed to be bounded as well.

According to Lemma 15 (we take in Lemma 15 $K = 1$, $m = 0$) for any $R > 0$, $\varepsilon \in (0,1)$

$$\begin{aligned} \int_0^T \mathsf{M}|h_t^n(x_t^n)|\, dt &\le \int_0^T \mathsf{P}\{|x_t^n| > R - 1\}\, dt \\ &\quad + \int_0^T \mathsf{M} w_t^n(R,\varepsilon)\, dt + N(d_1)\varepsilon^{-d_1} \\ &\quad \times \int_0^T \mathsf{M} \int_{|y| \le R} |h_t^n(y)|\, dy\, dt. \end{aligned} \tag{10}$$

We make use of the fact that the convergence in measure is equivalent to the convergence in the mean for uniformly bounded sequences. Thus we have that the sequence

$$\int_0^T \mathsf{M}|h_t^n(y)|\,dt \to 0$$

as $n \to \infty$ for any $y \in E_{d_1}$. Furthermore, each term of this sequence does not exceed T. This implies that the last expression in (10) tends to zero as $n \to \infty$ for any $\varepsilon > 0$, $R > 0$. Letting $n \to \infty$ in (10), next, $\varepsilon \downarrow 0$, $R \to \infty$, and in addition, using (7), (8) and the fact mentioned above that the convergence in the mean and the convergence in measure are related, we complete the proof of Lemma 16. □

8. The Dependence of Solutions of a Stochastic Equation on a Parameter

Let E be a Euclidean space, let a region $D \subset E$ (D denotes a region of parameter variation), and let T, K, m be fixed nonnegative constants; $(\mathbf{w}_t, \mathscr{F}_t)$ is a d_1-dimensional Wiener process. Furthermore, for $t \in [0, T]$, $x \in E_d$, $p \in D$, $n = 0, 1, 2, \ldots$ we are given: $\sigma_t(x)$, $\sigma_t^n(x)$, $\sigma_t(p,x)$ are random matrices of dimension $d \times d_1$ and $b_t(x)$, $b_t^n(x)$, $b_t(p,x)$, ξ_t^n, $\xi_t(p)$ are random d-dimensional vectors which are progressively measurable with respect to $\{\mathscr{F}_t\}$. Assume that for all t, ω, x, y

$$\|\sigma_t(x) - \sigma_t(y)\| + |b_t(x) - b_t(y)| \le K|x - y|. \tag{1}$$

Also, assume that $\sigma_t^n(x)$, $b_t^n(x)$ satisfy (1) for each $n \ge 0$ and, in addition, $\sigma_t(p,x)$, $b_t(p,x)$ satisfy (1) for each $p \in D$.

Suppose that all the processes in question belong to $\mathscr{L}$ for all values of x, n, p. Recall that the space $\mathscr{L}$ was introduced in Section 7. We shall frequently use further on other concepts and results given in Section 7.

We define the processes x_t^x, x_t^n, x_t^p as the solutions of the following equations:

$$x_t^x = x + \int_0^t \sigma_s(x_s^x)\,d\mathbf{w}_s + \int_0^t b_s(x_s^x)\,ds;$$

$$x_t^n = \xi_t^n + \int_0^t \sigma_s^n(x_s^n)\,d\mathbf{w}_s + \int_0^t b_s^n(x_s^n)\,ds;$$

$$x_t^p = \xi_t(p) + \int_0^t \sigma_s(p,x_s^p)\,d\mathbf{w}_s + \int_0^t b_s(p,x_s^p)\,ds.$$

Note that by Theorem 5.7 the above equations have solutions. We also note that by Corollary 5.6 these solutions belong to $\mathscr{L}$. If ξ_t^n, $\xi_t(p) \in \mathscr{L}B$ for all n, p, according to Corollary 5.10 x_t^x, x_t^n, $x_t^p \in \mathscr{L}B$ as well for all n, p, x.

1. Theorem. *Let $\sigma_t^n(x) \to \sigma_t^0(x)$, $b_t^n(x) \to b_t^0(x)$ in $\mathscr{L}$ as $n \to \infty$ for each $x \in E_d$ and let $\xi_t^n \to \xi_t^0$ in $\mathscr{L}$ as $n \to \infty$. Then $x_t^n \to x_t^0$ in $\mathscr{L}$ as $n \to \infty$.*

If $\xi_t^n \to \xi_t^0$ in $\mathscr{L}B$ as well as $n \to \infty$, then $x_t^n \to x_t^0$ in $\mathscr{L}B$ as $n \to \infty$.

PROOF. Let $\hat{\sigma}_t^n(x) = \sigma_t^n(x) - \sigma_t^n(0)$. It is seen that $\hat{\sigma}_t^n(x)$ satisfies the Lipschitz condition (1) and the growth condition $\|\hat{\sigma}_t^n(x)\| \le K|x|$. Furthermore, $\hat{\sigma}_t^n(x) \to \hat{\sigma}_t^0(x)$ in $\mathscr{L}$ for all x. From this, using Theorem 7.17 and Remark 7.18, we conclude that $\hat{\sigma}_t^n(x_t^0) \to \hat{\sigma}_t^0(x_t^0)$ in $\mathscr{L}$. Adding the last relation with $\sigma_t^n(0) \to \sigma_t^0(0)$ in $\mathscr{L}$, we obtain: $\sigma_t^n(x_t^0) \to \sigma_t^0(x_t^0)$ in $\mathscr{L}$. Similarly, $b_t^n(x_t^0) \to b_t^0(x_t^0)$.

Applying Corollary 5.5 and Theorem 5.9 for $(\tilde{x}_t,\tilde{\sigma}_t,\tilde{b}_t) = (x_t^0,\sigma_t^0,b_t^0)$, $(x_t,\sigma_t,b_t) = (x_t^n,\sigma_t^n,b_t^n)$, we immediately arrive at the assertions of the theorem. □

2. Corollary. *If the process $\xi_t(p)$ is $\mathscr{L}$-continuous ($\mathscr{L}B$-continuous), and, in addition, for each $x \in E_d$ the processes $\sigma_t(p,x)$, $b_t(p,x)$ are $\mathscr{L}$-continuous in p at the point $p_0 \in D$, the process x_t^p is $\mathscr{L}$-continuous ($\mathscr{L}B$-continuous) at the point p_0.*

3. Lemma. *Suppose that for each $t \in [0,T]$, $p \in D$, ω the functions $\sigma_t(p,x)$, $b_t(p,x)$ are linear with respect to x. Let the process $\xi_t(p)$ and for each $x \in E_d$ the processes $\sigma_t(p,x)$ and $b_t(p,x)$ be i times ($\mathscr{L}$-continuously) $\mathscr{L}$-differentiable at the point $p_0 \in D$. Then, the process x_t^p is i times ($\mathscr{L}$-continuously) $\mathscr{L}$-differentiable at p_0. If, in addition, $\xi_t(p)$ is i times ($\mathscr{L}B$-continuously) $\mathscr{L}B$-differentiable at the point p_0, the process x_t^p will be the same as the process $\xi_t(p)$.*

PROOF. Due to the linearity of $\sigma_t(p,x)$, $b_t(p,x)$

$$x_t^p = \xi_t(p) + \sum_{j=1}^{d} \int_0^t (x_s^p)^j \sigma_s(p,e_j)\, d\mathbf{w}_s + \sum_{j=1}^{d} \int_0^t (x_s^p)^j b_s(p,e_j)\, ds,$$

where $(x_s^p)^j$ is the jth coordinate of the vector x_s^p in the basis $\{e_j\}$. This implies that the last assertion of the lemma is a corollary of the first assertion as well as the results, which were proved in Section 7, related to the $\mathscr{L}B$-differentiability of integrals and the $\mathscr{L}$-differentiability of products of $\mathscr{L}$-differentiable processes.

We prove the first assertion. To this end, we make use of the induction with respect to i and, in addition, assume that $i = 1$. We take a unit vector $l \in E$ and, in accord with what was said in Section 7, let the processes

$$\mathscr{L}\text{-}\frac{\partial}{\partial l}\, \xi_t(p_0), \qquad \mathscr{L}\text{-}\frac{\partial}{\partial l}\, \sigma_s(p_0,x), \qquad \mathscr{L}\text{-}\frac{\partial}{\partial l}\, b_s(p_0,x)$$

be progressively measurable for $x \in E_d$.

By Corollary 2, we conclude that the process x_t^p is $\mathscr{L}$-continuous at the point p_0. It is not hard to see that for $p = p_0$ the process

$$\eta_t(p) \equiv \mathscr{L}\text{-}\frac{\partial}{\partial l}\, \xi_t(p) + \sum_{j=1}^{d} \int_0^t (x_s^p)^j \mathscr{L}\text{-}\frac{\partial}{\partial l}\, \sigma_s(p,e_j)\, d\mathbf{w}_s$$

$$+ \sum_{j=1}^{d} \int_0^t (x_s^p)^j \mathscr{L}\text{-}\frac{\partial}{\partial l}\, b_s(p,e_j)\, ds$$

exists, is progressively measurable (and $\mathscr{L}$-continuous with respect to p, if

$$\mathscr{L}\text{-}\frac{\partial}{\partial l}\,\xi_t(p),\qquad \mathscr{L}\text{-}\frac{\partial}{\partial l}\,\sigma_s(p,x),\qquad \mathscr{L}\text{-}\frac{\partial}{\partial l}\,b_s(p,x)$$

are $\mathscr{L}$-continuous with respect to p). Furthermore, $\eta_t(p_0) \in \mathscr{L}$.

According to Theorem 5.7, the solution of the equation

$$y_t^p = \eta_t(p) + \int_0^t \sigma_s(p,y_s^p)\,d\mathbf{w}_s + \int_0^t b_s(p,y_s^p)\,ds \tag{2}$$

exists and is unique for $p = p_0$.

Let us show that $y_t^p = \mathscr{L}\text{-}(\partial/\partial l)x_t^p$ for $p = p_0$. To this end, we take a sequence $r_n \to 0$ and assume that $y_t^p(n) = r_n^{-1}(x_t^{p+r_n l} - x_t^p)$. It can easily be seen that

$$y_t^p(n) = \eta_t(p,n) + \int_0^t \sigma_s(p + r_n l, y_s^p(n))\,d\mathbf{w}_s + \int_0^t b_s(p + r_n l, y_s^p(n))\,ds, \tag{3}$$

where

$$\eta_t(p,n) = r_n^{-1}[\xi_t(p + r_n l) - \xi_t(p)] + \int_0^t r_n^{-1}[\sigma_s(p + r_n l, x_s^p) - \sigma_s(p,x_s^p)]\,d\mathbf{w}_s$$
$$+ \int_0^t r_n^{-1}[b_s(p + r_n l, x_s^p) - b_s(p,x_s^p)]\,ds.$$

We are given the expression

$$\mathscr{L}\text{-}\lim_{n\to\infty} r_n^{-1}[\xi_t(p_0 + r_n l) - \xi_t(p_0)] = \mathscr{L}\text{-}\frac{\partial}{\partial l}\,\xi_t(p_0).$$

In addition, since the $\mathscr{L}$-limit of the product (sum) equals the product (sum) of $\mathscr{L}$-limits, we have that in $\mathscr{L}$

$$r_n^{-1}[\sigma_s(p_0 + r_n l, x_s^{p_0}) - \sigma_s(p_0,x_s^{p_0})] = \sum_{j=1}^d (x_s^{p_0})^j r_n^{-1}[\sigma_s(p_0 + r_n l, e_j) - \sigma_s(p_0,e_j)]$$
$$\to \sum_{j=1}^d (x_s^{p_0})^j\,\mathscr{L}\text{-}\frac{\partial}{\partial l}\,\sigma_s(p_0,e_j).$$

Similarly, in $\mathscr{L}$

$$r_n^{-1}[b_s(p_0 + r_n l, x_s^{p_0}) - b_s(p_0,x_s^{p_0})] \to \sum_{j=1}^d (x_s^{p_0})^j\,\mathscr{L}\text{-}\frac{\partial}{\partial l}\,b_s(p_0,e_j).$$

Thus, $\eta_t(p_0,n) \to \eta_t(p_0)$ in $\mathscr{L}$ as $n \to \infty$. Comparing (2) with (3), we have from Theorem 3 that $y_t^{p_0}(n) \to y_t^{p_0}$ in $\mathscr{L}$. Hence

$$y_t^p = \mathscr{L}\text{-}\frac{\partial}{\partial l}\,x_t^p \tag{4}$$

for $p = p_0$, proving thereby that x_t^p is $\mathscr{L}$-differentiable.

It is clear that (4) is satisfied at any point p at which there exist $\mathscr{L}$-derivatives $\xi_t(p)$, $\sigma_t(p,x)$, $b_t(p,x)$. Further, if the foregoing derivatives are continuous at the point p_0, they are defined in some neighborhood in which (4) is satisfied. In this case, as we noted above, $\eta_t(p)$ is $\mathscr{L}$-continuous at the point p_0. Also, by Corollary 2, it follows from Eq. (2) that the process y_t^p is $\mathscr{L}$-continuous at the point p_0. This fact implies that the process x_t^p is $\mathscr{L}$-continuously $\mathscr{L}$-differentiable at p_0.

Suppose that our lemma is proved for $i = i_0$ and that the assumptions of the lemma are satisfied for $i = i_0 + 1$. We shall complete proving our lemma if we show that each first $\mathscr{L}$-derivative of x_t^p is i_0 times ($\mathscr{L}$-continuously) $\mathscr{L}$-differentiable at the point p_0. We consider, for instance, $\mathscr{L}$-$(\partial/\partial l)x_t^p$. This process exists and satisfies Eq. (2) for p close to p_0.

Since the assumptions of Lemma 3 are satisfied for $i = i_0$ (even for $i = i_0 + 1$), by the induction assumption, the process x_t^p is i_0 times ($\mathscr{L}$-continuously) $\mathscr{L}$-differentiable at p_0. From this, it follows that the process $\eta_t(p)$ is i_0 times ($\mathscr{L}$-continuously) $\mathscr{L}$-differentiable at p_0. Applying the induction assumption to (2), we convince ourselves that the process y_t^p is i_0 times ($\mathscr{L}$-continuously) $\mathscr{L}$-differentiable at the point p_0. The lemma is proved. □

4. Theorem. *Suppose that the process $\xi_t(p)$ is i times ($\mathscr{L}$-continuously) $\mathscr{L}$-differentiable at a point $p_0 \in D$, and that the functions $\sigma_s(p,x)$, $b_s(p,x)$ for each s, ω are i times continuously (with respect to p, x) differentiable with respect to p, x for $p \in D$, $x \in E_d$. Furthermore, assume that all derivatives of the foregoing functions, up to order i inclusive, do not exceed $K(1 + |x|)^m$ with respect to the norm for any $p \in D$, s, ω, x. Then the process x_t^p is i times ($\mathscr{L}$-continuously) $\mathscr{L}$-differentiable at the point p_0. If, in addition, the process $\xi_t(p)$ is i times ($\mathscr{L}B$-continuously) $\mathscr{L}B$-differentiable at the point p_0, the process x_t^p will be the same as the process $\xi_t(p)$.*

PROOF. Because the notion of the $\mathscr{L}$-derivative is local, it suffices to prove the theorem in any subregion D' of a region D, which together with its closure lies in D. We construct an infinitely differentiable function $w(p)$ in such a way that $w(p) = 1$ for $p \in D'$, $w(p) = 0$ for $p \notin D$. Let $\bar{\xi}_t(p) = \xi_t(p)w(p)$, $\bar{\sigma}_s(p,x) = \sigma_s(p,x)w(p)$, $\bar{b}_s(p,x) = b_s(p,x)w(p)$. Then ξ_t, $\bar{b}_s$, $\bar{\sigma}_s$ satisfy the conditions of the theorem for $D = E$. Further, since the assertions of the theorem hold for $\bar{\xi}$, $\bar{\sigma}$, $\bar{b}$ in E, they hold as well for ξ, b, σ in the region D'. This reasoning shows that in proving our theorem, we can assume that the assumptions of the theorem are satisfied for $D = E$.

In this case we use the induction over i. First, let $i = 1$. Further, we take a unit vector $l \in E$ and a sequence of numbers $r_n \to 0$. Let

$$y_t^p(n) = r_n^{-1}(x_t^{p+r_n l} - x_t^p),$$
$$x_t^p(n,u) = ux_t^{p+r_n l} + (1-u)x_t^p.$$

Using the Newton–Leibniz formula, we easily obtain

$$y_t^p(n) = \eta_t(p,n) + \int_0^t \tilde{\sigma}_s^n(p,y_s^p(n))\,d\mathbf{w}_s + \int_0^t \tilde{b}_s^n(p,y_s^p(n))\,ds, \tag{5}$$

where

$$\eta(p,n) = r_n^{-1}[\xi_t(p + r_n l) - \xi_t(p)]$$
$$+ \int_0^t \left[\sum_j l^j \int_0^1 \sigma_{s,p^j}(p + ur_n l, x_s^p(n,u))\, du \right] d\mathbf{w}_s$$
$$+ \int_0^t \left[\sum_j l^j \int_0^1 b_{s,p^j}(p + ur_n l, x_s^p(n,u))\, du \right] ds,$$

$$\tilde{\sigma}_s^n(p,x) = \sum_{j=1}^d x^j \int_0^1 \sigma_{s,x^j}(p + ur_n l,\ x_s^p(n,u))\, du,$$

$$\tilde{b}_s^n(p,x) = \sum_{j=1}^d x^j \int_0^1 b_{s,x^j}(p + ur_n l, x_s^p(n,u))\, du,$$

We look upon the pair $(p + ur_n l, x_t^p(n,u))$ as a process in $E \times E_d$ with a time parameter t. It is seen that

$$|(p + ur_n l, x_t^p(n,u)) - (p,x_t^p)| \le |(p + r_n l, x_t^{p+r_n l}) - (p,x_t^p)|.$$

Furthermore, by Corollary 2 and the $\mathscr{L}$-continuity of $\mathscr{L}$-differentiable functions, $x^{p_0+r_n l} \to x_t^{p_0}$ in $\mathscr{L}$. In order to apply Lemma 7.8, we note that, for instance, $|b_{s,x^j}(p,x)| \le K(1 + \sqrt{|x|^2 + |p|^2})^m$ for all ω, s, p, x. By this lemma, for $p = p_0$

$$\tilde{\sigma}_s^n(p,x) \to \tilde{\sigma}_s(p,x), \qquad \tilde{b}_s^n(p,x) \to \tilde{b}_s(p,x), \qquad \eta_t(p,n) \to \eta_t(p)$$

in the sense of convergence in the space $\mathscr{L}$ in which

$$\tilde{\sigma}_s(p,x) \equiv \sum_{j=1}^d x^j \sigma_{s,x^j}(p,x_s^p),$$

$$\tilde{b}_s(p,x) \equiv \sum_{j=1}^d x^j b_{s,x^j}(p,x_s^p).$$

$$\eta_t(p) \equiv \mathscr{L}\text{-}\frac{\partial}{\partial l}\xi_t(p) + \int_0^t \sum_j l^j \sigma_{s,p^j}(p,x_s^p)\, dw_s + \int_0^t \sum_j l^j b_{s,p^j}(p,x_s^p)\, ds$$

Note that $\tilde{\sigma}_s$, $\tilde{b}_s$, η_t, $\tilde{\sigma}_s^n$, $\tilde{b}_s^n$, $\eta_t(p,n)$ are progressively measurable for those p, x, for which they exist. In fact, one can take the derivative $\mathscr{L}\text{-}(\partial/\partial l)\xi_t(p)$ to be progressively measurable. Also, for example, $\sigma_{s,x^j}(p,x)$ is progressively measurable (ordinary derivative with respect to the parameter of a progressively measurable process) and continuous with respect to p, x. Hence the process $\sigma_{s,x^j}(p + ur_n l, x_s^p(n,u))$ is progressively measurable and continuous with respect to u, which, in turn, implies the progressive measurability of the Riemann integral

$$\int_0^1 \sigma_{s,x^j}(p + ur_n l, x_s^p(n,u))\, du$$

and the progressive measurability of the process $\tilde{\sigma}_s^n(p,x)$.

Further, since $\sigma_s(p,x)$, $b_s(p,x)$ satisfy the Lipschitz condition (1) with respect to x, $\sigma_{s,x^j}(p,x)$, $b_{s,x^j}(p,x)$ are bounded variables. This implies that the functions $\tilde{\sigma}_s(p,x)$ and $\tilde{b}_s(p,s)$, linear with respect to x, satisfy the Lipschitz condition (1). By Theorem 5.7, for $p = p_0$ there exists a solution of the equation

$$y_t^p = \eta_t(p) + \int_0^t \tilde{\sigma}_s(p,y_s^p)\, d\mathbf{w}_s + \int_0^t \tilde{b}_s(p,y_s^p)\, ds. \tag{6}$$

By Theorem 1, comparing (5) with (6), we conclude that

$$\mathscr{L}\text{-}\lim_{n\to\infty} r_n^{-1}(x_t^{p+r_nl} - x_t^p) = \mathscr{L}\text{-}\lim_{n\to\infty} y_t^p(n) = y_t^p$$

for $p = p_0$. This shows that $y_t^p = \mathscr{L}\text{-}(\partial/\partial l)x_t^p$ for $p = p_0$, and therefore the process x_t^p is $\mathscr{L}$-differentiable at the point p_0. It is also seen that $y_t^p = \mathscr{L}\text{-}(\partial/\partial l)x_t^p$ at each point p at which $\mathscr{L}\text{-}(\partial/\partial l)\xi_t(p)$ exists.

Next, let $\xi_t(p)$ be $\mathscr{L}$-continuously $\mathscr{L}$-differentiable at the point p_0. Then $\mathscr{L}\text{-}(\partial/\partial l)\xi_t(p)$ exists in some neighborhood of the point p_0, y_t^p being the $\mathscr{L}$-derivative of x_t^p along the l direction in this neighborhood. In addition, the process (p,x_t^p) is $\mathscr{L}$-continuous at the point p_0, and the functions $\sigma_{s,p^j}(p,x)$, $b_{s,p^j}(p,x)$ are continuous with respect to (p,x) and do not exceed $K(1 + |x|)^m$ with respect to the norm. Therefore, by Theorem 7.9, the processes $\sigma_{s,p^j}(p,x_s^p)$, $b_{s,p^j}(p,x_s^p)$ are $\mathscr{L}$-continuous, and consequently, the process $\eta_t(p)$ is $\mathscr{L}$-continuous at p_0. Similarly, the fact that the functions $\sigma_{s,x^j}(p,x)$, $b_{s,x^j}(p,x)$ are bounded and continuous with respect to p, x, implies that the processes $\tilde{\sigma}_s(p,x)$ and $\tilde{b}_s(p,x)$ are $\mathscr{L}$-continuous at p_0 for each x. To conclude our reasoning, we observe that, by Corollary 2, the process y_t^p interpreted as the solution of Eq. (6) is $\mathscr{L}$-continuous at the point p_0.

Thus, we have proved the first assertion of the theorem for $i = 1$. Further, suppose that this theorem has been proved for $i = i_0$, and, in addition, the assertions of the theorem are satisfied for $i = i_0 + 1$. Consider the derivative $\mathscr{L}\text{-}(\partial/\partial l)x_t^p$. As was shown above, we may assume that this process is y_t^p and that it satisfies Eq. (6). By the induction assumption, x_t^p is i_0 times $\mathscr{L}$-differentiable at p_0. Therefore the pair (p,x_t^p) is i_0 times $\mathscr{L}$-differentiable as well. By Theorem 7.9, the processes $\sigma_{s,p^j}(p,x_s^p)$, $b_{s,p^j}(p,x_s^p)$, $\sigma_{s,x^j}(p,x_s^p)$, $b_{s,x^j}(p,x_s^p)$ are i_0 times $\mathscr{L}$-differentiable at the point p_0. Hence, in Eq. (6) the processes $\eta_t(p)$, $\tilde{\sigma}_s(p,x)$, $\tilde{b}_s(p,x)$ are i_0 times $\mathscr{L}$-differentiable with respect to p. Since $\tilde{\sigma}_s(p,x)$, $\tilde{b}_s(p,x)$ are linear functions of x, according to the preceding lemma the process y_t^p is i_0 times $\mathscr{L}$-differentiable at the point p_0. We have thus proved that the derivative $(\partial/\partial l)x_t^p$ is i_0 times $\mathscr{L}$-differentiable at the point p_0. Since l is an arbitrary unit vector from E, this implies, by definition, that x_t^p is $i_0 + 1$ times $\mathscr{L}$-differentiable at the point p_0.

In addition, if $\xi_t(p)$ is $i_0 + 1$ times $\mathscr{L}$-continuously $\mathscr{L}$-differentiable at the point p_0, we can prove that x_t^p is $i_0 + 1$ times $\mathscr{L}$-continuously $\mathscr{L}$-differentiable at the point p_0 if we put the word "$\mathscr{L}$-continuously" in the appropriate places in the above arguments. This completes the proof of the first assertion of Theorem 4.

For proving the second assertion of the theorem, we need only to prove, due to the equality

$$x_t^p = \xi_t(p) + \int_0^t \sigma_s(p,x_s^p)\, d\mathbf{w}_s + \int_0^t b_s(p,x_s^p)\, ds,$$

that the processes $\sigma_s(p,x_s^p)$, $b_s(p,x_s^p)$ are i times ($\mathscr{L}$-continuously) $\mathscr{L}$-differentiable at the point p_0. It is obvious that a process which is identically equal to $(p,0)$ is i times $\mathscr{L}$-continuously $\mathscr{L}$-differentiable. It is also seen that, since the function $\sigma_s(p,0)$ is i times continuously differentiable with respect to p and, in addition, the derivatives of this function are bounded, the process $\sigma_s(p,0)$ is i times $\mathscr{L}$-continuously $\mathscr{L}$-differentiable in accord with Theorem 7.9. Furthermore, the process (p,x_s^p) is i times ($\mathscr{L}$-continuously) $\mathscr{L}$-differentiable at the point p_0, the function $\sigma_s(p,x) - \sigma_s(p,0)$, with respect to the norm, does not exceed $K|x|$, and, in addition, the derivatives of this function satisfy the necessary restrictions on the growth. By Theorem 7.9, the process $\sigma_s(p,x_s^p) - \sigma_s(p,0)$ is i times ($\mathscr{L}$-continuously) $\mathscr{L}$-differentiable at the point p_0; the same holds for the process $\sigma_s(p,x_s^p) = \sigma_s(p,0) + [\sigma_s(p,x_s^p) - \sigma_s(p,0)]$. The process in $b_s(p,x_s^p)$ can be considered in a similar way. The theorem is proved. □

5. Remark. For $i \geq 1$ we have proved that for any unit vector $l \in E$ the solution of Eq. (6) is the $\mathscr{L}$-derivative of x_t^p along the l direction:

$$\begin{aligned} y_t^p = \mathscr{L}\text{-}\frac{\partial}{\partial l}\,\xi_t(p) &+ \int_0^t \sigma_{s,(l)}(p,x_s^p)\, d\mathbf{w}_s \\ &+ \int_0^t b_{s,(l)}(p,x_s^p)\, ds + \int_0^t \sigma_{s,(y_s^p)}(p,x_s^p)|y_s^p|\, d\mathbf{w}_s \\ &+ \int_0^t b_{s,(y_s^p)}(p,x_s^p)|y_s^p|\, ds. \end{aligned}$$

We have seen that the last equation is linear with respect to y_s^p; also, we applied Lemma 3 to this equation for $i \geq 2$. In Lemma 3 we derived Eq. (2), according to which the solution of the equation which follows is an $\mathscr{L}$-derivative of y_t^p along the l direction, that is, a second $\mathscr{L}$-derivative of x_t^p along the l direction. This equation is the following:

$$z_t^p = \eta_t(p) + \int_0^t \sigma_{s,(z_s^p)}(p,x_s^p)|z_s^p|\, d\mathbf{w}_s + \int_0^t b_{s,(z_s^p)}(p,x_s^p)|z_s^p|\, ds,$$

where, according to the rules of $\mathscr{L}$-differentiation of a composite function (see (2)),

$$\begin{aligned} \eta_t(p) = \mathscr{L}\text{-}\frac{\partial}{\partial l}&\left[\mathscr{L}\text{-}\frac{\partial}{\partial l}\,\xi_t(p) + \int_0^t \sigma_{s,(l)}(p,x_s^p)\, d\mathbf{w}_s + \int_0^t b_{s,(l)}(p,x_s^p)\, ds\right] \\ &+ \sum_{j=1}^d \int_0^t (y_s^p)^j[\sigma_{s,x^j(l)}(p,x_s^p) + \sigma_{s,x^j(y_s^p)}(p,x_s^p)|y_s^p|]\, d\mathbf{w}_s \\ &+ \sum_{j=1}^d \int_0^t (y_s^p)^j[b_{s,x^j(l)}(p,x_s^p) + b_{s,x^j(y_s^p)}(p,x_s^p)|y_s^p|]\, ds. \end{aligned}$$

Note that the above equations as well as equations for the highest $\mathscr{L}$-derivatives of x_t^p can be obtained proceeding from the fact that x_t^p is $\mathscr{L}$-differentiable the desired number of times, if we differentiate the equality

$$x_t^p = \xi_t(p) + \int_0^t \sigma_s(p,x_s^p)\, d\mathbf{w}_s + \int_0^t b_s(p,x_s^p)\, ds,$$

interchange the order of the derivatives with those of the integrals, and, in addition, make use of the formula for an $\mathscr{L}$-derivative of a composite function.

The following assertion is a simple consequence of Theorem 4 and Corollary 2 in the case where $D = E_d$, $\xi_t(p) \equiv p$, $\sigma_t(p,x) = \sigma_t(x)$, $b_t(p,x) = b_t(x)$.

6. Theorem. *The process x_t^x is $\mathscr{L}B$-continuous. If $\sigma_s(x)$, $b_s(x)$ are i times continuously differentiable with respect to x for all ω, s, and if, in addition, each derivative of these functions up to order i inclusively does not exceed $K(1 + |x|)^m$ with respect to the norm for any s, x, ω, the process x_t^x is i times $\mathscr{L}B$-continuously $\mathscr{L}B$-differentiable.*

In concluding this section, we give two theorems on estimation of moments of derivatives of a solution of a stochastic equation. Since, as we saw in Remark 5, it was possible to write equations for such derivatives, it is reasonable to apply Corollaries 5.6 and 5.10–5.12 for estimating the moments of these derivatives. The reader can easily prove the theorems which follow.

7. Theorem. *Let there be a constant K_1 such that for all s, x, p, ω*

$$|b_s(p,x)| + \|\sigma_s(p,x)\| \le K_1(1 + |x|).$$

Suppose that the process $\xi_t(p)$ is $\mathscr{L}B$-differentiable at a point $p_0 \in D$. Further, suppose that $\mathscr{L}B$-derivatives of the process $\xi_t(p)$ have modifications which are progressively measurable and separable at the same time. Let the functions $\sigma_s(p,x)$, $b_s(p,x)$ for each s, ω be continuously differentiable with respect to p, x for $p \in D$, $x \in E_d$. In addition, let the matrix norms of the derivatives of the function $\sigma_s(p,x)$ and the norms of the derivatives of the function $b_s(p,x)$ be smaller than $K(1 + |x|)^m$ $(m \ge 1)$ along all directions for all $p \in D$, s, ω, x. Then for any unit vector $l \in E$, $q \ge 1$, $t \in [0,T]$

$$\mathsf{M} \sup_{s \le t} \left| \mathscr{L}B\text{-}\frac{\partial}{\partial l} x_s^{p_0} \right|^{2q} \le N e^{Nt} \left(1 + \mathsf{M} \sup_{s \le t} \left| \mathscr{L}B\text{-}\frac{\partial}{\partial l} \xi_s(p_0) \right|^{2q} + \mathsf{M} \int_0^t |\xi_s(p_0)|^{2qm}\, ds \right),$$

where $N = N(q,K,m,K_1)$.

8. Theorem. (a) *Let the functions $\sigma_s(x)$, $b_s(x)$ be continuously differentiable with respect to x for each s, ω. Then for any unit vector $l \in E_d$, $q \ge 1$, $t \in [0,T]$, $x \in E_d$*

$$\mathsf{M} \sup_{s \le t} \left| \mathscr{L}B\text{-}\frac{\partial}{\partial l} x_s^x \right|^q \le N e^{Nt},$$

where $N = (q,K)$.

(b) *Let the functions* $\sigma_s(x)$, $b_s(x)$ *be twice continuously differentiable for each* s, ω. *Further, for each* x, s, ω *and unit vectors* $l \in E_d$ *let*

$$\|\sigma_{s(l)(l)}(x)\| + |b_{s(l)(l)}(x)| \leq K(1 + |x|)^m.$$

Also, suppose that $\|\sigma_s(x)\| + |b_s(x)| \leq K_1(1 + |x|)$ *for all* x, s, ω *for some constant* K_1. *Then for any* $q \geq 1$, $t \in [0,T]$, $x \in E_d$ *and the unit vector* $l \in E_d$

$$\mathsf{M} \sup_{s \leq t} \left| \mathscr{L}B\text{-}\frac{\partial^2}{\partial l^2} x_s^x \right|^q \leq N(1 + |x|)^{qm} e^{Nt},$$

where $N = N(q,K,m,K_1)$.

9. The Markov Property of Solutions of Stochastic Equations

The Markov property of solutions of a stochastic equation with non random coefficients is well known (see [9, 11, 24]). In this section, we shall prove a similar property for random coefficients of the equation (Theorem 4), and moreover, deduce some consequences from this property.

We fix two constants $T, K > 0$. In this section we repeatedly assume about $(\mathbf{w}_t, \mathscr{F}_t)$, ξ_t, $\sigma_t(x)$, $b_t(x)$, with indices and tildes or without them, the following: $(\mathbf{w}_t, \mathscr{F}_t)$ is a d_1-dimensional Wiener process, $\sigma_t(x)$ is a random matrix of dimension $d \times d_1$, $b_t(x)$, ξ_t are random d-dimensional vectors; $\sigma_t(x)$, $b_t(x)$, ξ_t are defined for $t \in [0,T]$, $x \in E_d$, progressively measurable with respect to $\{\mathscr{F}_t\}$, and

$$\mathsf{M} \int_0^T [|\xi_t|^2 + \|\sigma_t(x)\|^2 + |b_t(x)|^2]\, dt < \infty,$$

$$\|\sigma_t(x) - \sigma_t(y)\| + |b_t(x) - b_t(y)| \leq K|x - y|$$

for all possible values of the indices and arguments.

We can now specify the objective of this section. It consists in deriving formulas for a conditional expectation under the condition $\mathscr{F}_0$ of functionals of solutions of the stochastic equation

$$x_t = \xi_t + \int_0^t \sigma_s(x_s)\, d\mathbf{w}_s + \int_0^t b_s(x_s)\, ds. \tag{1}$$

Note that if the assumptions made above are satisfied, in accord with Theorem 5.7 the solution of Eq. (1) on an interval $[0,T]$ exists and is unique.

1. Lemma. *Suppose that for all integers* $i, j > 0$, $t_1, \ldots, t_i \in [0,T]$, $z_1, \ldots, z_j \in E_d$ *the vector*

$$\{\mathbf{w}_{t_p}, \xi_{t_p}, \sigma_{t_p}(z_q), b_{t_p}(z_q) : p = 1, \ldots, i,\ q = 1, \ldots, j\}$$

does not depend on $\mathscr{F}_0$. *Then the process* x_t, *which is a solution of Eq.* (1), *does not depend on* $\mathscr{F}_0$ *either.*

Proof. As we did in proving Theorem 5.7, we introduce here an operator I using the formula

$$Iy_t = \int_0^t \sigma_s(y_s)\,d\mathbf{w}_s + \int_0^t b_s(y_s)\,ds.$$

In proving Theorem 5.7 we said that the operator I is defined on a set of progressively measurable functions in $\mathscr{L}_2([0,T] \times \Omega)$ and also that this operator maps this set into itself.

Let a function $y_t(\omega)$ from the set indicated (for example, $y_t \equiv 0$) be such that the totality of random variables

$$\{\mathbf{w}_t,\xi_t,y_t,\sigma_t(x),b_t(x):t \subset [0,T], x \subset E_d\} \tag{2}$$

does not depend on $\mathscr{F}_0$. We prove that in this case the totality of random variables

$$\{\mathbf{w}_t, \xi_t + Iy_t, \sigma_t(x), b_t(x): t \in [0,T], x \in E_d\} \tag{3}$$

does not depend on $\mathscr{F}_0$ either.

We denote by Σ the completion of a σ-algebra of subsets Ω, which is generated by the totality of random variables (2). By assumption, Σ does not depend on $\mathscr{F}_0$. It is seen that for proving that (3) is independent of $\mathscr{F}_0$, it suffices to prove that random variables Iy_t are Σ-measurable for $t \in [0,T]$.

For real a let $x_n(a) = 2^{-n}[2^n a]$, where $[a]$ is the greatest integer less than or equal to a. If $y \in E_d$, we assume that $x_n(y) = (x_n(y^1), \ldots, x_n(y^d))$, and, in addition, that Γ_n is a set of values of the function $x_n(y)$, $y \in E_d$. Due to the continuity of $\sigma_t(x)$ with respect to x we have

$$\sigma_t(y_t) = \lim_{n\to\infty} \sigma_t(\kappa_n(y_t)) = \lim_{n\to\infty} \sum_{y\in\Gamma_n} \sigma_t(y)\chi_{\kappa_n}(y_t) = y. \tag{4}$$

Therefore, the variable $\sigma_t(y_t)$ is Σ-measurable. The Σ-measurability of $b_t(y_t)$ can be proved in a similar way. Further (see Appendix 1), for almost all $s \in [0,1]$ for some sequence of integers n' in probability

$$\lim_{n'\to\infty} \left[\int_0^t \sigma_{\kappa_{n'}(r+s)-s}(y_{\kappa_{n'}(r+s)-s})\,d\mathbf{w}_r + \int_0^t b_{\kappa_{n'}(r+s)-s}(y_{\kappa_{n'}(r+s)-s})\,dr \right] = Iy_t. \tag{5}$$

Since the function $\kappa_n(r + s) - s$ assumes only a finite number of values on an interval $[0,t]$, the integrals in a limiting expression are integrals of step functions. The former integrals are to be written as finite sums which consist of the product of values of $\sigma_r(y_r)$ and an increment $\mathbf{w}_r$ and the product of values of $b_r(y_r)$ and increments r. The foregoing sums are Σ-measurable. Hence the limiting expressions are Σ-measurable, which implies the Σ-measurability of Iy_t.

As we did in proving Theorem 5.7, we define here the sequence x_t^n using the recurrence formula

$$x_t^0 \equiv 0, \qquad x_t^{n+1} = \xi_t + Ix_t^n, \qquad n \geq 0.$$

By induction, it follows from what has been proved above that the processes x_t^n do not depend on $\mathscr{F}_0$ for $n \geq 0$, $t \in [0,T]$. According to Remark 5.13, for $t \in [0,T]$

$$\underset{n \to \infty}{\text{l.i.m.}}\ x_t^n = x_t.$$

Therefore, the process x_t does not depend on $\mathscr{F}_0$. The lemma is proved. □

In the next lemma we consider $(\tilde{\mathbf{w}}_t, \tilde{\mathscr{F}}_t)$, $\tilde{\xi}_t$, $\tilde{\sigma}_t(x)$, $\tilde{b}_t(x)$ as well as $(\mathbf{w}_t, \mathscr{F}_t)$, ξ_t, $\sigma_t(x)$, $b_t(x)$. As we agreed above, we assume here that these elements satisfy the same conditions. Let $\tilde{x}_t$ be a solution of the equation

$$\tilde{x}_t = \tilde{\xi}_t + \int_0^t \tilde{\sigma}_r(\tilde{x}_r)\, d\tilde{\mathbf{w}}_r + \int_0^t \tilde{b}_r(\tilde{x}_r)\, dr.$$

2. Lemma. *Suppose that for all integers i, $j > 0$ and $t_1, \ldots, t_i \in [0,T]$, $z_1, \ldots, z_j \in E_d$ the following vectors are identically distributed:*

$$\{\mathbf{w}_{t_p}, \xi_{t_p}, \sigma_{t_p}(z_q), b_{t_p}(z_q) : p = 1, \ldots, i, q = 1, \ldots, j\},$$
$$\{\tilde{\mathbf{w}}_{t_p}, \tilde{\xi}_{t_p}, \tilde{\sigma}_{t_p}(z_q), \tilde{b}_{t_p}(z_q) : p = 1, \ldots, i, q = 1, \ldots, j\}.$$

Then the finite-dimensional distribution of the process x_t is equivalent to that of the process $\tilde{x}_t$.

PROOF. We make use again of the operator I from the previous proof. Let

$$\tilde{I}\tilde{y}_t = \int_0^t \tilde{\sigma}_s(\tilde{y}_s)\, d\tilde{\mathbf{w}}_s + \int_0^t \tilde{b}_s(\tilde{y}_s)\, ds$$

and let the processes y_t, $\tilde{y}_t$ be progressively measurable with respect to $\{\mathscr{F}_t\}$, $\{\tilde{\mathscr{F}}_t\}$, respectively:

$$\mathsf{M} \int_0^t |y_t|^2\, dt < \infty, \qquad \mathsf{M} \int_0^t |\tilde{y}_t|^2\, dt < \infty.$$

Further, for any $i, j > 0$, $t_1, \ldots, t_i \in [0,T]$, $z_1, \ldots, z_j \in E_d$ let the vectors

$$\begin{gathered}\{\mathbf{w}_{t_p}, \xi_{t_p}, y_{t_p}, \sigma_{t_p}(z_q), b_{t_p}(z_q) : p = 1, \ldots, i, q = 1, \ldots, j\},\\ \{\tilde{\mathbf{w}}_{t_p}, \tilde{\xi}_{t_p}, \tilde{y}_{t_p}, \tilde{\sigma}_{t_p}(z_q), \tilde{b}_{t_p}(z_q) : p = 1, \ldots, i, q = 1, \ldots, j\}\end{gathered} \tag{6}$$

have identical distributions. Note that if two random vectors have identical distributions, any (Borel) function of one vector has the same distribution as the other has. From this it follows, in accord with Eq. (4), that for any $i, j > 0$, $t_1, \ldots, t_i \in [0,T]$, $z_1, \ldots, z_j \in E_d$ the vectors

$$\begin{gathered}\{\mathbf{w}_{t_p}, \xi_{t_p}, y_{t_p}, \sigma_{t_p}(y_{t_p}), \sigma_{t_p}(z_q), b_{t_p}(y_{t_p}), b_{t_p}(z_q) : p = 1, \ldots, i, q = 1, \ldots, j\},\\ \{\tilde{\mathbf{w}}_{t_p}, \tilde{\xi}_{t_p}, \tilde{y}_{t_p}, \tilde{\sigma}_{t_p}(\tilde{y}_{t_p}), \tilde{\sigma}_{t_p}(z_q), \tilde{b}_{t_p}(\tilde{y}_{t_p}), \tilde{b}_{t_p}(z_q) : p = 1, \ldots, i, q = 1, \ldots, j\}\end{gathered} \tag{7}$$

have the same distributions. It is useful to draw the reader's attention to the fact that in order to prove the proposition made above, we need to use vectors of type (6) at the values of z_q different from those which appear in (7).

We choose $s \in [0,1]$ so that Eq. (5) holds for $t = t_1, \ldots, t_i$, and, in addition, that similar representations hold for $\tilde{I}$. Having done this, we can see that the vectors

$$\begin{aligned}&\{\mathbf{w}_{t_p}, \xi_{t_p}, \xi_{t_p} + Iy_{t_p}, \sigma_{t_p}(z_q), b_{t_p}(z_q): p = 1, \ldots, i, q = 1, \ldots, j\},\\&\{\tilde{\mathbf{w}}_{t_p}, \tilde{\xi}_{t_p}, \tilde{\xi}_{t_p} + \tilde{I}\tilde{y}_{t_p}, \tilde{\sigma}_{t_p}(z_q), \tilde{b}_{t_p}(z_q): p = 1, \ldots, i, q = 1, \ldots, j\}\end{aligned} \tag{8}$$

are representable as the limits in probability of identical functions of vectors of type (7). Therefore, the vectors (8) have identical distributions for any $i, j > 0$, $t_1, \ldots, t_i \in [0,T]$, $z_1, \ldots, z_j \in E_d$.

Next, we compare the vectors (6) and (8). Also, we find sequences of the processes

$$x_t^0 \equiv 0, \qquad \tilde{x}_t^0 \equiv 0, \qquad x_t^{n+1} = \xi_t + Ix_t^n, \qquad \tilde{x}_t^{n+1} = \tilde{\xi}_t + \tilde{I}\tilde{x}_t^n.$$

Passing from vectors of type (6) to vectors of type (8), we prove by induction that the finite-dimensional distribution of x_t^n is equivalent to that of $\tilde{x}_t^n$. Therefore, the finite-dimensional distributions of the limits of these processes in the mean square coincide, i.e., x_t and $\tilde{x}_t$. The lemma is proved. □

3. Corollary. *If ξ_t, $\sigma_t(x)$, $b_t(x)$ are nonrandom and if, in addition, they are equal to $\tilde{\xi}_t$, $\tilde{\sigma}_t(x)$, $\tilde{b}_t(x)$, respectively, for all $t \in [0,T]$, $x \in E_d$, the processes x_t, $\tilde{x}_t$ have identical finite-dimensional distributions. Furthermore, the process x_t does not depend on $\mathscr{F}_0$, and the process $\tilde{x}_t$ does not depend on $\tilde{\mathscr{F}}_0$.*

This corollary follows from Lemmas 1 and 2 and the fact that all Wiener processes have identical finite-dimensional distributions and that, for example, $\mathbf{w}_t = \mathbf{w}_t - \mathbf{w}_0$ does not depend on $\mathscr{F}_0$.

The formula mentioned at the beginning of the section can be found in the next theorem. In order not to complicate the formulation of the theorem, we list the conditions under which we shall prove the theorem.

Let Z be a separable metric space with metric ρ and let $(\mathbf{w}_t^z, \mathscr{F}_t^z) \equiv (\mathbf{w}_t, \mathscr{F}_t)$, $\sigma_t^z(x)$, $b_t^z(x)$ be defined for $z \in Z$. We assume (in addition to the assumption mentioned at the beginning of the section) that the functions $\sigma_t^z(x,\omega)$, $b_t^z(x,\omega)$ are continuous with respect to z for all t, ω, x and

$$\mathsf{M} \int_0^T \left[\sup_z \|\sigma_t^z(x)\|^2 + \sup_z |b_t^z(x)|^2\right] dt < \infty$$

for all x.

4. Theorem. *Suppose that the assumptions made before proving the theorem are satisfied. Let the totality of variables*

$$\{\mathbf{w}_t, \sigma_t^z(x), b_t^z(x): t \in [0,T], x \in E_d\}$$

be independent of $\mathscr{F}_0$ for all $z \in Z$. Further, let ξ be an $\mathscr{F}_0$-measurable random variable with values in E_d and a finite second moment, let ζ be an $\mathscr{F}_0$-measurable *random function with values in Z. Finally, let y_t be a solution of the equation*

$$y_t = \xi + \int_0^t \sigma_r^\zeta(y_r)\, d\mathbf{w}_r + \int_0^t b_r^\zeta(y_r)\, dr. \tag{9}$$

We denote by $x_t^{z,x}$ a solution of the equation

$$x_t^{z,x} = x + \int_0^t \sigma_r^z(x_r^{z,x})\, d\mathbf{w}_r + \int_0^t b_r^z(x_r^{z,x})\, dr. \tag{10}$$

Let $F(z,x_{[0,T]})$ be a nonnegative measurable function on $Z \times C([0,T],E_d)$. Then

$$\mathsf{M}\{F(\zeta,y_{[0,T]})\,|\,\mathscr{F}_0\} = \Phi(\zeta,\xi) \qquad \text{(a.s.)}, \tag{11}$$

where

$$\Phi(z,x) \equiv \mathsf{M}F(z,x_{[0,T]}^{z,x}).$$

Proof. First we note that due to the conditions imposed, Eq. (9) and Eq. (10) are solvable and, in addition, are continuous with respect to t. Further, it suffices to prove Eq. (11) for functions of the form $F(z,x_{t_1}, \ldots ,x_{t_n})$, where $t_1, \ldots, t_n \in [0,T]$ and $F(z,x_1, \ldots ,x_n)$ is a bounded continuous function of $(z,x_1, \ldots ,x_n)$. In fact, in this case Eq. (11) extends in a standard manner to all nonnegative functions $F(z,x_{[0,T]})$, which are measurable with respect to a product of a σ-algebra of Borel sets in z and the smallest σ-algebra which contains cylinder sets of the space $C([0,T],E_d)$. It is a well-known fact that the latter σ-algebra is equivalent to the σ-algebra of Borel sets of the metric space $C([0,T],E_d)$.

In future, we shall consider functions F only of the type indicated. Let $A = \{z^{(i)}; i \geq 1\}$ be a countable everywhere dense subset in Z. For $z \in Z$ we denote by $\bar{\kappa}_n(z)$ the first member of the sequence $\{z^{(i)}\}$ for which $\rho(z,z^{(i)} \leq 2^{-n}$. It is easily seen that $\bar{\kappa}_n(z)$ is the measurable function of z and that $\rho(z,\bar{\kappa}_n(z)) \leq 2^{-n}$ for all $z \in Z$. In addition, we define the function $\kappa_n(x)$ as in the proof of Lemma 1.

By Lemma 1, almost surely

$$\begin{aligned}
&\mathsf{M}\,\{F(\bar{\kappa}_n(\zeta), x_{[0,T]}^{\bar{\kappa}_n(\zeta),\kappa_n(\xi)}\,|\,\mathscr{F}_0\} \\
&\quad = \sum_{z \in A} \sum_{x \in \Gamma_n} \chi_{\bar{\kappa}_n(\zeta)=z,\kappa_n(\xi)=x} \mathsf{M}\,\{F(z,x_{[0,T]}^{z,x})|\mathscr{F}_0\} \\
&\quad = \sum_{z \in A} \sum_{x \in \Gamma_n} \chi_{\bar{\kappa}_n(\zeta)=z,\kappa_n(\xi)=x} \Phi(z,x) = \Phi(\bar{\kappa}_n(\zeta),\kappa_n(\xi))
\end{aligned} \tag{12}$$

where we take the limit as $n \to \infty$. We agreed to consider only bounded continuous functions $F(z,x_{[0,T]})$ (moreover, of special type). Hence, the left side of (12) yields the left side of (11), if we show that for some subsequence $\{n'\}$

$$\mathsf{P}\left\{\lim_{n' \to \infty} \sup_{t \leq T} |x_t^{\bar{\kappa}_{n'}(\zeta),\kappa_{n'}(\xi)} - y_t| = 0\right\} = 1. \tag{13}$$

In this case the right side of (12) yields the right side of (11) if we prove that $\Phi(z,x)$ is a continuous function of (z,x).

Since the variables $\bar{\kappa}_n(\zeta)$, $\kappa_n(\xi)$ are $\mathscr{F}_0$-measurable, we can bring an indicator of the set $\{\bar{\kappa}_n(\zeta) = z, \kappa_n(\xi) = x\}$ under the sign of a stochastic integral. Multiplying (10) by the indicator of the above set, bringing this indicator under the integral signs, replacing the values z, x by values $\bar{\kappa}_n(\zeta)$, $\kappa_n(\xi)$, which are equal to z, x on the set considered, and, finally, bringing the indicator out, we have that on each set $\{\bar{\kappa}_n(\zeta) = z, \kappa_n(\xi) = x\}$ the process $x^{\kappa_n(\zeta),\kappa_n(\xi)}$ satisfies the equation

$$x_t = \kappa_n(\xi) + \int_0^t \sigma_r^{\bar{\kappa}_n(\zeta)}(x_r)\,d\mathbf{w}_r + \int_0^t b_r^{\bar{\kappa}_n(\zeta)}(x_r)\,dr. \tag{14}$$

The combination of the sets $\{\bar{\kappa}_n(\zeta) = z, \kappa_n(\xi) = x\}$ with respect to $z \in A$, $x \in \Gamma_n$ produces all Ω. Hence $x^{\kappa_n(\zeta)\kappa_n(\xi)}$ satisfies Eq. (14) on Ω. Comparing (9) with (14), we have in accord with Theorem 5.9 that

$$\begin{aligned} \mathsf{M} \sup_{t \le T} \left|x_t^{\bar{\kappa}_n(\zeta),\kappa_n(\xi)} - y_t\right|^2 &\le N\mathsf{M}|\xi - \kappa_n(\xi)|^2 \\ &\quad + N\mathsf{M} \int_0^T \big[\left|b_t^{\bar{\kappa}_n(\zeta)}(y_t) - b_t^{\zeta}(y_t)\right|^2 \\ &\quad + \left\|\sigma^{\bar{\kappa}_n(\zeta)}(y_t) - \sigma_t^{\zeta}(y_t)\right\|^2\big]\,dt. \end{aligned}$$

Here $|\xi - \kappa_n(\xi)| \to 0$ uniformly on Ω, $b_t^{\bar{\kappa}_n(\zeta)}(y_t) \to b_t^{\zeta}(y_t)$ for each t, due to continuity of $b_t^z(x)$ with respect to z. Furthermore, $|b_t^{\bar{\kappa}_n(\zeta)}(y_t)|^2 + |b_t^{\zeta}(y_t)|^2$ does not exceed $4 \sup_z |b_t^z(0)|^2 + 4K^2|y_t|^2$.

The last expression is summable over $d\mathsf{P} \times dt$. Investigating $\sigma_t^z(x)$ in a similar way, we conclude using the Lebesgue theorem that

$$\mathsf{M} \sup_{t \le T} \left|x_t^{\bar{\kappa}_n(\zeta),\kappa_n(\xi)} - y_t\right|^2 \to 0.$$

This implies (13). For proving the continuity of $\Phi(z,x)$ with respect to (z,x) it suffices to prove that for any sequence $(z_n,x_n) \to (z,x)$ there is a subsequence $(z_{n'},x_{n'})$ for which $\Phi(z_{n'},x_{n'}) \to \Phi(z,x)$. From a form of $\Phi(z,x)$ we easily find that it is enough to have

$$\mathsf{P}\left\{\lim_{n' \to \infty} \sup_{t \le T} \left|x_t^{z_{n'},x_{n'}} - x_t^{z,x}\right| = 0\right\} = 1.$$

The existence of such a subsequence $\{n'\}$ for any sequence (z_n,x_n) converging to (z,x) follows from the considerations which are very similar to the preceding considerations concerning Eq. (13). The theorem is proved. □

5. Remark. The function $\mathsf{M}F(z,x_{[0,T]}^{z,x})$ is measurable with respect to (z,x).

Indeed, the set of functions $F(z,x_{[0,T]})$ for which $\Phi(z,x)$ is measurable contains all continuous and bounded functions F. For these functions F, $\Phi(z,x)$ is continuous even with respect to (z,x). From this we derive in a usual way that the set mentioned contains all nonnegative Borel functions $F(z,x_{[0,T]})$.

6. Exercise

Prove that the assumptions of Theorem 4 about the finiteness of

$$\mathsf{M}\int_0^T\left[\sup_z\|\sigma_t^z(x)\|^2+\sup_z|b_t^z(x)|^2\right]dt$$

can be weakened, and that it is possible to require instead uniform integrability of the values $\|\sigma_t^z(0)\|^2$, $|b_t^z(0)|^2$ over $d\mathsf{P}\times dt$ for z which run through each bounded subset Z.

Further, we consider the problem of computing a conditional expectation under the condition $\mathscr{F}_s$, where $s\in[0,T]$. We shall reduce this problem to that of computing a conditional expectation under the condition $\tilde{\mathscr{F}}_0$ using a time shift. If the function $F(x_{[0,T-s]})$ is defined on $C([0,T-s],E_d)$ and $x_{[0,T-s]}\in C([0,T-s],E_d)$, we denote by $F(x_{[s,T]})$ a value of F on the function $\theta_s x$ which is given by the formula $(\theta_s x)_t=x_{t+s}$ for $t\in[0,T-s]$. Sometimes $F(x_{[s,T]})$ is written as $\theta_s F(x_{[0,T-s]})$. Similar notation can be used for the functions $F(x_{[0,\infty)})$.

7. Theorem. *Let the assumptions of Theorem 4 be satisfied. Further, let $s\in[0,T]$, and let $\zeta=\zeta(\omega)$, $\xi=\xi(\omega)$ be $\mathscr{F}_s$-measurable variables with values in Z and E_d, respectively. Finally, let $\sigma_{s+t}^z(x)$ and $b_{s+t}^z(x)$ be independent of ω for all $t\geq 0$.*

Suppose the process y_t satisfies the equation

$$y_t=\xi+\int_s^t\sigma_r^\zeta(y_r)\,d\mathbf{w}_r+\int_s^t b_r^\zeta(y_r)\,dr$$

for $t\in[s,T]$.

We define the process $x_t^{z,s,x}$ for $t\in[0,T-s]$ as a solution of the equation

$$x_t=x+\int_0^t\sigma_{s+r}^z(x_r)\,d\mathbf{w}_r+\int_0^t b_{s+r}^z(x_r)\,dr.$$

Then for any nonnegative measurable function $F(z,x_{[0,T-s]})$ given on $Z\times C([0,T-s],E_d)$,

$$\mathsf{M}\{F(\zeta,y_{[s,T]})\,|\,\mathscr{F}_s\}=\Phi(\zeta,\xi)\qquad\text{(a.s.)},$$

where

$$\Phi(z,x)=\mathsf{M}F(z,x_{[0,T-s]}^{z,s,x}).$$

PROOF. Let $\tilde{\mathbf{w}}_t=\mathbf{w}_{t+s}-\mathbf{w}_s$, $\tilde{\mathscr{F}}_t=\mathscr{F}_{t+s}$, $\tilde{y}_t=y_{t+s}$, $\tilde{\sigma}_t^z(x)=\sigma_{t+s}^z(x)$, $\tilde{b}_t^z(x)=b_{t+s}^z(x)$. It is seen that

$$\tilde{y}_t=\xi+\int_0^t\tilde{\sigma}_r^\zeta(\tilde{y}_r)\,d\tilde{\mathbf{w}}_r+\int_0^t\tilde{b}_r^\zeta(\tilde{y}_r)\,dr,$$

in this case ξ, ζ are $\tilde{\mathscr{F}}_0$-measurable, and $\tilde{\mathbf{w}}_t$ is a Wiener process with respect to $\tilde{\mathscr{F}}_t$. By Theorem 4

$$\mathsf{M}\{F(\zeta,y_{[s,T]})\,|\,\mathscr{F}_s\}=\mathsf{M}\{F(\zeta,\tilde{y}_{[0,T-s]})\,|\,\tilde{\mathscr{F}}_0\}=\tilde{\Phi}(\zeta,\xi)\qquad\text{(a.s.)},$$

where $\tilde{\Phi}(z,x) = \mathsf{M}F(z,\tilde{x}^{z,x}_{[0,T-s]})$ and $\tilde{x}^{z,x}_t$ is a solution of the equation

$$x_t = x + \int_0^t \tilde{\sigma}^z_r(x_r)\, d\tilde{\mathbf{w}}_r + \int_0^t \tilde{b}^z_r(x_r)\, dr.$$

It remains to note that, by Corollary 3, the processes $x^{z,s,x}_t$, $\tilde{x}^{z,x}_t$ have identical finite-dimensional distributions. Therefore $\tilde{\Phi}(z,x) = \Phi(z,x)$, thus proving the theorem. □

The technique involving a time shift can be applied in the case where s is a Markov time. The following fact, which we suggest the reader should prove using the above technique, leads to the so-called "strong Markovian" property of solutions of stochastic equations.

8. Exercise

Let $\sigma_t(x) \equiv \sigma(x)$, $b_t(x) \equiv b(x)$ be independent of t and ω, let τ be a Markov time with respect to $\{\mathscr{F}_t\}$, and let x^x_t be a solution (it is given for each t) of the equation

$$dx_t = \sigma(x_t)\, d\mathbf{w}_t + b(x_t)\, dt, \qquad x_0 = x.$$

Prove that in this situation for any $x \in E_d$ and a nonnegative measurable function $F = F(x_{[0,\infty)})$ given on $C([0,\infty),E_d)$,

$$\mathsf{M}_x\{\theta_\tau F \mid \mathscr{F}_\tau\} = \mathsf{M}_{x^x_\tau} F \qquad (\{\tau < \infty\}\text{-a.s.}),$$

where x indicates that in computing the conditional expectation one needs to take $x^x_{[0,\infty)}$ for the argument F, and x^x_τ indicates that first $\mathsf{M}_y F \equiv \mathsf{M}F(x^y_{[0,\infty)})$ is to be found and, second, y is to be replaced by x^x_τ.

9. Remark. The assertions of Theorems 4 and 7 hold not only for nonnegative functions F. This property of F was necessary to make the expressions we dealt with meaningful. For example, Theorem 7 holds for any measurable function F for which $\mathsf{M}|F(\zeta, y_{[s,T]})| < \infty$. In fact, by Theorem 7

$$\mathsf{M}\{F_{\pm}(\zeta, y_{[s,T]}) \mid \mathscr{F}_s\} = \Phi_{(\pm)}(\zeta,\xi) \qquad \text{(a.s.)}, \tag{15}$$

where $\Phi_{(\pm)}(z,x) = \mathsf{M}F_{\pm}(z,x^{z,s,x}_{[0,T-s]})$. In this case the left side of (15) is finite with probability 1 for both the sign $+$ and the sign $-$. In particular, the functions $\Phi_{(+)}(z,x)$, $\Phi_{(-)}(z,x)$ are finite for those (z,x) which are values of $(\zeta(\omega),\xi(\omega))$ on some subset Ω which has complete probability. Having subtracted from (15) with the $+$ sign, the same with the $-$ sign, we find

$$\mathsf{M}\{F(\zeta, y_{[s,T]}) \mid \mathscr{F}_s\} = \Phi(\zeta,\xi) \qquad \text{(a.s.)}, \tag{16}$$

where $\Phi(z,x) = \mathsf{M}F(z,x^{z,s,x}_{[0,T-s]})$; in this case the function $\Phi(z,x)$ exists at any rate for those (z,x) which are necessary for Eq. (16) to be satisfied.

Theorem 7 enables us to deduce the well-known Kolmogorov's equation for the case where $\sigma_t(x)$ and $b_t(x)$ do not depend on ω.

Denote by $x_t^{s,x}$ a solution of the equation

$$x_t = x + \int_0^t \sigma_{s+r}(x_r)\,d\mathbf{w}_r + \int_0^t b_{s+r}(x_r)\,dr, \tag{17}$$

$$(a_t^{ij}(x)) = \frac{1}{2}\sigma_t(x)\sigma_t^*(x),$$

$$L = L(t,x) = \sum_{i,j=1}^d a_t^{ij}(x)\frac{\partial^2}{\partial x^i\,\partial x^j} + \sum_{i=1}^d b_t^i(x)\frac{\partial}{\partial x^i} - c_t(x),$$

$$F(s,x_{[0,T-s]}) = \int_0^{T-s} f_{s+t}(x_t)\exp\left[-\int_0^t c_{s+r}(x_r)\,dr\right]dt$$

$$+ g(x_{T-s})\exp\left[-\int_0^{T-s} c_{s+r}(x_r)\,dr\right],$$

$$v(s,x) = \mathsf{M}F(s,x_{[0,T-s]}^{s,x}).$$

10. Theorem. *Let $c_t(x)$, $f_t(x)$, $g(x)$ be nonrandom real-valued functions, $c_t(x) \geq 0$. Let $\sigma_t(x)$, $b_t(x)$, $c_t(x)$, $f_t(x)$, $g(x)$ be twice differentiable in x, where neither $\sigma_t(x)$ nor $b_t(x)$ depends on ω. Furthermore, let the foregoing functions and their first and second derivatives with respect to x be continuous with respect to (t,x) in a strip $[0,T] \times E_d$. In addition, let the product of the functions $\sigma_t(x)$, $b_t(x)$, $c_t(x)$, $f_t(x)$, $g(x)$ and their first and second derivatives and the function $(1 + |x|)^{-m}$ (functions and their derivatives) be bounded in this strip. Then the function $v(t,x)$ has the following properties*:

1. $|v(t,x)| \leq N(1 + |x|)^m$ *for all $x \in E_d$, $t \in [0,T]$, where N does not depend on (t,x)*;
2. *$v(t,x)$ is once differentiable with respect to t, is twice differentiable with respect to x, and, in addition, the derivatives are continuous in the strip $[0,T] \times E_d$*;
3. *for all $t \in [0,T]$, $x \in E_d$*

$$\frac{\partial}{\partial t}v(t,x) + Lv(t,x) + f_t(x) = 0, \qquad v(T,x) = g(x). \tag{18}$$

Moreover, any function which has properties (1)–(3) *coincides with v in the strip $[0,T] \times E_d$.*

PROOF. By assumption, $\|\sigma_t(0)\|$, $|b_t(0)|$ are continuous. Therefore they are bounded on $[0,T]$ and

$$\|\sigma_t(x)\| + |b_t(x)| \leq \|\sigma_t(0)\| + |b_t(0)| + K|x| \leq N(1 + |x|),$$

where N does not depend on t, x. Furthermore, $F(s,x_{[0,T-s]}^{s,x})$ is a random variable since $F(s,x_{[0,T-s]})$ is a measurable (even continuous) function on

$C([0, T - s], E_d)$. From this and the assumptions $|f_t(x)| \le N(1 + |x|)^m$, $|g(x)| \le N(1 + |x|)^m$ and $c_t(x) \ge 0$ we deduce the first property of the function v if we use estimates of moments of solutions of a stochastic equation (see Corollary 5.12).

Equation (17) makes sense, in general, only for $t \in [0, T - s]$. It will be convenient to assume further that the process $x_t^{s,x}$ is defined for $t \in [0,T]$ for all $s \in (-\infty,\infty)$, $x \in E_d$. As before, we define the process $x_t^{s,x}$ as a solution of Eq. (17), in which, having redefined the functions $\sigma_t(x)$, $b_t(x)$ if necessary, we extend these functions from the interval $[0,T]$ to $(-\infty,\infty)$ defining $\sigma_t(x) = \sigma_T(x)$, $b_t(x) = b_T(x)$ for $t \ge T$ and $\sigma_t(x) = \sigma_0(x)$, $b_t(x) = b_0(x)$ for $t \le 0$. By Theorem 8.6, the process $x_t^{s,x}$ is twice $\mathscr{L}B$-differentiable with respect to x. By virtue of the results obtained in Section 7 (see Lemmas 7.11 and 7.12), the above proves that the random variable $F(s,x_{[0,T-s]}^{s,x})$ is twice $\mathscr{L}$-differentiable with respect to x for each $s \in [0,T]$, and also that the function $v(s,x)$ has all second derivatives with respect to x for each $s \in [0,T]$.

In order to prove that the function $v(s,x)$ is continuous with respect to (s,x), we need only assume in (17) that $p = (s,x)$, $x = \xi_t(p)$, $\sigma_{s+t}(y) = \sigma_t(p,y)$, $b_{s+t}(y) = b_t(p,y)$, write $c_{s+t}(y) = c_t(p,y)$, $f_{s+t}(y) = f_t(p,y)$ in the expression for F, and, in addition, make use of Corollary 8.2 as well as the results from Section 7. Using similar notation, taking the first and second $\mathscr{L}B$-derivatives of $x_t^{s,x}$ with respect to x (see Remark 8.5), and the $\mathscr{L}$-derivatives of $F(x,s_{[0,T-s]}^{s,x})$ and applying Corollary 8.2 as well as the results from Section 7, we prove that the first and second derivatives of $v(s,x)$ with respect to x are continuous with respect to (s,x).

This implies continuity of $Lv(s,x) + f_s(x)$ with respect to (s,x). Hence, if the first relation in (18) has been proved, we have continuity of $(\partial/\partial t)v(t,x)$. It should be mentioned that the second relation in (18) is obvious. Therefore, it remains only to prove that the derivative $(\partial/\partial t)v(t,x)$ exists and the first equality in (18) is satisfied. Furthermore, it suffices to prove this fact not for $(\partial/\partial t)v(t,x)$ but only for the right derivative of the function $v(t,x)$ with respect to t for $t \in [0,T)$. Indeed, as is well known in analysis, if $f(t)$, $g(t)$ are continuous on $[0,T]$ and if the right derivative $f(t)$ is equal to $g(t)$ on $[0,T)$, then $f'(t) = g(t)$ on $[0,T]$. We fix x and take $t_2 > t_1$, $t_1, t_2 \in [0,T]$. Further, let $s = t_2 - t_1$. By Theorem 7 (see Remark 9),

$$\mathsf{M}\{F(t_2,x_{[s,T-t_1]}^{t_1,x})|\mathscr{F}_s\} = \Phi(x_s^{t_1,x}) \qquad \text{(a.s.)}, \tag{19}$$

where $\Phi(y) = \mathsf{M}F(t_2,x_{[0,T-t_2]}^{t_2,y}) = v(t_2,y)$. Furthermore, simple computations show that

$$F(t_1,x_{[0,T-t_1]}) = \int_0^s f_{t_1+t}(x_t)\exp\left[-\int_0^t c_{t_1+r}(x_r)\,dr\right]dt$$

$$+ F(t_2,x_{[s,T-t_1]})\exp\left[-\int_0^s c_{t_1+r}(x_r)\,dr\right].$$

From this and (19) we find

$$v(t_1,x) = \mathsf{M} \int_0^s f_{t_1+t}(x_t^{t_1,x})\Psi_t^{t_1,x}\,dt + \mathsf{M}v(t_2,x_s^{t_1,x})\Psi_s^{t_1,x}, \tag{20}$$

where $\Psi_t^{t_1,x} \equiv \exp[-\int_0^t c_{t_1+r}(x_r^{t_1,x})\,dr]$.

Next, let $w(y)$ be a smooth function with compact support equal to 1 for $|y - x| \le 1$. Also, let $v_1(t_2,y) = v(t_2,y)w(y)$, $v_2(t_2,y) = v(t_2,y) - v_1(t_2,y)$. We represent the second term in (20) as the sum of two expressions starting from the equality $v = v_1 + v_2$. Using Ito's formula, we transform the expression which contains v_1. Note that derivatives of $v_1(t_2,y)$ are continuous and have compact support, and therefore bounded. We have

$$v(t_1,x) = v_1(t_2,x) + \mathsf{M}v_2(t_2,x_s^{t_1,x})\Psi_s^{t_1,x} + \mathsf{M}h_s^{t_1,x}, \tag{21}$$

where

$$h_s^{t_1,x} = \int_0^s \{f_{t_1+t}(x_t^{t_1,x}) + L(t_1 + t, x_t^{t_1,x})v_1(t_2,x_t^{t_1,x})\}\Psi_t^{t_1,x}\,dt.$$

It is seen that $v = v_1$ at a point x. We replace the expression $v_1(t_2,x)$ in (21) by the expression $v(t_2,x)$ and carry the latter into the left-hand side of (21). Further, we divide both sides of the equality by $s = t_2 - t_1$ and, in addition, we let $t_2 \downarrow t_1$. By the mean-value theorem, due to continuity of the expressions considered

$$\frac{1}{s}\,h_s^{t_1,x} \to f_{t_1}(x) + L(t_1,x)v_1(t_1,x) = f_{t_1}(x) + L(t_1,x)v(t_1,x).$$

Moreover, $|(1/s)h_s^{t_1,x}|$ does not exceed the summable quantity

$$N\left(1 + \sup_{t\in[0,T]} |x_t^{t_1,x}|\right)^q \tag{22}$$

for some suitable values of the constant N, q. Finally, $v_2(t_2,y) = 0$ for $|y - x| \le 1$, and, by property 1, $|v_2(t_2,y)| \le N(1 + |y|)^m$. Hence $|v_2(t_2,y)| \le N|y - x|^{m+4}$ and by Corollary 5.12,

$$\left|\frac{1}{s}\,\mathsf{M}v_2(t_2,x_s^{t_1,x})\Psi_s^{t_1,x}\right| \le \frac{N}{s}\,\mathsf{M}\sup_{t\le s}|x_t^{t_1,x} - x|^{m+4} \le Ns^{(m/2)+1} \to 0.$$

The arguments carried out above enable us to derive from (21) that the right derivative of the function $v(t,x)$ exists at a point $t = t_1$, and also prove that the derivative equals $[-f_{t_1}(x) - Lv(t_1,x)]$ for all $t_1 \in [0,T)$. As was explained above, this suffices to complete the demonstration of properties 1–3 for the function v.

We prove the last assertion of the theorem concerning uniqueness of solution of (18). Let $u(t,x)$ be a function having properties 1–3. In accord with Ito's formula for any $R > 0$

$$u(s,x) = \mathsf{M}\left\{u(s + \tau_R, x_{\tau_R}^{s,x})\Psi_{\tau_R}^{s,x} - \int_0^{\tau_R} \Psi_t^{s,x}\left[\frac{\partial}{\partial t} u(s + t, x_t^{s,x}) + L(s + t, x_t^{s,x})u(s + t, x_t^{s,x})\right]dt\right\}$$

$$= \mathsf{M}\left\{u(s + \tau_R, x_{\tau_R}^{s,x})\Psi_{\tau_R}^{s,x} + \int_0^{\tau_R} f_{s+t}(x_t^{s,x})\Psi_t^{s,x}\,dt\right\}, \tag{23}$$

where τ_R equals the minimum of $T - s$ and the first exit time of $x_t^{s,x}$ from S_R. It is seen that $\tau_R \to T - s$ for $R \to \infty$. Moreover, the expression in the curly brackets under the sign of the last mathematical expectation in (23) is continuous with respect to τ_R and, in addition, it does not exceed a summable quantity of the type (22). Therefore, assuming in (23) that $R \to \infty$, using the Lebesgue theorem, we can interchange the sign of the limit and the sign of the expectation. Having done this and, further, having noted that $u(T,x) = g(x)$, we immediately obtain $u(s,x) = v(s,x)$, thus proving the theorem. □

11. Remark. The last assertion of the theorem shows that $v(s,x)$ depends neither on an initial probability space nor on a Wiener process. The function $v(s,x)$ can be defined uniquely by the functions $a_t(x)$, $b_t(x)$, $c_t(x)$, $f_t(x)$, $g(x)$, i.e., by the elements which belong to (18). The function $v(s,x)$ does not change if we replace the probability space, or take another Wiener process, perhaps, even a d_2-dimensional process with $d_2 \neq d_1$, or, finally, take another matrix $\sigma_t(x)$ of dimension $d \times d_2$, provided only that the matrix $\sigma_t(x)\sigma_t^*(x)$ does not change.

10. Ito's Formula with Generalized Derivatives

Ito's formula is an essential tool of stochastic integral theory. The classical formulation of the theorem on Ito's formula involves the requirement that the function to which this formula can be applied be differentiable a sufficient number of times. However, in optimal control theory there arises a necessity to apply Ito's formula to nonsmooth functions (see Section 1.5).

In this section, we prove that in some cases Ito's formula remains valid for functions whose generalized derivatives are ordinary functions. Moreover, we prove some relationships between functions having generalized derivatives and mathematical expectations. These relationships will be useful for our further discussion.

We fix two bounded regions $D \subset E_d$, $Q \subset E_{d+1}$ in spaces E_d and E_{d+1}, respectively. Let d_1 be an integer, $d_1 \geq d$, let $(\mathbf{w}_t, \mathscr{F}_t)$ be a d_1-dimensional Wiener process, let $\sigma_t = \sigma_t(\omega)$ be a matrix of dimension $d \times d_1$, let $b_t = b_t(\omega)$ be a d-dimensional vector, and, finally, let $c_t = c_t(\omega)$ be real-valued. Furthermore, let

$$a_t = \frac{1}{2}\sigma_t\sigma_t^*, \qquad \varphi_t = \int_0^t c_t\,dr,$$

$$L_t = \sum_{i,j=1}^{d} a_t^{ij}\frac{\partial^2}{\partial x^i\,\partial x^j} + \sum_{i=1}^{d} b_t^i\frac{\partial}{\partial x^i} - c_t.$$

Assume that σ_t, b_t, c_t are progressively measurable with respect to $\{\mathscr{F}_t\}$ and, in addition, for all $t \geq 0$

$$\mathsf{M}\int_0^t \|\sigma_r\|^2\,dr < \infty, \qquad \mathsf{M}\int_0^t |b_r|\,dr < \infty, \qquad c_t \geq 0.$$

Under the assumption made above, for each $x_0 \in E_d$ the process

$$x_t = x_0 + \int_0^t \sigma_r\,d\mathbf{w}_r + \int_0^t b_r\,dr$$

is well-defined.

1. Theorem. *Let s, x_0 be fixed, $x_0 \in E_d$, $s \in (-\infty,\infty)$. Also, let τ_Q be the first exit time of the process $(s+t, x_t)$ from a region Q, let τ be some Markov time (with respect to $\{\mathscr{F}_t\}$) such that $\tau \leq \tau_Q$, let τ_D be the first exit time of the process x_t from a region D, and, finally, let τ' be a Markov time not exceeding τ_D. Suppose that there exist constants K, $\delta > 0$ such that $\|\sigma_t(\omega)\| + |b_t(\omega)| + c_t(\omega) \leq K$, $(a_t\lambda,\lambda) \geq \delta|\lambda|^2$ for all $\lambda \in E_d$ and (ω,t), which satisfy the inequality $t < \tau \vee \tau'$.*

Then for any $u \in \bar{W}^2(D)$, $v \in \bar{W}^{1,2}(Q)$, $t \geq 0$

$$e^{-\varphi_{\tau'}}u(x_{\tau'}) - e^{-\varphi_t}u(x_t) = \int_t^{\tau'} e^{-\varphi_r}L_r u(x_r)\,dr + \int_t^{\tau'} e^{-\varphi_r}\operatorname{grad}_x u(x_r)\sigma_r\,d\mathbf{w}_r,$$

$$e^{-\varphi_\tau}v(s+\tau, x_\tau) - e^{-\varphi_t}v(s+t, x_t) = \int_t^{\tau} e^{-\varphi_r}\left(\frac{\partial}{\partial r} + L_r\right)v(s+r, x_r)\,dr + \int_t^{\tau} e^{-\varphi_r}\operatorname{grad}_x v(s+r, x_r)\sigma_r\,d\mathbf{w}_r \quad (1)$$

almost surely on the sets $\{\tau' \geq t\}$, $\{\tau \geq t\}$, respectively. Furthermore, for any $u \in W^2(D)$, $v \in W^{1,2}(Q)$

$$u(x_0) = -\mathsf{M}\int_0^{\tau'} e^{-\varphi_r}L_r u(x_r)\,dr + \mathsf{M}e^{-\varphi_{\tau'}}u(x_{\tau'}),$$

$$v(s,x_0) = -\mathsf{M}\int_0^{\tau} e^{-\varphi_r}\left(\frac{\partial}{\partial r} + L_r\right)v(s+r, x_r)\,dr + \mathsf{M}e^{-\varphi_{\tau}}v(s+\tau,x_\tau).$$

PROOF. We prove both the assertions of Theorem 1 in the same way via approximation of u, v by smooth functions. Hence we prove the first assertion only.

Let a sequence $v^n \in C^{1,2}(\bar{Q})$ be such that

$$\|v - v^n\|_{B(Q)} \to 0, \qquad \|v - v^n\|_{W^{1,2}(Q)} \to 0,$$
$$\| |\operatorname{grad}_x (v - v^n)|^2 \|_{d+1,Q} \to 0.$$

Further, let

$$y_t = x_0 + \int_0^t \chi_{r<\tau} \sigma_r \, d\mathbf{w}_r + \int_0^t \chi_{r<\tau} b_r \, dr.$$

We note that $y_t = x_t$ for $t \le \tau < \infty$, which can easily be seen for $t < \tau$, and which follows from the continuity property of y_t and x_t for $t = \tau < \infty$. We prove that the right side of Eq. (1) makes sense. Obviously, for $r < \tau$

$$\left| \left(\frac{\partial}{\partial r} + L_r \right) v(s + r, x_r) \right| \le N \left[\sum_{i,j=1}^{d} |v_{x^i x^j}(s + r, x_r)| + \left| \frac{\partial}{\partial r} v(s + r, x_r) \right| \right.$$
$$\left. + \sum_{i=1}^{d} |v_{x^i}(s + r, x_r)| + |v(s + r, x_r)| \right],$$

where N depends only on d, K. From this, using Theorem 2.4[11] we obtain

$$\mathsf{M} \int_0^\tau \left| \left(\frac{\partial}{\partial r} + L_r \right) v(s + r, x_r) \right| dr = \mathsf{M} \int_0^\tau \chi_Q(s + r, y_r) \left| \left(\frac{\partial}{\partial r} + L_r \right) v(s + r, y_r) \right| dr$$
$$\le N \|v\|_{W^{1,2}(Q)}. \tag{2}$$

Similarly,

$$\mathsf{M} \left| \int_0^\tau e^{-\varphi_r} \operatorname{grad}_x v(s + r, x_r) \sigma_r \, d\mathbf{w}_r \right|^2 \le N \mathsf{M} \int_0^\tau \chi_Q(s + r, y_r) |\operatorname{grad}_x v(s + r, y_r)|^2 \, dr$$
$$\le N \| |\operatorname{grad}_x v|^2 \|_{d+1,Q}. \tag{3}$$

Further, we apply Ito's formula to the expression $v^n(t, y_t) e^{-\varphi_t}$. Then, we have on the set $\{t \le \tau\}$ almost surely

$$e^{-\varphi_\tau} v^n(s + \tau, x_\tau) - e^{-\varphi_t} v^n(s + t, x_t) = \int_t^\tau e^{-\varphi_r} \left(\frac{\partial}{\partial r} + L_r \right) v^n(s + r, x_r) \, dr$$
$$+ \int_t^\tau e^{-\varphi_r} \operatorname{grad}_x v^n(s + r, x_r) \sigma_r \, d\mathbf{w}_r. \tag{4}$$

We pass to the limit in equality (4) as $n \to \infty$. Using estimates similar to estimates (2) and (3), we easily prove that the right side of (4) tends to the right side of (1).

The first assertion of Theorem 1 can be proved for the function u by an almost word-for-word repetition of the proof given. The slight difference is

[11] In Theorem 2.4, we need take for D any region such that $(-\infty,\infty) \times D \supset Q$.

that if for v^n the existence of the terms in (4) follows from the obvious boundedness of $\tau(\omega)$, then a similar formula for proving the first assertion of the theorem for u is valid since $\tau'(\omega) < \infty$ (a.s.) and even $M\tau' < \infty$ (in Theorem 2.4, assume that $s = 0$, $g \equiv 1$). The theorem is proved. □

Henceforth, when we mention this theorem we shall call the assertions of the theorem Ito's formulas.

The assumption that the process x_t is nondegenerate is the most restrictive assumption of Theorem 1. However, we note that the formulation of the well-known Ito formula imposes no requirement for a process to be nondegenerate when only differentiable functions are being considered. In the next theorem the assumption about nondegeneracy will be dropped, and in Ito's formula instead of an equality an inequality will be proved.

Consider the case where σ_t, b_t, and c_t depend on the parameter $x \in E_d$. We fix $s \in E_1$. Furthermore, for $t \geq s$, $x \in E_d$ let there be given: $\sigma_t(x)$, a random matrix of dimension $d \times d_1$; $b_t(x)$, a random d-dimensional vector; $c_t(x)$ and $f_t(x)$, random variables. Assume that $\sigma_{s+t}(x)$, $b_{s+t}(x)$, $c_{s+t}(x)$, $f_{s+t}(x)$ are progressively measurable with respect to $\{\mathscr{F}_t\}$ for each x, and that $c_t(x)$, $f_t(x)$ are continuous with respect to x and bounded for $(\omega,t,x) \in \Omega \times Q$, where Q, as before, is a bounded region in E_{d+1}. Also, for all $t \geq s$, x and $y \in E_d$ let

$$\|\sigma_t(x) - \sigma_t(y)\| + |b_t(x) - b_t(y)| \leq K|x - y|,$$
$$\|\sigma_t(x)\| + |b_t(x)| \leq K(1 + |x|),$$

where K is a constant.

Under the above assumptions, for each $x \in E$ the solution $x_t^{s,x}$ of the equation

$$x_t = x + \int_0^t \sigma_{s+r}(x_r)\, d\mathbf{w}_r + \int_0^t b_{s+r}(x_r)\, dr$$

exists and is unique (see Theorem 5.7).

We denote by $\tau_Q^{s,x}$ the first exit time of $(s + t, x_t^{s,x})$ from the region Q;

$$a_t(x) = \frac{1}{2}\sigma_t(x)\sigma_t^*(x);$$

$$L_t(x) = \sum_{i,j=1}^{d} a_t^{ij}(x)\frac{\partial^2}{\partial x^i\, \partial x^j} + \sum_{i=1}^{d} b_t^i(x)\frac{\partial}{\partial x^i} - c_t(x);$$

$$\varphi_t^{s,x} = \int_0^t c_{s+r}(x_r^{s,x})\, dr.$$

2. Theorem. *Let* $(s,x) \in Q$ *and, in addition, let a function* $v \in C(\bar{Q})$ *belong to* $W^{1,2}(Q')$ *for each region* Q'*, which together with its closure lies in* Q*. Assume that the derivatives of* v *can be chosen so that for some set* $\Gamma \subset Q$*, for which* meas $(Q\backslash\Gamma) = 0$*, for all* ω *and* $(t,y) \in \Gamma$ *the inequality*

$$-\left[\frac{\partial}{\partial t} + L_t(y)\right]v(t,y) \geq f_t(y) \tag{5}$$

can be satisfied. Then for any Markov time τ (with respect to $\{\mathscr{F}_t\}$) not exceeding $\tau_Q^{s,x}$,

$$v(s,x) \geq \mathsf{M}e^{-\varphi_\tau}v(s+\tau, x_\tau) + \mathsf{M} \int_0^\tau e^{-\varphi_t} f_{s+t}(x_t)\,dt, \tag{6}$$

where $\varphi_t = \varphi_t^{s,x}$, $x_t = x_t^{s,x}$.

PROOF. In proving Theorem 2, we drop the superscripts s, x. First, we note that in proving this theorem we can assume that $\tau \leq \tau_{Q'}$, where $Q' \subset \bar{Q}' \subset Q$. Indeed, for all such Markov times let our theorem have been proved. We take an arbitrary time $\tau \leq \tau_Q$. It is seen that $\tau_{Q'} \uparrow \tau_Q$ and $\tau \wedge \tau_{Q'} \uparrow \tau$ when the regions Q', while expanding, converge to Q. Substituting in (6) the variable $\tau \wedge \tau_{Q'}$ for τ, taking the limit as $Q' \uparrow Q$, and, finally, noting that v is continuous in $\bar{Q}$, φ_t and x_t are continuous with respect to t, and, in addition, τ and $f_{s+t}(x_t)$ for $t \leq \tau$ are bounded, we have proved the assertion of the theorem in the general case.

Thus, let $\tau \leq \tau_{Q'}$. Further, we apply a rather well-known method of perturbation of an initial stochastic equation (see Exercise 1.1.1). We consider some d-dimensional Wiener process $\tilde{\mathbf{w}}_t$ independent of $\{\mathscr{F}_t\}$. Formally, this can be done by considering a direct product of two probability spaces: an initial space and a space on which a d-dimensional Wiener process is defined.

We denote by x_t^n a solution of the equation

$$x_t^n = x + \int_0^t \sigma_{s+r}(x_r^n)\,d\mathbf{w}_r + \varepsilon_n \tilde{w}_t + \int_0^t b_{s+r}(x_r^n)\,dr,$$

where $\varepsilon_n \neq 0$, $\varepsilon_n \to 0$ as $n \to \infty$.

It is convenient to rewrite the last equation in a different form. Let $\sigma_t^n(x)$ be a matrix of dimension $d \times (d_1 + d)$, such that the first d_1 columns of the matrix $\sigma_t^n(x)$ form a matrix $\sigma_t(x)$, and also the columns numbered $d_1 + 1, \ldots, d_1 + d$ form a matrix $\varepsilon_n I$, where I is a unit matrix of dimension $d \times d$. Furthermore, we take a $(d_1 + d)$-dimensional Wiener process $\bar{\mathbf{w}}_t = (\mathbf{w}_t^1, \ldots, \mathbf{w}_t^{d_1}, \tilde{\mathbf{w}}_t^1, \ldots, \tilde{\mathbf{w}}_t^d)$. Then

$$x_t^n = x + \int_0^t \sigma_{s+r}^n(x_r^n)\,d\bar{\mathbf{w}}_r + \int_0^t b_{s+r}(x_r^n)\,dr. \tag{7}$$

By Theorem 8.1, $\sup_{r \leq t} |x_r^n - x_r| \to 0$ as $n \to \infty$ in probability for each t. Therefore, there exists a subsequence $\{n_i\}$ such that $\sup_{r \leq t} |x_r^{n_i} - x_r| \to 0$ (a.s.) as $i \to \infty$ and for each t. In order not to complicate the notation, we assume that $\{n_i\} = \{n\}$.

Let $\tau_{Q'}^n$ be the first exit time of $(s + t, x_t^n)$ from Q'. It is not hard to show that $\underline{\lim}_{n \to \infty} \tau_{Q'}^n \geq \tau_{Q'}$ (a.s.). Hence, if we assume that

$$\tau^i = \tau \wedge \inf_{n \geq i} \tau_{Q'}^n,$$

then $\tau^i \leq \tau_{Q'}$ and $\tau^i \to \tau$ as $i \to \infty$ (a.s.).

Further, we apply Theorem 1 to v, Q', x_t^n, τ^i for $n \geq i$. Note that $\tau^i \leq \tau_Q^n$ for $n \geq i$. Moreover, $v \in W^{1,2}(Q')$. Next, it is seen that

$$\chi_{t < \tau_{Q'}^n}(|\sigma_{s+t}^n(x_t^n)| + |b_{s+t}(x_t^n)| + |c_{s+t}(x_t^n)|) \leq N,$$

where N does not depend on t, ω, n. Finally,

$$a_t^n \equiv \frac{1}{2}\sigma_t^n(x_t^n)[\sigma_t^n(x_t^n)]^* = a_t(x_t^n) + \frac{\varepsilon_n^2}{2} I,$$

$$(a_t^n\lambda,\lambda) \geq \frac{\varepsilon_n^2}{2}|\lambda|^2.$$

All the assumptions of Theorem 1 have been satisfied. Therefore, computing for the process x_r^n (see (7)) the operator L_r appearing in Theorem 1, and in addition, assuming that

$$\varphi_t^n = \int_0^t c_{s+r}(x_r^n)\,dr,$$

$$g_t^n(x) = \left[\frac{\partial}{\partial t} + L_t(x)\right]v(t,x) + \frac{\varepsilon_n^2}{2}\Delta v(t,x),$$

for $n \geq i$, we have

$$v(s,x) = -\mathsf{M}\int_0^{\tau^i} e^{-\varphi_r^n} g_{s+r}^n(x_r^n)\,dr + \mathsf{M}e^{-\varphi_{\tau^i}^n} v(s+\tau^i, x_{\tau^i}^n). \tag{8}$$

By the hypothesis of the theorem,

$$-\chi_\Gamma(t,x)g_t^n(x) \geq \chi_\Gamma(t,x)f_t(x) - \chi_\Gamma(t,x)\frac{\varepsilon_n^2}{2}\Delta v(t,x).$$

Furthermore, by Theorem 2.4,

$$\mathsf{M}\int_0^{\tau^i} \chi_{Q\setminus\Gamma}(s+r, x_r^n)\,dr \leq N\|\chi_{Q\setminus\Gamma}\|_{d+1,H_\infty} = 0.$$

Therefore, in integrating over r in the first expression in the right side of (8), we can assume that $(s+r, x_r^n) \in \Gamma$. From (8) we find

$$v(s,x) \geq \mathsf{M}\int_0^{\tau^i} e^{-\varphi_r^n} f_{s+r}(x_r^n)\,dr + \mathsf{M}e^{-\varphi_{\tau^i}^n} v(s+\tau^i, x_{\tau^i}^n)$$
$$-\frac{\varepsilon_n^2}{2}\mathsf{M}\int_0^{\tau^i} e^{-\varphi_r^n}\Delta v(s+r, x_r^n)\,dr.$$

Because τ^i does not exceed the diameter T of the region Q', $\sup_{r\leq T} |x_r^n - x_r| \to 0$ as $n \to \infty$, $f_{s+r}(y)$ and $c_{s+r}(y)$ are continuous with respect to y, and $\tau^i \uparrow \tau$ as $i \to \infty$, we conclude that in the last expression for $v(s,x)$ the first two terms in the right side as $n \to \infty$, then as $i \to \infty$, yield the right side of Eq. (6).

Therefore, for proving the theorem it remains only to show that

$$\lim_{n\to\infty} \varepsilon_n^2 \mathsf{M}\int_0^{\tau^n} |\Delta v(s+r, x_r^n)|\,dr = 0.$$

Making use of Theorem 2.2, we assume $s = 0$, $c_t \equiv 1$, $F(c,a) = c$, $b_t = b_{s+t}(x_t^n)$, $r_t \equiv 1$, $p = d$, $\sigma_t = \sigma_{s+t}^n(x_t^n)$. Note that, as was noted before, $|b_t| \leq$

$N \cdot 1 = Nc_t$ for $t < \tau^n_{Q'}$, where N does not depend on n, and, moreover,

$$(\det a_t)^{1/(d+1)} = (\det a^n_{s+t}(x^n_t))^{1/(d+1)} \geq \left(\frac{\varepsilon_n^2}{2}\right)^{d/(d+1)}.$$

Therefore

$$\begin{aligned}
\varepsilon_n^2 \mathsf{M} \int_0^{\tau^n} |\Delta v(s+r, x^n_t)|\, dr &\leq \varepsilon_n^2 e^T \mathsf{M} \int_0^{\tau^n} e^{-r} |\Delta v(s+r, x^n_r)|\, dr \\
&\leq \varepsilon_n^2 e^T \mathsf{M} \int_0^{\tau^n_{Q'}} e^{-r} |\Delta v(s+r, x^n_r)|\, dr \\
&\leq 2^{d/(d+1)} \varepsilon_n^{2/(d+1)} e^T \mathsf{M} \int_0^{\tau^n_{Q'}} e^{-r} (\det a^n_{s+r}(x^n_r))^{1/(d+1)} \\
&\quad \times (\chi_{Q'} |\Delta v|)(s+r, x^n_r)\, dr \\
&\leq 2^{d/(d+1)} \varepsilon_n^{2/(d+1)} e^T N \| \chi_{Q'} |\Delta v| \|_{d+1, H_\infty},
\end{aligned}$$

where N does not depend on n. The last expression tends to zero as $n \to \infty$, since $v \in W^{1,2}(Q')$. Therefore, the norm of that expression is finite. The theorem is proved. □

3. Remark. It is seen from the proof that if for all (t,ω) the function $f_t(x)$ is upper semicontinuous, $\underline{\lim}_{x_n \to x} f_t(x_n) \geq f_t(x)$, the assertion of the theorem still holds.

4. Corollary. *If $\sigma_t(x)$, $b_t(x)$, $c_t(x)$ do not depend on ω and in addition, $L_t(x)v(t,x) + \partial v(t,x)/\partial t$ is a bounded continuous function of $(t,x) \in Q$, we have in the notation of the theorem*

$$v(s,x) = \mathsf{M} e^{-\varphi_\tau} v(s+\tau, x_\tau) - \mathsf{M} \int_0^\tau e^{-\varphi_r} \left[L_{s+r}(x_r) v(s+r, x_r) + \frac{\partial v}{\partial \tau}(s+r, x_r) \right] dr.$$

5. Exercise to Theorem 1

(Compare [44, p. 39].) Let $d \geq 2$, $\alpha \in (0,1)$, $\mu = [(d-1)/(1-\alpha)] - 1$, $u(x) = |x|^\alpha$, $\sigma(x) = \sqrt{2a(x)}$, where $a^{ij}(x) = \delta^{ij} + \mu(x^i x^j / |x|^2)$. We take as D a sphere S_R, and also, we take as x_t some (possibly "weak") solution of the equation $dx_t = \sigma(x_t)\, d\mathbf{w}_t$, $x_0 = 0$. Let $\sigma_t = \sigma(x_t)$, $b_t = 0$, $c_t = 0$.

Show that second derivatives of u are summable with respect to D to the power $p = \alpha d/(2-\alpha)$. (Note that $p \to d$ as $\alpha \to 1$.) Also, show that $Lu(x_t) = 0$ (a.s.) and that Ito's formula is not applicable to $u(x_t)$.

6. Remark. In the case where $Q = (0,T) \times S_R$, we have $\tau^{s,x} = 0$ for $s = 0$ in the notations introduced before Theorem 2. This suggests that it would be useful to have in mind that if $Q = (0,T) \times S_R$, one can take in Theorem 2 instead of $\tau^{s,x}$ (in Theorem 1 instead of τ_Q) the minimum between $T - s$ and the first exit time of the process $x^{s,x}_t$ (respectively, the first exit time of the process x_t) from S_R. For $s = 0$ this minimum is not in general equal to zero. Thus we can derive meaningful assertions from Theorems 1 and 2.

In order to prove the validity of the remark made above, it suffices to repeat word-for-word the proof of Theorems 1 and 2.

Notes

Section 1. The notations and definitions given in this section are of common usage. Definition 2 as well as the concept of an exterior norm are somewhat special.

Sections 2, 3, 4. The results obtained in these sections generalize the corresponding results obtained in [32, 34, 36, 40]. Estimates of stochastic integrals having a jumplike part can be found in Pragarauskas [62].

Sections 5, 7, 8, 9. These sections contain more or less well-known results related to the theory of Ito's stochastic integral equations; see Dynkin [11], Liptser and Shiryayev [51], and Gikhman and Skorokhod [24]. The introduction of the spaces $\mathscr{L}$, $\mathscr{L}B$ is our idea.

Section 6. The existence of a solution of a stochastic equation containing measurable coefficients not depending on time was first proved in [28] by the method due to Skorokhod [70]. In this section we use Skorokhod's method in the case when the coefficients may depend on time. For the problem of uniqueness of a solution of a stochastic equation as well as the problem of constructing the corresponding Markov process, see [24, 28, 38]; also see S. Anoulova and G. Pragarauskas: On weak Markov solutions of stochastic equations, *Litovsky Math. Sb.* 17(2) (1977), 5–26, also see the references listed in this paper.

Section 10. The results obtained in this section are related to those in [28, 34].

General Properties of a Payoff Function 3

In this chapter we study general properties of a payoff function—the properties that a payoff function possesses under minimal assumptions on the system. Our attention is focused on proving the continuity property of a payoff function as well as proving different versions of Bellman's principle, and, furthermore, proving the fact that strategies which are close to optimal ones can be found among natural strategies. In Chapter 5 we shall discuss the feasibility of further reduction of a set of strategies to Markov strategies without decreasing the payoff. Considering the problem of optimal stopping of a controlled process, we describe in this chapter a subclass of stopping rules, which yield the same payoff performance as the class of all possible stopping rules.

1. Basic Results

Let A be a separable metric space (a set of admissible controls), let E_d be a Euclidean space of dimension d, and let T be a nonnegative number. We consider a controlled process in the space E_d in an interval of time $[0,T]$. Taking an integer d_1, we assume that $(\mathbf{w}_t,\mathscr{F}_t)$ is a d_1-dimensional Wiener process.

Suppose that for all $\alpha \in A$, $t \geq 0$, $x \in E_d$ we are given $\sigma(\alpha,t,x)$, which is a matrix of dimension $d \times d_1$, $b(\alpha,t,x)$, which is a d-dimensional vector, and, in addition, we are given real-valued functions $c^\alpha(t,x) \geq 0$, $f^\alpha(t,x)$, $g(x)$. As in Chapter 1, σ characterizes here the diffusion component of the process, b characterizes a deterministic component of the process, $f^\alpha(t,x)\,\Delta t$ plays the role of the payoff during the interval of time from t to $t + \Delta t$, if the controlled process is near a point x at time t and if, in addition, a control α is used, and $g(x)$ is the gain at time T. The function $c^\alpha(t,x)$ is the measure of "discounting." This function is introduced, first, for greater generality, second, because we consider the problems of optimal stopping of a controlled process and investigate them, as we did in Section 1.2 and 1.5, using the method of randomized stopping.

We assume that the functions σ, b, c, f, g are continuous with respect to (α,x) and continuous with respect to x uniformly over α for each t. Also, we assume that the above functions are Borel with respect to (α,t,x). Furthermore, for some constants m, $K \geq 0$ for all x, $y \in E_d$, $t \geq 0$, $\alpha \in A$, let

$$\|\sigma(\alpha,t,x) - \sigma(\alpha,t,y)\| + |b(\alpha,t,x) - b(\alpha,t,y)| \leq K|x - y|, \tag{1}$$

$$\|\sigma(\alpha,t,x)\| + |b(\alpha,t,x)| \leq K(1 + |x|), \tag{2}$$

$$|c^\alpha(t,x)| + |f^\alpha(t,x)| + |g(x)| \leq K(1 + |x|)^m. \tag{3}$$

We shall consider below a function $g(t,x)$ as well, the assumptions about which are formulated before the proof of Theorem 8.

As was done in Section 1.4, we introduce here the concepts of a strategy, a natural strategy, and Markov strategy.

1. Definition. By a *strategy* we mean a process $\alpha_t(\omega)$ progressively measurable with respect to a system of σ-algebras of $\{\mathscr{F}_t\}$, having values in A. We denote by $\mathfrak{A}$ the set of all strategies.

To each strategy $\alpha \in \mathfrak{A}$, $s \in [0,T]$, $x \in E_d$ we set into correspondence a solution $x_t^{\alpha,s,x}$ of the equation

$$x_t = x + \int_0^t \sigma(\alpha_r, s + r, x_r)\, d\mathbf{w}_r + \int_0^t b(\alpha_r, s + r, x_r)\, dr, \qquad t \geq 0.$$

Note that due to the assumptions about σ and b, the solution of the last equation exists and is unique. It is always convenient to represent the process $x_t^{\alpha,s,x}$ as a set of the last d coordinates of a $(d + 1)$-dimensional process $z_t^{\alpha,s,x} = (y_t^{\alpha,s,x}, x_t^{\alpha,s,x})$ which is a solution of the following system of equations:

$$y_t = s + \int_0^t 1\, dr;$$

$$x_t = x + \int_0^t \sigma(\alpha_r, y_r, x_r)\, d\mathbf{w}_r + \int_0^t b(\alpha_r, y_r, x_r)\, dr.$$

In this case s appears as one of the components of the given initial process. Furthermore, if we consider the process $x_t^{\alpha,s,x}$ on an interval of time $[0, T - s]$, this implies that the process $z_t^{\alpha,s,x}$ is considered before the first exit from a strip $[0,T) \times E_d$.

For $s \le T$ let

$$\varphi_t^{\alpha,s,x} = \int_0^t c^{\alpha_r}(s + r, x_r^{\alpha,s,x})\,dr,$$

$$v^\alpha(s,x) = \mathsf{M}\left[\int_0^{T-s} f^{\alpha_t}(s + t, x_t^{\alpha,s,x})e^{-\varphi_t^{\alpha,s,x}}\,dt + g(x_{T-s}^{\alpha,s,x})e^{-\varphi_{T-s}^{\alpha,s,x}}\right],$$

$$v(s,x) = \sup_{\alpha \in \mathfrak{A}} v^\alpha(s,x).$$[1]

For convenience we shall use the superscripts α and the subscripts s, x on the expectation sign to indicate expectations of quantities which depend on s, x, and a strategy α. For example,

$$\mathsf{M}_{s,x}^\alpha \int_0^{T-s} f^{\alpha_t}(s + t, x_t)e^{-\varphi_t}\,dt \equiv \mathsf{M} \int_0^{T-s} f^{\alpha_t}(s + t, x_t^{\alpha,s,x})e^{-\varphi_t^{\alpha,s,x}}\,dt,$$

$$\mathsf{M}_{s,x}^\alpha g(x_{T-s})e^{-\varphi_{T-s}} \equiv \mathsf{M}g(x_{T-s}^{\alpha,s,x})e^{-\varphi_{T-s}^{\alpha,s,x}}.$$

We treat the probabilities of events in a similar way. For example, $\mathsf{P}_{s,x}^\alpha\{|x_t| \ge R\} = \mathsf{P}\{|x_t^{\alpha,s,x}| \ge R\}$.

In the above notation

$$v(s,x) = \sup_{\alpha \in \mathfrak{A}} \mathsf{M}_{s,x}^\alpha\left[\int_0^{T-s} f^{\alpha_t}(s + t, x_t)e^{-\varphi_t}\,dt + g(x_{T-s})e^{-\varphi_{T-s}}\right].$$

Let $C([0,\infty),E_d)$ be a space of all continuous functions x_t with values in E_d defined on $[0,\infty)$. Also, let $\mathcal{N}_t$ be the smallest σ-algebra of the subsets $C([0,\infty),E_d)$ which contains all sets of the form $\{x_{[0,\infty)}: x_r \in \Gamma\}$ for $r \in [0,t]$ and Borel $\Gamma \subset E_d$.

2. Definition. *A function* $\alpha_t(x_{[0,\infty)}) = \alpha_t(x_{[0,t]})$ with values in A given for $t \in [0,\infty)$, $x_{[0,\infty)} \in C([0,\infty),E_d)$ is said to be *a natural strategy admissible at a point* (s,x) if this function is progressively measurable with respect to $\{\mathcal{N}_t\}$ and, if, in addition, there exists at least one solution of the stochastic equation

$$x_t = x + \int_0^t \sigma(\alpha_r(x_{[0,r]}), s + r, x_r)\,d\mathbf{w}_r + \int_0^t b(\alpha_r(x_{[0,r]}), s + r, x_r)\,dr, \tag{4}$$

which is progressively measurable with respect to $\mathcal{F}_t$. We denote by $\mathfrak{A}_E(s,x)$ the set of all natural strategies admissible at the point (s,x).

To each strategy $\alpha \in \mathfrak{A}_E(s,x)$ we set into correspondence one (fixed) solution $x_t^{\alpha,s,x}$ of Eq. (4).

3. Definition. The natural strategy $\alpha_t(x_{[0,t]})$ is said to be a (nonstationary) *Markov* strategy if $\alpha_t(x_{[0,t]}) = \alpha_t(x_t)$ for some Borel function $\alpha_t(x)$. We denote by $\mathfrak{A}_M(s,x)$ the set of all Markov strategies admissible at the point (s,x).

[1] The finiteness of the last two expressions will be proved after the statement of Theorem 7 (see below).

As was done in Section 1.4, we establish here the natural embedding of $\mathfrak{A}_E(s,x)$ in $\mathfrak{A}$. One essential characteristic of this embedding should be noted. To the natural strategy $\alpha \in \mathfrak{A}_E(s,x)$ we set into correspondence a strategy $\beta \in \mathfrak{A}$ according to the formula $\beta_t(\omega) = \alpha_t(x_{[0,t]}^{\alpha,s,x}(\omega))$. In this case $\beta_t(\omega)$ *depends* on (s,x). Furthermore, if this same strategy α belongs to $\mathfrak{A}_E(s',x')$ as well for $(s',x') \neq (s,x)$, the strategy $\beta'_t(\omega) = \alpha_t(x_{[0,t]}^{\alpha,s',x'}(\omega))$ does not, generally speaking, coincide with β. Therefore, the operation of embedding of $\mathfrak{A}_E(s,x)$ in $\mathfrak{A}$ depends on (s,x).

Further, we note that $\mathfrak{A}_M(s,x) \subset \mathfrak{A}_E(s,x)$ and $\mathfrak{A}_M(s,x) \neq \varnothing$ ($\mathfrak{A}_M(s,x)$ contains strategies of the form $\alpha_t(x_{[0,T]}) \equiv \alpha$, where α is a fixed element of A). For $s \leq T$ let

$$v_{(E)}(s,x) = \sup_{\alpha \in \mathfrak{A}_E(s,x)} v^\alpha(s,x), \qquad v_{(M)}(s,x) = \sup_{\alpha \in \mathfrak{A}_M(s,x)} v^\alpha(s,x).$$

It is seen that $v_{(M)} \leq v_{(E)} \leq v$.

4. Definition. Let $\varepsilon \geq 0$. A strategy $\alpha \in \mathfrak{A}$ is said to be ε-optimal for a point (s,x) if $v(s,x) \leq v^\alpha(s,x) + \varepsilon$. A 0-optimal strategy is said to be optimal.

We formulate those results related to v, Bellman's principle, and ε-optimal strategies, which we shall prove in the subsequent sections of Chapter 3.

5. Theorem. *The function $v(s,x)$ is continuous with respect to (s,x) on $[0,T] \times E_d$, $v(T,x) = g(x)$. There exists a constant $N = N(m,K,T)$ such that for all $s \in [0,T]$, $x \in E_d$*

$$|v(s,x)| \leq N(1 + |x|)^m. \tag{5}$$

6. Theorem. *Suppose that $s \in [0,T]$, $x \in E_d$. Furthermore, for each $\alpha \in \mathfrak{A}$ we are given a time $\tau^\alpha \leq T - s$ which is Markov with respect to $\{\mathcal{F}_t\}$ and, in addition, we are given a nonnegative process r_t^α which is progressively measurable with respect to $\{\mathcal{F}_t\}$ and bounded with respect to (t,ω). Then*

$$v(s,x) = \sup_{\alpha \in \mathfrak{A}} \mathsf{M}_{s,x}^\alpha \left\{ \int_0^\tau [f^{\alpha_t}(s + t, x_t) + r_t v(s + t, x_t)] \exp\left(-\varphi_t - \int_0^t r_u \, du \right) dt \right.$$
$$\left. + \, v(s + \tau, x_\tau) \exp\left(-\varphi_\tau - \int_0^\tau r_u \, du \right) \right\}. \tag{6}$$

7. Theorem. $v(s,x) = v_{(E)}(s,x)$. *In the previous theorem one may take the upper bound over $\alpha \in \mathfrak{A}_E(s,x)$.*

Let us discuss the assertions of Theorems 5, 6, and 7. In Theorem 5 the equality $v(T,x) = g(x)$ is obvious. Inequality (5) follows from the fact that from Corollary 2.5.12

$$\mathsf{M}_{s,x}^\alpha \sup_{t \leq T-s} |x_t|^m \leq N(m,K,T)(1 + |x|)^m, \tag{7}$$

and also from the fact that

$$|v(s,x)| \le \sup_{\alpha \in \mathfrak{A}} \mathsf{M}_{s,x}^{\alpha} \left\{ \int_0^{T-s} |f^{\alpha_t}(s+t, x_t)| \, dt + |g(x_{T-s})| \right\}$$

$$\le K(T - s + 1) \sup_{\alpha \in \mathfrak{A}} \mathsf{M}_{s,x}^{\alpha} \sup_{t \le T-s} (1 + |x_t|)^m.$$

Thus, the main assertion of Theorem 5 is the one about the continuity of $v(s,x)$. We shall see that the continuity of $v(s,x)$ with respect to x follows from the continuity of σ, b, c, f, g with respect to x, and, finally, the continuity of $v(s,x)$ with respect to s follows from the specific nature of the problem. In this connection, we recall that σ, b, c, f are only measurable with respect to t.

Theorem 6 for $r_t^\alpha \equiv 0$ is a usual Bellman's principle. In a certain sense the assertion of Theorem 6 is Bellman's principle in the general case as well. We explain this for $r_t^\alpha = \lambda$, using a method close in essence to the method of randomized stopping described in Section 1.2. We introduce a random variable ξ which is exponentially distributed with parameter λ and is independent of $\{\mathscr{F}_t\}$. Bellman's principle implies the equality

$$v(s,x) = \sup_{\alpha \in \mathfrak{A}} \mathsf{M}_{s,x}^{\alpha} \left[\int_0^{\tau \wedge \xi} f^{\alpha_t}(s+t, x_t) e^{-\varphi_t} \, dt + v(s + \tau \wedge \xi, x_{\tau \wedge \xi}) e^{-\varphi_{\tau \wedge \xi}} \right]$$

which can easily be transformed into (6) using Fubini's theorem and writing the mathematical expectation given above as follows:

$$\mathsf{M}_{s,x}^{\alpha} \int_0^{\infty} \lambda e^{-\lambda p} \left[\int_0^{\tau \wedge p} f^{\alpha_t}(s+t, x_t) e^{-\varphi_t} \, dt + v(s + \tau \wedge p, x_{\tau \wedge p}) e^{-\varphi_{\tau \wedge p}} \right] dp.$$

Theorem 7 shows that one should seek ε-optimal strategies among natural strategies. We shall see from the proof of Theorem 7 (see Remark 3.4) that one can take ε-optimal natural strategies of a very specific type.

We can prove some general results for the optimal stopping problem as well. Let $g(t,x)$ be a continuous function of (t,x) $(x \in E_d, t \ge 0)$ such that $|g(t,x)| \le K(1 + |x|)^m$ for all t, x. For $s \in [0,T]$ we denote by $\mathfrak{M}(T-s)$ the set of all Markov times (with respect to $\{\mathscr{F}_t\}$) not exceeding $T - s$. For $\alpha \in \mathfrak{A}$, $\tau \in \mathfrak{M}(T-s)$ let

$$v^{\alpha,\tau}(s,x) = \mathsf{M}_{s,x}^{\alpha} \left[\int_0^{\tau} f^{\alpha_t}(s+t, x_t) e^{-\varphi_t} \, dt + g(s+\tau, x_\tau) e^{-\varphi_\tau} \right],$$

$$w(s,x) = \sup_{\alpha \in \mathfrak{A}} \sup_{\tau \in \mathfrak{M}(T-s)} v^{\alpha,\tau}(s,x).$$

Using a similar formula, we introduce $w_{(E)}$ $(w_{(M)})$, replacing the upper bound over $\alpha \in \mathfrak{A}$ by the upper bound over $\mathfrak{A}_E(s,x)$ (over $\mathfrak{A}_M(s,x)$).

8. Theorem. *The function $w(s,x)$ is continuous with respect to (s,x) on $[0,T] \times E_d$, $w(s,x) \ge g(s,x)$, $w(T,x) = g(T,x)$. There exists a constant $N = N(m,K,T)$*

such that for all $s \in [0,T]$, $x \in E_d$

$$|w(s,x)| \le N(1 + |x|)^m. \tag{8}$$

9. Theorem. *Let* $s \in [0,T]$, $x \in E_d$, *and furthermore, let a Markov time* $\tau^\alpha \in \mathfrak{M}(T - s)$ *be defined for each* $\alpha \in \mathfrak{A}$. *Then*

$$w(s,x) = \sup_{\alpha \in \mathfrak{A}} \sup_{\gamma \in \mathfrak{M}(T-s)} \mathsf{M}_{s,x}^\alpha \left\{ \int_0^{\tau \wedge \gamma} f^{\alpha_t}(s + t, x_t) e^{-\varphi_t}\, dt \right.$$

$$\left. + g(s + \gamma, x_\gamma) e^{-\varphi_\gamma} \chi_{\gamma \le \tau} + w(s + \tau, x_\tau) e^{-\varphi_\tau} \chi_{\tau < \gamma} \right\}; \tag{9}$$

in this situation the upper bound over $\mathfrak{A}$ *can be replaced by the upper bound over* $\mathfrak{A}_E(s,x)$.

Let $\varepsilon > 0$ and let

$$\tau_e^{\alpha,s,x} = \inf\{t \ge 0 : w(s + t, x_t^{\alpha,s,x}) \le g(s + t, x_t^{\alpha,s,x}) + \varepsilon\}.$$

10. Theorem. $w(s,x) = w_{(E)}(s,x)$. *Furthermore, for* $\varepsilon > 0$, $s \in [0,T]$, $x \in E_d$ *the inequality*

$$w(s,x) \le \sup_{\alpha \in \mathfrak{A}_E(s,x)} \mathsf{M}_{s,x}^\alpha \left\{ \int_0^{\tau_\varepsilon} f^{\alpha_t}(s + t, x_t) e^{-\varphi_t}\, dt + g(s + \tau_\varepsilon, x_{\tau_\varepsilon}) e^{-\varphi_{\tau_\varepsilon}} \right\} + \varepsilon$$

holds. If A consists of a single point, the last inequality becomes an equality for $\varepsilon = 0$.

11. Theorem. *In the notation of Theorem 6*

$$w(s,x) \ge \sup_{\alpha \in \mathfrak{A}} \mathsf{M}_{s,x}^\alpha \left\{ \int_0^\tau [f^{\alpha_t}(s + t, x_t) + r_t w(s + t, x_t)] \right.$$

$$\left. \times \exp\left(-\varphi_t - \int_0^t r_p\, dp \right) dt + w(s + \tau, x_\tau) \exp\left(-\varphi_\tau - \int_0^\tau r_p\, dp \right) \right\}. \tag{10}$$

If $\tau^\alpha \le \tau_\varepsilon^{\alpha,s,x}$ *for some* $\varepsilon > 0$ *and all* $\alpha \in \mathfrak{A}$, *we have equality in* (10). *In any case, the upper bound over* $\alpha \in \mathfrak{A}$ *can be replaced by the upper bound over* $\alpha \in \mathfrak{A}_E(s,x)$. *Finally, if A consists of a single point,* $\tau^\alpha \le \tau_0^{\alpha,s,x}$, *then we have equality in* (10).

As in Theorem 5, the first assertion is the strongest one in Theorem 8. In fact, the equality $w(T,x) = g(T, x)$ is obvious and the inequality $w(s,x) \ge g(s,x)$ follows immediately from the definition of $w(s,x)$ and the fact that $\tau \equiv 0$ is a Markov time. Moreover, (8) can be proved with the aid of (7) in the same way as (5) was proved.

Theorem 9 is Bellman's principle for the optimal stopping problem of a controlled process. Further, we note that $\tau_\varepsilon^{\alpha,s,x}$ is the time of first exit of the process $(s + t, x_t^{\alpha,s,x})$ from an open (in relative topology of $[0,T] \times E_d$) set $Q_\varepsilon = \{(s,x): w(s,x) > g(s,x) + \varepsilon\}$. Since $w(T,x) = g(T,x)$, $\tau_\varepsilon^{\alpha,s,x} \le T - s$. It is seen that $\tau_\varepsilon^{\alpha,s,x} \in \mathfrak{M}(T - s)$. Theorem 10 shows that $\tau_\varepsilon^{\alpha,s,x}$ is an ε-optimal stopping time of the controlled process. If we deal with the stopping of a diffusion process, Theorem 10 asserts optimality of a time $\tau_0^{s,x}$. In regard to this, we note that Theorems 5–11 can be used in investigating solutions of stochastic equations in the case when there is no control.

Theorem 11 includes one more formulation of Bellman's principle, which is more convenient than Theorem 9 for the deduction of differential equations. Theorem 11 is the central theorem in the precise sense. Note that Theorem 10 follows immediately from Theorem 11 if in Theorem 11 we take $r_t^\alpha \equiv 0$, $\tau^\alpha = \tau_\varepsilon^{\alpha,s,x}$, and take advantage of the fact that

$$w(s + \tau^\alpha, x_{\tau^\alpha}^{\alpha,s,x}) \le g(s + \tau^\alpha, x_{\tau^\alpha}^{\alpha,s,x}) + \varepsilon.^2$$

Also, we show how Theorem 9 can be deduced from Theorem 11. To this end, we write the right side of (9) as $w_1(s,x)$. It follows from the inequality

$$w_1(s,x) \ge \sup_{\alpha \in \mathfrak{A}} \mathsf{M}_{s,x}^\alpha \left\{ \int_0^{\tau \wedge \tau_\varepsilon} f^{\alpha_t}(s + t, x_t) e^{-\varphi_t}\, dt \right.$$
$$\left. + g(s + \tau_\varepsilon, x_{\tau_\varepsilon}) e^{-\varphi_{\tau_\varepsilon}} \chi_{\tau_\varepsilon \le \tau} + w(s + \tau, x_\tau) e^{-\varphi_\tau} \chi_{\tau \le \tau_\varepsilon} \right\}$$

and the inequality

$$g(s + \tau_\varepsilon, x_{\tau_\varepsilon}) \ge w(s + \tau_\varepsilon, x_{\tau_\varepsilon}) - \varepsilon$$

that

$$w_1(s,x) \ge \sup_{\alpha \in \mathfrak{A}} \mathsf{M}_{s,x}^\alpha \left\{ \int_0^{\tau \wedge \tau_\varepsilon} f^{\alpha_t}(s + t, x_t) e^{-\varphi_t}\, dt \right.$$
$$\left. + w(s + \tau \wedge \tau_\varepsilon, x_{\tau \wedge \tau_\varepsilon}) e^{-\varphi_{\tau \wedge \tau_\varepsilon}} \right\} - \varepsilon.$$

Since $\tau \wedge \tau_\varepsilon \le \tau_\varepsilon$, by Theorem 11 for $r_t^\alpha \equiv 0$ we have that the last upper bound is equal to $w(s,x)$. Therefore, $w_1(s,x) \ge w(s,x)$. On the other hand, $g(s,x) \le w(s,x)$. Therefore,

$$w_1(s,x) \le \sup_{\alpha \in \mathfrak{A}} \sup_{\gamma \in \mathfrak{M}(T-s)} \mathsf{M}_{s,x}^\alpha \left\{ \int_0^{\tau \wedge \gamma} f^{\alpha_t}(s + t, x_t) e^{-\varphi_t}\, dt \right.$$
$$\left. + w(s + \tau \wedge \gamma, x_{\tau \wedge \gamma}) e^{-\varphi_{\tau \wedge \gamma}} \right\}.$$

It remains only to assume in Theorem 11 that $r_t^\alpha \equiv 0$, and, in addition, to note that the last upper bound does not exceed $w(s,x)$. Hence $w_1(s,x) \le w(s,x)$. Similar reasoning is possible if one takes in (9) instead of $\mathfrak{A}$ a set $\mathfrak{A}_E(s,x)$.

[2] The inequality is strict if $(s, x) \notin Q_0$.

Approximating the initial σ, b, c, f, g by means of differentiable functions and passing to the limit is a crucial point in proving the results formulated above and many others. In this section we prove one theorem on the passage to the limit of the indicated.

Suppose the functions $h_n^\alpha(t,x)$ $(n = 0,1,2,\ldots)$ are given. We write $h_n^\alpha(t,x) \to h_0^\alpha(t,x)$ in $\mathscr{L}_1([0,T],B)$ if for each $R > 0$

$$\lim_{n\to\infty} \int_0^T \sup_{\alpha\in A} \sup_{|x|\le R} |h_n^\alpha(t,x) - h_0^\alpha(t,x)|\, dt = 0.$$

12. Theorem. *Let $\sigma_n(\alpha,t,x)$ be a matrix of dimension $d \times d_1$, let $b_n(\alpha,t,x)$ be a d-dimensional vector, let $c_n^\alpha(t,x)$ be nonnegative, and, finally, let $f_n^\alpha(t,x)$, $g_n(t,x)$ be real, defined for $n = 1, 2, \ldots, \alpha \in A$, $t \in [0,T]$, $x \in E_d$. Assume that σ_n, b_n, c_n, f_n are measurable with respect to (α,t,x) and that, in addition, they converge to σ, b, c, f in $\mathscr{L}_1([0,T],B)$ as $n \to \infty$. Furthermore, for each n let the functions σ_n, b_n, c_n, f_n, g_n satisfy inequalities (1)–(3) with identical constants K and m, let $g_n(t,x)$ be measurable with respect to (t,x), and, finally, for each $R > 0$ let*

$$\lim_{n\to\infty} \sup_{t\in[0,T]} \sup_{|x|\le R} |g_n(t,x) - g(t,x)| = 0. \tag{11}$$

For $\alpha \in \mathfrak{A}$, $s \in [0,T]$, $x \in E_d$ we denote by $x_t^{\alpha,s,x}(n)$ a solution of the equation

$$x_t = x + \int_0^t \sigma_n(\alpha_r, s + r, x_r)\, d\mathbf{w}_r + \int_0^t b_n(\alpha_r, s + r, x_r)\, dr.$$

Further, let $\varphi_t^{\alpha,s,x}(n) = \int_0^t c_n^{\alpha_r}(s + r, x_r^{\alpha,s,x}(n))\, dr$. Then for any $q \ge 1$, $R > 0$

$$\begin{aligned} &\mathsf{M}_{s,x}^\alpha \int_0^{T-s} |f_n^{\alpha_t}(s + t, x_t(n))e^{-\varphi_t(n)} - f^{\alpha_t}(s + t, x_t)e^{-\varphi_t}|^q\, dt \to 0,\\ &\mathsf{M}_{s,x}^\alpha \sup_{t\le T-s} |g_n(s + t, x_t(n))e^{-\varphi_t(n)} - g(s + t, x_t)e^{-\varphi_t}|^q \to 0 \end{aligned} \tag{12}$$

as $n \to \infty$ uniformly over $\alpha \in \mathfrak{A}$, $s \in [0,T]$, $x \in S_R$.

Proof. We assume that each of the functions σ, b, c, f, g, σ_n, b_n, c_n, f_n, g_n is equal to zero for $t \ge T$. Then, in assertions (12) we can replace *T-s by T*. Further, as can easily be seen, the assertion that the left sides of (12) tend uniformly to zero is equivalent to the fact that relations (12) will still hold if we permit the values α, s, x in these relations to depend on n in an arbitrary way, provided $\alpha = \alpha_n \in \mathfrak{A}$, $s = s_n \in [0,T]$, $x = x_n \in S_R$. In the future, we shall assume that in (12) the values α, s, x are replaced by similar values α_n, s_n, x_n and also that $R > 0$ is fixed.

Let

$$x_t^n = x_t^{\alpha_n,s_n,x_n}(n), \qquad y_t^n = x_t^{\alpha_n,s_n,x_n},$$
$$\xi_t^n = |x_t^n| + |y_t^n|, \qquad \eta_t^n = |x_t^n - y_t^n|.$$

By Corollary 2.5.12 for any $q \ge 1$

$$\sup_n \mathsf{M} \sup_{t\le T} (\xi_t^n)^q < \infty. \tag{13}$$

Furthermore, let

$$h_t^n(x) = \sup_{\alpha \in A} \left| f_n^\alpha(s_n + t, x) - f^\alpha(s_n + t, x) \right|,$$

$$w_t^n(r) = \sup_{|x| \le r} h_t^n(x).$$

Obviously, as $n \to \infty$

$$\int_0^T w_t^n(r)\, dt \le \int_0^T \sup_{\alpha \in A} \sup_{|x| \le r} \left| f_n^\alpha(t,x) - f^\alpha(t,x) \right| dt \to 0.$$

In addition, $|h_t^n(x) - h_t^n(y)| \le 2w_t^n(|x| \vee |y|)$, which implies in accord with Theorem 2.7.17 as $n \to \infty$ in $\mathscr{L}$ that

$$\left| f_n^{\alpha_{nt}}(s_n + t, x_t^n) - f^{\alpha_{nt}}(s_n + t, x_t^n) \right| \le h_t^n(x_t^n) \to 0. \tag{14}$$

Replacing in the above arguments f by σ, b, and using Theorem 2.5.9, we have that $\eta_t^n \to 0$ in $\mathscr{L}B$ as $n \to \infty$. Further, the function

$$w_t(r,\delta) \equiv \sup_{\alpha \in A} \sup_{|x-y| \le \delta, |x|, |y| \le r} \left| f^\alpha(t,x) - f^\alpha(t,y) \right|$$

tends to zero as $\delta \downarrow 0$ due to the uniform continuity (with respect to α) of $f^\alpha(t,x)$ with respect to x. In addition, the above function does not exceed $2K(1 + r)^m$. By Lebesgue's theorem

$$\lim_{\delta \to 0} \overline{\lim_{n \to \infty}} \int_0^T w_{s_n+t}(r,\delta)\, dt \le \lim_{\delta \to 0} \int_0^T w_t(r,\delta)\, dt = 0.$$

By Lemma 2.7.5, $w_{s_n+t}(r, a_t^n) \to \infty$ for any $r > 0$ as $n \to \infty$ in measure $d\mathsf{P} \times dt$. This yields, by virtue of (13) and Chebyshev's inequality,

$$\lim_{n \to \infty} \int_0^T \mathsf{P}\{w_{s_n+t}(\xi_t^n, \eta_t^n) > \varepsilon\}\, dt \le \lim_{r \to \infty} \overline{\lim_{n \to \infty}} \int_0^T \mathsf{P}\{|\xi_t^n| > r\}\, dt$$

$$+ \lim_{r \to \infty} \overline{\lim_{n \to \infty}} \int_0^T \mathsf{P}\{w_{s_n+t}(r, \eta_t^n) > \varepsilon\}\, dt = 0.$$

In other words, $w_{s_n+t}(\xi_t^n, \eta_t^n) \to 0$ in measure $d\mathsf{P} \times dt$. Since, obviously,

$$\left| f^{\alpha_{nt}}(s_n + t, x_t^n) - f^{\alpha_{nt}}(s_n + t, y_t^n) \right| \le w_{s_n+t}(\xi_t^n, \eta_t^n),$$

the first of the above expressions tends to zero as $n \to \infty$ in measure $d\mathsf{P} \times dt$. Using (13), we can easily prove that this first expression indicated tends to zero in $\mathscr{L}$ as well (see the deduction of Lemma 2.7.6 from Lemma 2.7.5). Comparing the above with (14), we conclude that

$$f_n^{\alpha_{nt}}(s_n + t, x_t^n) - f^{\alpha_{nt}}(s_n + t, y_t^n) \to 0$$

as $n \to \infty$ in $\mathscr{L}$. We would prove this theorem if the functions c_n^α, c^α, g_n, g were equal to zero.

If the functions c_n^α, c^α, g_n, g are not equal to zero, the reader will easily complete proving the theorem by noting that

$$\left| f_1 e^{-\varphi_1} - f_2 e^{-\varphi_2} \right| \le |f_1 - f_2| + |f_1 + f_2| \, |\varphi_1 - \varphi_2|,$$

if $\varphi_1, \varphi_2 \geq 0$, and, in addition, applying the previous results as well as Hölder's inequality. The theorem is proved. □

13. Corollary. *Suppose that the assumptions of Theorem 12 are satisfied. Also, suppose that we are given measurable functions $g_n(x)$ satisfying the inequality $|g_n(x)| \leq K(1 + |x|)^m$ and such that for each $R > 0$*

$$\lim_{n\to\infty} \sup_{|x|\leq R} |g_n(x) - g(x)| = 0. \tag{15}$$

Using the functions σ_n, b_n, c_n, f_n, $g_n(t,x)$, $g_n(x)$, we construct functions $v_n^{\alpha,\tau}$, v_n^α, w_n, v_n in the same way as we constructed above the functions $v^{\alpha,\tau}$, v^α, w, v on the basis of the functions σ, b, c, f, $g(t,x)$, $g(x)$.

Then, $w_n(s,x) \to w(s,x)$, $v_n(s,x) \to v(s,x)$ as $n \to \infty$ uniformly over $s \in [0,T]$, $x \in S_R$ for each $R > 0$. Moreover, $v_n^{\alpha,\tau\wedge(T-s)}(s,x) \to v^{\alpha,\tau\wedge(T-s)}(s,x)$, $v_n^\alpha(s,x) \to v^\alpha(s,x)$ as $n \to \infty$ uniformly over $\alpha \in \mathfrak{A}$, $\tau \in \mathfrak{M}(T)$, $s \in [0,T)$, $x \in S_R$ for each $R > 0$.

Indeed, for example,

$$\begin{aligned}|w_n(s,x) - w(s,x)| &\leq \sup_{\alpha\in\mathfrak{A}} \sup_{\tau\in\mathfrak{M}(T-s)} |v_n^{\alpha,\tau}(s,x) - v^{\alpha,\tau}(s,x)| \\ &= \sup_{\alpha\in\mathfrak{A}} \sup_{\tau\in\mathfrak{M}(T)} |v_n^{\alpha,\tau\wedge(T-s)}(s,x) - v^{\alpha,\tau\wedge(T-s)}(s,x)|.\end{aligned}$$

Furthermore, it is seen that

$$\begin{aligned}&|v_n^{\alpha,\tau\wedge(T-s)}(s,x) - v^{\alpha,\tau\wedge(T-s)}(s,x)| \\ &\leq \mathsf{M}_{s,x}^\alpha \int_0^{T-s} |f_n^{\alpha_t}(s+t, x_t(n))e^{-\varphi_t(n)} - f^{\alpha_t}(s+t, x_t)e^{-\varphi_t}|\, dt \\ &\quad + \mathsf{M}_{s,x}^\alpha \sup_{t\leq T-s} |g_n(s+t, x_t(n))e^{-\varphi_t(n)} - g(s+t, x_t)e^{-\varphi_t}|.\end{aligned}$$

The last expression tends to zero as $n \to \infty$ by the theorem, uniformly over $\alpha \in \mathfrak{A}$, $s \in [0,T]$, $x \in S_R$.

We once approximated the given functions with the aid of infinitely differentiable functions, using convolutions with smooth kernels. Let us see what this method of approximation can give in the case considered. We shall show that $\sigma_n(\alpha,t,x)$, etc., can be taken to be infinitely differentiable with respect to x.

14. Theorem. *Let a sequence $\varepsilon_n \to 0$ as $n \to \infty$. Then the assertions of Theorem 12 and Corollary 13 hold for $\sigma_n(\alpha,t,x) = \sigma^{(0\varepsilon_n)}(\alpha,t,x)$, $b_n(\alpha,t,x) = b^{(0,\varepsilon_n)}(\alpha,t,x)$, etc. (For the notation, see Section 2.1).*

Proof. We assert that the functions σ_n, b_n, c_n, f_n satisfy, in general, inequalities (1)–(3) with identical constants K and m, and that, in addition, these functions converge to σ, b, c, f in $\mathscr{L}_1([0,T],B)$. Furthermore, we need verify (11), (15) and the fact that $|g_n(x)| + |g_n(t,x)| \leq N(1 + |x|)^m$, where N does not depend on n, t, x.

The desired estimates for σ_n, b_n, c_n, f_n, g_n follow from the fact that, for example,

$$\|\sigma_n(\alpha,t,x)\| = \left\| \int_{|z|\le 1} \sigma(\alpha, t, x - \varepsilon_n z)\zeta(z)\,dz \right\|$$
$$\le \sup_{|z|\le 1} \|\sigma(\alpha, t, x - \varepsilon_n z)\| \le K(1 + \varepsilon_n + |x|)$$
$$\le N(1 + |x|).$$
$$\|\sigma_n(\alpha,t,x) - \sigma_n(\alpha,t,y)\| \le \sup_{|z|\le 1} \|\sigma(\alpha, t, x - \varepsilon_n z) - \sigma(\alpha, t, y - \varepsilon_n z)\| \le K|x - y|.$$

Further, as in proving Theorem 12, we introduce here $w_t(r,\delta)$. We mentioned that $w_t(R,\varepsilon) \to 0$ as $\varepsilon \to 0$ for each t, R. From this, for each t we obtain

$$\sup_{\alpha\in A}\sup_{|x|\le R} |f_n^\alpha(t,x) - f^\alpha(t,x)| = \sup_{\alpha\in A}\sup_{|x|\le R} \left| \int_{|y|\le 1} [f^\alpha(t, x - \varepsilon_n y) - f^\alpha(t,x)]\zeta(y)\,dy \right|$$
$$\le w_t(R + \varepsilon_n, \varepsilon_n) \to 0.$$

By Lebesgue's theorem,

$$\int_0^T \sup_{\alpha\in A}\sup_{|x|\le R} |f_n^\alpha(t,x) - f^\alpha(t,x)|\,dt \to 0.$$

In a similar way we prove relations (11), (15) and the fact that the functions σ_n, b_n, $c_n \to \sigma$, b, c in $\mathscr{L}_1([0,T],B)$. The theorem is proved. □

In some cases one can take the functions σ_n, b_n, c_n, f_n, g_n to be infinitely differentiable with respect to (t,x).

15. Theorem. *Suppose that the set A consists of a finite number of points only and, in addition, that the sequence $\varepsilon_n \to 0$ as $n \to \infty$, $|\varepsilon_n| \le 1$. Then the assertions of Theorem 12 and Corollary 13 hold for $\sigma_n(\alpha,t,x) = \sigma^{(\varepsilon_n)}(\alpha,t,x)$, $b_n(\alpha,t,x) = b^{(\varepsilon_n)}(\alpha,t,x)$, etc. (in computing the convolution with respect to (t,x) we assume for $t \le 0$ that $\sigma(\alpha,t,x) = \sigma(\alpha,0,x)$, etc.).*

PROOF. Estimates of the growth of the functions σ_n, b_n, c_n, f_n, g_n can be obtained in the same way as in the preceding proof. Furthermore, Eqs. (11) and (15), as was mentioned in Section 2.1, are known. Therefore, it remains to show that σ_n, b_n, c_n, $f_n \to \sigma$, b, c, f in $\mathscr{L}_1([0,T],B)$. We shall prove only the convergence of f_n. For proving the convergence of σ_n, b_n, c_n, we repeat word-for-word the corresponding arguments.

Note that the definition of convergence in $\mathscr{L}_1[0,T]$, $B)$ involves the upper bound with respect to A. Since A consists of a finite number of points, the upper bound mentioned does not exceed the sum (with respect to $\alpha \in A$) of expressions standing under the sign of the upper bound. Therefore, we prove that $f_n \to f$ in $\mathscr{L}_1([0,T],B)$ if we show that for each $\alpha \in A$ for all $R > 0$

$$\lim_{n\to\infty} \int_0^T \sup_{|x|\le R} |f_n^\alpha(t,x) - f^\alpha(t,x)|\, dt = 0. \tag{16}$$

Let us take the function $w_t(r,\delta)$ from the proof of Theorem 12. Writing the convolution $f_n^\alpha(t,x)$ in a complete form and recalling that $\zeta(t,x) = \zeta_1(t)\zeta(x)$ (see Section 2.1), we easily prove that for $|x|$, $|y| \le R$, $|x - y| \le \varepsilon$

$$|f_n^\alpha(t,x) - f_n^\alpha(t,y)| \le w_t(R + 1, \varepsilon_n) * \varepsilon_n^{-1}\zeta_1(t\varepsilon_n^{-1}) \equiv w_t^n(R + 1, \varepsilon_n),$$

which implies convolution with respect to t. Let $h_n^\alpha = f_n^\alpha - f^\alpha$. For $|x|$, $|y| \le R$, $|x - y| \le \varepsilon$

$$|h_n^\alpha(t,x) - h_n^\alpha(t,y)| \le w_t^n(R + 1, \varepsilon) + w_t(R,\varepsilon).$$

Next, we apply Lemma 2.7.15, taking in this lemma $\xi \equiv x \in S_R$ and $R + 1$ instead of R. For each $\varepsilon > 0$

$$|h_n^\alpha(t,x)| \le w_t^n(R + 2, \varepsilon) + w_t(R + 1, \varepsilon) + N\varepsilon^{-d} \int_{|y|\le R+1} |h_n^\alpha(t,y)|\, dy.$$

From this it follows that the limiting expression in (16) is smaller than

$$\int_0^T [w_t^n(R + 2, \varepsilon) + w_t(R + 1, \varepsilon)]\, dt + N\varepsilon^{-d} \int_0^T dt \int_{|y|\le R+1} |h_n^\alpha(t,y)|\, dy.$$

In the last expression the second term tends to zero as $n \to \infty$ since the mean of any function from $\mathscr{L}_1$ converges in $\mathscr{L}_1$. For the same reason the first term as $n \to \infty$ tends to

$$\int_0^T w_t(R + 2, \varepsilon)\, dt + \int_0^T w_t(R + 1, \varepsilon)\, dt. \tag{17}$$

Thus, (17) estimates the left side of (16) for any $\varepsilon > 0$. In proving Theorem 12, we saw that (17) tended to zero as $\varepsilon \to 0$. We have thus proved Theorem 15. □

2. Some Preliminary Considerations

We shall prove the assertions of Theorems 1.5–1.11 by approximating an arbitrary strategy with the aid of step strategies, i.e., using strategies which are constant on each interval of a subdivision I of an interval of time $[0,T]$. We would expect that the upper bound of the payoffs given by all step strategies which have been constructed on the basis of subdivisions $I = \{0 = t_0, t_1, \ldots, t_n = T\}$ will tend to the corresponding payoff function as $\max_i(t_{i+1} - t_i) \to 0$. In this section, we prepare the proof of the above fact in a special formulation (see Theorem 3.2), and, furthermore, we prove that $v(s,x)$ and $w(s,x)$ are continuous with respect to x.

Using the definitions, assumptions, and notations given in Section 1, we introduce here some new objects. Let us take $\beta \in A$, $0 \le s \le t$, and a function

$u(x)$. Also, let us define a strategy $\beta_t \equiv \beta$ and assume, in addition, that

$$G_{s,t}^{\beta}u(x) = \mathsf{M}_{s,x}^{\beta}\left[\int_0^{t-s} f^{\beta}(s+r, x_r)e^{-\varphi_r}\,dr + u(x_{t-s})e^{-\varphi_{t-s}}\right],$$

$$G_{s,t}u(x) = \sup_{\beta \in A} G_{s,t}^{\beta}u(x).$$

In order to make oneself familiar with the operator $G_{s,t}$, we suggest the reader should work out the following exercise.

1. Exercise

Let $0 \le s_0 \le s_1 \le \cdots \le s_n = T$. Show that $G_{s_0,s_1}\, G_{s_1,s_2}, \ldots, G_{s_{n-1},s_n}g(x)$ is the upper bound of $v^{\alpha}(s_0,x)$ with respect to all strategies $\alpha \in \mathfrak{A}$ for which α_t is constant on each semiinterval of time $[s_i - s_0, s_{i+1} - s_0)$.

We shall repeatedly assume about the functions $u(x)$ to be substituted in the operators G^{β}, G, that for some constants K and $m \ge 0$ for all $x \in E_d$

$$|u(x)| \le K(1 + |x|)^m. \tag{1}$$

In this case $G_{s,t}^{\beta}|u|(x) \le N(1 + |x|)^m$, where N does not depend on β, s, t, x. As was seen in the discussion of Theorem 1.5–1.7, such inequalities readily follow from estimates of the moments of solutions of stochastic equations.

2. Theorem. *Let the continuous function $u(x)$ satisfy inequality* (1). *Then the function $G_{s,t}^{\beta}u(x)$ is continuous in x uniformly with respect to $\beta \in A$ and s, t such that $0 \le s \le t \le T$. The functions $v^{\alpha}(s,x)$, $v^{\alpha,\tau\wedge(T-s)}(s,x)$ are continuous in x uniformly with respect to $\alpha \in \mathfrak{A}$, $s \in [0,T]$, $\tau \in \mathfrak{M}(T)$. In particular, the functions $G_{s,t}u(x)$, $v(s,x)$, $w(s,x)$ are continuous in x uniformly with respect to s, t such that $0 \le s \le t \le T$.*

Proof. The last assertion follows from the fact that, for example,

$$\sup_{0 \le s \le t \le T} |G_{s,t}u(x_n) - G_{s,t}u(x_0)| \le \sup_{0 \le s \le t \le T}\,\sup_{\beta \in A} |G_{s,t}^{\beta}u(x_n) - G_{s,t}^{\beta}u(x_0)|.$$

Furthermore, the right side of the last expression tends to zero as $x_n \to x_0$ according to the first assertion of the theorem.

Next, we take a point $x_0 \in E_d$, a sequence $x_n \to x_0$, and also we assume that $h_n = x_n - x_0$, $\sigma_n(\alpha,t,x) = \sigma(\alpha, t, x + h_n)$. In a similar way, we introduce b_n, c_n, f_n, g_n, u_n. For instance, $u_n(x) = u(x + h_n)$. Since $c^{\alpha}(t,x)$ is continuous in x uniformly with respect to α, for each t $\sup_{\alpha \in A}\sup_{|x| \le R}|c_n^{\alpha}(t,x) - c^{\alpha}(t,x)| \to 0$ as $n \to \infty$. By Lebesgue's theorem,

$$\lim_{n\to\infty}\int_0^T \sup_{\alpha \in A}\,\sup_{|x| \le R} |c_n^{\alpha}(t,x) - c^{\alpha}(t,x)|\,dt = 0.$$

It is not hard to verify that the remaining assumptions of Theorem 1.12 are satisfied. Therefore, Theorem 1.12 is applicable in our case. Furthermore, we note that the process $x_t^{\alpha,s,x_0}(n)$ from Theorem 1.12 as well as the process $x_t^{\alpha,s,x_n} - h_n$ satisfy the same equation in an obvious way. Hence, $x_t^{\alpha,s,x_0}(n) = x_t^{\alpha,s,x_n} - h_n$, $c_n^{\alpha_t}(s + t, x_t^{\alpha,s,x_0}(n)) = c^{\alpha_t}(s + t, x_t^{\alpha,s,x_n})$, etc. By Theorem 1.12 we have that

$$\mathsf{M} \int_0^{T-s} \left| f^{\alpha_t}(s + t, x_t^{\alpha,s,x_n})e^{-\varphi_t^{\alpha,s,x_n}} - f^{\alpha_t}(s + t, x_t^{\alpha,s,x_0})e^{-\varphi_t^{\alpha,s,x_0}}\right| dt \to 0,$$

$$\mathsf{M} \sup_{t\le T-s} \left|g(s + t, x_t^{\alpha,s,x_n})e^{-\varphi_t^{\alpha,s,x_n}} - g(s + t, x_t^{\alpha,s,x_0})e^{-\varphi_t^{\alpha,s,x_0}}\right| \to 0$$

as $n \to \infty$ uniformly with respect to $\alpha \in \mathfrak{A}$, $s \in [0,T]$. Taking instead of $g(s,x)$ the function $u(x)$ in the last relation (this can be done because of continuity of $u(x)$ and by virtue of (1)), we find

$$\mathsf{M} \sup_{t\le T-s} \left|u(x_t^{\alpha,s,x_n})e^{-\varphi_t^{\alpha,s,x_n}} - u(x_t^{\alpha,s,x_0})e^{-\varphi_t^{\alpha,s,x_0}}\right| \to 0$$

as $n \to \infty$ uniformly with respect to $\alpha \in \mathfrak{A}$, $s \in [0,T]$. We derive the assertions of the theorem from the limiting relations proved above, in an elementary way. This completes the proof of Theorem 2. □

Further, we need continuity of v^α, G^α with respect to α.

Let a metric in the set A be given by a function $\rho(\alpha_1,\alpha_2)$. We assume that $\rho(\alpha_1,\alpha_2) < 1$ for all α_1, $\alpha_2 \in A$. We can easily satisfy this inequality if we replace, when needed, the initial metric by an equivalent metric, using the formula

$$\rho'(\alpha_1,\alpha_2) = \frac{2}{\pi}\arctan \rho(\alpha_1,\alpha_2).$$

3. Definition. For α^1, $\alpha^2 \in \mathfrak{A}$ let

$$\tilde{\rho}(\alpha^1,\alpha^2) = \mathsf{M} \int_0^T \rho(\alpha_t^1,\alpha_t^2)\, dt.$$

If $\alpha^n \in \mathfrak{A}$ $(n = 0,1, \ldots)$ and $\tilde{\rho}(\alpha^n,\alpha^0) \to 0$ as $n \to \infty$, we write $\alpha^n \to \alpha^0$.

Since $\rho(\alpha_1,\alpha_2) < 1$, $\tilde{\rho}(\alpha^1,\alpha^2)$ will be defined for each α^1, $\alpha^2 \in \mathfrak{A}$.

4. Exercise

Using Theorem 2.8.1, prove that if $\tilde{\rho}(\alpha^1,\alpha^2) = 0$, then

$$\sup_{t\le T} \left|x_t^{\alpha^1,s,x} - x_t^{\alpha^2,s,x}\right| = 0 \qquad \text{(a.s.)}$$

for all (s,x).

By hypothesis, the set A is separable. We fix a countable subset $\{\alpha(i)\}$ dense everywhere in A.

5. Definition. Let $I = \{0 = t_0, t_1, \ldots, t_n = T\}$ be a subdivision of an interval of time $[0,T]$, $\alpha \in \mathfrak{A}$, and let N be an integer. We write $\alpha \in \mathfrak{A}_{cT}(I,N)$, if

$\alpha_t(\omega) \in \{\alpha(1), \ldots, \alpha(N)\}$ for all $\omega \in \Omega$, $t \in [0,T]$, and $\alpha_t = \alpha_{t_i}$ for $t \in [t_i, t_{i+1})$, $i = 0, 1, \ldots, n-1$. Let $\mathfrak{A}_{cT}(I) = \bigcup_N \mathfrak{A}_{cT}(I,N)$, $\mathfrak{A}_{cT} = \bigcup_I \mathfrak{A}_{cT}(I)$. Strategies of the class $\mathfrak{A}_{cT}$ are said to be *step strategies*.

6. Lemma. *Suppose that the diameter of the subdivision I_n of the interval $[0,T]$ tends to zero as $n \to \infty$. Then for each strategy $\alpha \in \mathfrak{A}$ there is a sequence of strategies $\alpha^n \in \mathfrak{A}_{cT}(I_n)$ converging to the strategy $\alpha \in \mathfrak{A}$.*

PROOF. The distance $\tilde{\rho}$ satisfies a triangle inequality. Hence it suffices to show that:

a. in the sense of the distance $\tilde{\rho}$ the set $\bigcup_n \mathfrak{A}_{cT}(I_n)$ is dense in $\mathfrak{A}_{cT}$;
b. the set $\mathfrak{A}_{cT}$ is dense in a set of all the strategies each of which assumes only a finite number of values from $\{\alpha(i)\}$;
c. the latter set is dense in $\mathfrak{A}$.

Proof of (a). If $\alpha \in \mathfrak{A}_{cT}$, for some subdivision $I = \{0 = t_0, t_1, \ldots, t_p = T\}$ the equalities $\alpha_t = \alpha_{t_i}$ can be satisfied for $t \in [t_i, t_{i+1})$. Using the strategy α and the subdivision I_n, we construct a strategy α^n so that α_t^n will be right continuous, constant on each interval of the subdivision I_n, and, in addition, coincide with α_t at the left end points of the foregoing intervals. In this situation α_t^n differs from α_t only on those intervals of the subdivision I_n each of which contains at least one point t_i. It is seen that $\rho(\alpha_t^n, \alpha_t) \to 0$ as $n \to \infty$ everywhere, except for, perhaps, the points t_i. Hence $\tilde{\rho}(\alpha^n, \alpha) \to 0$.

Proof of (b). We take a strategy α_t that assumes values in $\{\alpha(1), \ldots, \alpha(N)\}$. In a Euclidean space E_N let us choose N arbitrary points $x_1, \ldots, x_N$ so that $|x_i - x_j| \geq 1$ for $i \neq j$. Let $\beta_t(\omega) = x_i$, if $\alpha_t(\omega) = \alpha(i)$, $t \in [0,T]$, $\beta_t(\omega) = 0$ for $t > T$. It is easily seen that for $s, t \in [0,T]$

$$\rho(\alpha_t, \alpha_s) \leq |\beta_t - \beta_s|^2. \tag{2}$$

Assuming $\alpha_t = \alpha_0$, $\beta_t = \beta_0$, we define completely the functions α_t, β_t for negative t. Let $\kappa(n,t) = j2^{-n}$ for $j2^{-n} \leq t < (j+1)2^{-n}$, $j = 0, \pm 1, \pm 2, \ldots$ It is a well-known fact (see, for instance, the proof of Lemma 4.4 in [51]) that there exists a number s and a sequence consisting of integers $n' \to \infty$, such that

$$\lim_{n' \to \infty} \mathsf{M} \int_0^T |\beta_t - \beta_{\kappa(n', t-s)+s}|^2 \, dt = 0.$$

By virtue of (2), we have for the functions $\alpha_t^n = \alpha_{\kappa(n,t-s)+s}$: $\tilde{\rho}(\alpha^n, \alpha) \to 0$ as $n' \to \infty$. Furthermore, it is not hard to see that $\kappa(n, t-s) + s$ is a step function of t, $\kappa(n, t-s) + s \leq t$. Hence α_t^n is $\mathscr{F}_t$-measurable and $\alpha^n \in \mathfrak{A}_{cT}$.

Proof of (c). Let us introduce on A the following functions:

$$i_n(\alpha) = \min\left\{i : \rho(\alpha, \alpha(i)) \leq \frac{1}{n}\right\}, \qquad \kappa_n(\alpha) = \alpha(i_n(\alpha)).$$

It is seen that $\kappa_n(\alpha)$ is equal to that $\alpha(i)$, which is at a distance from α of not more than $1/n$, and, which in addition, has the smallest possible index. Since

$\{\alpha(i)\}$ is dense everywhere in A, the functions $i_n(\alpha)$, $\kappa_n(\alpha)$ are defined on A and $\rho(\kappa_n(\alpha),\alpha)) \le 1/n$ for all $\alpha \in A$. Further, let $\kappa_{n,N}(\alpha) = \alpha(N \wedge i_n(\alpha))$. Obviously, $\kappa_{n,N}(\alpha) \to \alpha$, if we let first $N \to \infty$, second, $n \to \infty$. Hence $\tilde{\rho}(\kappa_{n,N}(\alpha),\alpha) \to 0$ under the same conditions for each strategy $\alpha \in \mathfrak{A}$. It remains only to note that the strategy $\kappa_{n,N}(\alpha_t)$ takes on values only in the set $\{\alpha(1), \ldots, \alpha(N)\}$. We have thus proved the lemma.

7. Lemma. *Let $s \in [0,T]$, let τ_1, τ_2 be random variables with values in $[0, T - s]$, and, finally, let $u(x)$ be a continuous function satisfying the condition* (1). *Then the random variable*

$$\int_{\tau_1}^{\tau_2} f^{\alpha_t}(s + t, x_t^{\alpha,s,x})e^{-\varphi_t^{\alpha,s,x}}\,dt + u(x_{\tau_2}^{\alpha,s,x})e^{-\varphi_{\tau_2}^{\alpha,s,x}} \tag{3}$$

is an $\mathscr{L}$-continuous function of (α,x) for $\alpha \in \mathfrak{A}$, $x \in E_d$.

Proof. Note first that if $\alpha^n \to \alpha$, $x^n \to x$, the $\mathscr{L}B$-limit $x_t^{\alpha^n,s,x^n}$ is equal to $x_t^{\alpha,s,x}$, which fact follows, according to Theorem 2.8.1, from the continuity of $\sigma(\alpha,t,y)$ and $b(\alpha,t,y)$ with respect to α, boundedness of $\sigma(\alpha,t,y)$ and $b(\alpha,t,y)$ for fixed y, and, finally, convergence of $\alpha_t^n(\omega)$ to $\alpha_t(\omega)$ in measure $dP \times dt$.

Further, reasoning in the same way as in the proof of Lemma 2.7.6, and, moreover, using condition (1.3) as well as the continuity of $c^\alpha(t,x)$ and $f^\alpha(t,x)$ with respect to (α,x), we can prove that the processes $c^{\alpha_t}(s + t, x_t^{\alpha,s,x})$, $f^{\alpha_t}(s + t, x_t^{\alpha,s,x})$ are $\mathscr{L}$-continuous with respect to (α,x). Also, applying the results obtained in Section 2.7 related to the $\mathscr{L}B$-continuity of integrals and the $\mathscr{L}$-continuity of products of $\mathscr{L}$-continuous processes, we immediately arrive at the assertion of the lemma. Lemma 7 is proved. □

8. Corollary. *For $s \in [0,T]$ and $\tau \in \mathfrak{M}(T - s)$ the functions $v^\alpha(s,x)$ and $v^{\alpha,\tau}(s,x)$ are continuous with respect to (α,x) for $\alpha \in \mathfrak{A}$, $x \in E_d$. For $0 \le s \le t \le T$ the function $G_{s,t}^\beta u(x)$ is continuous with respect to (β,x) on $A \times E_d$.*

Combining Corollary 8 with Lemma 6, we have

9. Corollary

$$v(s,x) = \lim_{n \to \infty} \sup_{\alpha \in \mathfrak{A}_{ct}(I_n)} v^\alpha(s,x)$$

for any sequence of subdivisions whose diameter tends to zero.

10. Exercise

Prove that (3) is $\mathscr{L}$-continuous with respect to s and, next, deduce from that the continuity of $v^\alpha(s,x)$ with respect to s. This together with Theorem 2 enables one to conclude that $v^\alpha(s,x)$ is continuous with respect to (s,x) and that, in addition, $v(s,x)$ is a Borel function of (s,x).

We prove some other properties of the operators $G^{\beta}_{s,t}$. Letting the sequence $\varepsilon_n \to 0$, $\varepsilon_n \neq 0$, we consider the mean functions for the functions σ, b, c, f. Let $\sigma_n(\alpha,t,x) = \sigma^{(\varepsilon_n)}(\alpha,t,x)$ (see the notation in Section 2.1), etc. In other words, we take σ_n, b_n, c_n, f_n from Theorem 1.15. We denote by $x_t^{\alpha,s,x}(n)$ a solution of the equation

$$dx_t = \sigma_n(\alpha_t, s + t, x_t)\, d\mathbf{w}_t + b_n(\alpha_t, s + t, x_t)\, dt, \qquad x_0 = x.$$

Furthermore, let

$$\varphi_t^{\alpha,s,x}(n) = \int_0^t c_n^{\alpha_r}(s + r, x_r^{\alpha,s,x}(n))\, dr$$

and, in addition, for a constant strategy $\beta_t \equiv \beta$ for $0 \leq s \leq t \leq T$ let

$$G^{\beta,n}_{s,t}u(x) = \mathsf{M}^{\beta}_{s,x}\left[\int_0^{t-s} f_n^{\beta}(s + r, x_r(n))e^{-\varphi_r(n)}\, dr + u(x_{t-s}(n))e^{-\varphi_{t-s}(n)}\right].$$

Regarding β as a single point of A and using Theorem 1.15 as well as the estimates of moments of solutions of stochastic equations, we have the following assertion.

11. Lemma. *Let a continuous function $u(x)$ satisfy condition* (1), *and let $u_n(x) = u^{(\varepsilon_n)}(x)$. Then*

$$|G^{\beta,n}_{s,t}u_n(x)| \leq N(1 + |x|)^m$$

for all $s \leq t \leq T$, $x \in E_d$, $n > 0$, $\beta \in A$, where N does not depend on s, t, x, n, β. Furthermore,

$$G^{\beta,n}_{s,t}u_n(x) \to G^{\beta}_{s,t}u(x)$$

as $n \to \infty$ for each $\beta \in A$ uniformly on each set of the form $\{(s,t,x): 0 \leq s \leq t \leq T, |x| \leq R\}$.

The functions σ_n, b_n, c_n, f_n, u_n are smooth with respect to (t,x). In addition, their derivatives grow not more rapidly than $(1 + |x|)^m$. For example,

$$\left|\frac{\partial^i}{\partial l^i} u_n(x)\right| = \left|\varepsilon_n^{-i} \int_{|z| \leq 1} u(x - \varepsilon_n z) \frac{\partial^i}{\partial l^i} \zeta(z)\, dz\right|$$

$$\leq \varepsilon_n^{-i} K(1 + \sup|\varepsilon_n| + |x|)^m \int_{|z| \leq 1} \left|\frac{\partial^i}{\partial l^i} \zeta(z)\right| dz.$$

By Theorem 2.9.10, the function $G^{\beta,n}_{s,t}u_n(x)$ is the unique solution of a certain equation. By Remark 2.9.11 the foregoing function is uniquely determined by the functions $a_n = \frac{1}{2}\sigma_n\sigma_n^*$, b_n, c_n, f_n, u_n, for finding which it suffices, obviously, to give σ, b, c, f, u. This, by Lemma 11, implies

12. Corollary. *The function $G^{\beta}_{s,t}u(x)$ does not change if we change the probability space and, furthermore, take another d_1-dimensional Wiener process. The function $G^{\beta}_{s,t}u(x)$ can be determined uniquely by σ, b, c, f, u.*

Let us use the properties of the function $G_{s,t}^{\beta,n}u_n(x)$ to a greater extent.

13. Corollary. *Suppose that $\beta \in A$, and that the function $\sigma(\beta,t,x)(b(\beta,t,x))$ for each $t \in [0,T]$ is twice (once) continuously differentiable over x. Furthermore, suppose that second (first) derivatives of the foregoing function with respect to x are bounded on any set of the form $[0,T] \times S_R$. Let $t \in [0,T]$, and let $\eta(s,x)$ be an infinitely differentiable function on E_{d+1}, which is equal to zero outside a certain cylinder $[0,t] \times S_R$. Then*

$$\int_0^t ds \int [G_{s,t}^{\beta}u(x)L^{\beta *}\eta(s,x) + f^{\beta}(s,x)\eta(s,x)]\,dx = 0,$$

where

$$L^{\beta *}\eta(s,x) \equiv -\frac{\partial}{\partial s}\eta + \sum_{i,j=1}^{d} (a^{ij}(\beta,s,x)\eta)_{x^i x^j} - \sum_{i=1}^{d} (b^i(\beta,s,x)\eta)_{x^i} - c^{\beta}(s,x)\eta,$$

$$a(\beta,s,x) = \frac{1}{2}\sigma(\beta,s,x)\sigma^*(\beta,s,x).$$

In fact, let

$$L_n^{\beta}(s,x) = \frac{\partial}{\partial s} + \sum_{i,j=1}^{d} a_n^{ij}(\beta,s,x)\frac{\partial^2}{\partial x^i\,\partial x^j} + \sum_{i=1}^{d} b_n^i(\beta,s,x)\frac{\partial}{\partial x^i} - c_n^{\beta}(s,x).$$

By Theorem 2.9.10,

$$L_n^{\beta}(s,x)G_{s,t}^{\beta,n}u_n(x) + f_n^{\beta}(s,x) = 0$$

in a strip $[0,t] \times E_d$. Multiplying the last equality by η, integrating by parts, and, in addition, introducing an operator $L_n^{\beta *}$ in the usual way, we have

$$\int_0^t ds \int [G_{s,t}^{\beta,n}u_n(x)L_n^{\beta *}\eta(s,x) + f_n^{\beta}(s,x)\eta(s,x)]\,dx = 0.$$

It remains only to let $n \to \infty$ and to note that the integration is to be carried out over a bounded set and, for example,

$$b_{nx^i}(\beta,s,x) = b_{x^i}(\beta,s,x) * \varepsilon_n^{-(d+1)}\zeta(\varepsilon_n^{-1}s,\varepsilon_n^{-1}x) \to b_{x^i}(\beta,s,x)$$

for almost all s, x. (For the properties of mean functions, see Section 2.1).

The final property of the operator $G_{s,t}^{\beta}$ which we give in this section follows immediately from Theorem 2.9.7 and Remark 2.9.9.

14. Lemma. *Suppose that $s \in [0,T]$, $0 \le t_1 \le t_2 \le T - s$ and a strategy $\alpha \in \mathfrak{A}$ is such that $\alpha_t = \alpha_{t_1}$ for $t \in [t_1,t_2)$. Let the continuous function $u(x)$ satisfy condition (1). Then almost surely*

$$\mathsf{M}_{s,x}^{\alpha}\left\{\int_{t_1}^{t_2} f^{\alpha_t}(s + t, x_t)e^{-\varphi_t}\,dt + u(x_{t_2})e^{-\varphi_{t_2}} \,\middle|\, \mathscr{F}_{t_1}\right\} = e^{-\varphi_{t_1}^{\alpha,s,x}} G_{s+t_1,s+t_2}^{\alpha_{t_1}} u(x_{t_1}^{\alpha,s,x}).$$

Note that for proving the lemma we should take in Theorem 2.9.7, A, t_1, t_2, $x_{t_1}^{\alpha,s,x}$, α_{t_1}, $\sigma_{s+r}^{\alpha}(x)$, $b_{s+r}^{\alpha}(x)$, instead of Z, s, T, ξ, ζ, $\sigma_r^z(x)$, $b_r^z(x)$ respectively.

3. The Proof of Theorems 1.5–1.7

In the preceding section we proved that some mathematical expectations of the form $M_{s,x}^{\alpha}F^{\alpha}$ are continuous with respect to (α,x) on $\mathfrak{A} \times E_d$. Furthermore, we learned how to approximate any strategy by means of step strategies. Also, we introduced the operators $G_{s,t}^{\beta}$, $G_{s,t}$ which are crucial for the discussion in this section. Having thus completed the technicalities, we proceed now to prove Theorems 1.5–1.7.

1. Lemma. *Let* $s_0 < s_1 < \cdots < s_n = T$. *Then*

$$v_{(E)}(s_0,x) \geq G_{s_0,s_1}G_{s_1,s_2} \cdots G_{s_{n-1},s_n}g(x).$$

PROOF. Let $u_i(x) = G_{s_i,s_{i+1}} \cdots G_{s_{n-1}s_n}g(x)$ $(i = 0,1,\ldots,n-1)$, $u_n(x) = g(x)$. Also, we fix $\varepsilon > 0$. By Theorem 2.2, the function $u_{n-1}(x) = G_{s_{n-1},s_n}g(x)$ is continuous and furthermore, it satisfies the inequality $|u_{n-1}(x)| \leq N(1 + |x|)^m$. This implies in accord with Theorem 2.2 that the function $u_{n-2}(x) = G_{s_{n-2},s_{n-1}}u_{n-1}(x)$ is continuous. Arguing in the same way, we convince ourselves that all the functions $u_i(x)$ are continuous.
Further,

$$u_I(x) = G_{s_i,s_{i+1}}u_{i+1}(x) = \sup_{\beta \in A} G_{s_is_{i+1}}^{\beta}u_{i+1}(x), \qquad i = 0, 1, \ldots, n-1.$$

By Corollary 2.8, the functions $G_{s_i,s_{i+1}}u_{i+1}(x)$ are continuous with respect to β. Hence the last upper bound can be computed on any countable set everywhere dense in A. Noting in addition that $G_{s_i,s_{i+1}}^{\beta}u_{i+1}(x)$ is continuous with respect to x according to Corollary 2.8, we conclude that there exists a (countable-valued) Borel function $\beta_i(x)$ such that for all x

$$u_i(x) \leq G_{s_i,s_{i+1}}^{\beta_i(x)}u_{i+1}(x) + \varepsilon, \qquad i = 0, 1, \ldots, n-1.$$

In a space of continuous functions $x_{[0,\infty)}$ with values in E_d we define the function $\alpha_t(x_{[0,\infty)}) = \alpha_t(x_{[0,t]})$ using the formula $\alpha_t(x_{[0,t]}) = \beta_i(x_{s_i-s_0})$ for $t \in [s_i - s_0, s_{i+1} - s_0)$, $i = 0, \ldots, n-1$, $\alpha_t(x_{[0,t]}) = \beta_0(0)$ for $t \geq T - s_0$. It is seen that the function α_t is progressively measurable with respect to $\{\mathcal{N}_t\}$ and also it is seen that the equation

$$x_t = x + \int_0^t \sigma(\alpha_r(x_{[0,r]}), s + r, x_r)\, d\mathbf{w}_r + \int_0^t b(\alpha_r(x_{[0,r]}), s + r, x_r)\, dr$$

is equivalent to a sequence of equations

$$x_t = x + \int_0^t \sigma(\beta_0(x), s + r, x_r)\, d\mathbf{w}_r + \int_0^t b(\beta_0(x), s + r, x_r)\, dr, \qquad t \in [0, s_1 - s_0);$$

$$x_t = x_{s_1-s_0} + \int_{s_1-s_0}^t \sigma(\beta_1(x_{s_1-s_0}), s + r, x_r)\, d\mathbf{w}_r$$
$$+ \int_{s_1-s_0}^t b(\beta_1(x_{s_1-s_0}), s + r, x_r)\, dr, \qquad t \in [s_1 - s_0, s_1 - s_0);$$

etc. Each of the equations given is solvable. Therefore, α_i is a natural strategy admissible at each point (s,x).

Finally, by Lemma 2.14 for $i = 0, 1, \ldots, n-1$

$$\begin{aligned} \mathsf{M}^{\alpha}_{s_0,x}&\left\{\int_{s_i-s_0}^{s_{i+1}-s_0} f^{\alpha_t}(s_0+t,x_t)e^{-\varphi_t}\,dt + u_{i+1}(x_{s_{i+1}-s_0})e^{-\varphi_{s_{i+1}-s_0}}\right\} \\ &= \mathsf{M}^{\alpha}_{s_0,x}e^{-\varphi_{s_i-s_0}}G^{\alpha_{s_i-s_0}}_{s_i,s_{i+1}}u_{i+1}(x_{s_i-s_0}) \\ &\geq \mathsf{M}^{\alpha}_{s_0,x}e^{-\varphi_{s_i-s_0}}u_i(x_{s_i-s_0}) - \varepsilon. \end{aligned}$$

Adding up all such inequalities and collecting like terms, we find

$$v_{(E)}(s_0,x) \geq v^{\alpha}(s_0,x) \geq u_0(x) - n\varepsilon,$$

thus proving our lemma. □

In the theorem which follows we prove the first assertion of Theorem 1.7.

2. Theorem. (a) $v_{(E)} = v$. (b) *Let* $s_0 = s_0^i \leq s_1^i < \cdots < s_{n(i)}^i = T$ $(i = 1,2,\ldots)$, $\max_j(s_{j+1}^i - s_j^i) \to 0$ *for* $i \to \infty$. *Then*

$$\begin{aligned} v(s_0,x) &= \lim_{i\to\infty} G_{s_0^i,s_1^i}G_{s_1^i,s_2^i}\cdots G_{s_{n(i)-1}^i,s_{n(i)}^i}g(x) \\ &= \sup_i G_{s_0^i,s_1^i}G_{s_1^i,s_2^i}\cdots G_{s_{n(i)-1}^i,s_{n(i)}^i}g(x). \end{aligned} \tag{1}$$

Proof. Assertion (a) follows from (b), Lemma 1, and the obvious inequality $v_{(E)} \leq v$. Furthermore, it follows from Lemma 1 that the upper bound in (1) does not exceed $v(s_0,x)$. Since the upper limit is smaller that the upper bound, to prove (b) we need only to show that

$$v(s_0,x) \leq \varliminf_{i\to\infty} G_{s_0^i,s_1^i}\cdots G_{s_{n(i)-1}^i,s_{n(i)}^i}g(x). \tag{2}$$

Using Corollary 2.9, we construct step strategies so that $v^{\alpha^i}(s_0,x) \to v(s_0,x)$ as $i \to \infty$ and $\alpha_t^i = \alpha^i_{s_j^i-s_0}$ for $t \in [s_j^i - s_0, s_{j+1}^i - s_0)$.

Also, we introduce functions u_j^i according to the formulas $u_{n(i)}^i(x) = g(x)$, $u_j^i(x) = G_{s_j^i,s_{j+1}^i}u_{j+1}^i(x)$ $(j = 0,1,\ldots,n(i)-1)$.

By Lemma 2.14

$$\begin{aligned} \mathsf{M}^{\alpha^i}_{s_0,x}&\left[\int_{s_j^i-s_0}^{s_{j+1}^i-s_0} f^{\alpha_r}(s_0+r,x_r)e^{-\varphi_r}\,dr + u_{j+1}^i(x_{s_{j+1}^i-s_0})e^{-\varphi_{s_{j+1}^i-s_0}}\right] \\ &= \mathsf{M}^{\alpha^i}_{s_0,x}e^{-\varphi_{s_j^i-s_0}}G^{\alpha^i_{s_j^i-s_0}}_{s_j^i,s_{j+1}^i}u_{j+1}^i(x_{s_j^i-s_0}) \\ &\leq \mathsf{M}^{\alpha^i}_{s_0,x}e^{-\varphi_{s_j^i-s_0}}u_j^i(x_{s_j^i-s_0}). \end{aligned} \tag{3}$$

Adding up such inequalities with respect to j from $j = 0$ to $j = n(i) - 1$, and, in addition, collecting like terms, we obtain: $v^{\alpha^i}(s_0,x) \leq u_0^i(x)$. Therefore, $v(s_0,x) \leq \varliminf_{i\to\infty} u_0^i(x)$, which is completely equivalent to (2). The theorem is proved. □

3. Exercise

Prove that if the subdivisions $\{s_j^i\}$ are embedded, the functions under the limit in (1) converge monotonically to $v(s_0,x)$.

4. Remark. The theorem proved above together with the constructions made in Lemma 1 provides a technique for finding ε-optimal strategies in the class of step natural strategies.

5. Lemma. (a) *Let* $s \in [0,T]$, $x \in E_d$, $\alpha \in \mathfrak{A}$. *Then the processes*

$$\delta_t^{\alpha,s,x} \equiv v(s + t, x_t^{\alpha,s,x})e^{-\varphi_t^{\alpha,s,x}} - \mathsf{M}_{s,x}^{\alpha}\{g(x_{T-s})e^{-\varphi_{T-s}} + \int_t^{T-s} f^{\alpha_r}(s + r, x_r)e^{-\varphi_r}\, dr \,|\, \mathscr{F}_t\},$$

$$\kappa_t^{\alpha,s,x} \equiv v(s + t, x_t^{\alpha,s,x})e^{-\varphi_t^{\alpha,s,x}} + \int_0^t f^{\alpha_r}(s + r, x_r^{\alpha,s,x})e^{-\varphi_t^{\alpha,s,x}}\, dr,$$

defined for $t \in [0, T - s]$, *are supermartingales with respect to* $\{\mathscr{F}_t\}$, *the first process being nonnegative* (a.s.).

(b) $G_{s,t}v(t,x) \le v(s,x)$ *for* $x \in E_d$, $0 \le s \le t \le T$.

PROOF. It is seen that

$$\delta_t^{\alpha,s,x} - \kappa_t^{\alpha,s,x} = -\mathsf{M}_{s,x}^{\alpha}\left\{\int_0^{T-s} f^{\alpha_r}(s + r, x_r)e^{-\varphi_r}\, dr + g(x_{T-s})e^{-\varphi_{T-s}} \middle| \mathscr{F}_t\right\},$$

where the right side is a martingale. Hence $\delta_t^{\alpha,s,x}$ is a supermartingale if $\kappa_t^{\alpha,s,x}$ is a supermartingale. The nonnegativity of $\delta_t^{\alpha,s,x}$ follows from the fact that by the definition of a supermartingale, $\delta_t^{\alpha,s,x} \ge \mathsf{M}_{s,x}^{\alpha}\{\delta_{T-s}|\mathscr{F}_t\}$ and $\delta_{T-s}^{\alpha,s,x} = 0$.

Further, by Theorem 2.2 the function $v(s + t, x)$ is continuous with respect to x. In addition, $|v(s + t, x)| \le N(1 + |x|)^m$. Therefore, by Lemma 2.7

$$\mathscr{L}\text{-}\lim_{n\to\infty} \kappa_t^{\alpha^n,s,x} = \kappa_t^{\alpha,s,x}$$

for each $t \in [0, T - s]$ if $\alpha^n \to \alpha$. By Lemma 2.6, we can choose step $\alpha^n \to \alpha$, which implies that the supermartingaleness of $x_t^{\alpha,s,x}$ needs to be proved only for step strategies. Since the constancy segments of α_t can be considered one by one, it suffices to prove that $\mathsf{M}_{s,x}^{\alpha}\{x_{t_2}|\mathscr{F}_{t_1}\} \le \kappa_{t_1}^{\alpha,s,x}$ (a.s.) for $t_2 \ge t_1$ if $\alpha_t = \alpha_{t_1}$ for $t \in [t_1,t_2)$. By Lemma 2.14, for such a strategy

$$\mathsf{M}_{s,x}^{\alpha}\{\kappa_{t_2}|\mathscr{F}_{t_1}\} = \int_0^{t_1} f^{\alpha_r}(s + r, x_r^{\alpha,s,x})e^{-\varphi_r^{\alpha,s,x}}\, dr + e^{-\varphi_{t_1}^{\alpha,s,x}} G_{s+t_1,s+t_2}^{\alpha_{t_1}} v(s + t_2, x_{t_1}^{\alpha,s,x}) \qquad \text{(a.s.)},$$

from which it is seen that it remains now to prove assertion (b) of the lemma. We assume $\beta_0 \in A$, $s_0 = t$, and, in addition, we construct a sequence of subdivisions $s_0 = s_0^i < s_j^i < \cdots < s_{n(i)}^i = T$ of an interval $[t,T]$ so that

$\max_j (s_{j+1}^i - s_j^i) \to 0$. By Theorem 2 and Lemma 2.14, it is not hard to obtain

$$\begin{aligned} N(1+|x|)^m &\geq v(t,x) \geq G_{s_0^i,s_1^i} \cdots G_{s_{n(i)-1}^i,s_{n(i)}^i} g(x) \\ &\geq G^{\beta_0}_{s_0^i,s_1^i} \cdots G^{\beta_0}_{s_{n(i)-1}^i,s_{n(i)}^i} g(x) \qquad \text{(see Ftn. 3)} \\ &= \mathsf{M}^{\beta_0}_{t,x}\left\{\int_0^{T-t} f^{\beta_0}(t+r,x_r)e^{-\varphi_r}\,dr + g(x_{T-t})e^{-\varphi_{T-t}}\right\} \\ &\geq -N(1+|x|)^m, \end{aligned}$$

where the constants N do not depend on x. This implies that for each $\beta \in A$ the magnitude of the sequence

$$G_{s_0^i,s_1^i} \cdots G_{s_{n(i)-1}^i,s_{n(i)}^i} g(x_{t-s}^{\beta,s,x})$$

does not exceed $N(1+|x_{t-s}^{\beta,s,x}|)^m$, the latter expression having a finite mathematical expectation. Therefore, recalling that

$$G^{\beta}_{s,t}u(x) \equiv \mathsf{M}^{\beta}_{s,x}\left[\int_0^{t-s} f^{\beta}(s+r,x_r)e^{-\varphi_r}\,dr + u(x_{t-s})e^{-\varphi_{t-s}}\right],$$

and also applying Lebesgue's theorem, we easily find

$$G^{\beta}_{s,t}v(t,x) = \lim_{i\to\infty} G^{\beta}_{s,t}G_{s_0^i,s_1^i} \cdots G_{s_{n(i)-1}^i,s_{n(i)}^i} g(x),$$

where the expression standing under the limit does not exceed $v(s,x)$ in accord with Lemma 1 or Theorem 2. We have thus proved Lemma 5. □

6. Theorem (Bellman's Principle). *For $s \leq t \leq T$*

$$v(s,x) = \sup_{\alpha\in\mathfrak{A}} \mathsf{M}^{\alpha}_{s,x}\left[\int_0^{t-s} f^{\alpha_r}(s+r,x_r)e^{-\varphi_r}\,dr + v(t,x_{t-s})e^{-\varphi_{t-s}}\right];$$

in this case we can take the upper bound with respect to $\alpha \in \mathfrak{A}_E(s,x)$ as well.

Proof. The properties of supermartingales imply that

$$v(s,x) = \mathsf{M}^{\alpha}_{s,x}\kappa_0 \geq \mathsf{M}^{\alpha}_{s,x}\kappa_{t-s} \geq \mathsf{M}^{\alpha}_{s,x}\kappa_{T-s} = v^{\alpha}(s,x).$$

Taking upper bounds with respect to $\alpha \in \mathfrak{A}$ or $\alpha \in \mathfrak{A}_E(s,x)$, we prove the required result. □

The following lemma proves Theorem 1.5.

7. Lemma. *The function $v(s,x)$ is continuous with respect to (s,x) for $s \in [0,T]$, $x \in E_d$.*

Proof. By Theorem 2.2, the function $v(s,x)$ is continuous in x uniformly with respect to $s \in [0,T]$. Therefore, it suffices to prove that $v(s,x)$ is continuous in s for each x. We fix x_0. We need to prove that if $s_n, t_n \in [0,T]$,

[3] Assuming that A consists of a single point β_0, we have in (3) equalities instead of inequalities.

$t_n - s_n \to 0$, then $v(s_n,x_0) - v(t_n,x_0) \to 0$. We consider without loss of generality that $t_n \geq s_n$. Further, we use Theorem 6 for $x = x_0$, $s = s_n$, $t = t_n$, and, in addition, choose $\alpha^n \in \mathfrak{A}$ such that the upper bound mentioned in the assertion of Theorem 6, attained for $\alpha = \alpha^n$ to within $1/n$. We have

$$\overline{\lim_{n\to\infty}} |v(s_n,x_0) - v(t_n,x_0)| \leq \overline{\lim_{n\to\infty}} \mathsf{M} \int_0^{t_n - s_n} |f^{\alpha_r^n}(s_n + r, x_r^n)|\, dr$$
$$+ \overline{\lim_{n\to\infty}} \mathsf{M}|v(t_n, x_{t_n - s_n}^n)e^{-\varphi_{t_n - s_n}^n} - v(t_n,x_0)|, \quad (4)$$

where the superscript n attached to x, φ stands for (α^n,s_n,x_0). By Corollary 2.5.12, for any $q \geq 1$

$$\sup_n \mathsf{M} \sup_{t \leq T} |x_t^n|^q < \infty, \quad (5)$$

from which it follows due to (1.3) that the limiting expression in the first term of (4) does not exceed $N(t_n - s_n)$, and also that this term itself is equal to zero. If we replace f by c in the above arguments, and if, furthermore, we use Chebyshev's inequality, we can see that $\varphi_{t_n - s_n}^n \to 0$ in probability.

By Corollary 2.5.12, $x_{t_n - s_n}^n \to x_0$ in probability. Due to the uniform continuity of $v(t,x)$ with respect to x,

$$h(y) \equiv \sup_{t \in [0,T]} |v(t, y + x_0) - v(t,x_0)| \to 0$$

as $y \to 0$. It follows, in turn, that $h(x_{t_n - s_n}^n - x_0) \to 0$ in probability. In particular, $v(t_n,x_{t_n - s_n}^n) - v(t_n,x_0) \to 0$ in probability. We can now easily prove that the expression standing under the sign of mathematical expectation in the second term in (4) tends to zero in probability. From (5) we conclude that the mathematical expectation of (5) as well tends to zero (compare with the deduction of Lemma 2.7.6 from Lemma 2.7.5). The lemma is proved. □

8. Proof of Theorem 1.6. We shall drop the superscripts (α,s,x). Further, we take the supermartingale $\kappa_t = \kappa_t^{\alpha,s,x}$ from Lemma 5, which is, according to Lemma 7, continuous in t. Therefore, by the lemma given in Appendix 2, the processes

$$\rho_t = \int_0^t [f^{\alpha_u}(s + u, x_u) + r_u v(s + u, x_u)] \exp\left(-\varphi_u - \int_0^u r_u\, du\right) du$$
$$+ v(s + t, x_t) \exp\left(-\varphi_t - \int_0^t r_u\, du\right), \qquad \kappa_t - \rho_t,$$

are supermartingales for $t \in [0, T - s]$. Therefore,

$$0 = \mathsf{M}_{s,x}^\alpha[\kappa_0 - \rho_0] \geq \mathsf{M}_{s,x}^\alpha[\kappa_\tau - \rho_\tau].$$

Applying the properties of supermartingales one more time, we obtain

$$v(s,x) = \mathsf{M}_{s,x}^\alpha \rho_0 \geq \mathsf{M}_{s,x}^\alpha \rho_t \geq \mathsf{M}_{s,x}^\alpha \kappa_\tau \geq \mathsf{M}_{s,x}^\alpha \kappa_{T-s} = v^\alpha(s,x).$$

It remains to take in the above inequalities upper bounds with respect to $\alpha \in \mathfrak{A}$, which completes the proof of Theorem 1.6. □

If we take in the above inequalities upper bounds with respect to $\alpha \in \mathfrak{A}_E(s,x)$ and if, in addition, we use Theorem 2a, we arrive at the second assertion of Theorem 1.7, thus completing the proof of Theorem 1.7.

9. Exercise

In proving Lemma 7, we introduced the function $h(y)$. With the aid of h, we define a convex modulus of continuity of $v(t,x)$ at a point x_0 according to the formula

$$\omega(\varepsilon) = \sup\{\mathsf{M}\bar{h}(\xi) : \mathsf{M}\xi \le \varepsilon, 0 \le \xi \le 1\},$$

where $\bar{h}(r) = \sup_{|y| \le r} h(y)$.

Prove that

$$|v(s,x_0) - v(t,x_0)| \le N|t - s|(1 + |x_0|)^{2m+1} + \omega(N\sqrt{|t - s|})$$

if $N\sqrt{|t - s|} \le 1$, where $N = N(K,T,m)$.

10. Remark. From Theorem 2 and Corollary 2.12 it follows that the function $v(s,x)$ can be defined uniquely after the functions σ, b, c, f, g have been given. The function $v(s,x)$ depends on neither the probability space nor the Wiener process involved.

4. The Proof of Theorems 1.8–1.11 for the Optimal Stopping Problem

In this section we shall use the method of randomized stopping (see Section 1.2). Recall that this method consists in the introduction of the multiplier $\exp(-\int_0^t r_u\,du)$ into the functional which characterizes the payoff, and also in the replacement of the function f^α by $f^\alpha + rg$. In accord with this remark we shall carry out the following construction. For $n > 0$ let $B_n = A \times [0,n]$. Furthermore, for $\beta = (\alpha,\bar{r}) \in B_n$ let

$$\sigma(\beta,t,x) = \sigma(\alpha,t,x), \qquad b(\beta,t,x) = b(\alpha,t,x),$$
$$c^\beta(t,x) = c^\alpha(t,x) + \bar{r},$$
$$f^\beta(t,x) = f^\alpha(t,x) + \bar{r}g(t,x), \qquad g(x) = g(T,x).$$

It is clear that for each n for $\beta \in B_n$, $t \ge 0$, x and $y \in E_d$ the functions $\sigma(\beta,t,x)$, $b(\beta,t,x)$ satisfy conditions (1.1) and (1.2) with the same constant K, and, in additon, the functions $c^\beta(t,x)$, $f^\beta(t,x)$, $g(x)$ satisfy the growth condition (1.3) with the same constant m and a different constant K. Hence as in Section 1 where we introduced the concepts of a strategy, a natural strategy, and a payoff function with respect to A, $\sigma(\alpha,t,x)$, $b(\alpha,t,x)$, $c^\alpha(t,x)f^\alpha(t,x)$, $g(x)$, we can introduce here analogous quantities with respect to B_n, $\sigma(\beta,t,x)$, $b(\beta,t,x)$, $c^\beta(t,x)$, $f^\beta(t,x)$, $g(x) = g(T,x)$. We denote by $\mathfrak{B}_n$ the set of corresponding strategies, and denote by $\mathfrak{B}_{n,E}(s,x)$ the set of natural strategies.

Let $\mathfrak{R}_n$ be a set of nonnegative processes $\bar{r}_t$ which are progressively measurable with respect to $\{\mathscr{F}_t\}$ and such that $\bar{r}_t(\omega) \le n$ for all (t,ω), $\mathfrak{B} = \bigcup \mathfrak{B}_n$, $\mathfrak{B}_E(s,x) = \bigcup \mathfrak{B}_{n,E}(s,x)$, $\mathfrak{R} = \bigcup \mathfrak{R}_n$.

Each strategy $\beta \in \mathfrak{B}_n$ is, obviously, a pair of processes $(\alpha_t,\bar{r}_t)$, with $\alpha = \{\alpha_t\} \in \mathfrak{A}$, $\bar{r} = \{\bar{r}_t\} \in \mathfrak{R}_n$. Conversely, each pair of this kind yields a strategy in $\mathfrak{B}_n$. It is easily seen that if $\beta = (\alpha,\bar{r}) \in \mathfrak{B}_n$, $x_t^{\alpha,s,x}$ is a solution of the equation

$$x_t = x + \int_0^t \sigma(\beta_{t_1}, s + t_1, x_{t_1})\, d\mathbf{w}_{t_1} + \int_0^t b(\beta_{t_1}, s + t_1, x_{t_1})\, dt_1.$$

In other words, $x_t^{\beta,s,x} = x_t^{\alpha,s,x}$.

Let

$$\tilde{v}_n(s,x) = \sup_{\beta \in \mathfrak{B}_n} \mathsf{M}_{s,x}^{\beta}\left[\int_0^{T-s} f^{\beta_t}(s + t, x_t)e^{-\varphi_t}\, dt + g(T, x_{T-s})e^{-\varphi_{T-s}}\right].$$

Here, as well as above, the indices attached to the sign of the mathematical expectation imply the mathematical expectation of a expression in which these indices are used wherever possible. Theorems 1.5–1.7, as well as the results obtained in Sections 2 and 3, are applicable to the function $\tilde{v}_n(s,x)$ as well as to a payoff function in the control problem without stopping. In particular $\tilde{v}_n(s,x)$ is continuous with respect to (s,x), $\tilde{v}_n(T,x) = g(T,x)$ (Theorem 1.5).

1. Lemma. *Let $s \in [0,T]$, $x \in E_d$, $\beta = (\alpha,\bar{r}) \in \mathfrak{B}$. Then the process*

$$\tilde{v}_n(s + t, x_t^{\alpha,s,x})e^{-\varphi_t^{\beta,s,x}} + \int_0^t [f^{\alpha_p}(s + p, x_p^{\alpha,s,x}) + \bar{r}_p\tilde{v}_n(s + p, x_p^{\alpha,s,x})]e^{-\varphi_p^{\beta,s,x}}\, dp,$$

defined for $t \in [0, T - s]$, is a continuous supermartingale.

PROOF. By Lemma 3.5a for $\beta \in \mathfrak{B}_n$ the process

$$\tilde{v}_n(s + t, x_t^{\beta,s,x})e^{-\varphi_t^{\beta,s,x}} + \int_0^t f^{\beta_p}(s + p, x_p^{\beta,s,x})e^{-\varphi_t^{\beta,s,x}}\, dp$$

is a supermartingale. In particular, $(\beta = (\alpha,0))$,

$$\tilde{v}_n(s + t, x_t^{\alpha,s,x})e^{-\varphi_t^{\alpha,s,x}} + \int_0^t f^{\alpha_p}(s + p, x_p^{\alpha,s,x})e^{-\varphi_p^{\alpha,s,x}}\, dp$$

is a supermartingale. It remains to apply the lemma from Appendix 2 to the last expression, thus completing the proof of our lemma.

2. Lemma. *Let $s \in [0,t]$, $x \in E_d$, $\gamma^i \in \mathfrak{M}(T - s)$, $\beta^i = (\alpha^i,\bar{r}^i) \in \mathfrak{B}(i = 1,2,\ldots)$. Further, let a Borel function $u(t,x)$ satisfy the inequality $|u(t,y)| \le N(1 + |y|)^m$ with the same constant N for all $t \ge 0$, $y \in E_d$. In addition, let*

$$\lim_{i\to\infty} \mathsf{M}_{s,x}^{\alpha^i} \int_0^{\gamma^i} \bar{r}_t^i \exp\left(-\varphi_t - \int_0^t \bar{r}_p^i\, dp\right) dt = 0.$$

Then

$$\lim_{i\to\infty}\left|\mathsf{M}^{\beta^i}_{s,x}\left\{\int_0^{\gamma^i} f^{\beta^i_t}(s+t,x_t)e^{-\varphi_t}\,dt + u(s+\gamma^i,x_{\gamma^i})e^{-\varphi_{\gamma^i}}\right\}\right.$$
$$\left.-\,\mathsf{M}^{\alpha^i}_{s,x}\left\{\int_0^{\gamma^i} f^{\alpha^i_t}(s+t,x_t)e^{-\varphi_t}\,dt + u(s+\gamma^i,x_{\gamma^i})e^{-\varphi_{\gamma^i}}\right\}\right| = 0.$$

PROOF. It can easily be seen that

$$\left|f^{\beta^i_t}(s+t,x_t^{\alpha^i,s,x})e^{-\varphi_t^{\beta^i,s,x}} - f^{\alpha^i_t}(s+t,x_t^{\alpha^i,s,x})e^{-\varphi_t^{\alpha^i,s,x}}\right|$$
$$\le \left|f^{\alpha^i_t}(s+t,x_t^{\alpha^i,s,x})\right|e^{-\varphi_t^{\alpha^i,s,x}}\left(1-\exp\left(-\int_0^t \bar r^i_p\,dp\right)\right)$$
$$+\, r^i_t\left|g(s+t,x_t^{\alpha^i,s,x})\right|\exp\left(-\int_0^t \bar r^i_p\,dp\right).$$

Integrating the both sides of the last expression over $t\in[0,\gamma^i]$, introducing the notation

$$h^i = \sup_{t\in[0,T-s]}(1+|x_t^{\alpha^i,s,x}|)^m$$

and finally, noting that $|f^\alpha(t,x)|\le K(1+|x|)^m$, $|g(t,x)|\le K(1+|x|)^m$, we find

$$\left|\mathsf{M}^{\beta^i}_{s,x}\int_0^{\gamma^i} f^{\beta^i_t}(s+t,x_t)e^{-\varphi_t}\,dt - \mathsf{M}^{\alpha^i}_{s,x}\int_0^{\gamma^i} f^{\alpha^i_t}(s+t,x_t)e^{-\varphi_t}\,dt\right|$$
$$\le K\mathsf{M}h^i\left[\int_0^{\gamma^i}\left(1-\exp\left(-\int_0^t \bar r^i_p\,dp\right)\right)dt + \int_0^{\gamma^i}\bar r^i_t\exp\left(-\int_0^t \bar r^i_p\,dp\right)dt\right]$$
$$\le K(T-s+1)\mathsf{M}h^i\left(1-\exp\left(-\int_0^{\gamma^i}\bar r^i_t\,dt\right)\right).$$

It is also seen that

$$\left|\mathsf{M}^{\beta^i}_{s,x}u(s+\gamma^i,x_{\gamma^i})e^{-\varphi_{\gamma^i}} - \mathsf{M}^{\alpha^i}_{s,x}u(s+\gamma^i,x_{\gamma^i})e^{-\varphi_{\gamma^i}}\right|$$
$$\le N\mathsf{M}h^i\left(1-\exp\left(-\int_0^{\gamma^i}\bar r^i_t\,dt\right)\right).$$

Therefore, it suffices to show that the last expression tends to zero. Since $c^\alpha(t,x)\le K(1+|x|)^m$, then $\varphi_t^{\alpha^i,s,x}\le KRT$ for $h^i\le R$. Hence

$$I \equiv \overline{\lim_{i\to\infty}}\,\mathsf{M}h^i\left(1-\exp\left(-\int_0^{\gamma^i}\bar r^i_t\,dt\right)\right) \le \overline{\lim_{i\to\infty}}\,\mathsf{M}h^i\chi_{h^i>R}\left(1-\exp\left(-\int_0^{\gamma^i}\bar r^i_t\,dt\right)\right)$$
$$+\,R\,\overline{\lim_{i\to\infty}}\,\mathsf{M}\chi_{h^i\le R}\int_0^{\gamma^i}\bar r^i_t\exp\left(-\int_0^t \bar r^i_p\,dp\right)dt$$
$$\le \frac{1}{R}\sup_i \mathsf{M}(h^i)^2 + Re^{KRT}\,\overline{\lim_{i\to\infty}}\,\mathsf{M}\chi_{h^i\le R}\int_0^{\gamma^i}\bar r^i_t\exp\left(-\varphi_t^{\alpha^i,s,x}-\int_0^t \bar r^i_p\,dp\right)dt.$$

By hypothesis, the last term is equal to zero. Furthermore, it follows from estimates of moments of solutions of stochastic equations (see Corollary 2.5.12) that $\sup_i \mathsf{M}(h^i)^2 < \infty$. Therefore, letting $R \to \infty$ in the inequality

$$I \leq \frac{1}{R} \sup_i \mathsf{M}(h^i)^2,$$

we obtain $I = 0$, which proves the lemma. □

3. Lemma. (a) *Let* $s \in [0,T]$, $x \in E_d$ *and for each* $\alpha \in \mathfrak{A}$ *let* $\tau^\alpha \in \mathfrak{M}(T-s)$, $r^\alpha \in \mathfrak{R}$ *be defined. Then*

$$\tilde{v}_n(s,x) = \sup_{\alpha \in \mathfrak{A}} \mathsf{M}^\alpha_{s,x}\left\{\tilde{v}_n(s+\tau, x_\tau)\exp\left(-\varphi_\tau - \int_0^\tau r_p\,dp\right)\right.$$
$$\left. + \int_0^\tau [f^{\alpha_t} + n(g-\tilde{v}_n)_+ + r_t\tilde{v}_n](s+t, x_t)\exp\left(-\varphi_t - \int_0^t r_p\,dp\right)dt\right\}. \tag{1}$$

(b) *Let* $g_n = g \wedge \tilde{v}_n$. *Then*

$$\tilde{v}_n(s,x) = \sup_{\alpha \in \mathfrak{A}} \sup_{\tau \in \mathfrak{M}(T-s)} \mathsf{M}^\alpha_{s,x}\left[\int_0^\tau f^{\alpha_t}(s+t, x_t)e^{-\varphi_t}\,dt + g_n(s+\tau, x_\tau)e^{-\varphi_\tau}\right].$$

Furthermore, we can replace in (a) *and* (b) *the upper bound with respect to* $\alpha \in \mathfrak{A}$ *by an upper bound with respect to* $\alpha \in \mathfrak{A}_E(s,x)$.

PROOF. By Theorem 1.6, Eq. (1) holds for any τ^α, r^α if it holds for $\tau^\alpha = T-s$, $r^\alpha = 0$. We deduce assertion (b) from (a) for $r^\alpha = 0$ in the same way as we deduced the corresponding assertion from Eq. (1.5.2) in proving Lemma 1.5.2. Theorem 1.7 implies that it is possible to replace $\mathfrak{A}$ by $\mathfrak{A}_E(s,x)$ in the preceding considerations.

Thus, it remains only to prove (1) for $\tau^\alpha = T-s$, $r^\alpha = 0$. Let $\beta = (\alpha,\bar{r}) \in \mathfrak{B}_n$. Furthermore, let

$$\kappa = \tilde{v}_n(s+t, x_t^{\beta,s,x})e^{-\varphi_t^{\beta,s,x}} + \int_0^t f^{\beta_p}(s+p, x_p^{\beta,s,x})e^{-\varphi_p^{\beta,s,x}}\,dp,$$

$$\Phi_t = \exp\left(\int_0^t \bar{r}_p\,dp\right).$$

According to Lemma 3.5 the process κ_t is a supermartingale. According to the lemma given in Appendix 2, the process $\rho_t \equiv \kappa_t\Phi_t - \int_0^t \kappa_s\,d\Phi_s$ is a supermartingale and, in addition,

$$\tilde{v}_n(s,x) = \mathsf{M}\rho_0 \geq \mathsf{M}\rho_{T-s} \geq e^{n(T-s)}[\mathsf{M}\kappa_{T-s} - \tilde{v}_n(s,x)] + \tilde{v}_n(s,x). \tag{2}$$

Using Fubini's theorem, we easily prove that

$$\mathsf{M}\rho_{T-s} = \mathsf{M}^\alpha_{s,x}\left\{g(T,x_{T-s})e^{-\varphi_{T-s}} + \int_0^{T-s}[f^{\alpha_t} + \bar{r}_t(g-\tilde{v}_n)](s+t, x_t)e^{-\varphi_t}\,dt\right\}.$$

Obviously, the upper bound of the last expression with respect to $\bar{r} \in \mathfrak{R}_n$ is equal to

$$\mathsf{M}_{s,x}^{\alpha}\left\{\int_0^{T-s} [f^{\alpha_t} + n(g - \tilde{v}_n)_+](s + t, x_t)e^{-\varphi_t}\, dt + g(T,x_{T-s})e^{-\varphi_{T-s}}\right\}.$$

Taking this fact into consideration, recalling the definition of $\tilde{v}_n$, and, finally, computing the upper bounds in (2) with respect to $\alpha \in \mathfrak{A}$, $\bar{r} \in \mathfrak{R}_n$, we arrive at (1) for $\tau^{\alpha} = T - s$, $r^{\alpha} = 0$. The lemma is proved. □

4. Corollary. *Since* $g_n \le g$, $\tilde{v}_n \le w$.

5. Lemma. (a) *The function* $w(s,x)$ *is continuous with respect to* s, x.

(b) *There exists a constant* N *such that* $|\tilde{v}_n(s,x)| \le N(1 + |x|)^m$ *for all* n, s, x.

(c) $\tilde{v}_n(s,x) \uparrow w(s,x)$ *uniformly on each set of the form* $\{(s,x): s \in [0,T], |x| \le R\}$.

PROOF. Assertion (a) follows from (b) as well as the continuity property of $\tilde{v}_n(s,x)$. Since $\mathfrak{B}_n \subset \mathfrak{B}_{n+1}$, the sequence $\tilde{v}_n(s,x)$ increases. Moreover, in accord with Corollary 4, $\tilde{v}_n \le w$ and, obviously, $\tilde{v}_0(s,x) \le \tilde{v}_n(s,x)$; in this case the function $\tilde{v}_0$ does not differ essentially from the function v considered in Section 1. All this together with the estimates of v, w given in Section 1 proves assertion (b).

Let $\bar{w}(s,x) = \lim_{n\to\infty} \tilde{v}_n(s,x)$. By Corollary 4, $\bar{w}(s,x) \le w(s,x)$. On the other hand, for $\alpha \in \mathfrak{A}$, $\tau \in \mathfrak{M}(T - s)$ let $\bar{r}_t = n\chi_{\tau \le t}$, $\beta_t = (\alpha_t, \bar{r}_t)$. Then, using Fubini's theorem, it is not hard to obtain

$$\begin{aligned}\tilde{v}_n(s,x) &\ge \mathsf{M}_{s,x}^{\beta}\left\{\int_0^{T-s} f^{\beta_t}(s + t, x_t)e^{-\varphi_t}\, dt + g(T,x_{T-s})e^{-\varphi_{T-s}}\right\}\\ &= \mathsf{M}_{s,x}^{\alpha}\int_{\tau}^{T-s} ne^{-n(t-\tau)}\left[\int_0^t f^{\alpha_p}(s + p, x_p)e^{-\varphi_p}\, dp + g(s + t, x_t)e^{-\varphi_t}\right] dt\\ &\quad + \mathsf{M}_{s,x}^{\alpha}e^{-n(T-s-\tau)}\left[\int_0^{T-s} f^{\alpha_t}(s + t,x_t)e^{-\varphi_t}\, dt + g(T,x_{T-s})e^{-\varphi_{T-s}}\right].\end{aligned}$$

Let us write last relation in a different form. Let

$$\eta^{\alpha,s,x}(t) = \int_0^t f^{\alpha_t}(s + p, x_p^{\alpha,s,x})e^{-\varphi_p^{\alpha,s,x}}\, dp + g(s + t, x_t^{\alpha,s,x})e^{-\varphi_t^{\alpha,s,x}}$$

for $t \le T - s$, $\eta^{\alpha,s,x}(t) = \eta^{\alpha,s,x}(T - s)$ for $t > T - s$. Furthermore, we introduce a random variable ξ which has an exponential distribution with a parameter equal to unity and which, in addition, does not depend on $\{\eta^{\alpha,s,x}(t)\}$. What we have obtained can be written as follows: $\tilde{v}_n(s,x) \ge \mathsf{M}_{s,x}^{\alpha}\eta(\tau + (1/n)\xi)$. $\bar{w}(s,x) \ge \mathsf{M}_{s,x}^{\alpha}\eta(\tau + (1/n)\xi)$. Letting n tend to infinity, we note that the process $\eta^{\alpha,s,x}(t)$ is continuous with respect to t,

$$|\eta^{\alpha,s,x}(t)| \le K(T - s + 1)\left(1 + \sup_{t \le T-s} |x_t^{\alpha,s,x}|\right)^m,$$

and, finally, we note that the last quantity is summable. Therefore, by Lebesgue's theorem,

$$\bar{w}(s,x) \geq \mathsf{M}_{s,x}^{\alpha}\eta(\tau) = v^{\alpha,\tau}(s,x), \qquad \bar{w}(s,x) \geq w(s,x).$$

We conclude that $\bar{w}(s,x) = w(s,x)$.

From the last equality and the inequality $w(s,x) \geq g(s,x)$ it follows, in particular, that the decreasing sequence of nonnegative continuous functions

$$g(s,x) - g_n(s,x) = g(s,x) - g(s,x) \wedge \tilde{v}_n(s,x)$$
$$\to g(s,x) - g(s,x) \wedge w(s,x) = 0.$$

By Dini's theorem, $g(s,x) - g_n(s,x) \to 0$ uniformly on each cylinder $\bar{C}_{T,R}$. In view of Lemma 3 (and Corollary 1.13), in order to prove (c) it suffices to show that $|g_n(s,x)| \leq N(1 + |x|)^m$ with the same constant N for all n, s, x. The last inequality follows from assertion (b), thus proving the lemma. □

6. Remark. The proof of (a) completes the proof of Theorem 1.8.

Theorems 1.9 and 1.10, as was seen in Section 1, follow from Theorem 1.11. For proving Theorem 1.11 we need an analog of Lemma 1, which is a combination of Lemma 1 and Lemma 5 (and Theorem 1.12).

7. Corollary. *Let $s \in [0,T]$, $x \in E_d$, $\beta = (\alpha,\bar{r}) \in \mathfrak{B}$. Then the process*

$$\rho_t^{\beta,s,x} \equiv w(s + t, x_t^{\alpha,s,x})e^{-\varphi_t^{\beta,s,x}}$$
$$+ \int_0^t [f^{\alpha_p}(s + p, x_p^{\alpha,s,x}) + \bar{r}_p w(s + p, x_p^{\alpha,s,x})]e^{-\varphi_p^{\beta,s,x}}\, dp,$$

defined for $t \in [0, T - s]$, is a continuous supermartingale.

The process $\rho_{t\wedge\tau}^{\beta,s,x}$ is also a supermartingale for each $\tau \in \mathfrak{M}(T - s)$. Subtracting from the process $\rho_{t\wedge\tau}^{\beta,s,x}$ for $\beta = (\alpha,0)$ the martingale

$$\mathsf{M}_{s,x}^{\alpha}\left\{\int_0^\tau f^{\alpha_p}(s + p, x_p)e^{-\varphi_p}\, dp + g(s + \tau, x_\tau)e^{-\varphi_\tau} \middle| \mathscr{F}_t\right\},$$

we arrive at a supermartingale which for $t = T - s$ is equal to

$$[w(s + \tau, x_\tau^{\alpha,s,x}) - g(s + \tau, x_\tau^{\alpha,s,x})]e^{-\varphi_\tau^{\alpha,s,x}}.$$

The last expression is nonnegative. Further, from the definition of the supermartingale it immediately follows that a supermartingale which is positive (a.s.) at a certain moment of time, is positive (a.s.) at each preceding moment of time. Summing up what has been said above, we have the following result.

8. Corollary. *Let $s \in [0,T]$, $\tau \in \mathfrak{M}(T - s)$, $x \in E_d$, $\alpha \in \mathfrak{A}$. Then the process*

$$w(s+t\wedge\tau, x_{t\wedge\tau}^{\alpha,s,x})e^{-\varphi_{t\wedge\tau}^{\alpha,s,x}} - \mathsf{M}_{s,x}^{\alpha}\left\{g(s+\tau, x_\tau)e^{-\varphi_\tau} + \int_{\tau\wedge t}^{\tau} f^{\alpha_p}(s+p, x_p)e^{-\varphi_p}\, dp \middle| \mathscr{F}_t\right\}$$

is a nonnegative supermartingale for $t \in [0, T - s]$.

9. Proof of Theorem 1.11. Corollary 7 and the properties of supermartingales imply

$$w(s,x) \geq \mathsf{M}_{s,x}^{\alpha}\left\{\int_0^\tau [f^{\alpha_t}(s+t,x_t) + r_t w(s+t,x_t)] \exp\left(-\varphi_t - \int_0^t r_p\,dp\right) dt \right.$$

$$\left. + w(s+\tau,x_\tau)\exp\left(-\varphi_\tau - \int_0^\tau r_p\,dp\right)\right\}, \tag{3}$$

which proves inequality (1.10). Next, let $\tau^\alpha \leq \tau_\varepsilon^{\alpha,s,x}$, $\varepsilon > 0$. For $\beta = (\alpha,\bar{r}) \in \mathfrak{B}$ let $\tau^\beta = \tau^\alpha$. According to Bellman's principle (see Theorem 1.7) for each n

$$\tilde{v}_n(s,x) = \sup_{\beta \in \mathfrak{B}_{n,E}(s,x)} \mathsf{M}_{s,x}^{\alpha}\left\{\int_0^\tau f^{\beta_t}(s+t,x_t)e^{-\varphi_t}\,dt + \tilde{v}_n(s+\tau,x_\tau)e^{-\varphi_\tau}\right\}$$

$$\leq \sup_{\alpha \in \mathfrak{A}_E(s,x)} \sup_{\bar{r} \in \mathfrak{R}} \mathsf{M}_{s,x}^{\alpha}\left\{\int_0^\tau [f^{\alpha_t}(s+t,x_t)\right.$$

$$+ \bar{r}_t g(s+t,x_t)] \exp\left(-\varphi_t - \int_0^t \bar{r}_p\,dp\right) dt$$

$$\left. + w(s+\tau,x_\tau)\exp\left(-\varphi_\tau - \int_0^\tau \bar{r}_p\,dp\right)\right\}.$$

Taking the limit in the last expression as $n \to \infty$ and, furthermore, using the inequality $g(s,x) \leq w(s,x)$, (3) and the fact that $\tilde{v}_n \uparrow w$, we have

$$w(s,x) = \sup_{\alpha \in \mathfrak{A}_E(s,x)} \mathrm{up}_{\bar{r} \in \mathfrak{R}} \mathsf{M}_{s,x}^{\alpha}\left\{\int_0^\tau [f^{\alpha_t}(s+t,x_t)\right.$$

$$+ \bar{r}_t g(s+t,x_t)] \exp\left(-\varphi_t - \int_0^t \bar{r}_p\,dp\right) dt$$

$$\left. + w(s+\tau,x_\tau)\exp\left(-\varphi_\tau - \int_0^\tau \bar{r}_p\,dp\right)\right\}.$$

Further, we take a sequence $\alpha^i \in \mathfrak{A}_E(s,x)$, $\bar{r}^i \in \mathfrak{R}$, for which

$$w(s,x) = \lim_{i\to\infty} \mathsf{M}_{s,x}^{\alpha^i}\left\{\int_0^\tau [f^{\alpha_t^i}(s+t,x_t) + \bar{r}_t^i g(s+t,x_t)] \exp\left(-\varphi_t - \int_0^t \bar{r}_p^i\,dp\right) dt \right.$$

$$\left. + w(s+\tau,x_\tau)\exp\left(-\varphi_\tau - \int_0^\tau \bar{r}_p^i\,dp\right)\right\}. \tag{4}$$

From the inequality $g(s+t,x_t^{\alpha,s,x}) < w(s+t,x_t^{\alpha,s,x}) - \varepsilon$ for $t < \tau_\varepsilon^{\alpha,s,x}$, and also from (3) and (4), we find

$$\varepsilon \lim_{i\to\infty} \mathsf{M}_{s,x}^{\alpha^i} \int_0^\tau \bar{r}_t^i \exp\left(-\varphi_t - \int_0^t \bar{r}_p^i\,dp\right) dt = 0.$$

By Lemma 2, the last expression and (4) yield

$$w(s,x) = \lim_{i\to\infty} \mathsf{M}_{s,x}^{\alpha^i}\left\{\int_0^\tau f^{\alpha_t^i}(s+t,x_t)e^{-\varphi_t}\,dt + w(s+\tau,x_\tau)e^{-\varphi_\tau}\right\}. \tag{5}$$

By Corollary 7, the process

$$\kappa_t^{\alpha,s,x} \equiv w(s+t,x_t^{\alpha,s,x})e^{-\varphi_t^{\alpha,s,x}} + \int_0^t f^{\alpha_p}(s+p,x_p^{\alpha,s,x})e^{-\varphi_p^{\alpha,s,x}}\,dp$$

is a continuous supermartingale. Therefore, according to the lemma given in Appendix 2, the process $\kappa_t^{\alpha,s,x} - \rho_t^{\beta,s,x}$ is a supermartingale for each $\beta = (\alpha,r) \in \mathfrak{B}$. In particular, $\mathsf{M}_{s,x}^{\alpha}\kappa_\tau \le \mathsf{M}_{s,x}^{\beta}\rho_\tau$, which together with (5) and (3) yields

$$\begin{aligned} w(s,x) &\le \lim_{i\to\infty} \mathsf{M}_{s,x}^{\alpha^i}\left\{\int_0^\tau [f^{\alpha_t^i}(s+t,x_t) + r_t w(s+t,x_t)]\exp\left(-\varphi_t - \int_0^t r_p\,dp\right)dt \right.\\ &\quad \left. + w(s+\tau,x_\tau)\exp\left(-\varphi_\tau - \int_0^\tau r_p\,dp\right)\right\} \\ &\le \sup_{\alpha\in\mathfrak{A}_E(s,x)} \mathsf{M}_{s,x}^{\alpha^i}\left\{w(s+\tau,x_\tau)\exp\left(-\varphi_\tau - \int_0^\tau r_p\,dp\right)\right. \\ &\quad \left. + \int_0^\tau [f^{\alpha_t}(s+t,x_t) + r_t w(s+t,x_t)]\exp\left(-\varphi_t - \int_0^t r_p\,dp\right)dt\right\} \\ &\le w(s,x). \end{aligned}$$

It only remains to prove that in (1.10) we have equality if A consists of a single point and if, in addition, $\tau^\alpha \le \tau_0^{\alpha,s,x}$. In this case we do not write the superscript α since we deal with only one strategy. Let $\tau \le \tau_0^{s,x}$. For $\varepsilon > 0$, as can be seen, $\tau \wedge \tau_\varepsilon^{s,x} \le \tau_\varepsilon^{s,x}$, and, further, in accord with what has been proved,

$$\begin{aligned} w(s,x) &= \mathsf{M}_{s,x}\left\{\int_0^{\tau\wedge\tau_\varepsilon} [f(s+t,x_t) + r_t w(s+t,x_t)]\exp\left(-\varphi_t - \int_0^t r_p\,dp\right)dt\right. \\ &\quad \left. + w(s+\tau\wedge\tau_\varepsilon, x_{\tau\wedge\tau_\varepsilon})\exp\left(-\varphi_{\tau\wedge\tau_\varepsilon} - \int_0^{\tau\wedge\tau_\varepsilon} r_p\,dp\right)\right\}. \end{aligned}$$

Letting $\varepsilon \downarrow 0$ and noting that $\tau \wedge \tau_\varepsilon^{s,x} \uparrow \tau \wedge \tau_0^{s,x} = \tau \le T - s$, the function $w(t,x)$ is continuous with respect to (t,x) and also, noting that the quantity

$$\int_0^{T-s}\left[|f(s+t,x_t)| + r_t \sup_{p\le T-s}|w(s+p,x_p)|\right]\exp\left(-\int_0^t r_p\,dp\right)dt,$$

$\sup_{p\le T-s}|w(s+p,x_p)|$ has a finite mathematical expectation, we have proved what was required. This completes the proof of Theorem 1.11. □

10. Remark. Applying Remark 3.10 to the functions $\tilde{v}_n(s,x)$, we see that these functions are defined uniquely by the functions $\sigma(\beta,t,y)$, $b(\beta,t,y)$, $c^\beta(t,y)$, $f^\beta(t,y)$

and $g(T,y)$. The latter functions are expressible in terms of $\sigma(\alpha,t,y)$, $b(\alpha,t,y)$, $c^\alpha(t,y)$, $f^\alpha(t,y)$ and $g(t,y)$, which together with Lemma 5c proves that in order to compute the function $w(s,x)$ it is sufficient to give the functions $\sigma(\alpha,t,y)$, $b(\alpha,t,y)$, $c^\alpha(t,y)$, $f^\alpha(t,y)$, and $g(t,y)$.

11. Exercise

Examine the possibility of "pasting" to the strategy which "serves well" until a moment of time τ_ε a strategy which "serves well" during an interval between τ_ε and $\tau_{\varepsilon/2}$, and, furthermore, of extending this procedure, and show that the final assertion of Theorem 1.11 holds in every case, not merely in the case where A consists of a single point.

12. Exercise

Prove that for $u \equiv w$ and for $u \equiv g$

$$w(s,x) = \sup_{\alpha \in \mathfrak{A}} \sup_{r \in \mathfrak{R}} \mathsf{M}_{s,x}^{\alpha} \left\{ \int_0^{T-s} [f^{\alpha_t}(s+t, x_t) + r_t u(s+t, x_t)] \exp\left(-\varphi_t - \int_0^t r_p\, dp \right) dt + u(T, x_{T-s}) \exp\left(-\varphi_{T-s} - \int_0^{T-s} r_p\, dp \right) \right\}.$$

We conclude the discussion in this section by formulating two theorems. In the first theorem we shall estimate the rate of convergence of $\tilde{v}_n$ to w. In the second theorem we shall give one connectivity property of a set Q_0.

Both theorems mentioned will be proved in Section 5.3. However, here we note that the expressions appearing in the formulations of Theorems 13 and 14 which follow will be determined in the introduction to Chapter 4. Also, we note that the spaces $W^{1,2}_{\text{loc}}(H_T)$ will be introduced in Section 5.3 (see Definition 5.3.1).

13. Theorem. *Let* $g \in W^{1,2}_{\text{loc}}(H_T) \cap C(\bar{H}_T)$, $F[g] \geq -K(1 + |x|)^m$ (H_T-a.s.). *Then in* H_T

$$|w(s,x) - \tilde{v}_n(s,x)| \leq \frac{1}{n} N(K,m,T)(1 + |x|)^m.$$

14. Theorem. *Let* $g \in W^{1,2}_{\text{loc}}(H_T) \cap C(H_T)$, $s \in [0,T]$. *Furthermore, let there exist a function* $h(t,x)$ *which is continuous in* $(s,T) \times E_d$ *and which coincides with* $F[g](t,x)$ *almost everywhere in* $(s,T) \times E_d$. *Further, let*

$$Q = \{(t,x) : t \in (s,T), x \in E_d, h(t,x) > 0\},$$
$$Q_0' = \{(t,x) : t \in (s,T), x \in E_d, w(t,x) > g(t,x)\}.$$

Then $Q \subset Q_0'$ *and also each connected component of the region* Q_0' *contains at least one connected component of the region* Q. *In particular, if the set* Q *is connected, that is, if it consists of a single connected component, then the set* Q_0' *is connected as well.*

Notes

Section 1. The results valid for the general case appear here for the first time. Some of these results for particular cases can be found in Krylov [36] and Portenko and Skorokhod [61].

Sections 2, 3. Some of the results in these sections can be found but without detailed proofs, in Portenko and Skorokhod [61]. The step strategies are considered in Fleming [14].

Section 4. The methods for investigating the optimal stopping problem used in this section have been borrowed from [29–31, 36]. Theorem 14 is the generalization of a result obtained in [56].

The Bellman Equation 4

In Chapter 3 we investigated general properties of controlled processes, such as continuity of a payoff function, the feasibility of passing to the limit from one process to another, the validity of Bellman's principle of various forms, etc. The assumptions we have made are rather weak. In this chapter we shall see that by making additional assumptions on the smoothness of initial objects, we can prove some smoothness of the payoff functions as well as the fact that the payoff functions satisfy the Bellman equation.

The assumptions, definitions, and notations given in Section 3.1 are used throughout this chapter. Section 5 dealing with a passage to the limit in the Bellman equation is an exception, however, and self-contained in terms of assumptions and definitions. In addition to the main assumptions taken from Section 3.1, each section of this chapter contains assumptions which will be formulated or referred to at the beginning of each section, and which will be of use only in that section. We wish to give particular attention to one peculiarity of our assumptions. We make an assumption about a parameter $m \geq 0$, which yields the rate of growth of functions as $|x| \to \infty$. The simple case when $m = 0$ is not excluded from our assumptions, and in this case the functions in question satisfy the boundedness assumption. For a first reading of Chapter 4, we therefore recommend the reader assume that $m = 0$. Furthermore, it will be easier to comprehend the material of this chapter under the assumption that $c^\alpha(t,x) = 0$.

Let

$$a(\alpha,t,x) = \frac{1}{2}\sigma(\alpha,t,x)\sigma^*(\alpha,t,x),$$

$$L^\alpha u = L^\alpha(t,x)u = \frac{\partial u}{\partial t} + \sum_{i,j=1}^{d} a^{ij}(\alpha,t,x)u_{x^ix^j} + \sum_{i=1}^{d} b^i(\alpha,t,x)u_{x^i} - c^\alpha(t,x)u,$$

$$L^{\alpha *}u = L^{\alpha *}(t,x)u = -\frac{\partial u}{\partial t} + \sum_{i,j=1}^{d} [a^{ij}(\alpha,t,x)u]_{x^i x^j} - \sum_{i=1}^{d} [b^i(\alpha,t,x)u]_{x^i} - c^\alpha(t,x)u,$$

$$F(u_0,u_{ij},u_i,u,t,x) = \sup_{\alpha\in A}\left[u_0 + \sum_{i,j=1}^{d} a^{ij}(\alpha,t,x)u_{ij}\right.$$

$$\left. + \sum_{i=1}^{d} b^i(\alpha,t,x)u_i - c^\alpha(t,x)u + f^\alpha(t,x)\right],$$

$$F_1(u_{ij},t,x) = \sup_{\alpha\in A}\sum_{i,j=1}^{d} a^{ij}(\alpha,t,x)u_{ij},$$

$$F[u] = F[u](t,x) = F\left(\frac{\partial}{\partial t}u(t,x), u_{x^i x^j}(t,x), u_{x^i}(t,x), u(t,x), t, x\right),$$

$$F_1[u] = F_1[u](t,x) = F_1(u_{x^i x^j}(t,x),t,x).$$

Note some properties of the quantities introduced. Since for each (t,x) the functions $a(\alpha,t,x)$, $b(\alpha,t,x)$, $c^\alpha(t,x)$, $f^\alpha(t,x)$ are uniformly bounded with respect to α (see (3.1.2) and (3.1.3)), the functions F, F_1 are finite. Since $a(\alpha,t,x)$, $b(\alpha,t,x)$, $c^\alpha(t,x)$, $f^\alpha(t,x)$ are continuous with respect to $\alpha \in A$ and, since, in addition, the set A is separable, in determining F, F_1, one can take the upper bound with respect to any countable set everywhere dense in A. This implies, for example, that the functions $F(u_0,u_{ij},u_i,u,t,x)$, $F_1(u_{ij},t,x)$, $F[u](t,x)$ are measurable with respect to their arguments. Furthermore, if in a region Q for each $\alpha \in A$ and a function $u(t,x)$

$$L^\alpha u + f^\alpha \le 0 \qquad \text{(a.e. on } Q), \tag{1}$$

one can remove a set Γ_α of measure zero from Q for each $\alpha \in A$, so that the expression $L^\alpha u + f^\alpha$ does not exceed zero on the remaining set. The union of Γ_α with respect to α from any countable subset $\tilde{A}$ of the set A has measure zero, in addition, outside this union $L^\alpha u + f^\alpha \le 0$ on Q for each $\alpha \in \tilde{A}$. If we take in the last inequality the upper bound with respect to $\alpha \in \tilde{A}$, and if we take the set $\tilde{A}$ to be everywhere dense in A, it turns out that $F[u](t,x) \le 0$ on $Q \bigcup_{\alpha\in A} \Gamma_\alpha$. In particular, $F[u] \le 0$ (Q-a.e.). This reasoning shows that if (1) is satisfied for each $\alpha \in A$, then $F[u] \le 0$ (Q-a.e.). It is seen that the converse holds true as well.

Finally, we mention that the function F_1 can be computed on the basis of the function F immediately according to the following simple formula:

$$F_1(u_{ij},t,x) = \lim_{r\to\infty}\frac{1}{r}F(u_0,ru_{ij},u_i,u,t,x).$$

1. Estimation of First Derivatives of Payoff Functions

In addition to the assumptions made in Section 3.1, we assume here that for all $t \in [0,T]$, $\alpha \in A$, $R > 0$, $x, y \in S_R$

$$|c^\alpha(t,x) - c^\alpha(t,y)| + |f^\alpha(t,x) - f^\alpha(t,y)| + |g(x) - g(y)| + |g(t,x) - g(t,y)| \\ \leq K(1 + R)^m|x - y|. \tag{1}$$

1. Theorem. *The functions $v(s,x)$, $w(s,x)$ have, for each $s \in [0,T]$, first-order generalized derivatives with respect to x. Furthermore, there exists a constant $N = N(K,m)$ such that for each $s \in [0,T]$ for almost all x*

$$|\operatorname{grad}_x v(s,x)| + |\operatorname{grad}_x w(s,x)| \leq N(1 + |x|)^{2m}e^{N(T-s)}.$$

PROOF. First, we prove the theorem under the assumption that for each $t \in [0,T]$, $\alpha \in A$ the functions σ, b, c, f, g are once continuously differentiable in x. Then it follows from our assumptions (see (1) and (3.1.1)) that for $l \in E_d$

$$\|\sigma_{(l)}(\alpha,t,x)\| + |b_{(l)}(\alpha,t,x)| \leq K,$$
$$|c^\alpha_{(l)}(t,x)| + |f^\alpha_{(l)}(t,x)| + |g_{(l)}(x)| + |g_{(l)}(t,x)| \leq K(1 + |x|)^m.$$

Relying upon the results obtained in Sections 2.7 and 2.8, we obtain that for any strategy $\alpha \in \mathfrak{A}$, $s \in [0,T]$ and $\tau \in \mathfrak{M}(T - s)$ the functions $v^\alpha(s,x)$ and $v^{\alpha,\tau}(s,x)$ are continuously differentiable in x. In this case, for example,

$$v^{\alpha,\tau}_{(l)}(s,x) = \int_0^{T-s} \mathsf{M}\chi_{t<\tau}\mathscr{L}\text{-}\frac{\partial}{\partial l}[f^{\alpha_t}(s + t, x_t^{\alpha,s,x})e^{-\varphi_t^{\alpha,s,x}}]\,dt$$
$$+ \mathsf{M}\mathscr{L}\text{-}\frac{\partial}{\partial l}[g(s + \tau, x_\tau^{\alpha,s,x})e^{-\varphi_\tau^{\alpha,s,x}}].$$

In order to estimate $v^{\alpha,\tau}_{(l)}(s,x)$, we put $y_t^{\alpha,s,x} = \mathscr{L}\text{-}(\partial/\partial l)x_t^{\alpha,s,x}$. We have

$$\mathsf{M}\chi_{t<\tau}\mathscr{L}\text{-}\frac{\partial}{\partial l}[f^{\alpha_t}(s + t, x_t^{\alpha,s,x})e^{-\varphi_t^{\alpha,s,x}}]$$
$$= \mathsf{M}^\alpha_{s,x}\chi_{t<\tau}f^{\alpha_t}_{(y_t)}(s + t, x_t)|y_t|e^{-\varphi_t}$$
$$- \mathsf{M}^\alpha_{s,x}\chi_{t<\tau}f^{\alpha_t}(s + t, x_t)e^{-\varphi_t}\int_0^t c^{\alpha_r}_{(y_r)}(s + r, x_r)|y_r|\,dr, \tag{2}$$

where the magnitude of the first term does not exceed

$$K\mathsf{M}^\alpha_{s,x}(1 + |x_t|)^m|y_t| \leq K[\mathsf{M}^\alpha_{s,x}(1 + |x_t|)^{2m}]^{1/2}[\mathsf{M}^\alpha_{s,x}|y_t|^2]^{1/2},$$

which yields in turn, by Corollary 2.5.12 and Theorem 2.8.8 on estimation of moments $x_t^{\alpha,s,x}$, $y_t^{\alpha,s,x}$,

$$[\mathsf{M}^\alpha_{s,x}\chi_{t<\tau}f^{\alpha_t}_{(y_t)}(s + t, x_t)|y_t|e^{-\varphi_t}] \leq N(K,m)(1 + |x|)^m e^{N(K,m)t}.$$

In order to estimate the second term in (2), we apply the Cauchy inequality. The square of (2) is estimated in terms of the product of the quantity

$$\mathsf{M}^{\alpha}_{s,x}|f^{\alpha_t}(s+t,x_t)|^2 \le K^2\mathsf{M}^{\alpha}_{s,x}(1+|x_t|)^{2m} \le N(1+|x|)^{2m}e^{Nt}$$

and the quantity

$$\mathsf{M}^{\alpha}_{s,x}\left|\int_0^t c^{\alpha_r}_{(y_r)}(s+r,x_r)|y_r|\,dr\right|^2 \le K^2 t\mathsf{M}^{\alpha}_{s,x}\int_0^t (1+|x_r|)^{2m}|y_r|^2\,dr$$

$$\le K^2 t\int_0^t [\mathsf{M}^{\alpha}_{s,x}(1+|x_r|)^{4m}]^{1/2}[\mathsf{M}^{\alpha}_{s,x}|y_r|^4]^{1/2}\,dr$$

$$\le Nt(1+|x|)^{2m}\int_0^t e^{Nr}\,dr \le N(1+|x|)^{2m}e^{Nt}.$$

Therefore, the magnitude of the second term in (2) does not exceed $N(1+|x|)^{2m}e^{Nt}$. Estimating in a similar way the expression

$$\mathsf{M}\mathscr{L}\text{-}\frac{\partial}{\partial l}[g(s+\tau,x^{\alpha,s,x}_{\tau})e^{-\varphi^{\alpha,s,x}_{\tau}}],$$

which resembles rather closely the left side of (2), we finally find

$$|v^{\alpha,\tau}_{(l)}(s,x)| \le N(K,m)(1+|x|)^{2m}e^{N(K,m)(T-s)}.$$

Next, for $|x|$, $|y| < R$, according to the Lagrange theorem

$$|w(s,x)-w(s,y)| \le \sup_{\substack{\alpha\in\mathfrak{A},\\ \tau\in\mathfrak{M}(T-s)}} |v^{\alpha,\tau}(s,x)-v^{\alpha,\tau}(s,y)|$$

$$\le N(1+R)^{2m}|x-y|e^{N(T-s)}. \tag{3}$$

As in Section 2.1, a function satisfying a Lipschitz condition has generalized derivatives, and in addition, the gradient of such a function does not exceed the Lipschitz constant. Hence (3) implies the existence of first-order generalized derivatives of $w(s,x)$ and, in addition, the inequality

$$|\operatorname{grad}_x w(s,x)| \le N(1+R)^{2m}e^{N(T-s)}$$

for almost all $x \in S_R$. The last inequality implies precisely that

$$|\operatorname{grad}_x w(s,x)| \le N(1+|x|)^{2m}e^{N(T-s)} \qquad \text{(a.e.)}$$

with the same constant N. The function $v(s,x)$ can be considered in a similar manner.

We have thus proved the theorem for smooth functions σ, b, c, f, g. For proving this theorem in the general case we make use of Theorem 3.1.14 and Corollary 3.1.13. We approximate the functions α, b, c, f, $g(x)$, $g(t,x)$ using smooth functions σ_n, b_n, c_n, f_n, $g_n(x)$, $g_n(t,x)$, which we have obtained from the initial functions by means of convolution with a function $\varepsilon_n^{-d}\zeta(\varepsilon_n^{-1}x)$

(see Theorem 3.1.14). Let $\varepsilon_n = 1/n$. For $x, y \in S_R$, for example, let

$$|f_n^\alpha(t,x) - f_n^\alpha(t,y)| = \left|\int_{|z|\le 1} \zeta(z)\left[f^\alpha\left(t, x - \frac{1}{n}z\right) - f^\alpha\left(t, y - \frac{1}{n}z\right)\right]dz\right|$$
$$\le K\left(1 + R + \frac{1}{n}\right)^m |x - y| \le 2^m K(1 + R)^m |x - y|.$$

It is seen from the above that σ_n, b_n, c_n, f_n, $g_n(x)$, $g_n(t,x)$ satisfy our assumptions for the same constant $\tilde{K}$, m. Hence, if we denote by $w_n(s,x)$ the payoff function constructed on the basis of σ_n, b_n, c_n, $g_n(t,x)$, then for $|x|$, $|y| < R$ (see (3))

$$|w_n(s,x) - w_n(s,y)| \le N(K,m)(1 + R)^{2m} e^{N(K,m)(T-s)} |x - y|.$$

Taking the limit in the last inequality as $n \to \infty$, we arrive at (3) in the general case using Corollary 3.1.13 and Theorem 3.1.14. As we have seen above, inequality (3) implies the assertions of the theorem for the function $w(s,x)$. Similar reasoning is suitable for $v(s,x)$, thus proving the theorem. □

2. Exercise

Using Bellman's principle, prove that for some constant $N = N(K,m)$ for $t \le s$

$$|v(t,x) - v(s,x)| \le N\sqrt{t - s}(1 + |x|)^{2m+1} e^{N(T-s)}.$$

3. Remark. If $c^\alpha(t,x) \equiv 0$, (2) contains no second term and, therefore

$$|\operatorname{grad}_x v(s,x)| + |\operatorname{grad}_x w(s,x)| \le N(1 + |x|)^m e^{N(T-s)} \qquad \text{(a.s.)}.$$

It is not in general permissible to assert that Bellman's equation $F[v] = 0$ holds for the function $v(t,x)$ having only first derivatives with respect to x. In fact, the equation $F[v] = 0$, in addition to first derivatives with respect to x, involves second derivatives as well as a derivative with respect to t. It turns out that although the foregoing derivatives enter into the inequality $F[v] \le 0$, we can make this inequality meaningful by integration by parts.

4. Theorem. *Let $R > 0$, and, in addition, let $\eta(t,x)$ be a nonnegative function which is infinitely differentiable on E_{d+1} and is equal to zero outside $[0,T] \times S_R$. Then for $u(t,x) \equiv v(t,x)$ and for $u(t,x) \equiv w(t,x)$ for each $\beta \in A$*

$$\int_{H_T}\left[u\frac{\partial\eta}{dt} + \sum_{i,j=1}^{d} (a^{ij}\eta)_{x^i} u_{x^j} + \sum_{j=1}^{d} (b^i\eta)_{x^i} u + \eta c u - \eta f\right] dx\,dt \ge 0,$$

where we assume for the sake of simplicity that $a^{ij} = a^{ij}(\beta,t,x)$, etc.

Note that the assertion of the theorem makes sense since the functions w_{x^i}, v_{x^i} exist, $\sigma(\beta,t,x)$, $b(\beta,t,x)$ satisfy a Lipschitz condition with respect to x, and, furthermore, they even have bounded first generalized derivatives. The first-order generalized derivatives of the function $a(\beta,t,x) = \frac{1}{2}\sigma(\beta,t,x)\sigma^*(\beta,t,x)$ are bounded in each cylinder C_{T,R_1}. We first prove the theorem for differentiable σ, b. In the lemma which follows assumption (1) will be absent.

5. Lemma. *Let $\beta \in A$ and let the function $\sigma(\beta,t,x)$ for each $t \in [0,T]$ be twice continuously differentiable in x. Also, let $b(\beta,t,x)$ be once continuously differentiable in x for each $t \in [0,T]$. Furthermore, let the corresponding derivatives of these functions be bounded in each cylinder C_{T,R_1}. Then for $u(t,x) \equiv v(t,x)$ and for $u(t,x) \equiv w(t,x)$*

$$\int_{H_T} [uL^{\beta *}\eta + f^\beta \eta]\, dx\, dt \le 0,$$

where η is a function having the same properties as that in Theorem 4.

PROOF. We introduce a constant strategy $\beta_t \equiv \beta$. For $\lambda \ge 0$ let

$$w_\lambda^\beta(s,x) = \mathsf{M}_{s,x}^\beta \int_0^{T-s} e^{-\varphi_t - \lambda t}[f^\beta(s+t, x_t) + \lambda u(s+t, x_t)]\, dt$$
$$+ \mathsf{M}_{s,x}^\beta u(T, x_{T-s}) e^{-\varphi_{T-s} - \lambda(T-s)}.$$

If in Corollary 3.2.13 we take $\tilde{c}^\beta(t,x) = c^\beta(t,x) + \lambda$ instead of $c^\beta(t,x)$ and a function $\tilde{f}^\beta(t,x) = f^\beta(t,x) + \lambda u(t,x)$ instead of $f^\beta(t,x)$, and if, in addition, we replace $u(x)$ by $u(T,x)$, we have

$$\int_{H_T} [w_\lambda^\beta(L^{\beta *}\eta - \lambda\eta]) + (f^\beta + \lambda u)\eta]\, dx\, dt = 0.$$

According to Bellman's principle (Theorems 3.1.6 and 3.1.11) $w_\lambda^\beta \le u$; therefore, $0 \le \lambda\eta(u - w_\lambda^\beta)$ and

$$\int_{H_T} [w_\lambda^\beta L^{\beta *}\eta + f^\beta\eta]\, dx\, dt \le 0 \tag{4}$$

for each $\lambda \ge 0$. Further, let us take the limit in (4) as $\lambda \to \infty$.

We note that due to the estimate

$$|u(s+t, x_t^{\beta,s,x})| \le N\left(1 + \sup_{t \le T-s} |x_t^{\beta,s,x}|\right)^m,$$

summability of the last quantity and, in addition, the continuity of $u(s+t, x_t^{\beta,s,x})$ in t, the function

$$h_{s,x}(t) \equiv \mathsf{M}_{s,x}^\beta e^{-\varphi_t} u(s+t, x_t)$$

is a continuous function of t. Therefore, if ξ is a random variable having an exponential distribution with exponent equal to unity, then $\mathsf{M}h_{s,x}[(T-s) \wedge (\xi/\lambda)] \to h_{s,x}(0)$ as $\lambda \to \infty$. This implies precisely that for $s \in [0,T]$, $\lambda \to \infty$

$$\mathsf{M}_{s,x}^\beta\left[\int_0^{T-s} e^{-\varphi_t - \lambda t}\lambda u(s+t, x_t)\, dt + u(T, x_{T-s})e^{-\varphi_{T-s} - \lambda(T-s)}\right] \to u(s,x). \tag{5}$$

Furthermore, it follows from the estimates of moments of solutions of stochastic equations that $|h_{s,x}(t)| \le N(1 + |x|)^m$, where N does not depend on

s, x, t. Hence the left sides in (5) are bounded in $C_{T,R}$ uniformly with respect to λ. From the estimates mentioned above and the inequality $|f^\beta(t,s)| \leq K(1 + |x|)^m$ it follows that for $s \in [0,T]$

$$\left| \mathsf{M}_{s,x}^\beta \int_0^{T-s} e^{-\varphi_t - \lambda t} f^\beta(s + t, x_t)\, dt \right| \leq N(1 + |x|)^m \int_0^{T-s} e^{-\lambda t}\, dt \leq \frac{1}{\lambda} N(1 + |x|)^m,$$

where N does not depend on s, x, λ. Therefore, the totality of functions w_λ^β is bounded on $[0,T] \times S_R$ and $w_\lambda^\beta \to u$ as $\lambda \to \infty$. Since in (4) we can take an integral over the set $C_{T,R}$, we replace the function w_λ^β by u, letting $\lambda \to \infty$. The lemma is proved. □

Proof of Theorem 4. As in proving Theorem 1, we approximate σ, b, c, f, $g(x)$ using smooth functions σ_n, b_n, c_n, f_n, $g_n(x)$, and, in addition, taking the convolution of σ, b, c, f, $g(x)$ with the function $n^d\zeta(nx)$ (see Theorem 3.1.14). We denote by v_n the payoff function which has been constructed on the basis of σ_n, b_n, c_n, f_n, $g_n(x)$. According to Lemma 5,

$$\int_{H_T} [v_n L_n^{\beta *}\eta + f_n^\beta \eta]\, dx\, dt \leq 0, \tag{6}$$

where the operator $L_n^{\beta *}$ is constructed in the usual way on the basis of $a_n(\beta,t,x) = \frac{1}{2}\sigma(\beta,t,x)\sigma_n^*(\beta,t,x), b_n(\beta,t,x), c_n^\beta(t,x)$. Since the function v_n has a generalized derivative with respect to x, we have, integrating by parts in (6), that

$$\begin{aligned} &-\int_0^T dt \int \sum_{i,j=1}^d (a_n^{ij}\eta)_{x^i} v_{nx^j}\, dx \\ &\quad \leq \int_{H_T} \left[v_n \frac{\partial \eta}{\partial t} + \sum_{i=1}^d (b_n^i \eta)_{x^i} v_n + \eta c_n v_n - \eta f_n \right] dx\, dt. \end{aligned} \tag{7}$$

Let us take the limit in (7) as $n \to \infty$. According to Theorem 3.1.14 and Corollary 3.1.13, the functions $v_n(t,x)$ converge to $v(t,x)$ uniformly on $[0,T] \times S_R$. As was indicated in Section 2.1, $c_n^\beta(t,x) \to c^\beta(t,x)$ for all t, x due to the continuity of $c^\beta(t,x)$ in x and

$$(b_n^i\eta)_{x^i} = b_{nx^i}^i \eta + b_n^i \eta_{x^i} \to b_{x^i}^i \eta + b^i \eta_{x^i} = (b^i\eta)_{x^i}$$

for almost all (t,x) because the generalized derivative $b_{x^i}^i$ exists. Furthermore, $|b_{x^i}| \leq K$; hence $|b_{nx^i}| = |b_{x^i} * n^d\zeta(nx)| \leq K$. This reasoning shows that the right side of (7) tends to

$$\int_{H_T} \left[v \frac{\partial \eta}{\partial t} + \sum_{i=1}^d (b^i\eta)_{x^i} v + \eta c v - \eta f \right] dx\, dt$$

as $n \to \infty$.

Further, by Theorem 1,

$$|\operatorname{grad}_x v_n(t,x)| \leq N(1 + |x|)^{2m}$$

for $t \in [0,T]$, $x \in E_d$, where $N = N(K,T,m)$ does not depend on n. Then we obtain

$$\left| \int_{H_T} (a_n^{ij}\eta)_{x^i} v_{nx^j}\, dx\, dt - \int_{H_T} (a^{ij}\eta)_{x^i} v_{x^j}\, dx\, dt \right|$$
$$\le N \int_{H_T} |(a_n^{ij}\eta)_{x^i} - (a^{ij}\eta)_{x^i}|\, dx\, dt + \left| \int_{H_T} (a^{ij}\eta)_{x^i} [v_{nx^j} - v_{x^j}]\, dx\, dt \right|, \quad (8)$$

where N depends only on K, T, m, R. We know that $(a_n^{ij})_{x^i} \to a_{x^i}^{ij} \to a_{x^i}^{ij}$ almost everywhere. It can easily be seen as well that the union of the foregoing derivatives is bounded on $[0,T] \times S_R$. Hence the first term in the right side of (8) tends to zero as $n \to \infty$. The second term tends to zero as well, since $v_n \to v$ in $\mathscr{L}_2(C_{T,R})$, and, in addition, because the norms $|\text{grad}_x v_n|$ in $\mathscr{L}_2(C_{T,R})$ are bounded and therefore (see Section 2.1) $v_{nx^j} \to v_{x^j}$ weakly in $\mathscr{L}_2(C_{T,R})$. Therefore, the limit of the left side of (7) is equal to

$$-\int_{H_T} \sum_{i,j=1}^{d} (a^{ij}\eta)_{x^i} v_{x^j}\, dx\, dt$$

as $n \to \infty$.

In a similar way this theorem can be proved for $w(s,x)$, thus completing the proof of the theorem. □

In order to derive two corollaries from the theorem proved above, we need two simple facts. If in a region $Q \subset [0,T] \times E_d$ the bounded functions $\varphi(t,x)$, $\psi(t,x)$ have bounded generalized derivatives, $\varphi_{x^ix^j}$, φ_{x^i}, ψ_{x^j}, then for any $\eta \in C_0^\infty(Q)$

$$\int_Q (\varphi\eta)_{x^i} \psi_{x^j}\, dx\, dt = -\int_Q (\varphi\eta)_{x^ix^j} \psi\, dx\, dt, \quad (9)$$

$$\int_Q \varphi_{x^i} (\psi\eta)_{x^j}\, dx\, dt = -\int_Q \varphi_{x^ix^j} \psi\eta\, dx\, dt. \quad (10)$$

Inequalities (9) and (10) can be proved in the same way, namely, we need to take a function $\eta_1 \in C_0^\infty(Q)$ such that it is equal to unity everywhere where $\eta \neq 0$. Next, we replace ψ_{x^j} by $(\psi\eta_1)_{x^j}$ in (9) and, furthermore, we replace ψ by mean functions $\psi^{(\varepsilon)}$ in both inequalities. Since the products $\psi^{(\varepsilon)}\eta_1, \psi^{(\varepsilon)}\eta \in C_0^\infty(Q)$, by the definition of a generalized derivative, we can shift the derivatives from $\psi^{(\varepsilon)}\eta_1$, $\psi^{(\varepsilon)}\eta$ onto φ. Also, we pass to the limit as $\varepsilon \to 0$ using the theorem on bounded convergence. Finally, the presence of η_1 has obviously no effect on values of the resulted expressions, which fact allows us to remove η_1 in general.

6. Corollary. *Let a region $Q \subset H_T$, $\beta \in A$, and also let $a(\beta,t,x)$ as a function of the variables (t,x) have second generalized derivatives with respect to x. Then for each nonnegative function $\eta \in C_0^\infty(Q)$ for $u \equiv v$ and for $u \equiv w$*

$$\int_Q u L^{\beta *} \eta\, dx\, dt \le -\int_Q f^\beta \eta\, dx\, dt.$$

We have made use of the preceding remarks in order to remove the derivatives from the function u. Using the same remarks and shifting the derivatives on u whenever possible in the assertion of the theorem, we have

$$\int_{H_T} \eta(L^\beta u + f^\beta)\,dx\,dt \le 0.$$

Taking arbitrary $\eta \ge 0$, we arrive the following assertion.

7. Corollary. *Let a region $Q \subset H_T$. Furthermore, let a function $v(w)$, as a function of (t,x), have two generalized derivatives in x and one derivative in t in the region Q. Then for any $\beta \in A$ almost everywhere on Q*

$$L^\beta v + f^\beta \le 0 \qquad (L^\beta w + f^\beta \le 0).$$

In other words, almost everywhere on Q

$$F[v] \le 0 \qquad (F[w] \le 0).$$

Lemma 5 (or Corollary 6) has a rather unusual application for proving assertions of the type of Theorems 2.3.3 and 2.3.4.

We recall (see [40]) that for a fixed $\lambda > 0$ a nonnegative infinitely differentiable function $u(x)$ given on E_d is said to be λ-convex if the matrix $(\lambda u(x)\delta^{ij} - u_{x^ix^j}(x))$ is nonnegative definite for all x. According to Corollary 1 of Lemma 1 in [40], for each λ-convex function $|\text{grad}\, u(x)| \le \sqrt{\lambda}\, u(x)$ at each point x.

8. Theorem. *Suppose that on a probability space measurable processes x_t, a_t, b_t, c_t, φ_t are defined for $t \in [0,\infty)$, with $x_t \in E_d$. We assume that a_t is a nonnegative definite matrix of dimension $d \times d$, b_t is a d-dimensional vector, c_t, φ_t are nonnegative numbers. Assume that there exists a constant $\lambda > 0$ such that $c_t \ge \lambda \operatorname{tr} a_t + \sqrt{\lambda}|b_t|$ for all t, ω. Finally, assume that for any smooth bounded function $u(t,x)$ which decreases in t, is λ-convex with respect to x and which has bounded derivatives $(\partial/\partial t)u$, u_{x^i}, $u_{x^ix^j}$ on $[0,\infty) \times E_d$, the inequality*

$$\mathsf{M}u(0,x_0) \ge -\mathsf{M}\int_0^\infty e^{-\varphi_t} L_t u(t,x_t)\,dt^1 \tag{11}$$

is satisfied, where

$$L_t u(t,x_t) = \sum_{i,j=1}^d a_t^{ij} u_{x^ix^j}(t,x_t) + \sum_{i=1}^d b_t^i u_{x^i}(t,x_t) - c_t u(t,x_t) + \frac{\partial}{\partial t} u(t,x_t).$$

Then for any nonnegative Borel function $f(t,x)$

$$\mathsf{M}\int_0^\infty e^{-\varphi_t}(\det a_t)^{1/(d+1)} f(t,x_t)\,dt \le N(d,\lambda)\|f\|_{d+1,[0,\infty)\times E_d}.$$

PROOF. First we note that

$$c_t u \ge \lambda u \operatorname{tr} a_t + \sqrt{\lambda} u|b_t| \ge \lambda u \operatorname{tr} a_t + (\text{grad}\, u, b_t).$$

[1] This condition can be replaced by (12).

Hence

$$-L_t u \geq \lambda u \operatorname{tr} a_t - \operatorname{tr}[a_t(u_{x^ix^j})] - \frac{\partial}{\partial t} u = \operatorname{tr}[a_t(\lambda u \delta^{ij} - u_{x^ix^j})] - \frac{\partial}{\partial t} u$$

is greater than zero since the trace of the product of positive matrices is positive; furthermore, by assumption, $(\partial/\partial t)\ u \leq 0$. The above implies, in particular, that the right side of (11) is always definite. Next, we have from (11) that

$$\mathsf{M} u(0,x_0) \geq -\mathsf{M} \int_0^\infty e^{-\varphi_t} L_t^0 u(t,x_t)\, dt, \tag{12}$$

where

$$L_t^0 u = \sum_{i,j=1}^d a_t^{ij} u_{x^ix^j} - \lambda u \operatorname{tr} a_t + \frac{\partial}{\partial t} u.$$

We take a smooth function $f(t,x)$ with compact support and some $n > 0$. Let A_n be a set of all matrices α of dimension $d \times d$ such that $\operatorname{tr} \alpha\alpha^* \leq 2n$. For $\alpha \in A_n$ let $\sigma(\alpha,t,x) = \alpha$, $b(\alpha,t,x) = 0$, $c^\alpha = \lambda \operatorname{tr} a(\alpha)$, $f^\alpha = (\det a(\alpha))^{1/(d+1)} f(t,x)$. We take as T a number such that $f(t,x) = 0$ for $t \geq T - 2$. Let $g(x) = 0$. On the basis of the quantities introduced, using some d-dimensional Wiener process, we define the payoff function $v_n(t,x)$. It is seen that $v_n(t,x)$ increase as n increases. Let $v(t,x) = \lim_{n\to\infty} v_n(t,x)$. By Theorem 2.3.3, for all x, n

$$v_n(t,x) \leq N(d,\lambda)\left(\int_t^\infty \int_{E_d} f^{d+1}(r,x)\, dr\, dx\right)^{1/(d+1)} \tag{13}$$

Therefore, we shall prove the theorem for the function f chosen if we show that

$$\sup_{t\geq 0} \sup_{x\in E_d} v(t,x) \geq \mathsf{M} \int_0^\infty e^{-\varphi_t} (\det a_t)^{1/(d+1)} f(t,x_t)\, dt. \tag{14}$$

It should be mentioned that in accord with the results obtained in [54, Section 1, 2] it suffices to prove the assertion of the theorem for smooth nonnegative $f(t,x)$ with compact support only. Thus, it remains only to prove (14).

Using Lemma 5, we take as $\eta(t,x)$ the function

$$\varepsilon^{-(d+1)} \zeta\left(\frac{s-t}{\varepsilon}, \frac{y-x}{\varepsilon}\right), \qquad \varepsilon \in (0,1),$$

which, being a function of (t,x), satisfies the conditions of Lemma 5 if $s \in [\varepsilon, T - \varepsilon]$. Noting in addition that in our case the coefficients L^β do not depend on (t,x), we can easily find using Lemma 5 for $s \in [\varepsilon, T-\varepsilon]$, $n > 0$ that

$$\sum_{i,j=1}^d a^{ij} v_{n y^i y^j}^{(\varepsilon)}(s,y) - \lambda v_n^{(\varepsilon)}(s,y) \operatorname{tr} a + \frac{\partial}{\partial s} v_n^{(\varepsilon)}(s,y) + (\det a)^{1/(d+1)} f^{(\varepsilon)}(s,y) \leq 0 \tag{15}$$

for all $y \in E_d$ and for nonnegative symmetric matrices α such that $\operatorname{tr} a \leq n$. As can easily be seen, the functions $v(t,x)$, $v_n(t,x)$ are equal to zero for $t \in [T - \varepsilon, T]$. It is convenient to suppose that the functions $v(t,x)$, $v_n(t,x)$ are defined not only for $t \in [0,T]$ but for $t \geq T$ as well, and, furthermore, that $v(t,x) = v_n(t,x) = 0$ for $t \geq T$. Then $f^{(\varepsilon)}(s,y) = v_n^{(\varepsilon)}(s,y) = 0$ for $s \in [T-1, \infty)$ ($\varepsilon < 1$). Therefore, (15) holds not only for $s \in [\varepsilon, T-\varepsilon]$ but for $s \geq T - \varepsilon$ as well. By virtue of (13), according to the Lebesgue theorem, $v_n^{(\varepsilon)} \to v^{(\varepsilon)}$. Further, from (15) we have

$$\sum_{i,j=1}^{d} a^{ij} v_{y^i y^j}^{(\varepsilon)}(s,y) - \lambda v^{(\varepsilon)}(s,y) \operatorname{tr} a + \frac{\partial}{\partial s} v^{(\varepsilon)}(s,y) + (\det a)^{1/(d+1)} f^{(\varepsilon)}(s,y) \leq 0 \tag{16}$$

for $s \geq \varepsilon$ and $y \in E_d$ for any nonnegative symmetric matrix a. If in (16) we take $a^{ij} = n\xi^i\xi^j$, divide both sides of (16) by n, and, finally, let $n \to \infty$, we obtain

$$\sum_{i,j=1}^{d} \xi^i \xi^j v_{y^i y^j}^{(\varepsilon)}(s,y) - \lambda v^{(\varepsilon)}(s,y)\xi^2 \leq 0.$$

In short, the matrix $(\lambda v^{(\varepsilon)}\delta^{ij} - v_{x^i x^j}^{(\varepsilon)}) \geq 0$ and, furthermore, the function $v^{(\varepsilon)}$ is λ-convex with respect to y. Taking $a = 0$, we obtain from (16) that $v^{(\varepsilon)}$ decreases in s.

From (16) and (12) for $u(t,x) = v^{(\varepsilon)}(\varepsilon + t, x)$ we find

$$\mathsf{M} v^{(\varepsilon)}(\varepsilon,x_0) \geq \mathsf{M} \int_0^\infty e^{-\varphi_t} (\det a_t)^{1/(d+1)} f^{(\varepsilon)}(\varepsilon + t, x_t)\, dt.$$

It remains to note that the left side of the last inequality does not obviously exceed the left side of (14). Also, we need to let $\varepsilon \downarrow 0$ and to obtain the right side of (14) from the right side of the inequality given above, using Fatou's lemma as well as the fact that $f^{(\varepsilon)}(t,x) \to f(t,x)$ uniformly in (t,x). We have thus proved the theorem. □

2. Estimation from Below of Second Derivatives of a Payoff Function

In this section we shall estimate from below second derivatives of a payoff function. Also, using the estimates obtained, we shall represent the inequality

$$\int_{H_T} [uL^{\beta *}\eta + f^\beta \eta]\, dx\, dt \leq 0$$

in a local form (see Lemma 1.5 or Corollary 1.6).

Assume that the conditions given in Section 3.1 are satisfied. We also assume that the functions $\sigma(\alpha,t,x)$, $b(\alpha,t,x)$, $c^\alpha(t,x)$, $f^\alpha(t,x)$, $g(x)$, $g(t,x)$ for each $\alpha \in A$, $t \in [0,T]$ are twice continuously differentiable over x. For all $\alpha \in A$, $t \in [0,T]$, $x \in E_d$, $l \in E_d$, let

$$\|\sigma_{(l)}(\alpha,t,x)\| + |b_{(l)}(\alpha,t,x)| \leq K,$$
$$\|\sigma_{(l)(l)}(\alpha,t,x)\| + |b_{(l)(l)}(\alpha,t,x)| \leq K(1 + |x|)^m$$

and for

$$u^\alpha(t,x) \equiv c^\alpha(t,x), \qquad u^\alpha(t,x) \equiv f^\alpha(t,x),$$
$$u^\alpha(t,x) \equiv g(x), \qquad u^\alpha(t,x) \equiv g(t,x)$$

let

$$|u^\alpha_{(l)}(t,x)| + |u^\alpha_{(l)(l)}(t,x)| \le K(1 + |x|)^m.$$

We prove under the foregoing assumptions that second-order generalized derivatives of the payoff functions $v(s,x)$, $w(s,x)$ with respect to x and first-order derivatives of these functions with respect to s are countably additive functions of sets (see Definition 2.1.2).

As was done in Section 1, we rely here upon the estimates of derivatives of functions $v^\alpha(s,x)$, $v^{\alpha,\tau}(s,x)$, which, according to the results obtained in Sections 2.7 and 2.8, are twice continuously differentiable in x. If we let $l \in E_d$, $\alpha \in \mathfrak{A}$, $\tau \in \mathfrak{M}(T - s)$ and if, in addition, we write the derivative $v^{\alpha,\tau}_{(l)(l)}(s,x)$, using the differentiation rules of mathematical expectations, integrals, and composite functions, we obtain a rather cumbersome expression. In order to simplify it, we introduce the following notation:

$$y_t^{\alpha,s,x} = \mathscr{L}B\text{-}\frac{\partial}{\partial l}\, x_t^{\alpha,s,x}, \qquad z_t^{\alpha,s,x} = \mathscr{L}B\text{-}\frac{\partial^2}{\partial l^2}\, x_t^{\alpha,s,x},$$
$$\xi^{\alpha,\tau}_{s,x}(t) = \int_t^\tau e^{-(\varphi_r - \varphi_t)} f^{\alpha_r}(s + r, x_r)\, dr + e^{-(\varphi_\tau - \varphi_t)} g(s + \tau, x_\tau); \tag{1}$$

$$\xi^{\alpha,\tau}_{(l),s,x}(t) = e^{-(\varphi_\tau - \varphi_t)} g_{(y_\tau)}(s + \tau, x_\tau)|y_\tau| + \int_t^\tau e^{-(\varphi_r - \varphi_t)} |y_r| \{ f^{\alpha_r}_{(y_r)}(s + r, x_r) - c^{\alpha_r}_{(y_r)}(s + r, x_r)\xi^{\alpha,\tau}_{s,x}(r)\}\, dr; \tag{2}$$

$$\begin{aligned}\xi^{\alpha,\tau}_{(l),(l),s,x}(t) = {} & e^{-(\varphi_\tau - \varphi_t)} g_{(y_\tau)(y_\tau)}(s + \tau, x_\tau)|y_\tau|^2 \\ & + \int_t^\tau e^{-(\varphi_r - \varphi_t)} |y_r|^2 \{ f^{\alpha_r}_{(y_r)(y_r)}(s + r, x_r) \\ & - c^{\alpha_r}_{(y_r)(y_r)}(s + r, x_r)\xi^{\alpha,\tau}_{s,x}(r)\}\, dr \\ & + e^{-(\varphi_\tau - \varphi_t)} g_{(z_\tau)}(s + \tau, x_\tau)|z_\tau| \\ & + \int_t^\tau e^{-(\varphi_r - \varphi_t)} |z_r| \{ f^{\alpha_r}_{(z_r)}(s + r, x_r) \\ & - c^{\alpha_r}_{(z_r)}(s + r, x_r)\xi^{\alpha,\tau}_{s,x}(r)\}\, dr \\ & - 2\int_t^\tau e^{-(\varphi_r - \varphi_t)} |y_r| c^{\alpha_r}_{(y_r)}(s + r, x_r)\xi^{\alpha,\tau}_{(l),s,x}(r)\, dr,\end{aligned} \tag{3}$$

where $x_r = x_r^{\alpha,s,x}$, $y_r = y_r^{\alpha,s,x}$, $z_r = z_r^{\alpha,s,x}$, $\varphi_r = \varphi_r^{\alpha,s,x}$.

As can easily be seen, for each ω and for almost all $t \in [0,\tau]$

$$\frac{d}{dt}\,[\xi^{\alpha,\tau}_{s,x}(t)e^{-\varphi_t}] = -e^{-\varphi_t} f^{\alpha_t}(s + t, x_t).$$

Differentiating the last expression, we find

$$\frac{d}{dt}\,\xi^{\alpha,\tau}_{s,x}(t) = c^{\alpha_t}(s + t, x_t)\xi^{\alpha,\tau}_{s,x}(t) - f^{\alpha_t}(s + t, x_t).$$

Further, noting that $\xi_{s,x}^{\alpha,\tau}(\tau) = g(s + \tau, x_\tau)$, we conclude that for $t \in [0,\tau]$

$$\xi_{s,x}^{\alpha,\tau}(t) = g(s + \tau, x_\tau) + \int_t^\tau \{f^{\alpha_r}(s + r, x_r) - c^{\alpha_r}(s + r, x_r)\xi_{s,x}^{\alpha,\tau}(r)\} \, dr, \tag{4}$$

which constitutes an equation with respect to $\xi_{s,x}^{\alpha,\tau}(t)$ for $t \in [0,\tau]$. Since the reverse of the transformations we have carried out is true, $\xi_{s,x}^{\alpha,\tau}(t)$ is the unique solution of (4) on $[0,\tau]$.

Further, it is not hard to see that the process $\xi_{s,x}^{\alpha,\tau}(t \wedge \tau)$ is $\mathscr{L}B$-differentiable in x. According to the well-known rule of operations with derivatives, it follows from (4) that for almost all ω for $t \in [0,\tau]$

$$\begin{aligned}\mathscr{L}B\text{-}\frac{\partial}{\partial t} \xi_{s,x}^{\alpha,\tau}(t \wedge \tau) &= g_{(y_\tau)}(s + \tau, x_\tau)|y_\tau| \\ &\quad + \int_t^\tau \{|y_r|[f_{(y_r)}^{\alpha_r}(s + r, x_r) - c_{(y_r)}^{\alpha_r}(s + r, x_r)\xi_{s,x}^{\alpha,\tau}(r)] \\ &\quad - c^{\alpha_r}(s + r, x_r)\mathscr{L}B\text{-}\frac{\partial}{\partial l} \xi_{s,x}^{\alpha,\tau}(r \wedge \tau)\} \, dr. \end{aligned} \tag{5}$$

It is convenient to regard (5) as an equation with respect to an $\mathscr{L}B$-derivative. Comparing (1) and (4) with (2) and (5), one easily sees that $\xi_{(l),s,x}^{\alpha,\tau}(t)$ satisfies Eq. (5) for $t \in [0,\tau]$. Since the solution of Eq. (5) is unique for the same reasons the solution of Eq. (4) is unique, then for almost all ω for $t \in [0,\tau]$

$$\mathscr{L}B\text{-}\frac{\partial}{\partial l} \xi_{s,x}^{\alpha,\tau}(t \wedge \tau) = \xi_{(l),s,x}^{\alpha,\tau}(t). \tag{6}$$

Differentiating (5) over l and, furthermore, regarding the relation thus obtained as an equation for $\mathscr{L}B\text{-}(\partial^2/\partial l^2)\xi_{s,x}^{\alpha,\tau}(t)$, we prove in a similar way that for almost all ω for $t \in [0,\tau]$

$$\mathscr{L}B\text{-}\frac{\partial^2}{\partial l^2} \xi_{s,x}^{\alpha,\tau}(t \wedge \tau) = \xi_{(l),(l),s,x}^{\alpha,\tau}(t). \tag{7}$$

1. Exercise

Prove (6) and (7) by direct differentiation of (1).

We can deduce from (6) and (7) the estimates of

$$\left|\mathscr{L}B\text{-}\frac{\partial}{\partial l} \xi_{s,x}^{\alpha,\tau}(t \wedge \tau)\right|, \qquad \left|\mathscr{L}B\text{-}\frac{\partial^2}{\partial l^2} \xi_{s,x}^{\alpha,\tau}(t \wedge \tau)\right|.$$

First, from our assumptions, for $t \le \tau$

$$|\xi_{s,x}^{\alpha,\tau}(t)| \le K(T - s + 1) \sup_{t \le T-s} (1 + |x_t^{\alpha,s,x}|)^m.$$

Further,

$$|c_{(y_r)}^{\alpha_r}(s + r, x_r)\xi_{s,x}^{\alpha,\tau}(r)| \le K^2(T - s + 1) \sup_{t \le T-s} (1 + |x_t^{\alpha,s,x}|)^{2m}.$$

Hence from (6) and (2) we obtain

$$\left|\mathscr{L}B\text{-}\frac{\partial}{\partial l}\xi^{\alpha,\tau}_{s,x}(t\wedge\tau)\right| \le N_1 \sup_{t\le T-s}(1+|x^{\alpha,s,x}_t|)^{2m}\sup_{t\le T-s}|y^{\alpha,s,x}_t|, \tag{8}$$

where $N_1 = K(T-s+1) + K^2(T-s+1)^2$. Similarly,

$$\left|\mathscr{L}B\text{-}\frac{\partial^2}{\partial l^2}\xi^{\alpha,\tau}_{s,x}(t\wedge\tau)\right| \le N_2 \sup_{t\le T-s}(1+|x^{\alpha,s,x}_t|)^{3m}\sup_{t\le T-s}|y^{\alpha,s,x}_t|^2$$
$$+ N_1 \sup_{t\le T-s}(1+|x^{\alpha,s,x}_t|)^{2m}\sup_{t\le T-s}|z^{\alpha,s,x}_t|, \tag{9}$$

where $N_2 = N_1 + 2N_1K(T-s)$.

We also make use of the estimates of moments of derivatives of solutions of stochastic equations given in Theorem 2.8.8. In this case we obtain, for example,

$$\mathsf{M}^{\alpha}_{s,x}\sup_{t\le T-s}(1+|x_t|)^{2m}\sup_{t\le T-s}|z_t|$$
$$\le\left[\mathsf{M}^{\alpha}_{s,x}\sup_{t\le T-s}(1+|x_t|)^{4m}\right]^{1/2}\left[\mathsf{M}^{\alpha}_{s,x}\sup_{t\le T-s}|z_t|^2\right]^{1/2}$$
$$\le Ne^{N(T-s)}(1+|x|)^{3m},$$

where $N = N(K,m)$. Estimating other expressions in the right sides of (8) and (9) in a similar way, and, in addition, noting that

$$v^{\alpha,\tau}(s,x) = \mathsf{M}^{\alpha}_{s,x}\xi^{\alpha,\tau}_{s,x}(0),\ v^{\alpha,\tau}_{(l)}(s,x) = \mathsf{M}^{\alpha}_{s,x}\xi^{\alpha,\tau}_{(l),s,x}(0),$$
$$v^{\alpha,\tau}_{(l)(l)}(s,x) = \mathsf{M}^{\alpha}_{s,x}\xi^{\alpha,\tau}_{(l),(l),s,x}(0),$$

we prove that the following assertion holds for $v^{\alpha,\tau}(s,x)$.

2. Lemma. *For each $s\in[0,T]$, $\alpha\in\mathfrak{A}$ and $\tau\in\mathfrak{M}(T-s)$ the functions $v^{\alpha}(s,x)$ and $v^{\alpha,\tau}(s,x)$ have second continuous derivatives with respect to x. There exists a constant $N = N(K,m)$ such that for all $l\in E_d$*

$$|v^{\alpha}_{(l)(l)}(s,x)| + |v^{\alpha,\tau}_{(l)(l)}(s,x)| \le Ne^{N(T-s)}(1+|x|)^{3m},$$
$$|v^{\alpha}_{(l)}(s,x)| + |v^{\alpha,\tau}_{(l)}(s,x)| \le Ne^{N(T-s)}(1+|x|)^{2m},$$
$$|v^{\alpha}(s,x)| + |v^{\alpha,\tau}(s,x)| \le Ne^{N(T-s)}(1+|x|)^{m}.$$

The proof of this lemma for $v^{\alpha}(s,x)$ is the same as that for $v^{\alpha,\tau}(s,x)$. However, we need to take $\tau = T-s$ and, furthermore, replace $g(t,x)$ by $g(x)$.

3. Theorem. *There exists a constant $N = N(K,m)$ such that for each $s\in[0,T]$ the functions*

$$v(s,x) + Ne^{N(T-s)}(1+|x|^2)^{(3m/2)+1},$$
$$w(s,x) + Ne^{N(T-s)}(1+|x|^2)^{(3m/2)+1}$$

are convex downward with respect to x.

PROOF. Let $l \neq 0$. Simple calculations show that

$$[(1 + |x|^2)^{(3m/2)+1}]_{(l)(l)}$$
$$= \left(\frac{3m}{2} + 1\right)\frac{3m}{2}(1 + |x|^2)^{(3m/2)-1}4(xl)^2 + \left(\frac{3m}{2} + 1\right)$$
$$\times (1 + |x|^2)^{3m/2}2 \geq (1 + |x|)^{3m}2^{-3m/2}. \tag{10}$$

We take N from Lemma 2. Also, let $N_1 = 2^{3m/2}N$. By virtue of (10) and Lemma 2 a second directional derivative of the functions $v^\alpha(s,x) + N_1 e^{N_1(T-s)}(1 + |x|^2)^{(3m/2)+1}$ is positive for any $\alpha \in \mathfrak{A}$, $s \in [0,T]$. Therefore, these functions are convex downward. A fortiori, their upper bound with respect to α is convex downward:

$$v(s,x) + N_1 e^{N_1(T-s)}(1 + |x|^2)^{(3m/2)+1}.$$

We consider $w(s,x)$ in a similar way, thus completing the proof of the theorem. □

4. Corollary. *Each of the functions v and w is representable as the difference between two functions convex downward with respect to x. The subtrahend can be taken equal to*

$$Ne^{N(T-s)}(1 + |x|^2)^{(3m/2)+1}$$

for a constant $N = N(K,m)$.

In fact, let us, for example, take N given in Theorem 5, and in addition, let us write

$$v(s,x) = [v(s,x) + Ne^{N(T-s)}(1 + |x|^2)^{(3m/2)+1}] - Ne^{N(T-s)}(1 + |x|^2)^{(3m/2)+1}.$$

In the lemma which follows we list general properties of differences of convex functions.

5. Lemma. *Suppose that in a convex region $Q \subset H_T$ we are given a function $u(s,x) = u_1(s,x) - u_2(s,x)$, in which u_1 and u_2 are defined, measurable, locally bounded in Q, and, furthermore, convex downward with respect to x in $Q_s = \{x:(s,x) \in Q\}$ for each s. Then, for each $l_1, l_2 \in E_d$ in the region Q there exist derivatives $u_{(l_1)(l_2)}(s,x)(ds\,dx)$ (see Definition 2.1.2). In this case inside Q*

$$u_{(l_1)(l_1)}(s,x)(ds\,dx) \geq -u_{2(l_1)(l_1)}(s,x)(ds\,dx). \tag{11}$$

In addition, if the bounded function $\eta(s,x)$ is measurable with respect to s, is twice continuously differentiable with respect to x for each s, and, finally, if this function is equal to zero outside some compact set lying in Q, then for any $l_1, l_2 \in E_d$

$$\int_Q u\eta_{(l_1)(l_2)}\,ds\,dx = \int_Q \eta u_{(l_1)(l_2)}(ds\,dx).$$

PROOF. It is easy to obtain from the equality $u = u_1 - u_2$ and the properties of derivatives of u_1, u_2 analogous properties of derivatives of u. We shall see below that second derivatives along the l_1 direction of a function which is convex downward with respect to x are nonnegative. Hence inequality (11) readily follows from the equality $u = u_1 - u_2$. Therefore, it suffices to prove the lemma for the functions u which are convex downward with respect to x ($u = u_1, u_2 = 0$). Further, it suffices, obviously, to prove that the assertions of the lemma hold in any bounded region $Q' \subset \bar{Q}' \subset Q$. Note that by hypothesis the function u is bounded in any region Q'. Then, in proving the lemma it is possible to assume that the region Q is bounded, and that the function u is convex downward with respect to x and is bounded in Q. It will be convenient to assume as well that the function u is extended in some way outside Q.

Let us take a unit vector $l \in E_d$, $s \in (0,T)$, and also let us take a nonnegative $\eta \in C_0^\infty(Q_s)$. Furthermore, for a real r we introduce an operator Δ_r^s using the formula

$$\Delta_r^2 \xi(s,x) = \xi(s, x + rl) - 2\xi(s,x) + \xi(s, x - rl).$$

Integrating by parts, we easily prove that

$$\Delta_r^2 \eta(x) = r^2 \int_{-1}^{1} (1 - |r_1|)\eta_{(l)(l)}(x + r_1 rl)\, dr_1,$$

from which we have, by the mean value theorem, $\Delta_r^2 \eta(x) = r^2 \eta_{(l)(l)}(x + \theta rl)$ where $|\theta| \le 1$. In particular, for $|r| \le 1$ the collection of functions $(1/r^2)\,\Delta_r^2 \eta(x)$ is bounded and for $r \to 0$ it converges to $\eta_{(l)(l)}(x)$. Therefore,

$$\int_{Q_s} u(s,x)\eta_{(l)(l)}(x)\, dx = \lim_{r \to 0} \frac{1}{r^2} \int_{Q_s} u(s,x)\, \Delta_r^2 \eta(x)\, dx. \tag{12}$$

The function η is equal to zero near the boundary Q_s. Hence for a sufficiently small r the function $\Delta_r^2 \eta(x)$ is equal to zero near the boundary Q_s. For sufficiently small r the last integral in (12) can be extended to E_d. Further, if we write $\int u\, \Delta_r^2 \eta\, dx$ as the sum of three integrals and if, in addition, we make in these integrals the change of variables of the type $y = x + a$, we easily obtain that for sufficiently small r

$$\int_{Q_s} u(s,x)\, \Delta_r^2 \eta(x)\, dx = \int_{E_d} \eta(x)\, \Delta_r^2 u(s,x)\, dx = \int_{Q_s} \eta(x)\, \Delta_r^2 u(s,x)\, dx.$$

Because u is convex, $\Delta_r^2 u(s,x) \ge 0$ if the distance between x and ∂Q_s is larger than $|r|$. Furthermore, $\eta \ge 0$. Therefore, for small r

$$\int_{Q_s} \eta(x)\, \Delta_r^2 u(s,x)\, dx \ge 0, \qquad \int_{Q_s} u(s,x)\, \Delta_r^2 \eta(x)\, dx \ge 0.$$

By virtue of (12),

$$\int_{Q_s} u(s,x)\eta_{(l)(l)}(x)\, dx \ge 0.$$

Similarly, for any nonnegative function $\eta \in C_0^\infty(Q)$ we have

$$\int_Q u\eta_{(l)(l)}\, ds\, dx \ge 0.$$

By Lemma 2.1.3 the inequalities proved above imply the existence for each s of the measure $u_{(l)(l)}(s,x)(dx)$ on Q_s as well as the existence of the measure $u_{(l)(l)}(s,x)(ds\,dx)$ on Q. By Fubini's theorem, for $\eta \in C_0^\infty(Q)$

$$\int_Q \eta u_{(l)(l)}(ds\,dx) = \int_Q u\eta_{(l)(l)}\,ds\,dx$$
$$= \int_0^T \left[\int_{Q_s} u\eta_{(l)(l)}\,dx\right] ds = \int_0^T ds\left[\int_{Q_s} \eta u_{(l)(l)}(dx)\right].$$

In short,

$$\int_Q \eta(s,x)u_{(l)(l)}(s,x)(ds\,dx) = \int_0^T ds\left[\int_{Q_s} \eta(s,x)u_{(l)(l)}(s,x)(dx)\right].$$

This equality proved for $\eta \in C_0^\infty(Q)$ can be extended to all nonnegative Borel functions η if we apply usual arguments from measure theory.

Further, by definition,

$$\int_{Q_s} u(s,x)\eta_{(l)(l)}(x)\,dx = \int_{Q_s} \eta(x)u_{(l)(l)}(s,x)(dx) \tag{13}$$

for all $\eta \in C_0^\infty(Q_s)$. Approximating the function $\eta \in C^2(\bar{Q}_s)$ which is equal to zero near ∂Q_s uniformly in Q_s along with its second derivative using the functions from $C_0^\infty(Q_s)$, we can see that Eq. (13) holds for the functions $\eta \in C^2(\bar{Q}_s)$ as well.

Next, if the nonnegative function η is one taken from the formulation of the lemma, then

$$\int_Q u\eta_{(l)(l)}\,ds\,dx = \int_0^T ds\left[\int_{Q_s} u(s,x)\eta_{(l)(l)}(s,x)\,dx\right]$$
$$= \int_0^T ds\left[\int_{Q_s} \eta(s,x)u_{(l)(l)}(s,x)(dx)\right] = \int_Q \eta u_{(l)(l)}(ds\,dx).$$

It remains only to recall (see Section 2.1) simple relations between $u_{(l_1)(l_2)}$ and $u_{(l_1+l_2)(l_1+l_2)}$, $u_{(l_1-l_2)(l_1-l_2)}$, and also to represent a bounded function η which satisfies the conditions of the lemma as the difference between two nonnegative functions which also satisfy the conditions of the lemma. The lemma is proved. □

6. Theorem. *For any* $l_1, l_2 \in E_d$ *inside* H_T *there exist generalized derivatives* $v_{(l_1)(l_2)}(s,x)(ds\,dx)$, $w_{(l_1)(l_2)}(s,x)(ds\,dx)$ *(see Definition* 2.1.2*). There exists a constant* $N = N(K,m)$ *such that for each* $l \in E_d$ *inside* H_T *for* $u \equiv v$ *and for* $u \equiv w$

$$u_{(l)(l)}(s,x)(ds\,dx) \geq -Ne^{N(T-s)}(1 + |x|)^{3m}\,ds\,dx. \tag{14}$$

Proof. From Lemma 5 and Corollary 4 follows the existence of derivatives as well as the fact that

$$u_{(l)(l)}(s,x)(ds\,dx) \geq -Ne^{N(T-s)}[(1 + |x|^2)^{(3m/2)+1}]_{(l)(l)}\,ds\,dx. \tag{15}$$

Using equality in (10) we have that

$$[(1+|x|^2)^{(3m/2)+1}]_{(l)(l)} \leq 4\left(\frac{3m}{2}+1\right)\frac{3m}{2}(1+|x|^2)^{3m/2} + 2\left(\frac{3m}{2}+1\right)(1+|x|)^{3m}$$
$$\leq (3m+2)(3m+1)(1+|x|)^{3m},$$

from which and from (15) we have (14), thus completing the proof of the theorem. □

Further, we can write the integro-differential inequalities given in Section 1, in a local form. Let

$$L^\beta u(s,x)(ds\,dx) = \frac{\partial}{\partial s}u(s,x)(ds\,dx) + \sum_{i,j=1}^{d} a^{ij}(\beta,s,x)u_{x^ix^j}(s,x)(ds\,dx)$$
$$+ \sum_{i=1}^{d} b^i(\beta,s,x)u_{x^i}(s,x)\,ds\,dx - c^\beta(s,x)u(s,x)\,ds\,dx.$$

7. Theorem. *In H_T there exist generalized derivatives*

$$v_{x^i}(s,x),\ w_{x^i}(s,x),$$
$$v_{x^ix^j}(ds\,dx), \qquad w_{x^ix^j}(ds\,dx), \qquad \frac{\partial}{\partial s}v(ds\,dx), \qquad \frac{\partial}{\partial s}w(ds\,dx).$$

Furthermore, for $u \equiv v$ and for $u \equiv w$ inside H_T for all $\beta \in A$

$$-L^\beta u(s,x)(ds\,dx) - f^\beta(s,x)\,ds\,dx \geq 0.$$

In other words, $[-L^\beta u(s,x)(ds\,dx) - f^\beta(s,x)\,ds\,dx]$ is a (positive) measure.

PROOF. The existence of the derivatives v_{x^i} and w_{x^i} was proved in Section 1. We have proved the existence of $v_{x^ix^j}(ds\,dx)$ and $w_{x^ix^j}(ds\,dx)$ in the preceding theorem. By Lemma 1.5 for $\beta \in A$ and for any nonnegative $\eta \in C_0^\infty(H_T)$

$$\int_{H_T}[vL^{\beta *}\eta + f^\beta\eta]\,ds\,dx \leq 0.$$

By Lemma 5, it follows from the last inequality that

$$\int_{H_T} v\frac{\partial}{\partial s}\eta\,ds\,dx \geq \int_{H_T}\eta v^\beta(ds\,dx),$$

where

$$v^\beta(ds\,dx) = \sum_{i,j=1}^{d} a^{ij}(\beta,s,x)v_{x^ix^j}(s,x)(ds\,dx) + \sum_{i=1}^{d} b^i(\beta,s,x)v_{x^i}(s,x)\,ds\,dx$$
$$- c^\beta(s,x)v(s,x)\,ds\,dx + f^\beta(s,x)\,ds\,dx.$$

By Lemma 2.1.3, the above enables us to conclude that the derivative $(\partial/\partial s)v(s,x)(ds\,dx)$ exists and that it does not exceed $[-v^\beta(ds\,dx)]$ inside H_T. Hence the theorem is proved for the function v. We can prove the theorem for the function w in a similar way, thus completing the proof of our theorem. □

8. Remark. In proving Theorem 7 we used no assumptions about non-degeneracy of a controlled process. In particular, all the assertions made in this section hold in the case where $\sigma(\alpha,t,x) \equiv 0$.

3. Estimation from Above of Second Derivatives of a Payoff Function

Inequalities of the form

$$L^\alpha v(ds\,dx) + f^\alpha\,ds\,dx \le 0 \tag{1}$$

(see Theorem 2.7) enable us to estimate from above second derivatives $v_{(l)(l)}(ds\,dx)$. Such estimation consists of the fact that we preserve the derivative $v_{(l)(l)}(ds\,dx)$ on the left side of (1) but we carry all the remaining expressions over to the right side of (1). It is necessary that the derivative $v_{(l)(l)}(ds\,dx)$ be "actually" present in some inequality of type (1) or that the derivative $u_{(l)(l)}(s,x)$ "actually" belong to the operator $F[u]$.

We assume that in addition to the assumptions made in this chapter, the assumptions made in Section 4.2 concerning the derivatives σ, b, c, f, $g(x)$, $g(t,x)$ are satisfied. For $t \in [0,T]$, $x \in E_d$, $\alpha \in A$, $l \ne 0$ let

$$n^\alpha(t,x) = (1 + \operatorname{tr} a(\alpha,t,x) + |b(\alpha,t,x)| + c^\alpha(t,x) + |f^\alpha(t,x)|)^{-1},$$

$$\mu(l) = \mu(t,x,l) = \inf_{\lambda:\,l\lambda=1}\ \sup_{\alpha\in A} n^\alpha(t,x)(a(\alpha,t,x)\lambda,\lambda),$$

$$Q(l) = \{(t,x) \in H_T : \mu(t,x,l) > 0\}.$$

We note that due to continuity of $a(\alpha,t,x)$, n^α with respect to α and due to separability of A we can, in determining $\mu(t,x,l)$, compute the upper bound on a basis of a countable subset of A. Therefore, the upper bound mentioned is measurable with respect to (t,x). Furthermore, this upper bound is continuous with respect to λ; therefore $\mu(t,x,l)$ is measurable with respect to (t,x). In particular, $Q(l)$ is a Borel set.

Further, we introduced the function $n^\alpha(t,x)$ into the formula which gives $\mu(t,x,l)$, for the sake of convenience. Since for each (t,x) the functions $n^\alpha(t,x)$, $[n^\alpha(t,x)]^{-1}$ are bounded from above on the set A, $\mu(t,x,l) > 0$ if and only if

$$\inf_{\lambda:\,l\lambda=1}\ \sup_{\alpha\in A}(a(\alpha,t,x)\lambda,\lambda) > 0,$$

in other words, if

$$\inf_{\lambda:\,l\lambda=1} F_1(\lambda^i\lambda^j,t,x) > 0.$$

Therefore,

$$Q(l) = \left\{(t,x) \in H_T:\ \inf_{\lambda:\,l\lambda=1} F_1(\lambda^i\lambda^j,t,x) > 0\right\}.$$

Next, we shall explain in what sense $Q(l)$ is a set on which the derivative $u_{(l)(l)}$ actually belongs to the operators $F[u]$, $F_1[u]$. Let a point $(t_0,x_0) \notin Q(l)$. Then, it can easily be seen that there is a vector λ_0 such that $l\lambda_0 = 1$ and $F_1(\lambda_0^i\lambda_0^j,t_0,x_0) = 0$. We can assume without loss of generality that the direction of λ_0 is equivalent to that of a first coordinate vector. Then $a^{11}(\alpha,t_0,x_0) = 0$ for all $\alpha \in A$. The property of nonnegative definiteness of the matrices $a(\alpha,t,x)$ implies that $a^{1i}(\alpha,t_0,x_0) = a^{i1}(\alpha,t_0,x_0) = 0$ for all $\alpha \in A$, $i = 1, \ldots, d$.

Therefore,

$$F_1[u](t_0,x_0) = \sup_{\alpha \in A} \sum_{i,j=2}^{d} a^{ij}(\alpha,t_0,x_0)u_{x^ix^j}.$$

For computing $F_1[u](t_0,x_0)$ we need to know the derivatives $u_{x^ix^j}$ only for $i, j \geq 2$. At the same time, it is impossible to express

$$|l|^2 u_{(l)(l)} = \sum_{i,j=1}^{d} u_{x^ix^j} l^i l^j,$$

in terms of the derivatives $u_{x^ix^j}$ mentioned since $l^1 = l(\lambda_0/|\lambda_0|) \neq 0$. For example, the functions u, $u + (x^1)^2$ have identical derivatives with respect to x^ix^j $(i,j \geq 2)$; however, their derivatives with respect to $(l)(l)$ are distinct. Operators F_1 on the functions u, $u + (x^1)^2$ apparently coincide at the point (t_0,x_0). An arbitrary variation of $u_{(l)(l)}$ has no effect on the value of $F_1[u]$.

In the same sense, the derivatives $u_{x^1x^1}$, $u_{x^2x^2}$ do not belong to the operator

$$Lu \equiv u_{x^1x^1} + 2u_{x^1x^2} + u_{x^2x^2}.$$

In fact, assuming $\bar{l} = (1,1)$, we easily prove that $Lu = 2u_{(\bar{l})(\bar{l})}$. It is impossible, however, to express either $u_{x^1x^1}$ or $u_{x^2x^2}$ in terms of $u_{(\bar{l})(\bar{l})}$.

The reader will understand how the estimation of the type $v_{(l)(l)}(ds\,dx) \leq \psi\,ds\,dx$ depends on the equality $\mu(t,x,l) = 0$ from the following exercise.

1. Exercise

Let $d = d_1 = 2$, $T = 1$,

$$A = \left[-\frac{1}{2},\frac{1}{2}\right], \qquad \sigma^{ij} \equiv 1, \qquad b^1(\alpha,t,x) = \alpha, \qquad b^2(\alpha,t,x) = -\alpha,$$

$$g(x) = \frac{(x^1 - x^2)^2}{1 + (x^1 - x^2)^2}, \qquad v(s,x) = \sup_{\alpha \in \mathfrak{A}} \mathsf{M}_{s,x}^{\alpha} g(x_{1-s}).$$

Prove that $\mu(l) = 0$ for $l \nparallel \bar{l} = (1,1)$, $\mu(\bar{l}) > 0$, $v_{(\bar{l})(\bar{l})} = 0$, and that the function $v(s,x)$ is a smooth function of s and $v_{(l)(l)}(\Gamma) > 0$ for $l \nparallel \bar{l}$, where $\Gamma = [0,1] \times \{x : x^1 = x^2\}$. Note that $\int_\Gamma ds\,dx = 0$.

Thus, if $\mu(l) = 0$, we need not know the value of $u_{(l)(l)}$ in order to compute $F[u]$. The derivative $v_{(l)(l)}(ds\,dx)$ is not in general absolutely continuous with respect to Lebesgue measure, that is, the generalized derivative $v_{(l)(l)}(s,x)$

does not exist. We shall see below (see Theorem 5) that if $\mu(t,x,l) > 0$ in any region Q, the generalized derivative $v_{(l)(l)}(s,x)$ does exist in Q.

2. Lemma. *Let $u = (u^{ij})$ be a matrix of dimension $d \times d$, and let ψ be a number. Assume that for all $\lambda \in E_d$*

$$(u\lambda,\lambda) \geq \psi|\lambda|^2.$$

Then for all $(t,x) \in H_T$ and for all units l

$$\mu(t,x,l)(ul,l) \leq \sup_{\alpha \in A} n^\alpha(t,x) \operatorname{tr} a(\alpha,t,x)u + \psi_-.$$

Proof. We fix t, x. Furthermore, we denote by Γ the smallest closed convex set of matrices of dimension $d \times d$, which contains all matrices $n^\alpha(t,x)a(\alpha,t,x)$ $(\alpha \in A)$. We can obtain Γ, for example, of the closure as the set

$$\left\{\sum_{i=0}^{n} p_i n^{\alpha_i}(t,x)a(\alpha_i,t,x): n = 1,2,\ldots; p_i \geq 0, \sum_{i=1}^{n} p_i = 1, \alpha_i \in A\right\},$$

which implies that the set Γ is bounded and, in addition, that

$$\begin{aligned} \max_{\alpha \in \Gamma} (a\lambda,\lambda) &= \sup_{\alpha \in A} n^\alpha(t,x)(a(\alpha,t,x)\lambda,\lambda), \\ \max_{\alpha \in \Gamma} \operatorname{tr} au &= \sup_{\alpha \in A} n^\alpha(t,x) \operatorname{tr} a(\alpha,t,x)u. \end{aligned} \tag{2}$$

Let us prove that

$$\mu(t,x,l) = \inf_{\lambda: l\lambda = 1} \max_{a \in \Gamma} (a\lambda,\lambda) = \max_{a \in \Gamma} \inf_{\lambda: l\lambda = 1} (a\lambda,\lambda). \tag{3}$$

The first equality in (3) follows from the first equality in (2). In order to prove the second equality in (3), we apply the main theorem of game theory. Let $R > 0$. The function $(a\lambda,\lambda)$ is given on $\Gamma \times \{\lambda: l\lambda = 1, |\lambda| = R\}$, convex upward (linear) with respect to a, convex downward with respect to λ because the matrices in Γ are nonnegative definite. Furthermore, the sets Γ, $\{\lambda: l\lambda = 1, |\lambda| \leq R\}$ are convex, bounded, and closed. Therefore, for each $R > 0$

$$\begin{aligned} \mu_R(l) &\equiv \min_{\lambda: l\lambda = 1, |\lambda| \leq T} \max_{a \in \Gamma} (a\lambda,\lambda) \\ &= \max_{a \in R} \min_{\lambda: l\lambda = 1, |\lambda| \leq R} (a\lambda,\lambda), \end{aligned}$$

where the second equality implies that there exists a matrix $a^R \in \Gamma$ such that $(a^R\lambda,\lambda) \geq \mu_R(l)$ if $l\lambda = 1$, $|\lambda| \leq R$. Letting $R \to \infty$ and, in addition, taking a convergent sequence of the matrices a^R, we find a matrix $\bar{a} \in \Gamma$ such that $(\bar{a}\lambda,\lambda) \geq \lim_{R \to \infty} \mu_R(l)$ if $l\lambda = 1$. The definition of $\mu_R(l)$ shows that $\lim_{R \to \infty} \mu_R(l) = \mu(l)$. Therefore,

$$\inf_{\lambda: l\lambda = 1} (\bar{a}\lambda,\lambda) \geq \mu(l),$$

$$\mu(l) \leq \inf_{\lambda: l\lambda = 1} (\bar{a}\lambda,\lambda) \leq \sup_{a \in \Gamma} \inf_{\lambda: l\lambda = 1} (a\lambda,\lambda).$$

On the other hand,

$$\sup_{a\in\Gamma}\inf_{\lambda:l\lambda=1}(a\lambda,\lambda)\le\sup_{a\in\Gamma}\inf_{\lambda:l\lambda=1,|\lambda|\le R}(a\lambda,\lambda)=\mu_R(l)\to\mu(l),$$

which completes the proof of the second equality in (3).

Further, let $\mu(a,l)=\inf_{\lambda:l\lambda=1}(a\lambda,\lambda)$. By definition, $\mu(a,l)\le(a\lambda,\lambda)$ if $l\lambda=1$. Using in the last inequality the expression $\lambda(l\lambda)^{-1}$ instead of λ, we find $\mu(a,l)(l\lambda)^2\le(a\lambda,\lambda)$ for $l\lambda\ne 0$. Thus certainly $\mu(a,l)(l\lambda)^2\le(a\lambda,\lambda)$ for $l\lambda=0$, which implies that the matrix $a-\mu(a,l)(l^il^j)\ge 0$ for each $a\in\Gamma$. The matrix $\frac{1}{2}(u+u^*)-\psi I$, where I denotes the unit matrix of dimension $d\times d$, is nonnegative definite as well by definition. The trace of the product of nonnegative definite symmetric matrices is nonnegative. Hence for each $a\in\Gamma$

$$\begin{aligned}0&\le\operatorname{tr}[a-\mu(a,l)(l^il^j)][\tfrac{1}{2}(u+u^*)-\psi I]\\&=\operatorname{tr}[a-\mu(a,l)(l^il^j)]u-\psi[\operatorname{tr}a-\mu(a,l)]\\&\le\operatorname{tr}au-\mu(a,l)(ul,l)+\psi_-.\end{aligned}$$

The last inequality holds due to the fact that $n^\alpha(t,x)\operatorname{tr}a(\alpha,t,x)\le 1$, $1\ge\operatorname{tr}a\ge\mu(a,l)$ for $a\in\Gamma$. Finally, we have $\mu(a,l)(ul,l)\le\operatorname{tr}au+\psi_-$, where it remains only to take upper bounds with respect to $a\in\Gamma$ and, in addition, to make use of (3) and (2). The lemma is proved. □

We introduce some additional notations. If v is a σ-additive function of sets, $|v|$ is the variation of v; furthermore, $v_-=\frac{1}{2}(|v|-v)$ is the negative part of the measure v. It is a well-known fact that if v is absolutely continuous with respect to the measure v_1 and, in addition, $v(dx)/v_1(dx)=f(x)$, the measures $|v|$ and v_- are absolutely continuous as well with respect to v_1 and

$$\frac{|v|(dx)}{v_1(dx)}=|f(x)|,\qquad\frac{v_-(dx)}{v_1(dx)}=f_-(x).$$

The next theorem incorporates the objects whose existence was proved in Sections 1 and 2 (in particular, see Theorem 2.7).

3. Theorem. *Let $u_1=v$, $u_2=w$. Let measurable functions ψ_1, ψ_2 be such that $\psi_i\,dt\,dx\le u_{i(l)(l)}(dt\,dx)$ inside H_T for any $l\ne 0$, $i=1,2$. (By Theorem 2.6, we can take as ψ_i the right side of inequality (2.14).) Then for all unit l and $i=1,2$ inside H_T*

$$\begin{aligned}\mu(l)u_{i(l)(l)}(dt\,dx)\le(\psi_i)_-\,dt\,dx&+\left(\frac{\partial}{\partial t}u_i\right)(dt\,dx)\\&+|\operatorname{grad}_x u_i|\,dt\,dx+(u_i)_+\,dt\,dx+dt\,dx,\end{aligned}\tag{4}$$

$$\frac{\partial}{\partial t}u_i(dt\,dx)\le\inf_\beta|L^\beta|[(\psi_i)_-+|\operatorname{grad}_x u_i|+(u_i)_++1]\,dt\,dx,\tag{5}$$

where

$$|L^\beta|\equiv|L^\beta(t,x)|\equiv\operatorname{tr}a(\beta,t,x)+|b(\beta,t,x)|+c^\beta(t,x)+f^\beta_-(t,x).$$

PROOF. We introduce the measure ν, using the formula

$$\nu(dt\,dx) = \left|\frac{\partial}{\partial t} v\right|(dt\,dx) + \sum_{i,j=1}^{d} |v_{x^i x^j}|(dt\,dx) + dt\,dx.$$

All set functions

$$\int_\Gamma \frac{\partial}{\partial t} v(dt\,dx), \qquad \int_\Gamma v_{x^i x^j}(dt\,dx), \qquad \int_\Gamma dt\,dx$$

are absolutely continuous with respect to ν. We denote by $\bar{v}_t(t,x)$, $\bar{v}_{ij}(t,x)$, $\rho(t,x)$ the respective Radon–Nikodym derivatives with respect to the measure ν.

According to Theorem 2.7, $[-L^\beta v(dt\,dx) - f^\beta\,dt\,dx]$ is a measure which is, obviously, absolutely continuous with respect to the measure ν and whose Radon–Nikodym derivative with respect to the measure ν is nonnegative. From this, for each $\beta \in A$ we have

$$\begin{aligned} \bar{v}_t(t,x) + \sum_{i,j=1}^{d} a^{ij}(\beta,t,x)\bar{v}_{ij}(t,x) + \sum_{i=1}^{d} b^i(\beta,t,x)v_{x^i}(t,x)\rho(t,x) \\ - c^\beta(t,x)v(t,x)\rho(t,x) + f^\beta(t,x)\rho(t,x) \le 0 \end{aligned} \tag{6}$$

almost everywhere in H_T with respect to the measure ν. The set (t,x) on which (6) are satisfied has full measure and, generally speaking, depends on β. Taking advantage of the separability of A as well as the continuity of a, b, c, and f with respect to β, we can easily prove that the set of those (t,x) for which (6) are satisfied for all β at the same time has a full measure ν as well.

Further, by assumption for each unit $\lambda \in E_d$ inside H_T

$$\psi_1\,dt\,dx \le v_{(\lambda)(\lambda)}(dt\,dx) = \sum_{i,j=1}^{d} \lambda^i\lambda^j v_{x^i x^j}(dt\,dx).$$

Therefore,

$$\psi_1 \rho \le \sum_{i,j=1}^{d} \lambda^i \lambda^j \bar{v}_{ij}$$

almost everywhere in H_T with respect to the measure ν. Due to continuity of $\lambda^i\lambda^j$ with respect to λ, the last inequality holds for all unit λ at the same time on some set of full measure ν. Therefore, the conditions of Lemma 2 are satisfied on the set mentioned for $u^{ij} = \bar{v}_{ij}$, Also, we note that since

$$n^\alpha(t,x)|b(\alpha,t,x)| \le 1, \qquad n^\alpha(t,x)c^\alpha(t,x) \le 1, \qquad n^\alpha(t,x)|f^\alpha(t,x)| \le 1,$$

from (6) on the set of full measure ν for all $\beta \in A$ we obtain

$$\begin{aligned} &n^\beta(t,x) \sum_{i,j=1}^{d} a^{ij}(\beta,t,x)\bar{v}_{ij}(t,x) \\ &\qquad \le [\bar{v}_t(t,x)]_- + |\operatorname{grad}_x v(t,x)|\rho(t,x) + v_+\rho(t,x) + \rho(t,x). \end{aligned}$$

Applying Lemma 2, we conclude that almost everywhere in H_T with respect to the measure v

$$\mu(l) \sum_{i,j=1}^{d} l^i l^j \bar{v}_{ij} \leq (\psi_1)_- \rho + (\bar{v}_t)_- + |\operatorname{grad}_x v| \rho + v_+ \rho + \rho.$$

Multiplying the last inequality by $v(dt\,dx)$, we arrive at (4) for $i = 1$. Next, let us prove (5) for $i = 1$.

Noting that the matrix $(\bar{v}_{ij} - \psi_1 \rho \delta^{ij}) \geq 0$ and, in addition, the trace of the product of symmetric positive matrices is positive, we find

$$\sum_{i,j=1}^{d} a^{ij}(\beta,t,x)\bar{v}_{ij}(t,x) = \operatorname{tr} a(\beta,t,x)(\bar{v}_{ij}) \geq \psi_1 \rho \operatorname{tr} a(\beta,t,x).$$

We have from (6) that almost everywhere in H_T with respect to the measure v

$$\begin{aligned}\bar{v}_t(t,x) &\leq [(\psi_1)_- \operatorname{tr} a(\beta,t,x) + |\operatorname{grad}_x v|\,|b(\beta,t,x)| + c^\beta(t,x)v_+ + f^\beta_-(t,x)]\rho \\ &\leq |L^\beta|[(\psi_1)_- + |\operatorname{grad}_x v| + v_+ + 1]\rho.\end{aligned}$$

Computing the lower bound of the right side of the last inequality on a basis of a countable set β which is dense in A, using the continuity of $|L^\beta|$ with respect to β, and, finally, multiplying the inequality thus obtained by $v(dt\,dx)$, we complete the proof of (5) for $i = 1$. We prove inequalities (4) and (5) for $i = 2$ in the same way. The theorem is proved. □

4. Corollary. *Let a region $Q \subset H_T$, and let a measure $((\partial/\partial t)v)_-(dt\,dx)$ (respectively a measure $((\partial/\partial t)w)_-(dt\,dx)$) in a region Q be absolutely continuous with respect to a Lebesgue measure $dt\,dx$. Then, the restriction of the set functions $v_{(l)(l)}(dt\,dx)$ (respectively, $w_{(l)(l)}(dt\,dx)$) on a set $Q \cap Q(l)$ is absolutely continuous with respect to Lesbesgue measure. Furthermore, in Q there exists a generalized derivative*

$$\frac{\partial}{\partial t} v(t,x) \left(\frac{\partial}{\partial t} w(t,x) \right)$$

in the sense of Definition 2.1.1.

In fact, we take as ψ_1 the right side of inequality (2.14). From (4) and Theorem 2.6 we have that inside Q

$$\mu(l)\psi_1\,dt\,dx \leq \mu(l)v_{(l)(l)}(dt\,dx) \leq h\,dt\,dx,$$

where

$$h = (\psi_1)_- + \frac{((\partial/\partial t)v)_-}{dt\,dx} + |\operatorname{grad}_x v| + v_+ + 1,$$

which, as can be seen, implies that inside Q

$$\mu(l)|v_{(l)(l)}|(dt\,dx) \leq [\mu(l)|\psi_1| + h]\,dt\,dx.$$

Therefore, if $\Gamma \subset Q$, $\int_\Gamma dt\,dx = 0$, then

$$\int_\Gamma \mu(l)|v_{(l)(l)}|(dt\,dl) = 0.$$

If, moreover, $\Gamma \subset Q(l)$, by virtue of the inequality $\mu(l) > 0$ on $Q(l)$

$$\int_\Gamma |v_{(l)(l)}|(dt\,dx) = 0.$$

In other words, $v_{(l)(l)}(dt\,dx)$ is absolutely continuous with respect to Lebesgue measure on $Q \cap Q(l)$.

Further, since

$$-\frac{\partial}{\partial t} v_-(dt\,dx) \le \frac{\partial}{\partial t} v(dt\,dx),$$

then inside Q

$$-\frac{((\partial/\partial t)v)_-}{dt\,dx} dt\,dx \le \frac{\partial}{\partial t} v(dt\,dx),$$

which yields the lower estimate of $(\partial/\partial t)v(dt\,dx)$ in terms of the Lebesgue measure $dt\,dx$. Next, taking the upper estimate from (5), we obtain, as above, that $(\partial/\partial t)v(dt\,dx)$ is absolutely continuous with respect to Lebesgue measure on Q. Therefore, if $\eta \in C_0^\infty(Q)$, then

$$\int_Q v \frac{\partial}{\partial t} \eta\,dt\,dx = -\int_Q \eta \frac{\partial}{\partial t} v(dt\,dx) = -\int_Q \eta \frac{(\partial/\partial t)v}{dt\,dx} dt\,dx$$

and therefore, the measure density $(\partial/\partial t)v(dt\,dx)$ with respect to Lebesgue measure is a generalized derivative of v with respect to t in the sense of Definition 2.1.1. The function w can be considered in the same way.

5. Theorem. *Let a region $Q \subset H_T$. Let a measure $((\partial/\partial t)v)_-(dt\,dx)$ in a region Q be absolutely continuous with respect to a Lebesgue measure $dt\,dx$. Then, in Q there exists a derivative $(\partial/\partial t)v(t,x)$ and furthermore;*

a. *if $Q \subset Q(l)$ for some unit vector l, a second generalized derivative $v_{(l)(l)}(t,x)$ exists, and also*

$$\mu(l)\psi_1 \le \mu(l)v_{(l)(l)} \le (\psi_1)_- + \left(\frac{\partial}{\partial t} v\right)_- + |\text{grad}_x v| + v_+ + 1 \qquad (Q\text{-a.e.}),$$

where ψ_1 is a function satisfying the assumption of Theorem 3;

b. *if for all $(t,x) \in Q$, $l \ne 0$*

$$\sup_{\alpha \in A} (a(\alpha,t,x)l,l) > 0, \tag{7}$$

then $Q \subset \bigcap_l Q(l)$, all generalized derivatives of the type $v_{x^i x^j}(t,x)$ exist in Q, and, in addition, $F[v] \le 0$ (Q-a.e.).

Proof. According to the preceding corollary, $(\partial/\partial t)v(t,x)$ exists in Q and, furthermore, under conditions (a) the function of sets $v_{(l)(l)}(dt\,dx)$ in the region

Q is absolutely continuous with respect to Lebesgue measure. The Radon–Nikodym derivative of the function with respect to Lebesgue measure is a generalized derivative of the type $v_{(l)(l)}(t,x)$. Estimates for the former derivative follow immediately from (4) and the assumption that

$$\psi_1 \, dt \, dx \leq v_{(l)(l)}(dt \, dx).$$

In order to prove (b), we note that the function

$$\sup_{\alpha \in A} n^\alpha(t,x)(a(\alpha,t,x)l,l)$$

is continuous with respect to l, which together with (7) implies that on Q

$$\mu(t,x) \equiv \inf_{|l|=1} \sup_{\alpha \in A} n^\alpha(t,x)(a(\alpha,t,x)l,l) > 0.$$

It is seen that for all $\lambda \in E_d$

$$\sup_{\alpha \in A} n^\alpha(t,x)(a(\alpha,t,x)\lambda,\lambda) \geq \mu(t,x)|\lambda|^2,$$

which yields for $|l| = 1$

$$\mu(t,x,l) \geq \inf_{\lambda : l\lambda = 1} \mu(t,x)|\lambda|^2 = \mu(t,x).$$

Therefore, $\mu(l) \geq \mu > 0$ on Q. In other words, $Q \subset \cap Q(l)$. Using assertion (a), we finally conclude that in Q there exist all generalized derivatives of the type $v_{(l)(l)}(t,x)$. As we have seen above, this implies the existence of all second mixed generalized derivatives in the region Q. Finally, the inequality $F[v] \leq 0$ (Q-a.e.) follows from Corollary 1.7, thus proving the theorem. □

6. Remark. Obviously, Theorem 5 will hold if we replace in its formulation v and ψ_1 by w and ψ_2, ψ_2 being a function satisfying the assumption of Theorem 3.

4. Estimation of a Derivative of a Payoff Function with Respect to t

We saw in Section 4.3 (see Theorem 3.5) that in order to prove the existence of second-order generalized derivatives of a payoff function with respect to space variables and to estimate them, we had to know how to estimate the derivatives of a payoff function with respect to t. In this section we estimate the absolute values of $(\partial/\partial t)v(t,x)$, $(\partial/\partial t)w(t,x)$, making more assumptions than in Sections 2 and 3. In addition to the main assumptions made in Chapter 4, we assume here that the functions $\sigma(\alpha,t,x)$, $b(\alpha,t,x)$, $c^\alpha(t,x)$, $f^\alpha(t,x)$ for each $\alpha \in A$ are once continuously differentiable with respect to (t,x) on $\bar{H}_T$, the function $g(x)$ is twice continuously differentiable with respect to x, and the function $g(t,x)$ is once differentiable with respect to t and twice differentiable with respect to x, with the derivatives $(\partial/\partial t)g(t,x)$, $g_{x^i}(t,x)$, $g_{x^ix^j}(t,x)$ being continuous in $\bar{H}_T$. Furthermore, for all $\alpha \in A$, $t \in [0,T]$, $l \in E_d$ let

$$\begin{aligned} &\left\|\frac{\partial}{\partial t}\sigma(\alpha,t,x)\right\| + \left|\frac{\partial}{\partial t}b(\alpha,t,x)\right| + \left|\frac{\partial}{\partial t}c^\alpha(t,x)\right| \\ &\quad + \left|\frac{\partial}{\partial t}f^\alpha(t,x)\right| + |c^\alpha_{(l)}(t,x)| + |f^\alpha_{(l)}(t,x)| \\ &\quad + |g_{(l)}(x)| + |g_{(l)(l)}(x)| + \left|\frac{\partial}{\partial t}g(t,x)\right| \\ &\quad + |g_{(l)}(t,x)| + |g_{(l)(l)}(t,x)| \le K(1+|x|)^m, \\ &|L^\alpha(T,x)g(x)| + |L^\alpha g(T,x)| \le K(1+|x|)^{3m}, \end{aligned} \tag{1}$$

where the constants K and m are the same as those in (3.1.1)–(3.1.3).

The last inequality follows readily from the foregoing and from (3.1.2) and (3.1.3) if in the right side of (1) we replace $K(1+|x|)^{3m}$ by $N(K,d)$ $(1+|x|)^{(m+2)\vee 2m}$. It is seen also that $(1+|x|)^m \le (1+|x|)^{(m+2)\vee 2m}$. Hence, if there exist constants K and $m \ge 0$ for which all the assumptions except (1) can be satisfied, there will exist (other) constants K and m for which all the assumptions including (1) can be satisfied. Therefore, we can easily do without (1). We shall not omit (1), however, because the estimate (1) is convenient in the case, for example, where $m = 0$ and, in addition, $\sigma(\alpha,t,x)$ and $b(\alpha,t,x)$ are bounded functions. Furthermore, the right side of (1) written in a special form will be convenient for our computations.

It is clear that we can always extend the functions σ, b, c, f, and $g(t,x)$ for $t > T$ so that our assumptions are satisfied for all $t \in [0,\infty)$. However, in this case we need, perhaps, to replace the constant K by $2K$ for $t > T$. Let us assume that we have carried out the extension as described above.

Due to the results obtained in Section 2.8 the process $x_t^{\alpha,s,x}$ is $\mathscr{L}B$-continuously $\mathscr{L}B$-differentiable over s for $s \in (0,T)$ on an interval $[0,T]$. Furthermore, if $q_t^{\alpha,s,x} = \mathscr{L}B\text{-}(\partial/\partial s)x_t^{\alpha,s,x}$, then for each $n \ge 1$ (see Theorem 2.8.7)

$$\mathsf{M}\sup_{t\le T}|q_t^{\alpha,s,x}|^{2n} \le Ne^{NT}(1+|x|)^{2nm}, \tag{2}$$

where $N = N(K,m,n)$. It is seen that the process $(s+t, x_t^{\alpha,s,x})$ is $\mathscr{L}B$-continuously $\mathscr{L}B$-differentiable as well. It follows from Section 2.7 that the function $v^{\alpha,\tau}(s,x)$ is continuously differentiable over s on $(0,T)$ for each $\alpha \in \mathfrak{A}$, $x \in E_d$, $\tau \in \mathfrak{M}(T)$. Our objective consists in finding the formula for $(\partial/\partial s)v^{\alpha,\tau\wedge(T-s)}(s,x)$.

1. Lemma. *Suppose that on a square $(0,T)\times(0,T)$ we are given a bounded function $\psi(s,r)$ which is measurable with respect to (s,r). Also, suppose that for almost all r the function $\psi(s,r)$ is absolutely continuous with respect to s*

and that the function

$$\frac{\partial}{\partial s}\psi(s,r) = \begin{cases} \lim\limits_{t\to s} \dfrac{\psi(t,r) - \psi(s,r)}{t - s} & \text{if the limit exists} \\ 0 & \text{otherwise} \end{cases}$$

satisfies the inequality

$$\int_0^T \int_0^T \left| \frac{\partial}{\partial s}\psi(s,r) \right| dr\, ds < \infty.$$

Then the function $\int_0^{T-s} \psi(s,r)\, dr$ *is absolutely continuous with respect to* s *and, in addition, a derivative of this function with respect to* s *coincides with*

$$\int_0^{T-s} \frac{\partial}{\partial s}\psi(s,r)\, dr - \psi(s, T - s)$$

for almost all s.

The proof of the lemma follows from the fact that for those r for which $\psi(s,r)$ is absolutely continuous with respect to s, the function $(\partial/\partial s)\psi(s,r)$ is a derivative of $\psi(s,r)$, is measurable with respect to (s,r), and, finally, as we can easily verify using Fubini's theorem,

$$\int_0^{T-s} \psi(s,r)\, dr = -\int_s^T \left[\int_0^{T-s_1} \frac{\partial}{\partial s_1}\psi(s_1,r)\, dr - \psi(s_1, T - s_1) \right] ds_1.$$

We can better describe the method of using this lemma if from the definition of $v^{\alpha,\tau}(s,x)$, applying Ito's formula, we derive the following formula:

$$v^{\alpha,\tau\wedge(T-s)}(s,x) = \int_0^{T-s} \mathsf{M}_{s,x}^{\alpha}\chi_{\tau\geq r}[f^{\alpha_r}(s + r, x_r) + L^{\alpha_r}g(s + r, x_r)]e^{-\varphi_r}\, dr + g(s,x).$$

If we assume that the function $g(t,y)$ is infinitely differentiable with its derivatives increasing not too rapidly for $|y| \to \infty$, the functions $L^{\alpha_r}g(s + r, y)$ are continuously differentiable with respect to (s,y). Further, since the process $(s + r, x_r^{\alpha,s,x})$ is $\mathscr{L}B$-continuously $\mathscr{L}B$-differentiable with respect to s, due to the results obtained in Section 2.7 the random variable

$$\chi_{\tau\geq r}[f^{\alpha_r}(s + r, x_r^{\alpha,s,x}) + L^{\alpha_r}g(s + r, x_r^{\alpha,s,x})]e^{-\varphi_r^{\alpha,s,x}}$$

is $\mathscr{L}$-continuously $\mathscr{L}$-differentiable with respect to s on $(0,T)$ for each $r \in [0,T]$. In this situation the mathematical expectation

$$\mathsf{M}_{s,x}^{\alpha}\chi_{\tau\geq r}[f^{\alpha_r}(s + r, x_r) + L^{\alpha_r}g(s + r, x_r)]e^{-\varphi_r} \tag{3}$$

is continuously differentiable with respect to s on $(0,T)$ for any α, x, r. If we compute the derivative of (3) with respect to s and, furthermore, if we make use of the familiar estimates of moments of $x_r^{\alpha,s,x}$ as well as inequality (2), we can easily prove (compare with the proof of Theorem 1.1) that the derivative of (3) with respect to s is bounded for $s \in (0,T)$, $r \in (0,T)$ for each x. Therefore, in this case the function $v^{\alpha,\tau\wedge(T-s)}(s,x)$ is absolutely continuous with respect to s according to Lemma 1. Also, ($(0,T)$-a.e.)

$$\frac{\partial}{\partial s} v^{\alpha,\tau\wedge(T-s)}(s,x) = -\mathsf{M}^{\alpha}_{s,x}\chi_{\tau\geq T-s}[f^{\alpha_{T-s}}(T,x_{T-s}) + L^{\alpha_{T-s}}g(T,x_{T-s})]e^{-\varphi_{T-s}}$$

$$+ \int_0^{T-s} \frac{\partial}{\partial s} \mathsf{M}^{\alpha}_{s,x}\chi_{\tau\geq r}[f^{\alpha_r}(s+r,x_r)$$

$$+ L^{\alpha_r}g(s+r,x_r)]e^{-\varphi_r}\,dr + \frac{\partial}{\partial s} g(s,x).$$

Let us transform the last expression. Using Ito's formula as well as the rules given in Section 2.7, which enable us to interchange the order of derivatives and integrals, we conclude that

$$\int_0^t \frac{\partial}{\partial s} \mathsf{M}^{\alpha}_{s,x}\chi_{\tau\geq r}e^{-\varphi_r}L^{\alpha_r}g(s+r,x_r)\,dr = \frac{\partial}{\partial s}\mathsf{M}^{\alpha}_{s,x}\int_0^{\tau\wedge t} e^{-\varphi_r}L^{\alpha_r}g(s+r,x_r)\,dr$$

$$= \frac{\partial}{\partial s}[\mathsf{M}^{\alpha}_{s,x}e^{-\varphi_{\tau\wedge t}}g(s+\tau\wedge t, x_{\tau\wedge t}) - g(s,x)].$$

Thus, if $g(t,y)$ is a sufficiently smooth function,

$$\frac{\partial}{\partial s} v^{\alpha,\tau\wedge(T-s)}(s,x) = -\mathsf{M}^{\alpha}_{s,x}\chi_{\tau\geq T-s}[f^{\alpha_{T-s}}(T,x_{T-s}) + L^{\alpha_{T-s}}g(T,x_{T-s})]e^{-\varphi_{T-s}}$$

$$+ \frac{\partial}{\partial s}\mathsf{M}\left\{\int_0^{\tau\wedge t} f^{\alpha_r}(s+r,x_r^{\alpha,s,x})e^{-\varphi_r^{\alpha,s,x}}\,dr\right.$$

$$\left.+ e^{-\varphi_{\tau\wedge t}^{\alpha,s,x}}g(s+\tau\wedge t, x_{\tau\wedge t}^{\alpha,s,x})\right\}\Bigg|_{t=T-s} \qquad ([0,T]\text{-a.e.}).$$

Immediate computation of the last derivative with respect to s and, next, the application of Fubini's theorem (or carrying out transformations identical to those given in Section 2) lead us to the following result.

2. Lemma. *For each $x\in E_d$, $\alpha\in\mathfrak{A}$, $\tau\in\mathfrak{M}(T)$ the function $v^{\alpha,\tau\wedge(T-s)}(s,x)$ is absolutely continuous with respect to s and, in addition, almost everywhere on $(0,T)$ its derivative coincides with $\theta^{\alpha,\tau\wedge(T-s)}(s,x)$ where*

$$\theta^{\alpha,\tau}(s,x) \equiv \mathsf{M}^{\alpha}_{s,x}\left\{e^{-\varphi_\tau}\left[\left(\frac{\partial}{\partial s}g\right)(s+\tau,x_\tau) + g_{(q_\tau)}(s+\tau,x_\tau)|q_\tau|\right]\right.$$

$$+ \int_0^\tau e^{-\varphi_t}\left\{\left(\frac{\partial}{\partial s}f^{\alpha_t}\right)(s+t,x_t) + f^{\alpha_t}_{(q_t)}(s+t,x_t)|q_t|\right.$$

$$\left.\left. - \xi^{\alpha,\tau}_{s,x}(t)\left[\left(\frac{\partial}{\partial s}c^{\alpha_t}\right)(s+t,x_t) + c^{\alpha_t}_{(q_t)}(s+t,x_t)|q_t|\right]\right\}dt\right\}$$

$$- \mathsf{M}^{\alpha}_{s,x}\chi_{\tau\geq T-s}[f^{\alpha_{T-s}}(T,x_{T-s}) + L^{\alpha_{T-s}}g(T,x_{T-s})]e^{-\varphi_{T-s}},$$

$$\xi^{\alpha,\tau}_{s,x}(t) = \int_t^\tau e^{-(\varphi_r-\varphi_t)}f^{\alpha_r}(s+r,x_r)\,dr + e^{-(\varphi_\tau-\varphi_t)}g(s+\tau,x_\tau).$$

The arguments adduced prior to the lemma included an assumption that $g(t,y)$ is very smooth. However, the lemma holds in the general case as well. In order to be convinced of this, one needs only to approximate $g(t,y)$ using convolutions with smooth kernels and, second, to pass to the limit in the formula

$$v^{\alpha,\tau\wedge(T-s_2)}(s_2,x) - v^{\alpha,\tau\wedge(T-s_1)}(s_1,x) = \int_{s_1}^{s_2} \theta^{\alpha,\tau\wedge(T-s)}(s,x)\,ds, \tag{4}$$

proved for smooth $g(t,y)$. Passage to the limit becomes possible due to the familiar estimates of the moments of $x_t^{\alpha,s,x}$, $q_t^{\alpha,s,x}$.

The foregoing estimates enable us to assert that

$$|\theta^{\alpha,\tau}(s,x)| \le Ne^{N(T-s)}(1+|x|)^{3m}, \tag{5}$$

where $N = N(K,m)$ (compare, for example, the proof of Theorem 1.1).

3. Theorem. *For each $x \in E_d$ the functions $v(s,x)$ and $w(s,x)$ are absolutely continuous with respect to s on $[0,T]$, have on this interval a generalized derivative with respect to s, and, finally, for some constant $N = N(K,m)$*

$$\left|\frac{\partial}{\partial s} v(s,x)\right| + \left|\frac{\partial}{\partial s} w(s,x)\right| \le Ne^{N(T-s)}(1+|x|)^{3m} \qquad ([0,T]\text{-a.e.}).$$

PROOF. From (4) and (5) for $s_2 > s_1$ we find

$$|v^{\alpha,\tau\wedge(T-s_2)}(s_2,x) - v^{\alpha,\tau\wedge(T-s_1)}(s_1,x)| \le Ne^{N(T-s_1)}(1+|x|)^{3m}(s_2-s_1).$$

Since

$$w(s,x) = \sup_{\alpha\in\mathfrak{A}} \sup_{\tau\in\mathfrak{M}(T)} v^{\alpha,\tau\wedge(T-s)}(s,x)$$

and the difference between the upper bounds is not greater than the upper bound of the differences, we have

$$|w(s_2,x) - w(s_1,x)| \le Ne^{N(T-s_1)}(1+|x|)^{3m}(s_2-s_1),$$

which implies the absolute continuity of $w(s,x)$ with respect to s. Dividing the last inequality by $s_2 - s_1$ and taking the limit as $s_i \to s$ we obtain the ordinary derivative $(\partial/\partial s)w(s,x)$. We complete the proof of the theorem for w invoking the fact that a generalized derivative of a function of a real variable coincides almost everywhere with the ordinary derivative of this function. In order to prove the theorem for the function $v(s,x)$, it suffices to assume that $\tau = T$ and to replace $g(s,x)$ by $g(x)$ in all the arguments in this section. The theorem is proved. □

4. Exercise

We change σ and b in Exercise 3.1. Let $\sigma(t,x)$ be a unit matrix of dimension 2×2 for $t\in[0,\frac{1}{2}]$; $\sigma^{ij}(t,x) = 1$ for $t\in(\frac{1}{2},1]$; $b(\alpha,t,x) = 0$ for $t\in[0,\frac{1}{2}]$;

$$b(\alpha,t,x) = (\alpha,-\alpha) \text{ for } t\in(\tfrac{1}{2},1].$$

Show that the derivative $(\partial/\partial s)v(s,x)$ is unbounded (near the point $(\frac{1}{2},0)$).

5. Passage to the Limit in the Bellman Equation

We know from Sections 1–4 the conditions imposed on σ, b, c, f, and g under which payoff functions have generalized second derivatives with respect to x and a generalized first derivative with respect to t. We shall show further on (see Theorems 7.1 and 7.2) that it is easy to deduce the Bellman equations for payoff functions from the existence of the derivatives mentioned and from the assumption about the nondegeneracy of all processes $x_t^{\alpha,s,x}$. In order to remove the condition that all the processes $x_t^{\alpha,s,x}$ are to be nondegenerate, we need theorems on passage to the limit in the Bellman equation.

Throughout this section Q denotes a bounded subregion of H_T, $\tilde{a}(\alpha,t,x)$ is a nonnegative symmetric matrix of dimension $d \times d$, $\tilde{b}(\alpha,t,x)$ is a d-dimensional vector, $\tilde{c}^\alpha(t,x)$, $\tilde{f}^\alpha(t,x)$ and $\tilde{r}^\alpha(t,x)$ are numbers. We assume that $\tilde{a}$, $\tilde{b}$, $\tilde{c}$, $\tilde{f}$, and $\tilde{r}$ are all definite for $(\alpha,t,x) \in A \times Q$, measurable with respect to (t,x) and also continuous with respect to α. Furthermore, we assume that $\tilde{c} \geq 0$, $\tilde{r} \geq 0$; $\tilde{a}$, $\tilde{b}$, $\tilde{c}$ *and* $\tilde{r}$ are bounded on $A \times Q$, $\sup_\alpha |\tilde{f}^\alpha(t,x)| \in \mathscr{L}_{d+1}(Q)$. Let

$$G(u_0,u_{ij},u_i,u,t,x) \equiv \sup_{\alpha \in A}\Bigg[\tilde{r}^\alpha(t,x)u_0 + \sum_{i,j=1}^{d} \tilde{a}^{ij}(\alpha,t,x)u_{ij}$$
$$+ \sum_{i=1}^{d} \tilde{b}^i(\alpha,t,x)u_i - \tilde{c}^\alpha(t,x)u + \tilde{f}^\alpha(t,x)\Bigg],$$

$$G[u] \equiv G[u](t,x) \equiv G\left(\frac{\partial}{\partial t}u(t,x), u_{x^ix^j}(t,x), u_{x^i}(t,x), u(t,x), t, x\right).$$

We denote by $\partial' Q$ the parabolic boundary of Q, that is, the set of those points (t_0,x_0) of the (usual) boundary of Q for each of which there exists a number $\delta > 0$ and, in addition, a continuous function x_t defined on $[t_0 - \delta, t_0]$ such that $x_{t_0} = x_0$, $(t,x_t) \in Q$ for $t \in [t_0 - \delta, t_0)$. It is easily seen that if the process $(s_0 + t, x_t)$ is continuous with respect to t and is inside Q at $t = 0$, this process $(s_0 + t, x_t)$ can leave the region Q only across the parabolic boundary of Q.

Two basic theorems of the present section are the following.

1. Theorem. *Let the functions* $u_n \in W^{1,2}(Q)$ $(n = 0,1,2,\ldots)$; *also,*

$$\sup_{n \geq 0} \|u^n\|_{B(\partial' Q)} < \infty, \qquad \lim_{n\to\infty} \|u_n - u_0\|_{d+1,Q} = 0.$$

Then:

a. *if* $\sup_{n\geq 0} G[u_n]$, $\sup_{n\geq 0}((\partial/\partial t) + \Delta)u_n \in \mathscr{L}_{d+1}(Q)$, *then*

$$\varliminf_{n\to\infty} G[u_n] \geq G[u_0] \qquad \text{(a.e. on } Q);$$

b. *if* $\inf_{n\geq 0} G[u_n]$, $\inf_{n\geq 0}((\partial/\partial t) + \Delta)u_n \in \mathscr{L}_{d+1}(Q)$, *then*

$$G[u_0] \geq \varlimsup_{n\to\infty} G[u_n] \qquad \text{(a.e. on } Q).$$

This theorem implies that under the appropriate conditions

$$\lim_{n\to\infty} G[u_n] \geq G\left[\lim_{n\to\infty} u_n\right] \geq \varliminf_{n\to\infty} G[u_n] \qquad \text{(a.e. on } Q).$$

2. Theorem. *For some constant* $\delta > 0$ *for all* $(\alpha,t,x) \in A \times Q$, $\lambda \in E_d$ let

$$(\tilde{a}(\alpha,t,x)\lambda,\lambda) \geq \delta|\lambda|^2, \qquad \tilde{r}^\alpha(t,x) \geq \delta.$$

Let $u_n \in W^{1,2}(Q)$ $(n = 0,1,2,\ldots)$; *also,*

$$\sup_{n \geq 0} \|u^n\|_{B(\partial' Q)} < \infty, \qquad \lim_{n\to\infty} \|u_n - u_0\|_{d+1,Q} = 0.$$

Then for any function $h \in \mathscr{L}_{d+1}(Q)$:

a. $\|(G[u_0] + h)_+\|_{d+1,Q} \leq N \varliminf_{n\to\infty} \|(G[u_n] + h)_+\|_{d+1,Q}$;
b. $\|(G[u_0] + h)_-\|_{d+1,Q} \leq N \varliminf_{n\to\infty} \|(G[u_n] + h)_-\|_{d+1,Q}$,

where N *depends only on* d, δ, *and the maximal values of the functions* $\tilde{b}^i(\alpha,t,x)$, $\tilde{a}^{ij}(\alpha,t,x)$, $\tilde{r}^\alpha(t,x)$ *with respect to* $i, j = 1, \ldots, d$, *and* $(\alpha,t,x) \in A \times Q$. *In particular, if* $G[u_n] \to -h$ *in the norm of* $\mathscr{L}_{d+1}(Q)$, *then* $G[u_0] = -h$ (a.e.).

It is essential to note that the hypotheses of the theorem include none on the convergence of derivatives of the functions u_n to derivatives of the function u_0. In this connection, let us bring the reader's attention to

3. Exercise

Let $d = T = 1$, $Q = (0,1) \times (-1,1)$,

$$G[u] = \sup_{|\alpha| < 1} \left(\frac{\partial}{\partial t} u + \alpha u_{xx}\right).$$

Let

$$\chi_n(x) = \operatorname{sgn} \sin(2^n \pi x),$$

$$u_n(t,x) = \int_{2^{-n_1}}^{x} (x - y)\chi_n(y)\,dy, \qquad n = 1, 2, \ldots; \qquad u_0(t,x) \equiv 0.$$

Prove that the totality of u_n, u_{nx}, u_{nxx} is bounded, $u_n \to u_0$ uniformly in Q, and, nevertheless,

$$0 = G[u_0] \geq \varliminf_{n\to\infty} G[u_n] = 1 \qquad \text{(a.e. on } Q).$$

Assertion (a) in Theorems 1 and 2 can sometimes be strengthened.

4. Exercise

For all $\alpha \in A$ let the functions $\tilde{a}(\alpha,t,x)(\tilde{b}(\alpha,t,x))$ be twice (once) continuously differentiable with respect to x, let $\tilde{r}^\alpha(t, x)$ be once continuously differentiable with respect to t in Q and, finally, let the respective derivatives of the foregoing functions be bounded in Q.

Prove that if $u_n \in W^{1,2}(Q)$ $(n = 0,1,2, \ldots)$, $h \in \mathscr{L}_{d+1}(Q)$, $\lim_{n\to\infty} \|u_n - u_0\|_{d+1,Q} = 0$, then $\varliminf_{n\to\infty} G[u_n] \geq G[u_0]$ (a.e. on Q) and if in addition $\sup_{n\geq 1} G[u_n \in \mathscr{L}_1(Q)$ (with the constant $N = 1$)

$$\|(G[u_0] + h)_+\|_{d+1,Q} \leqslant \varliminf_{n\to\infty} \|(G[u_n] + h)_+\|_{d+1,Q}.$$

Theorem 1 follows readily from Theorem 2. For example, let us prove Theorem 1b assuming that Theorem 2 has been proved. First we note that if $u \in W^{1,2}(Q)$, then $G[u] \in \mathscr{L}_{d+1}(Q)$. This fact follows immediately from the measurability of $G[u](t,x)$ (compare with Chapter 4, Introduction) and from the obvious inequality

$$|G[u]| \leq N\left(\left|\frac{\partial u}{\partial t}\right| + \sum_{i,j=1}^{d} |u_{x^i x^j}| + \sum_{i=1}^{d} |u_{x^i}| + |u| + 1\right), \tag{1}$$

where the constant N depends only on the upper bounds of the quantities $\tilde{a}^{ij}(\alpha,t,x)$, $\tilde{b}^i(\alpha,t,x)$, $\tilde{c}^\alpha(t,x)$, $\tilde{f}^\alpha(t,x)$, $\tilde{r}^\alpha(t,x)$. Further, for $\varepsilon > 0$ we put $\tilde{a}_\varepsilon(\alpha,t,x) = \tilde{a}(\alpha,t,x) + \varepsilon I$, where I denotes the unit matrix of dimension $d \times d$, $\tilde{r}_\varepsilon^\alpha(t,x) = \tilde{r}^\alpha(t,x) + \varepsilon$. We can construct the operator G_ε on the basis of $\tilde{a}_\varepsilon$, $\tilde{b}$, $\tilde{c}$, $\tilde{f}$, and $\tilde{r}_\varepsilon$ in the same way as we have constructed the operator G on the basis of $\tilde{a}$, $\tilde{b}$, $\tilde{c}$, $\tilde{f}$, and $\tilde{r}$. Obviously, $G_\varepsilon[u] = G[u] + \varepsilon((\partial/\partial t) + \Delta)u$ and

$$\tilde{r}_\varepsilon^\alpha(t,x) \geq \varepsilon, \qquad (\tilde{a}_\varepsilon(\alpha,t,x)\lambda,\lambda) \geq \varepsilon|\lambda|^2. \tag{2}$$

Let $h_{n_0} = -\inf_{n\geq n_0} G_\varepsilon[u_n]$ for $n_0 > 0$. Since

$$\inf_{n\geq 1} G_\varepsilon[u_n] \leq -h_{n_0} \leq G_\varepsilon[u_{n_0}],$$

the function $h_{n_0} \in \mathscr{L}_{d+1}(Q)$ due to inequality (1) and Theorem 1b. In addition, $(G_\varepsilon[u_0] + h_{n_0})_- = 0$ for $n \geq n_0$. Inequalities (2) enable us to use Theorem 2 and thus obtain $(G_\varepsilon[u_0] + h_{n_0})_- = 0$ (a.e. on Q), that is, $G_\varepsilon[u_0] \geq -h_{n_0}$ (a.e. on Q),

$$G[u_0] + \varepsilon\left(\frac{\partial}{\partial t} + \Delta\right)u_0 \geq \inf_{n\geq n_0} G_\varepsilon[u_n]$$
$$\geq \inf_{n\geq n_0} G[u_n] + \varepsilon \inf_{n\geq 0}\left(\frac{\partial}{\partial t} + \Delta\right)u_n \qquad \text{(a.e. on } Q\text{)},$$

For $n_0 \to \infty$, $\varepsilon \downarrow 0$ the foregoing proves Theorem 1b.

Therefore, we need to prove only Theorem 2, which we shall do at the end of this section. However, we investigate now an auxiliary problem assuming the conditions of Theorem 2 to be satisfied.

It is convenient to assume that $\tilde{a}(\alpha,t,x)$, $\tilde{b}(\alpha,t,x)$ are defined not only on Q but for all (t,x) in general. Redefining them, if necessary, outside Q, we can arrange so that $\tilde{b}(\alpha,t,x) = 0$, $\tilde{a}(\alpha,t,x) = \delta I$ for $(t,x) \notin Q$.

Let $\tilde{\sigma}(\alpha,t,x)$ be the positive symmetric square root of the matrix $2\tilde{a}(\alpha,t,x)$. We fix in a set A a countable everywhere dense subset $\{\alpha(i), i \geq 1\}$ and, furthermore, we denote by $\tilde{\mathfrak{A}}$ the set of all measurable functions $\tilde{\alpha}(t,x)$ given on $(-\infty,\infty) \times E_d$ and assuming values from $\{\alpha(i), i \geq 1\}$.

Since each eigenvalue of the matrix $\tilde{a}$ is greater than δ, each eigenvalue $\tilde{\sigma}$ is greater than $\sqrt{2\delta}$. Therefore, $(\tilde{\sigma}\lambda,\lambda) \geq \sqrt{2\delta}|\lambda|^2$. By Theorem 2.6.1, we have that for each $\tilde{\alpha} \in \tilde{\mathfrak{A}}$, $s \in [0,T]$, $x \in E_d$ there exists a probability space, a d-dimensional Wiener process $(\mathbf{w}_t, \mathscr{F}_t)$, and a continuous process $x_t = x_t^{\alpha,s,x}$ on this space such that

$$x_t = x + \int_0^t \tilde{\sigma}(\tilde{\alpha}(s+r,x_r), s+r, x_r)\, d\mathbf{w}_r + \int_0^t \tilde{b}(\tilde{\alpha}(s+r,x_r), s+r, x_r)\, dr.$$

Let

$$\tilde{c}^{\tilde{\alpha}}(t,x) = \tilde{c}^{\tilde{\alpha}(t,x)}(t,x),$$

$$R_\lambda^{\tilde{\alpha}} h(s,x) = \mathsf{M} \int_0^\tau h(s+t, x_t^{\tilde{\alpha},s,x}) \exp\left[-\lambda t - \int_0^t \tilde{c}^{\tilde{\alpha}}(s+r, x_r^{\tilde{\alpha},s,x})\, dr\right] dt,$$

$$\Pi_\lambda^{\tilde{\alpha}} h(s,x) = \mathsf{M} h(s+\tau, x_\tau^{\tilde{\alpha},s,x}) \exp\left[-\lambda\tau - \int_0^\tau \tilde{c}^{\tilde{\alpha}}(s+r, x_r^{\tilde{\alpha},s,x})\, dr\right],$$

where τ is the time of first exit of the process $(s+t, x_t^{\tilde{\alpha},s,x})$ from the region Q. Certainly, in this notation we should somehow incorporate the dependence on the choice of a probability space, a Wiener process, or $x_t^{\tilde{\alpha},s,x}$. However, we shall not do this but instead assume that for each $\tilde{\alpha} \in \tilde{\mathfrak{A}}$, $x \in E_d$, $s \in [0,T]$ one of the needed probability spaces, one Wiener process, and one of the processes $x_t^{\tilde{\alpha},s,x}$ is fixed. For shortening the notation we put also $\tilde{f}^{\tilde{\alpha}}(t,x) = \tilde{f}^{\tilde{\alpha}(t,x)}(t,x)$ for $\tilde{\alpha} \in \tilde{\mathfrak{A}}$.

5. Lemma. *Let* $u \in W^{1,2}(Q)$, $\tilde{r}^\alpha(t,x) \equiv 1$, $G[u] = -h$(a.e. *on* Q), $\lambda \geq 0$. *Then for all* $(t,x) \in Q$

$$u(t,x) = \sup_{\tilde{\alpha} \in \tilde{\mathfrak{A}}} [R_\lambda^{\tilde{\alpha}}(\lambda u + f^{\tilde{\alpha}} + h)(t,x) + \Pi_\lambda^{\tilde{\alpha}} u(t,x)] \tag{3}$$

PROOF. We denote by $\tilde{u}(s,x)$ the right side of (3) and note that it does not change, by Theorem 2.2.4, if we replace the function h by the equivalent function $\tilde{h}(t,x) \equiv -G[u](t,x)$. Using $\tilde{a}$, $\tilde{b}$, and $\tilde{c}$, we construct the operator $\tilde{L}^\alpha$ in the same way as we constructed the operator L^α on the basis of $a(\alpha,t,x)$, $b(\alpha,t,x)$, and $c^\alpha(t,x)$ in the introduction to this chapter. It is seen that $\tilde{L}^\alpha u + \tilde{f}^\alpha \leq -\tilde{h}$ everywhere in Q for $\alpha \in A$ and for $\tilde{\alpha} \in \tilde{\mathfrak{A}}$, $(t,x) \in Q$

$$\tilde{L}^{\tilde{\alpha}(t,x)} u(t,x) - \lambda u(t,x) \leq -\lambda u(t,x) - \tilde{f}^{\tilde{\alpha}}(t,x) - \tilde{h}(t,x).$$

Applying Ito's formula (Theorem 2.10.1), we obtain

$$\begin{aligned} u(t,x) &= R_\lambda^{\tilde{\alpha}}(\lambda u - \tilde{L}^{\tilde{\alpha}} u)(t,x) + \Pi_\lambda^{\tilde{\alpha}} u(t,x) \\ &\geq R_\lambda^{\tilde{\alpha}}(\lambda u + \tilde{f}^{\tilde{\alpha}} + \tilde{h})(t,x) + \Pi_\lambda^{\tilde{\alpha}} u(t,x), \end{aligned} \tag{4}$$

which holds for any function $\tilde{\alpha} \in \tilde{\mathfrak{A}}$. Therefore, $u \geq \tilde{u}$. For proving the converse we note first that due to the continuity of $\tilde{a}(\alpha,t,x)$, $\tilde{b}(\alpha,t,x)$, $\tilde{c}^\alpha(t,x)$, and $\tilde{f}^\alpha(t,x)$ with respect to α and, in addition, due to the density $\{\alpha(i)\}$ in the set A

$$G[u] = \sup_i [\tilde{L}^{\alpha(i)}u + \tilde{f}^{\alpha(i)}] = \lim_{n\to\infty} \max_{i\le n} [\tilde{L}^{\alpha(i)}u + \tilde{f}^{\alpha(i)}].$$

Therefore, for each $\varepsilon > 0$ and each $(t,x) \in Q$ there exists a number i such that

$$G[u](t,x) - \varepsilon \le \tilde{L}^{\alpha(i)}u(t,x) + \tilde{f}^{\alpha(i)}(t,x).$$

We denote by $i_\varepsilon(t,x)$ the smallest integer which satisfies the last inequality. It can easily be proved that the set $\{(t,x) \in Q : i_\varepsilon(t,x) = i\}$ is measurable for each i. Hence the function $i_\varepsilon(t,x)$ is measurable and, by this token, the function $\tilde{\alpha}_\varepsilon(t,x) = \alpha(i_\varepsilon(t,x))$ is measurable as well, for which

$$-\tilde{h}(t,x) - \varepsilon = G[u](t,x) - \varepsilon \le \tilde{L}^{\tilde{\alpha}_\varepsilon(t,x)}u(t,x) + \tilde{f}^{\tilde{\alpha}_\varepsilon}(t,x)$$

for $(t,x) \in Q$. The foregoing (compare with (4)) implies that

$$\begin{aligned} u(t,x) &\le R_\lambda^{\tilde{\alpha}_\varepsilon}(\lambda u + \tilde{f}^{\tilde{\alpha}_\varepsilon} + \tilde{h})(t,x) + \Pi_\lambda^{\alpha_\varepsilon} u(t,x) + R_\lambda^{\alpha_\varepsilon}\varepsilon(t,x) \\ &\le \tilde{u}(t,x) + R_\lambda^{\tilde{\alpha}_\varepsilon}\varepsilon(t,x) \le \tilde{u}(t,x) + \varepsilon(T-t). \end{aligned}$$

Here ε is an arbitrary positive number; therefore, $u \le \tilde{u}$. This completes the proof of the lemma. □

For $\lambda = 0$ we derive from the lemma a probabilistic representation of a solution of the equation $G[u] = -h$:

$$u = \sup_{\alpha\in\mathfrak{A}} [R_0^{\tilde{\alpha}}(\tilde{f}^{\tilde{\alpha}} + h) + \Pi_0^{\tilde{\alpha}} u]. \tag{5}$$

6. Exercise

Prove using (5) that if

$$\tilde{r}^\alpha(t,x) \equiv 1, \quad u_1,u_2 \in W^{1,2}(Q), \qquad G[u_1] \ge G[u_2] \qquad \text{(a.e. on } Q),$$

$u_1|_{\partial' Q} \le u_2|_{\partial' Q}$, then $u_1 = u_2$ everywhere in Q. In particular, if $F[u_1] = F[u_2]$(a.e. on Q), $u_1|_{\partial' Q} = u_2|_{\partial' Q}$, then $u_1 = u_2$ in the region Q.

We note another simple consequence of Eq. (5), which is, however, rather irrelevant to the discussion in this section.

7. Theorem. *Let (the first assumption of Theorem 2 be satisfied)*

$$\begin{aligned} &Q = C_{T,R}, \qquad \tilde{r}^\alpha(t,x) \equiv 1, \qquad u_1, u_2 \in W^{1,2}(C_{T,R}), \\ &h = \operatorname*{ess\,sup}_{C_{T,R}} |G[u_1] - G[u_2]|, \qquad u_1(T,x) = u_2(T,x). \end{aligned}$$

Then for each $n > 0$ there is a constant $N = N(K,n)$ such that for all $(s,x) \in C_{T,R}$

$$|u_1(s,x) - u_2(s,x)| \le h(T-s) + R^{-n}Ne^{N(T-s)}(1 + |x|)^n \|u_1 - u_2\|_{B(\partial' C_{T,R})}.$$

Proof. Let $h_i = -G[u_i]$. Further, we write representations (5) for u_1 and u_2 and, second, we subtract the representations obtained. Noting that the magnitude of the difference between the upper bounds does not exceed the upper bound of magnitudes of the differences, we have

$$|u_1 - u_2| \le \sup_{\tilde{\alpha} \in \tilde{\mathfrak{A}}} [R_0^{\tilde{\alpha}} |h_1 - h_2| + \Pi_0^{\tilde{\alpha}} |u_1 - u_2|].$$

Since $|h_1 - h_2| \le h$ (a.e.), $R_0^{\tilde{\alpha}} |h_1 - h_2|(s,x) \le R_0^{\tilde{\alpha}} h(s,x) \le h(T - s)$. Furthermore, since $u_1(T,x) = u_2(T,x)$,

$$\Pi_0^{\tilde{\alpha}} |u_1 - u_2|(s,x) \le \mathsf{M}|u_1 - u_2|(s + \tau, x_\tau^{\tilde{\alpha},s,x})\chi_{\tau < T-s}$$
$$\le \sup_{t \in [0,T]} \sup_{|y| = R} |u_1(t,y) - u_2(t,y)| \mathsf{P}\{|x_\tau^{\tilde{\alpha},s,x}| = R\}.$$

It remains only to estimate the last probability. It is seen that it equals

$$\mathsf{P}\left\{\sup_{t \le T-s} |x_t^{\tilde{\alpha},s,x}| \ge R\right\} \le \frac{1}{R^n} \mathsf{M} \sup_{t \le T-s} |x_t^{\tilde{\alpha},s,x}|^n.$$

According to Corollary 2.5.12 the last expression does not, in turn, exceed $R^{-n} N e^{N(T-s)} (1 + |x|)^n$. This completes the proof of the theorem. □

Using the probabilistic representation given in Lemma 5 in order to solve the equation $G[u] = -h$, we can give a probabilistic formula for the operator G.

8. Lemma. *Let* $\tilde{r}^\alpha(t,x) \equiv 1$ *and also let*

$$\mathscr{T}_\lambda^h u(t,x) = \sup_{\tilde{\alpha} \in \tilde{\mathfrak{A}}} [R_\lambda^{\tilde{\alpha}}(\lambda u + \tilde{f}^{\tilde{\alpha}} + h)(t,x) + \Pi_\lambda^{\tilde{\alpha}} u(t,x)]$$

Then for all

$$h \in \mathscr{L}_{d+1}(Q), \qquad u \in W^{1,2}(Q), \lambda \ge 1,$$
$$\lambda \| (\mathscr{T}_\lambda^h u - u)_+ \|_{d+1,Q} \le N \| (G[u] + h)_+ \|_{d+1,Q},$$

where N *depends only on* d, δ, *and the maximum of the moduli* $\tilde{a}^{ij}(\alpha,t,x)$, $\tilde{b}^i(\alpha,t,x)$. *Furthermore, for the same* h, u

$$\lim_{\lambda \to \infty} \| \lambda(\mathscr{T}_\lambda^h u - u) - G[u] - h \|_{d+1,Q} = 0.$$

Proof. By Lemma 5, $u = \mathscr{T}^{h_1} u$, where $h_1 = -G[u]$. Estimating the difference between the upper bounds, we find

$$\mathscr{T}_\lambda^h u - u = \mathscr{T}_\lambda^h u - \mathscr{T}_\lambda^{h_1} u \le \sup_{\tilde{\alpha} \in \tilde{\mathfrak{A}}} R_\lambda^{\tilde{\alpha}}(h - h_1) \le \sup_{\tilde{\alpha} \in \tilde{\mathfrak{A}}} R_\lambda^{\tilde{\alpha}}(h - h_1)_+,$$

$$\mathscr{T}_\lambda^h u - u \ge \inf_{\tilde{\alpha} \in \tilde{\mathfrak{A}}} R_\lambda^{\tilde{\alpha}}(h - h_1) = -\sup_{\tilde{\alpha} \in \tilde{\mathfrak{A}}} R_\lambda^{\tilde{\alpha}}(h_1 - h) \ge -\sup_{\tilde{\alpha} \in \tilde{\mathfrak{A}}} R_\lambda^{\tilde{\alpha}}(h - h_1)_-,$$

which together with Theorems 2.4.5 and 2.4.7 prove the assertions of the theorem. In fact, by Theorem 2.4.5

$$\lambda]\!](\mathscr{T}^h_\lambda u - u)_\pm\|_{p+1,Q} \le \lambda]\!]\sup_{\tilde\alpha\in\tilde{\mathfrak{A}}} R^{\tilde\alpha}_\lambda(h-h_1)_\pm\|_{p+1,Q} \le N\|(h-h_1)_\pm\|_{p+1,Q}.$$

By Theorem 2.4.7, the expressions

$$\lambda \sup_{\tilde\alpha\in\tilde{\mathfrak{A}}} R^{\tilde\alpha}_\lambda(h-h_1), \qquad -\lambda \sup_{\tilde\alpha\in\tilde{\mathfrak{A}}} R^{\tilde\alpha}_\lambda(h_1-h)$$

converge in the sense of the upper norm $\mathscr{L}_{d+1}(Q)$ to h-h_1. Since $\lambda(\mathscr{T}^h_\lambda u - u)$ is between the foregoing expressions, the former converges to $h - h_1$ as well. The lemma is proved. □

9. Proof of Theorem 2. First we consider the case $\tilde r^\alpha(t,x) \equiv 1$, $h = 0$. Let a region $Q' \subset \bar Q' \subset Q$. According to Lemma 8

$$\|G[u])_-\|_{d+1,Q'} = \lim_{\lambda\to\infty} \lambda]\!](\mathscr{T}^0_\lambda u - u)_-\|_{d+1,Q'}.$$

Further, it is seen that

$$|\mathscr{T}^0_\lambda u - \mathscr{T}^0_\lambda u_n| \le \lambda \sup_{\tilde\alpha\in\tilde{\mathfrak{A}}} R^{\tilde\alpha}_\lambda |u-u_n| + N_0 \sup_{\alpha\in\tilde{\mathfrak{A}}} \Pi^{\tilde\alpha}_\lambda 1,$$

where $N_0 = \sup_n \|u_n - u_0\|_{B(\partial' Q)}$. By Theorem 2.4.5

$$\lambda \left]\!\!\right]\sup_{\tilde\alpha\in\tilde{\mathfrak{A}}} R^{\tilde\alpha}_\lambda |u-u_n|\left\|_{d+1,Q}\right. \le N\|u-u_n\|_{d+1,Q},$$

Since the constant N does not depend on n, the right side of the last inequality tends to zero as $n \to \infty$.

Therefore,

$$\begin{aligned}|(G[u])_-\|_{d+1,Q'} &\le \overline{\lim_{\lambda\to\infty}}\ \lim_{n\to\infty} (\lambda]\!](\mathscr{T}^0_\lambda u_n - u_n)_-\|_{d+1,Q'} \\ &\quad + \lambda]\!]\mathscr{T}^0_\lambda u - \mathscr{T}^0_\lambda u_n\|_{d+1,Q'} + \lambda\|u-u_n\|_{d+1,Q'}) \\ &\le \lim_{\lambda\to\infty}\ \lim_{n\to\infty} \lambda]\!](\mathscr{T}^0_\lambda u_n - u_n)_-\|_{d+1,Q'} \\ &\quad + N_0 \lim_{\lambda\to\infty} \lambda \left\|\sup_{\tilde\alpha\in\tilde{\mathfrak{A}}} \Pi^{\tilde\alpha}_\lambda 1\right\|_{d+1,Q'},\end{aligned}$$

where the first term does not exceed $N\,\underline{\lim}_{n\to\infty} \|(G[u_n])_-\|_{d+1,Q}$, in accord with Lemma 8, and the second term is equal to zero in accord with Theorem 2.4.7. Finally,

$$\|(G[u])_-\|_{d+1,Q'} \le N \underline{\lim}_{n\to\infty} \|(G[u_n])_-\|_{d+1,Q} \tag{6}$$

for each region $Q' \subset \bar Q' \subset Q$, N depending only on d, δ as well as the maximal magnitudes of $\tilde a^{ij}(\alpha,t,x)$, $\tilde b^i(\alpha,t,x)$ with respect to α, t, x, i, j.

Next, we choose an increasing sequence of regions Q_i' whose union is Q. Putting instead of Q' the region Q_i' in the left side of (6) and, in addition, letting $i \to \infty$, we complete proving assertion (b) of the theorem for $h = 0$. In this case assertion (a) of the theorem can be proved in a similar manner.

Using formal transformations, we can derive the general assertion from the particular case considered. Let $h = 0$, and also let $\tilde{r}^\alpha(t,x)$ be an arbitrary function satisfying the conditions of the theorem. We construct the operator $\bar{G}[u]$ on a basis of the functions $(\tilde{r})^{-1}\tilde{a}$, $(\tilde{r})^{-1}\tilde{r}b$, $(\tilde{f})^{-1}\tilde{c}$, $(\tilde{r})^{-1}\tilde{f}$, 1 in the same way as we constructed the operator $G[u]$ on a basis of the functions $\tilde{a}$, $\tilde{b}$, $\tilde{c}$, $\tilde{f}$, $\tilde{r}$. Let

$$N_1 = \sup_{\alpha \in A} \sup_{(t,x) \in Q} \tilde{r}^\alpha(t,x).$$

Note that for any set of numbers l^α:

1. if $0 \le \sup_{\alpha \in A} l^\alpha$, then

$$\sup_{\alpha \in A} l^\alpha \le N_1 \sup_{\alpha \in A} (\tilde{r}^\alpha(t,x))^{-1} l^\alpha \le N_1 \delta^{-1} \sup_{\alpha \in A} l^\alpha;$$

2. if $\sup_{\alpha \in A} l^d \le 0$, then

$$\sup_{\alpha \in A} l^\alpha \ge N_1 \sup_{\alpha \in A} \tilde{r}^\alpha(t,x))^1 l^\alpha \ge N_1 \delta^{-1} \sup_{\alpha \in A} l^\alpha.$$

The foregoing implies that

$$(G[u])_+ \le N_1(\bar{G}[u])_+ \le N_1\delta^{-1}(G[u])_+,$$
$$(G[u])_- \le N_1(\bar{G}[u])_- \le N_1\delta^{-1}(G[u])_-.$$

These inequalities together with the assertions of the theorem which hold for the operator $\bar{G}$ and $h = 0$ immediately prove the theorem for G and $h = 0$. In order to prove the theorem for an arbitrary $h \in \mathscr{L}_{d+1}(Q)$, it suffices to note that $G[u] + h$ can be written as $\hat{G}[u]$ in an obvious way, if we construct $\hat{G}[u]$ on a basis of the functions $\tilde{a}$, $\tilde{b}$, $\tilde{c}$, $\tilde{f} + h$, and $\tilde{r}$ in the same way as we construct $G[u]$ on a basis of $\tilde{a}$, $\tilde{b}$, $\tilde{c}$, $\tilde{f}$, and $\tilde{r}$. This completes the proof of the theorem. □

6. The Approximation of Degenerate Controlled Processes by Nondegenerate Ones

Let $(\tilde{\mathbf{w}},\tilde{\mathscr{F}}_t)$ be a $(d_1 + d)$-dimensional Wiener process, let ε be a number, and, finally, let $\sigma_\varepsilon(\alpha,t,x)$ be a matrix of dimension $d \times (d_1 + d)$, in which the first d_1 columns coincide with the respective columns of the matrix $\sigma(\alpha,t,x)$ and, also, the block of last d columns gives εI, where I denotes a unit matrix of dimension $d \times d$. Denote by $\tilde{\mathfrak{A}}$ the set of all processes $\tilde{\alpha} = \tilde{\alpha}_t(\omega)$ which are progressively measurable with respect to $\{\tilde{\mathscr{F}}_t\}$ and which take on values from A. For $\tilde{\alpha} \in \tilde{\mathfrak{A}}$, $s \in [0,T]$, $x \in E_d$ we define the process $x_t^{\tilde{\alpha},s,x}(\varepsilon)$ to be a solution of the equation

$$x_t = x + \int_0^t \sigma_\varepsilon(\tilde{\alpha}_r, s + r, x_r)\, d\tilde{\mathbf{w}}_r + \int_0^t b(\tilde{\alpha}_r, s + r, x_r)\, dr. \tag{1}$$

Furthermore, let

$$\varphi_t^{\tilde{\alpha},s,x}(\varepsilon) = \int_0^t c^{\tilde{\alpha}_r}(s + r, x_r^{\tilde{\alpha},s,x}(\varepsilon))\,dr,$$

$$v_\varepsilon^{\tilde{\alpha}}(s,x) = \mathsf{M}_{s,x}^{\tilde{\alpha}}\left[\int_0^{T-s} f^{\tilde{\alpha}_t}(s + t, x_t(\varepsilon))e^{-\varphi_t(\varepsilon)}\,dt + g(x_{T-s}(\varepsilon))e^{-\varphi_{T-s}(\varepsilon)}\right],$$

$$v_\varepsilon(s,x) = \sup_{\tilde{\alpha}\in\tilde{\mathfrak{A}}} v_\varepsilon^{\tilde{\alpha}}(s,x).$$

For $s \in [0,T]$ we denote by $\tilde{\mathfrak{M}}(T - s)$ the set of all Markov times (with respect to $\{\tilde{\mathscr{F}}_t\}$) $\tilde{\tau}$ which do not exceed $T - s$,

$$v_\varepsilon^{\tilde{\alpha},\tilde{\tau}}(s,x) = \mathsf{M}_{s,x}^{\tilde{\alpha}}\left[\int_0^{\tilde{\tau}} f^{\tilde{\alpha}_t}(s + t, x_t(\varepsilon))e^{-\varphi_t(\varepsilon)}\,dt + g(s + \tilde{\tau}, x_{\tilde{\tau}}(\varepsilon))e^{-\varphi_{\tilde{\tau}}(\varepsilon)}\right],$$

$$w_\varepsilon(s,x) = \sup_{\tilde{\alpha}\in\tilde{\mathfrak{A}}}\ \sup_{\tilde{\tau}\in\tilde{\mathfrak{M}}(T-s)} v_\varepsilon^{\tilde{\alpha},\tilde{\tau}}(s,x).$$

The processes $x_t^{\tilde{\alpha},s,x}(\varepsilon)$ for $\varepsilon \neq 0$ are nondegenerate in the following sense. Let $a_\varepsilon(\alpha,t,x) = \frac{1}{2}\sigma_\varepsilon(\alpha,t,x)\sigma_\varepsilon^*(\alpha,t,x)$. It is seen that

$$a_\varepsilon(\alpha,t,x) = a(\alpha,t,x) + \tfrac{1}{2}\varepsilon^2 I \geq \tfrac{1}{2}\varepsilon^2 I. \tag{2}$$

Hence for any $\lambda \in E_d$

$$(a_\varepsilon(\alpha,t,x)\lambda,\lambda) \geq \tfrac{1}{2}\varepsilon^2|\lambda|^2.$$

The equality $a_\varepsilon = a + \frac{1}{2}\varepsilon^2 I$ immediately implies the following useful relations:

$$L_\varepsilon^\alpha u(t,x) \equiv \frac{\partial u(t,x)}{\partial t} + \sum_{i,j=1}^d a_\varepsilon^{ij}(\alpha,t,x)u_{x^ix^j}(t,x) + \sum_{i=1}^d b^i(\alpha,t,x)u_{x^i}(t,x) - c^\alpha(t,x)u(t,x)$$

$$= L^\alpha u(t,x) + \frac{1}{2}\varepsilon^2\,\Delta u(t,x), \tag{3}$$

$$F_\varepsilon[u] \equiv \sup_{\alpha\in A}[L_\varepsilon^\alpha u(t,x) + f^\alpha(t,x)] = F[u] + \frac{1}{2}\varepsilon^2\,\Delta u.$$

In the case where $\mathscr{F}_t \subset \tilde{\mathscr{F}}_t$ for all t, the set of strategies $\mathfrak{A} \subset \tilde{\mathfrak{A}}$. If, in addition, the first d_1 coordinates of the process $\tilde{\mathbf{w}}_t$ form a process $\mathbf{w}_t$, due to the uniqueness of a solution of Eq. (1) we have $x_t^{\alpha,s,x} = x_t^{\alpha,s,x}(0)$ for $\alpha \in \mathfrak{A}$. Hence we can say that the nondegenerate controlled process $x_t^{\alpha,s,x}(\varepsilon)$ as $\varepsilon \to 0$ approximates the (degenerate, in general) process $x_t^{\alpha,s,x}$.

1. Theorem. *As* $\varepsilon \to 0$

$$v_\varepsilon(t,x) \to v(t,x), \qquad w_\varepsilon(t,x) \to w(t,x)$$

uniformly on each cylinder $\bar{C}_{T,R}$.

PROOF. According to Corollary 3.1.13

$$v_\varepsilon(t,x) \to v_0(t,x), \qquad w_\varepsilon(t,x) \to w_0(t,x) \tag{4}$$

as $\varepsilon \to 0$ uniformly on each cylinder $\bar{C}_{T,R}$. It is seen that for $\varepsilon = 0$ the process $x_t^{\alpha,s,x}(\varepsilon)$ can be defined to be a solution of the equation

$$x_t = x + \int_0^t \sigma(\alpha_r, s + r, x_r)\, d\tilde{\mathbf{w}}_t' + \int_0^t b(\alpha_r, s + r, x_r)\, dr,$$

where $\tilde{\mathbf{w}}_t'$ is the vector composed of the first d_1 components of the vector $\tilde{\mathbf{w}}_t$. The last equation is equivalent to the equation for $x_t^{\alpha,s,x}$, in which, however, the Wiener process is (possibly) a different one and, in addition, it is allowed to choose strategies to be measurable with respect to rather large σ-algebras. However, as we know from Remarks 3.3.10 and 3.4.10, a payoff function does not depend on a probability space and the fact that one d_1-dimensional Wiener process is replaced by another d_1-dimensional Wiener (with respect to, possibly, very large σ-algebras) process. Therefore, $v_0 \equiv v$, $w_0 \equiv w$, which together with (4) proves the theorem. □

In some cases, for example, in finding numerical values of payoff functions, it is crucial to know how great the difference $|v_\varepsilon(s,x) - v(s,x)|$ is.

2. Theorem. *For all $s \in [0,T]$, $\alpha \in A$, $R > 0$, $x, y \in S_R$, let*

$$\begin{aligned} &|c^\alpha(s,x) - c^\alpha(s,y)| + |f^\alpha(s,x) - f^\alpha(s,y)| \\ &\qquad + |g(x) - g(y)| + |g(s,x) - g(s,y)| \le K(1 + R)^m|x - y|. \end{aligned}$$

Then there exists a constant $N = N(K,m)$ such that for all $(s,x) \in H_T$, $\varepsilon \in [-1,1]$

$$|v_\varepsilon(s,x) - v(s,x)| + |w_\varepsilon(s,x) - w(s,x)| \le |\varepsilon| N(1 + |x|)^{2m} e^{N(T-s)}.$$

PROOF. We can prove the theorem by differentiating Eq. (1) over the parameter ε. We prefer, however, a formal application of Theorem 1.1. We add the equation $\varepsilon_t = \varepsilon + \int_0^t 0\, d\tilde{\mathbf{w}}_r + \int_0^t 0\, dr$ to Eq. (1), replacing in (1) ε by ε_r, and furthermore, we regard ε_t as the last component of the controlled process $(x_t^{\alpha,s,(x,\varepsilon)}, \varepsilon_t^{\alpha,s,(x,\varepsilon)})$.

Note that for $s \in [0,T]$, $x, y \in E_d$, $\varepsilon_1, \varepsilon_2 \in E_1$, $\alpha \in A$,

$$\begin{aligned} \|\sigma_{\varepsilon_1}(\alpha,s,x) - \sigma_{\varepsilon_2}(\alpha,s,y)\|^2 &= \|\sigma(\alpha,s,x) - \sigma(\alpha,s,y)\|^2 + (\varepsilon_1 - \varepsilon_2)^2 \\ &\le K^2|x - y|^2 + (\varepsilon_1 - \varepsilon_2)^2 \\ &\le (K^2 + 1)|(x,\varepsilon_1) - (y,\varepsilon_2)|^2. \end{aligned}$$

In other words, the function $\sigma_\varepsilon(\alpha,s,x)$ satisfies a Lipschitz condition with respect to (x,ε) uniformly in α, s.

Therefore, the controlled process $(x_t^{\tilde{\alpha},s,(x,\varepsilon)}, \varepsilon_t^{\tilde{\alpha},s,(x,\varepsilon)})$ fits the scheme considered in Chapter 4. Theorem 1.1 estimates the gradient of the functions $v_\varepsilon(s,x)$, $w_\varepsilon(s,x)$ with respect to the variables (x,ε). In particular, the generalized derivatives v_ε, w_ε with respect to ε for $\varepsilon^2 + |x|^2 \le R^2$ do not exceed

$N(1 + R)^{2m}e^{N(T-s)}$. As was mentioned in Section 2.1, the boundedness of a generalized derivative yields a Lipschitz constant. Hence for $\varepsilon^2 + |x|^2 \le R^2$

$$|v_\varepsilon(s,x) - v_0(s,x)| + |w_\varepsilon(s,x) - w_0(s,x)| \le |\varepsilon| N(1 + R)^{2m}e^{N(T-s)},$$

where $N = N(K,m)$. It remains only to take $R^2 = x^2 + 1$ for $|\varepsilon| \le 1$. The theorem is proved. □

7. The Bellman Equation

The Bellman equation plays an essential role in finding a payoff function and ε-optimal strategies. It turns out that if the processes $x_t^{\alpha,s,x}$ are nondegenerate, we can obtain the Bellman equation under the assumption of the existence of generalized derivatives of a payoff function. First, we prove two results of the type mentioned, and second, we derive the Bellman equation imposing restrictions only on σ, b, c, f, g. These restrictions on σ, b, c, f, g will be formulated after Theorem 2. Here as well as everywhere else in this chapter, we assume that the assumptions made in Section 3.1 are satisfied.

1. Theorem. *Let a bounded region $Q \subset H_T$, let $w \in W^{1,2}(Q)$, and, finally, for each region Q' which together with its closure lies in Q let there exist a number $\delta = \delta(Q') > 0$ such that for all $(t,x) \in Q'$, $\alpha \in A$, $\lambda \in E_d$*

$$(a(\alpha,t,x)\lambda,\lambda) \ge \delta|\lambda|^2.$$

Then $F[w] \le 0$ (a.e. on Q), $F[w] = 0$ ($Q \cap \{(t,x): w(t,x) > g(t,x)\}$-a.e.), $w \ge g$ in the region Q. In short,

$$(F[w] + w - g)_+ + g - w = 0 \qquad (\text{a.e. } on\ Q).$$

PROOF. For $\beta \in A$ we introduce a constant strategy $\beta_t \equiv \beta$. Let the region $Q' \subset \bar{Q}' \subset Q$, let a point $(s,x) \in Q'$, and, finally, let τ' be the time of first exit of the process $(s + t, x_t^{\beta,s,x})$ from the region Q'. By Theorem 3.1.11, for each $\lambda \ge 0$

$$w(s,x) \ge \mathsf{M}_{s,x}^{\beta}\left\{\int_0^{\tau'} [f^\beta(s + t, x_t) + \lambda w(s + t, x_t)]e^{-\varphi_t - \lambda t}\, dt + w(s + \tau', x_{\tau'})e^{-\varphi_{\tau'} - \lambda\tau'}\right\}.$$

By Ito's formula (Theorem 2.10.1)

$$w(s,x) = \mathsf{M}_{s,x}^{\beta}\left\{\int_0^{\tau'} [\lambda w(s + t, x_t) - L^\beta w(s + t, x_t)]e^{-\varphi_t - \lambda t}\, dt + w(s + \tau', x_{\tau'})e^{-\varphi_{\tau'} - \lambda\tau'}\right\}.$$

Therefore, subtracting these two formulas, we obtain

$$0 \geq \mathsf{M}^{\beta}_{s,x} \int_0^{\tau'} [L^{\beta} w(s + t, x_t) + f^{\beta}(s + t, x_t)] e^{-\varphi_t - \lambda t} \, dt.$$

Multiplying the last inequality by λ and, in addition, letting $\lambda \to \infty$, we find according to Theorem 2.4.6 that $L^{\beta} w + f^{\beta} \leq 0$ (a.e. on Q'). Then, $F[w] \leq 0$ (a.e. on Q).

On the other hand, let $\varepsilon > 0$ and let the region

$$Q' \subset Q \cap \{(s,x): w(s,x) > g(s,x) + \varepsilon\}.$$

Then, according to the Bellman principle (according to Theorem 3.1.11)

$$0 = \sup_{\alpha \in \mathfrak{A}} \left\{ \mathsf{M}^{\alpha}_{s,x} \left[\int_0^{\tau'} f^{\alpha_t}(s + t, x_t) e^{-\varphi_t} \, dt + w(s + \tau', x_{\tau'}) e^{-\varphi_{\tau'}} \right] - w(s,x) \right\},$$

where τ' is the time of first exit of the process $(s + t, x_t^{\alpha,s,x})$ from the region Q'. By Ito's formula,

$$w(s,x) = \mathsf{M}^{\alpha}_{s,x} \left\{ \int_0^{\tau'} [-L^{\alpha_t} w(s + t, x_t)] e^{-\varphi_t} \, dt + w(s + \tau', x_{\tau'}) e^{-\varphi_{\tau'}} \right\},$$

which implies that

$$\begin{aligned} 0 &= \sup_{\alpha \in \mathfrak{A}} \mathsf{M}^{\alpha}_{s,x} \int_0^{\tau'} [L^{\alpha_t} w(s + t, x_t) + f^{\alpha_t}(s + t, x_t)] e^{-\varphi_t} \, dt \\ &\leq \sup_{\alpha \in \mathfrak{A}} \mathsf{M}^{\alpha}_{s,x} \int_0^{\tau'} F[w](s + t, x_t) e^{-\varphi_t} \, dt, \end{aligned} \tag{1}$$

where $F[w] \leq 0$ (a.e. on Q). Hence the right side of (1) is equal to zero. By virtue of Corollary 2.4.8, we have $F[w] = 0$ (a.e. on Q'). In view of arbitrariness of Q' it means that $F[w] = 0$ ($Q \cap \{(s,x): w(s,x) > g(s,x) + \varepsilon\}$-a.e.) for each $\varepsilon > 0$. The union of all these regions for all $\varepsilon > 0$ constitutes a region $Q \cap \{(s,x): w(s,x) > g(s,x)\}$. Therefore, in the last region $F[w] = 0$ almost everywhere.

Finally, the inequality $w \geq g$ is obvious (see, however, Theorem 3.1.8). We leave it as an exercise for the reader to prove the last assertion. The theorem is proved. □

2. Theorem. *Let a bounded region $Q \subset H_T$, let $v \in W^{1,2}(Q)$, and, finally, for each region Q' which together with its closure lies in Q let there exist a number $\delta = \delta(Q') > 0$ such that for all $(s,x) \in Q'$, $\alpha \in A$, $\lambda \in E_d$*

$$(a(\alpha,t,x)\lambda,\lambda) \geq \delta|\lambda|^2.$$

Then $F[v] = 0$ (a.e. on Q).

The proof of this theorem follows exactly the proof of Theorem 1. We need only, instead of Theorem 3.1.11, to use Theorem 3.1.6 where it is necessary.

We formulate the conditions which, in addition to the assumptions made in Section 3.1, we assume to be satisfied in the remaining part of this section.

Let us introduce a vector $\gamma^\alpha(t,x)$ of dimension $d \times (d_1 + d + 4)$, whose coordinates are given by the following variables: $\sigma^{ij}(\alpha,t,x)(i = 1,\ldots,d, j = 1,\ldots,d_1)$, $b^i(\alpha,t,x)$ $(i = 1,\ldots,d)$, $c^\alpha(t,x)$, $f^\alpha(t,x)$, $g(x)$, $g(t,x)$. For all $\alpha \in A$, $l \in E_d$ let the derivatives $\gamma^\alpha_{(l)}(t,x)$, $\gamma^\alpha_{(l)(l)}(t,x)$, $(\partial/\partial t)\gamma^\alpha(t,x)$ exist and be continuous with respect to (t,x) on $\bar{H}_T$. Assume that the derivatives mentioned (they are vectors) do not exceed $K(1 + |x|)^m$ in norm for all $\alpha \in A$, $l \in E_d$, $(t,x) \in \bar{H}_T$.

Also, it is convenient to assume that for all $\alpha \in A$, $x \in E_d$

$$|L^\alpha(T,x)g(x)| + |L^\alpha g(T,x)| \le K(1 + |x|)^{3m}.$$

We note that the relationship between this assumption and the preceding one was discussed in Section 4. We shall prove under the combined assumptions indicated that the functions v and w satisfy the corresponding Bellman equations in the region $Q^* = \{(t,x) \in H_T : \sup_{\alpha \in A}(a(\alpha,t,x)\lambda,\lambda) > 0 \text{ for all } \lambda \neq 0\}$.

First we show that the set Q^* is in fact a region. Let

$$n^\alpha(t,x) = (1 + \operatorname{tr} a(\alpha,t,x) + |b(\alpha,t,x)| + c^\alpha(t,x) + |f^\alpha(t,x)|)^{-1},$$

$$\mu = \mu(t,x) = \inf_{|\lambda|=1} \sup_{\alpha \in A} n^\alpha(t,x)(a(\alpha,t,x)\lambda,\lambda).$$

3. Lemma. *The function $\mu(t,x)$ is continuous in $[0,T] \times E_d$, the equality*

$$Q^* = \{(t,x) \in H_T : \mu(t,x) > 0\},$$

is satisfied, the set Q^* is open, and the function $\mu^{-1}(t,x)$ is locally bounded in Q^*.

PROOF. The third and fourth assertions follow from the first and second assertions and also from well-known properties of continuous functions.

Further, the derivatives (with respect to (t,x)) of the functions $a(\alpha,t,x)$, $b(\alpha,t,x)$, $c^\alpha(t,x)$, and $f^\alpha(t,x)$ are bounded on any set of the form $A \times [0,T] \times \{x : |x| \le R\}$. Therefore, these functions are continuous with respect to (t,x) uniformly with respect to α. By similar reasoning the function $(a(\alpha,t,x)\lambda,\lambda)$ is continuous with respect to (t,x) uniformly with respect to $\alpha \in A$, $\lambda \in S_1$. Then we have that the function $n^\alpha(t,x)(a(\alpha,t,x)\lambda,\lambda)$ is continuous with respect to (t,x) uniformly with respect to $\alpha \in A$, $\lambda \in S_1$. Furthermore, we note that the modulus (or absolute value) of the difference between the lower (upper) bounds does not exceed the upper bound of the moduli of the differences. Therefore, if $(t_n,x_n) \to (t_0,x_0)$, then

$$\begin{aligned}|\mu(t_n,x_n) - \mu(t_0,x_0)| &\le \sup_{|\lambda|=1} \Big| \sup_{\alpha \in A} n^\alpha(t_n,x_n)(a(\alpha,t_n,x_n)\lambda,\lambda) \\ &\qquad - \sup_{\alpha \in A} n^\alpha(t_0,x_0)(a(\alpha,t_0,x_0)\lambda,\lambda)\Big| \\ &\le \sup_{|\lambda|=1} \sup_{\alpha \in A} |n^\alpha(t_n,x_n)(a(\alpha,t_n,x_n)\lambda,\lambda) \\ &\qquad - n^\alpha(t_0,x_0)(a(\alpha,t_0,x_0)\lambda,\lambda)| \to 0\end{aligned}$$

by the definition of uniform continuity.

In order to prove the second assertion, we make use of the fact that due to the inequality $n^\alpha(t,x) \le 1$ for $|\lambda| = 1$ we have $\sup_{\alpha\in A}(a(\alpha,t,x)\lambda,\lambda) \ge \mu(t,x)$. Hence, if $(t,x) \in H_T$ and $\mu(t,x) > 0$, then $(t,x) \in Q^*$. If $\mu(t,x) = 0$, there will be a sequence $\lambda_n \in \partial S_1$ for which

$$\sup_{\alpha\in A} n^\alpha(t,x)(a(\alpha,t,x)\lambda_n,\lambda_n) \to 0.$$

Therefore, $(a(\alpha,t,x)\lambda_n,\lambda_n) \to 0$ for all $\alpha \in A$. We can consider without loss of generality that the sequence $\{\lambda_n\}$ converges to a limit. Ee denote this limit by λ_0. Then $(a(\alpha,t,x)\lambda_0,\lambda_0) = 0$ for all $\alpha \in A$. Therefore, $(t,x) \notin Q^*$, which completes the proof of the second assertion, thus proving the lemma. □

4. Theorem. *In H_T (in the region Q^*) the functions $v(t,x)$, $w(t,x)$ have all generalized first (respectively, second) derivatives with respect to x and a generalized first derivative with respect to t. The foregoing derivatives are locally bounded in H_T (respectively in Q^*). There exists a constant $N = N(K,m)$ such that for $u \equiv v$ and for $u \equiv w$ for any $l \in E_d$*

$$\left|\frac{\partial}{\partial t} u\right| + |\mathrm{grad}_x u| \le N(1 + |x|)^{3m} e^{N(T-t)} \qquad \text{(a.e. in } H_T), \tag{2}$$

$$-N(1 + |x|)^{3m} e^{N(T-t)} \le u_{(l)(l)} \le \frac{1}{\mu} N(1 + |x|)^{3m} e^{N(T-t)} \qquad \text{(a.e. on } Q^*). \tag{3}$$

This theorem is in some way a summary of the results obtained in Sections 1–4. The existence of $(\partial/\partial t)u$, $u_{(l)}$ follows immediately from Theorem 1.1 and Theorem 4.3, from which, in addition, we have estimates of the foregoing derivatives.

The existence of $(\partial/\partial t)u$ implies that the measures $(\partial/\partial t)u(dt\,dx)$ and $((\partial/\partial t)u)_-(dt\,dx)$ are absolutely continuous with respect to the Lebesgue measure, and, furthermore, their Radon–Nikodym derivatives are equal to $(\partial/\partial t)u$ and $((\partial/\partial t)u)_-$, respectively. By Theorem 3.5b and Remark 3.6, all generalized second derivatives of the functions $v(t,x)$, $w(t,x)$ with respect to x exist in Q^*.

Further, as was shown in proving Theorem 3.5, the function $\mu(l)$ appearing in assertion (a) of that theorem is greater than μ. Therefore, by Theorem 3.5 and Remark 3.6,

$$\psi \le u_{(l)(l)} \le \frac{1}{\mu}\left[|\psi| + \left|\frac{\partial}{\partial t} u\right| + |\mathrm{grad}_x u| + |u| + 1\right] \qquad \text{(a.e. on } Q^*),$$

where $\psi = -Ne^{N(T-t)}(1 + |x|)^{3m}$ is the right side of inequality (2.14). To complete the proof of inequality (3), it remains only to use inequality (2) and also, to recall (see Section 3.1) that $|u| \le N(1 + |x|)^m e^{N(T-t)}$. The theorem is proved. □

5. Theorem. $F[v] = 0$ (a.e. *on* Q^*), $F[w] \leq 0$ (a.e. *on* Q^*), $F[w] = 0$ ($Q^* \cap \{(s,x): w(s,x) > g(s,x)\}$-a.e.), $w(s,x) \geq g(s,x)$ *in the region* Q^*. *The assertion concerning* w *can be written in short as follows*:

$$(F[w] + w - g)_+ + g - w = 0 \qquad \text{(a.e. } on\ Q^*\text{)}.$$

PROOF. According to Corollary 1.7, $F[v] \leq 0$, $F[w] \leq 0$ (a.e. on Q^*). We prove that $F[w] = 0$ almost everywhere in any bounded region Q' which together with its closure lies in

$$Q^* \cap \{(t,x): w(t,x) > g(t,x)\},$$

which fact is obviously sufficient for proving the assertions of the theorem concerning w. Let us make use of the approximation of degenerate processes by means of the nondegenerate ones, which was described in Section 6. We take the matrix $\sigma_\varepsilon(\alpha,s,x)$, the process $x^{\tilde{\alpha},s,x}(\varepsilon)$, and the function $w_\varepsilon(s,x)$ from Section 6. As was indicated in Section 6, the matrix $a_\varepsilon(\alpha,t,x) \equiv \frac{1}{2}\sigma_\varepsilon(\alpha,t,x)\sigma_\varepsilon^*(\alpha,t,x)$ is equal to $a(\alpha,t,x) + \frac{1}{2}\varepsilon^2 I$ and it also satisfies inequality (6.2):

$$(a_\varepsilon(\alpha,t,x)\lambda,\lambda) \geq \tfrac{1}{2}\varepsilon^2|\lambda|^2. \tag{4}$$

Hence for $\varepsilon \neq 0$ the set Q^* associated with the matrix a_ε coincides with H_T, which implies, according to Theorem 4, the existence of generalized first and second derivatives of w_ε with respect to x, a generalized first derivative with respect to t. Furthermore, it implies that the foregoing derivatives are locally bounded in H_T. By Theorem 1, due to (4) for $\varepsilon \neq 0$ the function w_ε satisfies the equation (see (6.3)) $F[w_\varepsilon] + \frac{1}{2}\varepsilon^2 \Delta w_\varepsilon = 0$ almost everywhere in the region

$$\{(t,x) \in H_T: w_\varepsilon(t,x) > g(t,x)\}.$$

For all sufficiently small ε these regions contain Q'. In fact, since $\bar{Q}' \subset \{(s,x) \in H_T: w(s,x) > g(s,x)\}$, the continuous function $w(s,x) - g(s,x) > 0$ on $\bar{Q}'$. Since the set $\bar{Q}'$ is a compact, there exists a number $\delta > 0$ such that $w(s,x) - g(s,x) \geq \delta$ for $(s,x) \in \bar{Q}'$. By Theorem 6.1, the functions $w_\varepsilon(s,x) - g(s,x) \to w(s,x) - g(s,x)$ as $\varepsilon \to 0$ uniformly on $\bar{Q}'$. Therefore, for all sufficiently small ε on Q' (even on $\bar{Q}'$) the inequality $w_\varepsilon(s,x) - g(s,x) \geq \delta/2$ is satisfied.

From the above we conclude that for all sufficiently small ε

$$F[w_\varepsilon] \geq -\tfrac{1}{2}\varepsilon^2 \Delta w_\varepsilon \qquad \text{(a.e. on } Q'\text{)}. \tag{5}$$

Using Theorem 5.1b we take the limit in (5). However, before doing this, we need to estimate Δw_ε, $(\partial/\partial t)w_\varepsilon$.

Note that for $|\varepsilon| \leq 1$ the matrix $\sigma_\varepsilon(\alpha,t,x)$ satisfies the same conditions as those the matrix $\sigma(\alpha,t,x)$ satisfies, having, however, a different constant K. Indeed, the matrix norms of their derivatives with respect to t and x obviously coincide. Furthermore,

$$\|\sigma_\varepsilon(\alpha,t,x)\|^2 = \|\sigma(\alpha,t,x)\|^2 + \varepsilon^2 \leq (K^2 + 1)(1 + |x|)^2.$$

Hence, applying Theorem 4 to the function w_ε for $|\varepsilon| \le 1$, $\varepsilon \ne 0$, we find the constant N which depends only on K and m, for which

$$|w_{\varepsilon(l)(l)}| \le \left(1 + \frac{1}{\mu_\varepsilon}\right) N(1 + |x|)^{3m} e^{N(T-t)} \qquad \text{(a.e. in } H_T) \tag{6}$$

for all $l \in E_d$, where

$$\mu_\varepsilon(t,x) = \inf_{|\lambda|=1} \sup_{\alpha \in A} n_\varepsilon^\alpha(t,x)(a_\varepsilon(\alpha,t,x)\lambda,\lambda),$$

$$n_\varepsilon^\alpha(t,x) = (1 + \operatorname{tr} a_\varepsilon(\alpha,t,x) + |b(\alpha,t,x)| + c^\alpha(t,x) + |f^\alpha(t,x)|)^{-1}.$$

Further, it is seen that $(a_\varepsilon(\alpha,t,x)\lambda,\lambda) \ge (a(\alpha,t,x)\lambda,\lambda)$. Since

$$\operatorname{tr} a_\varepsilon(\alpha,t,x) = \operatorname{tr} a(\alpha,t,x) + \frac{\varepsilon^2}{2} d \le \operatorname{tr} a(\alpha,t,x) + \varepsilon^2 d,$$

$n_\varepsilon^\alpha(t,x) \ge (1/1 + \varepsilon^2 d))n^\alpha(t,x)$. Hence $\mu_\varepsilon \ge (1/(1 + \varepsilon^2 d))\mu$. From (6) we conclude that

$$|w_{\varepsilon(l)(l)}| \le \left[1 + \frac{1}{\mu}(1 + \varepsilon^2 d)\right] N(1 + |x|)^{3m} e^{N(T-t)} \qquad \text{(a.e. on } Q^*).$$

By virtue of Lemma 3 the last expression is bounded on Q' by a certain constant. Thus, there exists a constant N such that for any $\varepsilon \in [-1,1]$, $\varepsilon \ne 0$, the inequality $|\Delta w_\varepsilon| \le N$ can be satisfied almost everywhere on Q'. Theorem 4 also implies the uniform boundedness of $|(\partial/\partial t)w_\varepsilon|$ for $\varepsilon \in [-1,1]$.

Next, we take a sequence $w_{1/n}$. The above arguments and (5) yield

$$F[w_1] \ge \inf_{n \ge 1} F[w_{1/n}] \ge -N \qquad \text{(a.e. on } Q'),$$

$$\underline{\lim}_{n\to\infty} F[w_{1/n}] \ge -\frac{1}{2} \overline{\lim}_{n\to\infty} \frac{1}{n^2} \Delta w_{1/n} \ge 0 \qquad \text{(a.e. on } Q').$$

The former inequality allows us to assert that the function $\inf_{n\ge1} F[w_{1/n}] \in \mathscr{L}_{d+1}(Q')$ (it is bounded on Q').

The latter inequality together with Theorem 5.1b yields

$$F[w] \ge \underline{\lim}_{n\to\infty} F[w_{1/n}] \ge 0 \quad \text{(a.e. on } Q').$$

Recalling that $F[w] \le 0$ (a.e. on Q^*), we obtain $F[w] = 0$ (a.e. on Q'). We have proved the theorem for the function w.

It remains only to prove that $F[v] = 0$ (a.e. on Q^*). Let us consider the functions $v_\varepsilon(s,x)$ introduced in Section 6. By inequality (4) and Theorem 4, for $\varepsilon \ne 0$ generalized derivatives $(\partial/\partial t)v_\varepsilon(t,x)$, $v_{\varepsilon x^i}(t,x)$ $v_{\varepsilon x^i x^j}(t,x)$ exist and are locally bounded in H_T. By Theorem 2 for $\varepsilon \ne 0$

$$F[v_\varepsilon] + \tfrac{1}{2}\varepsilon^2 \Delta v_\varepsilon = 0 \qquad \text{(a.e. in } H_T).$$

We fix a certain bounded region $Q' \subset \bar{Q}' \subset Q^*$. In the same way as we did before, we estimate on Q' the derivatives $v_{\varepsilon(l)(l)}$, $(\partial/\partial t)v_\varepsilon$, using Theorem 4.

Then, using Theorem 5.1b, we can conclude that

$$F[v] \geq \varliminf_{n\to\infty} F[v_{1/n}] = \varliminf_{n\to\infty} \left(-\frac{1}{2n^2}\Delta v_{1/n}\right) \geq 0 \qquad \text{(a.e. on } Q'\text{)}.$$

On the other hand, since $F[v] \leq 0$ (a.e. on Q^*), $F[v] = 0$ (a.e. on Q'). Due to arbitrariness of Q', $F[v] = 0$ (a.e. on Q^*). The theorem is proved. □

6. Remark. Inequality (6) together with the estimate μ_ε given in the preceding proof, shows that for all $\varepsilon \in [-1,1]$, $\varepsilon \neq 0$, $l \in E_d$,

$$\frac{\mu}{1+\mu}|w_{\varepsilon(l)(l)}| \leq (1+\varepsilon^2 d)N(1+|x|)^{3m}e^{N(T-t)} \qquad \text{(a.e. on } Q^*\text{)}, \tag{7}$$

where $N = N(K,m)$. From (6) it follows in general that inequality (7) holds almost everywhere in H_T. However, the function $\mu = 0$ outside Q^*. By Theorem 4, inequality (7) holds as well for $\varepsilon = 0$. Absolutely similarly, for all $\varepsilon \in [-1,1]$, $l \in E_d$

$$\frac{\mu}{1+\mu}|v_{\varepsilon(l)(l)}| \leq (1+\varepsilon^2 d)N(1+|x|)^{3m}e^{N(T-t)} \qquad \text{(a.e. on } Q^*\text{)},$$

where $N = N(K,m)$.

In the case when for all $(t,x) \in \bar{H}_T$ and $\lambda \neq 0$

$$\sup_{\alpha\in A}(a(\alpha,t,x)\lambda,\lambda) > 0, \tag{8}$$

the set Q^* coincides with H_T, and, in addition, the continuous function $\mu(s,x) > 0$ at each point $[0,T] \times E_d$. Hence, the function μ^{-1} is bounded on each cylinder $\bar{C}_{T,R}$. Also, the above arguments show that the derivatives $w_{\varepsilon(l)(l)}$ and $v_{\varepsilon(l)(l)}$ are bounded (a.e.) in each cylinder $\bar{C}_{T,R}$ by a constant not depending on ε. The same remark applies to the mixed derivatives $w_{\varepsilon(l_1)(l_2)}$ and $v_{\varepsilon(l_1)(l_2)}$, which, as we know, can readily be expressed in terms of $w_{\varepsilon(l_1+l_2)(l_1+l_2)}$, $w_{\varepsilon(l_1-l_2)(l_1-l_2)}$, $v_{\varepsilon(l_1+l_2)(l_1+l_2)}$, and $v_{\varepsilon(l_1-l_2)(l_1-l_2)}$.

The next theorem follows immediately from Theorems 4 and 5, the results obtained in Section 3.1 on the continuity of v and w, as well as on the estimates of $|v|$ and $|w|$, and, finally, from the remarks made above about the properties of μ if condition (8) is satisfied. Recall that the assumptions made in Section 3.1 and the assumptions about the smoothness of σ, b, c, f, $g(x)$, and $g(t,x)$ which were formulated before Lemma 3 are assumed to be satisfied.

7. Theorem. *For all $(t,x) \in \bar{H}_T$ and $\lambda \neq 0$ let inequality (8) be satisfied (i.e., $F_1(\lambda^i\lambda^j,t,x) > 0$). Then the functions $v(t,x)$ and $w(t,x)$ are continuous in $\bar{H}_T$, have in H_T all generalized first and second derivatives with respect to x and a generalized first derivative with respect to t. These derivatives are bounded in each cylinder $C_{T,R}$. There exists a constant $N = N(K,m)$ such that for all $(t,x) \in \bar{H}_T$*

$$|v(t,x)| \leq N(1+|x|)^m e^{N(T-t)},$$
$$|w(t,x)| \leq N(1+|x|)^m e^{N(T-t)}.$$

Finally,
a. $F[v] = 0$ (a.e. *in* H_T), $v(T,x) = g(x)$;
b. $F[w] \leq 0$ (a.e. *in* H_T), $w(t,x) \geq g(t,x)$ *for* $(t,x) \in \bar{H}_T$, $F[w] = 0$ *almost everywhere on the set* $\{(t,x) \in H_T; w(t,x) > g(t,x)\}$, $w(T,x) = g(T,x)$.

It follows from this theorem that, in particular, the derivatives $(\partial/\partial t)v$, $(\partial/\partial t)w$, v_{x^i}, w_{x^i}, $v_{x^ix^j}$, and $w_{x^ix^j}$ are summable with respect to any cylinder $C_{T,R}$ to any power. Using theorems on embedding (see [47, Chapter II, Lemma 3.3]) we deduce from the foregoing the following.

8. Corollary. *Under the assumptions of Theorem* 7 $\text{grad}_x v(t,x)$ *and* $\text{grad}_x w(t,x)$ *are continuous in* $\bar{H}_T$. *Moreover, for any* $R > 0$, $\lambda \in (0,1)$ *there is a constant N such that for* $|x|$, $|x_1|$, $|x_2| \leq R$, *and* t, t_1, $t_2 \in [0,T]$ *the inequalities*

$$|\text{grad}_x u_i(t,x_1) - \text{grad}_x u_i(t,x_2)| \leq N|x_1 - x_2|^\lambda, \qquad i = 1,2,$$
$$|\text{grad}_x u_i(t_1,x) - \text{grad}_x u_i(t_2,x)| \leq N|t_1 - t_2|^{\lambda/2}, \qquad i = 1,2,$$

are satisfied, where $u_1 = v$, $u_2 = w$.

Further, since the nonnegative function $w(t,x) - g(t,x)$ is continuously differentiable with respect to x, its derivatives with respect to x vanish at the points of $\bar{H}_T$, at which this function vanishes. This implies

9. Corollary. *The smooth pasting condition*

$$\text{grad}_x w(t,x) = \text{grad}_x g(t,x)$$

is satisfied, under the assumptions of Theorem 7, *everywhere on the set* $\{(t,x) \in \bar{H}_T : w(t,x) = g(t,x)\}$ *and, in particular, on the boundary of this set.*

10. Remark. The assertions of Theorems 5 and 7 will hold if in formulating the conditions imposed on γ^α after Theorem 2 we do not require continuity of $\gamma^\alpha_{(l)}$, $\gamma^\alpha_{(l)(l)}$, and $(\partial/\partial t)\gamma^\alpha$ if by these derivatives we mean generalized derivatives, and if, finally, we drop the condition

$$|L^\alpha(T,x)g(x)| + |L^\alpha g(T,x)| \leq K(1 + |x|)^{3m}.$$

We shall explain this. In fact it suffices to show that Theorem 4 will still hold if we replace K, m by some other constants in the formulation of the theorem. Let us smooth the coordinates $\gamma^\alpha(t,x)$, assuming $\gamma^\alpha(t,x) = \gamma^\alpha(0,x)$ for $t \leq 0$. Furthermore, using the vector $\gamma^\alpha(t,x,\varepsilon) \equiv [\gamma^\alpha(t,x)]^{(\varepsilon)}$, let us construct the payoff functions $v(t,x,\varepsilon)$ and $w(t,x,\varepsilon)$.

For $0 < \varepsilon < 1$ the vector $\gamma^\alpha(t,x,\varepsilon)$ satisfies all the conditions formulated after Theorem 2 containing the constants K' and m', which do not depend on ε since, for example, $\gamma^\alpha_{(l)}(t,x,\varepsilon) = [\gamma^\alpha_{(l)}(t,x)]^{(\varepsilon)}$. Hence for the functions $v(t,x,\varepsilon)$ and $w(t,x,\varepsilon)$ Theorem 4 holds true in which K, m, μ, and Q^* are replaced by K', m', $\mu(t,x,\varepsilon)$ and $Q^*(\varepsilon)$, respectively, constructed on a basis of $\gamma^\alpha(t,x,\varepsilon)$.

Next, in each cylinder $\bar{C}_{T,R}$ the vector $\gamma^{\alpha}(t,x)$ satisfies a Lipschitz condition with respect to (t,x) with a constant not depending on α since estimates of the generalized derivatives $\gamma^{\alpha}_{(l)}$ and $(\partial/\partial t)\gamma^{\alpha}$ do not depend on α. This readily implies that $\gamma^{\alpha}(t,x,\varepsilon) \to \gamma^{\alpha}(t,x)$ as $\varepsilon \to 0$ uniformly on $A \times \bar{C}_{T,R}$ for each R. In particular, $\mu(t,x,\varepsilon) \to \mu(t,x)$. Furthermore, by Theorem 3.1.12 and Corollary 3.1.13, $v(t,x,\varepsilon) \to v(t,x)$ and $w(t,x,\varepsilon) \to w(t,x)$ as $\varepsilon \to 0$ uniformly on each cylinder $\bar{C}_{T,R}$. The convergence of the payoff functions and the estimate given in Theorem 4 of their generalized first derivatives with respect to (t,x), which is uniform with respect to $\varepsilon \in (0,1)$, enables us, as was mentioned in Section 2.1, to prove the existence and to estimate the generalized first derivatives of $v(t,x)$ and $w(t,x)$ with respect to (t,x). We can estimate $v_{(l)(l)}(t,x)$ and $w_{(l)(l)}(t,x)$ in the same way, if we take advantage of the fact that all $\mu(t,x,\varepsilon) \geq \frac{1}{2}\mu(t,x) > 0$ due to the uniform convergence of $\mu(t,x,\varepsilon)$ to $\mu(t,x)$ in each bounded region Q' which together with its closure lies in Q^*, starting from some instant of time.

Notes

This chapter uses the methods and results of [34, 36, 37], and [58, 59]. For control of jump processes, see Pragarauskas [63].

Section 1. If the set A consists of a single point, i.e., if we consider a diffusion process, it is possible to regard the functions v, $-v$ as payoff functions. Therefore, in Theorem 4 we then have equality instead of inequality. Similar assertions can be found in Freidlin [18]. Theorem 8 is the generalization of a result obtained in [38, 62].

Sections 2, 3. The method applied in these sections, involving the derivatives in the sense of Definition 2.1.2, enables us to do without the theorems on interior smoothness of solutions of elliptic as well as parabolic equations, that is, the theorems used by Krylov, Nisio, and Pragarauskas (see the references listed above).

Section 4. In Theorem 2.9.10 the differentiability of $v(t,x)$ with respect to t is derived from the existence of second derivatives of σ, b, c, f with respect to x. Exercise 4 shows that in the presence of control in order to estimate $(\partial/\partial t)v(t,x)$, we need to require that the derivatives of σ, b, c, f with respect to t exist.

Section 5. The results obtained in this section for the time homogeneous case, can be found in [34]. It is well known that the limit of harmonic functions is harmonic, and the associated theory has much in common with the theory developed in this section.

Section 6. The relationship between the payoff functions associated with a controlled process, as well as the nondegenerate approximation of this process, is investigated in Fleming [16], Krylov [37], Tobias [74].

Section 7. The fact that a payoff function satisfies the Bellman equation implies, in particular, that the Bellman equation is solvable. It is interesting to note that differential equations theory suggests no (other) methods for proving the solvability of the Bellman equations in question. The smooth pasting condition (Corollary 9) was first introduced by Shiryayev (see [69]).

The Construction of ε-Optimal Strategies 5

The main objective of investigating a controlled process approached from a practical point of view is to construct optimal strategies or strategies close to optimal. In this chapter we show how one can find ε-optimal strategies in optimal control problems which were discussed in Chapters 3 and 4. Recall that we proved in Chapter 3 that one can always find ε-optimal strategies in the class of natural strategies. In this chapter we focus our attention on constructing Markov (see Definition 3.1.3) ε-optimal strategies which is of interest from a practical point of view due to the simplicity of Markov strategies. Adjoint Markov strategies which are investigated in this chapter, are somewhat more complex, becoming thereby less applicable in engineering than Markov strategies (see Definition 3.17). However, from a theoretical point of view adjoint Markov strategies are more convenient in some respects than Markov strategies. Considering adjoint Markov strategies, we prove in Section 3 that a solution of the Bellman equation is a payoff function.

In the arguments of this chapter the results obtained in Chapter 4 on payoff functions satisfying the Bellman equation play an essential role.

Throughout Chapter 5 we use the assumptions, definitions, and notations given in Section 3.1.

1. ε-Optimal Markov Strategies and the Bellman Equation

We showed in Sections 1.1, 1.4, and 1.5 the way to construct ε-optimal strategies having the knowledge of the payoff function. In this section and the sequel we carry out construction of these strategies in the following cases:

a. for each $R > 0$ there exists a number $\delta_R > 0$ such that for all $\alpha \in A$, $(t,x) \in C_{T,R}$, $\lambda \in E_d$ the inequality
$$(a(\alpha,t,x)\lambda,\lambda) \geq \delta_R|\lambda|^2$$
is satisfied;

b. for all $t \in [0,T]$, $x \in E_d$, $\lambda \neq 0$
$$\sup_{\alpha \in A}(a(\alpha,t,x)\lambda,\lambda) > 0;$$

c. $a(\alpha,t,x)$ does not depend on x.

The technique for finding ε-optimal Markov strategies in the case (a) is given in this section. For (b) and (c) we shall prove the existence of ε-optimal Markov strategies and construct randomized ε-optimal Markov strategies in the subsequent sections.

The case (c) incorporates the control of a completely deterministic process when $\sigma(\alpha,t,x) = 0$.

In this section, in addition to the assumptions made in Section 3.1, we impose the following conditions. Let A be a convex set in a Euclidean space. Also, for each $(t,x) \in \bar{H}_T$ let the functions $\sigma(\alpha,t,x)$ and $b(\alpha,t,x)$ satisfy the Lipschitz condition with respect to α, namely, for all α, $\beta \in A$, $(t,x) \in \bar{H}_T$ let
$$\|\sigma(\alpha,t,x) - \sigma(\beta,t,x)\| + |b(\alpha,t,x) - b(\beta,t,x)| \leq K|\alpha - \beta|.$$

Furthermore, we introduce a vector $\gamma^\alpha(t,x)$ of dimension $d \times (d_1 + d + 4)$ whose coordinates are given by the following variables: $\sigma^{ij}(\alpha,t,x)$ $(i = 1, \ldots ,d,$ $j = 1, \ldots ,d_1)$, $b^i(\alpha,t,x)$ $(i = 1, \ldots ,d)$, $c^\alpha(t,x)$, $f^\alpha(t,x)$, $g(x)$, $g(t,x)$.

We assume that for each $\alpha \in A$, $l \in E_d$ the derivatives $\gamma^\alpha_{(l)}(t,x)$, $\gamma^\alpha_{(l)(l)}(t,x)$, $(\partial/\partial t)\gamma^\alpha(t,x)$ exist and are continuous with respect to (t,x) on $\bar{H}_T$. In addition, for all $\alpha \in A$, $l \in E_d$, $(t,x) \in \bar{H}_T$ let
$$\left|\frac{\partial}{\partial t}\gamma^\alpha(t,x)\right| + |\gamma^\alpha_{(l)}(t,x)| + |\gamma^\alpha_{(l)(l)}(t,x)| \leq K(1 + |x|)^m.$$

Finally, we assume for the sake of convenience that for all $\alpha \in A$, $x \in E_d$
$$|L^\alpha(T,x)g(x)| + |L^\alpha g(T,x)| \leq K(1 + |x|)^{3m}.$$

As was shown in Section 4.4, we can always get rid of the last assumption by choosing appropriate constants K and m in the other assumptions made above.

We regard the foregoing assumptions as satisfied throughout this section.

1. Lemma. *Let a bounded region $Q \subset H_T$ and let a function $u \in W^{1,2}(Q)$. For the function $\alpha = \alpha(t,x)$ with values in A let*
$$h^\alpha(t,x) = F[u](t,x) - [L^{\alpha(t,x)}(t,x)u(t,x) + f^{\alpha(t,x)}(t,x)].$$

We assert that for each $\varepsilon > 0$ one can find a function $\alpha(t,x)$ given on $(-\infty,\infty) \times E_d$ which is infinitely differentiable with respect to (t,x) and, in addition, has values in A and a constant N such that

$$\|h^{\alpha}\|_{d+1,Q} \leq \varepsilon, \qquad \sup_{t,x} \sup_{l \in E_d} |\alpha_{(l)}(t,x)| < \infty,$$

$$\|\sigma(\alpha(t,x),t,x) - \sigma(\alpha(t,y),t,y)\| + |b(\alpha(t,x),t,x) - b(\alpha(t,y),t,y)| \leq N|x - y|$$

for all $x, y \in E_d$, $t \geq 0$.

PROOF. We fix $\varepsilon > 0$. We proceed in the same way as in proving Lemma 1.4.9. We choose a countable subset $\{\alpha(i): i \geq 1\}$ which is everywhere dense in A. From the equality

$$F[u] = \sup_i [L^{\alpha(i)}u + f^{\alpha(i)}] = \lim_{n\to\infty} \max_{i \leq n} [L^{\alpha(i)}u + f^{\alpha(i)}]$$

and the boundedness of Q we readily derive the existence of the measurable function $\tilde{\alpha}(t,x)$ which assumes only a finite number of values from $\{\alpha(i)\}$ and is such that $\|h^{\tilde{\alpha}}\|_{d+1,Q} \leq \varepsilon/2$. Assume that $\tilde{\alpha}(t,x)$ is defined everywhere in E_{d+1} and equal to $\alpha(1)$ outside Q. We take the smoothing kernels $n^{d+1}\zeta(nt,nx)$. Let $\alpha_n(t,x) = n^{d+1}\zeta(nt,nx) * \tilde{\alpha}(t,x)$.

As has repeatedly been noted, $\alpha_n(t,x)$ are infinitely differentiable, $\alpha_n \to \tilde{\alpha}$ (a.e.). Moreover, $\alpha_n(t,x) \in A$ for all (t,x) *because A is convex.*

Further, it follows from the continuity of f^{α} and the coefficients of L^{α} with respect to α that $h^{\alpha_n} \to h^{\tilde{\alpha}}$ (a.e. on Q). Due to boundedness of f^{α} and the coefficients of L^{α} on Q there exists a constant N for which

$$|F[u]| + |L^{\alpha}u + f^{\alpha}| \leq N\left(\left|\frac{\partial}{\partial t}u\right| + \sum_{i,j=1}^{d} |u_{x^i x^j}| + \sum_{i=1}^{d} |u_{x^i}| + |u| + 1\right)$$

for all $\alpha \in A$ everywhere on Q. Hence the totality of the functions h^{α_n} is bounded by one function in $\mathscr{L}_{d+1}(Q)$. By the Lebesgue theorem, $\|h^{\alpha_n}\|_{d+1,Q} \to \|h^{\tilde{\alpha}}\|_{d+1,Q}$. Therefore, there exists a number $n(\varepsilon)$ such that

$$\|h^{\alpha_{n(\varepsilon)}}\|_{d+1,Q} \leq \|h^{\tilde{\alpha}}\|_{d+1,Q} + \frac{\varepsilon}{2} \leq \varepsilon.$$

Next, we let $\alpha(t,x) = \alpha_{n(\varepsilon)}(t,x)$ and prove that the function $\alpha(t,x)$ is the sought function. We embed the region Q into a cylinder $C_{T,R}$. The function $\zeta(t,x)$ is equal to zero for $|x| > 1$. Outside of Q, therefore, outside of $C_{T,R}\tilde{\alpha} = \alpha(1)$. It is easy to derive from the foregoing properties of the functions ζ and $\tilde{\alpha}$ that $\alpha_n(t,x) = \alpha(1)$ for $|x| > R + 1$ for all n. In particular, $\alpha(t,x) = \alpha(1)$ for $|x| > R + 1$. For reasons similar to those above, $\alpha(t,x) = \alpha(1)$ for $t < -1$ and for $t > T + 1$. Due to the continuous differentiability of $\alpha(t,x)$ we have

$$N_1 \equiv \sup_{t,x} \sup_{l \in E_d} |\alpha_{(l)}(t,x)| < \infty.$$

Finally, let us show that for all $x, y \in E_d, t \geq 0$

$$\|\sigma(\alpha(t,x),t,x) - \sigma(\alpha(t,y),t,y)\| \leq K(1 + N_1)|x - y|.$$

We have

$$\begin{aligned}\|\sigma(\alpha(t,x),t,x) - \sigma(\alpha(t,y),t,y)\| &\leq \|\sigma(\alpha(t,x),t,x) - \sigma(\alpha(t,y),t,x)\| + K|x - y| \\ &\leq K|\alpha(t,x) - \alpha(t,y)| + K|x - y| \\ &\leq KN_1|x - y| + K|x - y|.\end{aligned}$$

We can estimate the corresponding difference for the functions b in a similar way. The lemma is proved. □

Note that the function $\alpha(t,x)$ whose existence is asserted in the lemma depends on the combination Q, u, ε. In the case where Q is a subregion of H_T, $u \in W^{1,2}(Q)$, $\varepsilon > 0$, we denote the combination Q, u, ε by p:

$$p = (Q,u,\varepsilon).$$

It is convenient to write the function $\alpha(t,x)$ constructed on the basis of p as $\alpha[p](t,x)$. For a fixed $s_0 \in [0,T]$ and the function $\alpha[p](t,x)$ we can define the Markov strategy $\alpha[p]$ using the formula

$$\alpha_t[p](x_{[0,t]}) = \alpha[p](s_0 + t, x_t). \tag{1}$$

Since the functions $\sigma(\alpha[p](s_0 + t, x), s_0 + t, x)$, $b(\alpha[p](s_0 + t, x), s_0 + t, x)$ satisfy the Lipschitz condition with respect to x, the Markov strategy $\alpha[p]$ is admissible at a point (s_0,x) for each $x \in E_d$.

2. Theorem. *For each $R > 0$ let there exist a number $\delta_R > 0$ such that for all $\alpha \in A$, $(t,x) \in \bar{C}_{T,R}$, $\lambda \in E_d$*

$$(a(\alpha,t,x)\lambda,\lambda) \geq \delta_R|\lambda|^2.$$

Then $v_{(M)} = v$ on $\bar{H}_T$. Furthermore, we fix $s_0 \in [0,T]$, assume that $p = (C_{T,R},v,\varepsilon)$, and, finally, define the Markov strategy $\alpha[p]$ using Eq. (1). Then for each $x \in E_d$

$$\lim_{R\to\infty} \lim_{\varepsilon\downarrow 0} v^{\alpha[p]}(s_0,x) = v(s_0,x).$$

PROOF. First we note that since

$$v^{\alpha[p]}(s_0,x) \leq v_{(M)}(s_0,x) \leq v(s_0,x),$$

the first assertion follows from the second assertion. We have from Theorem 4.7.7 that $v \in W^{1,2}(C_{T,R})$ for all R. Therefore, we have defined the strategy $\alpha[p]$. Furthermore, $F[v] = 0$ (H_T-a.e.). Therefore, by the definition of the function $\alpha[p](t,x)$

$$\|h^p\|_{d+1,C_{T,R}} \leq \varepsilon,$$

where

$$h^p(t,x) = -L^{\alpha[p](t,x)}(t,x)v(t,x) - f^{\alpha[p](t,x)}(t,x).$$

According to Ito's formula

$$v(s_0,x) = \mathsf{M}_{s_0,x}^{\alpha[p]}\left[\int_0^{\tau_{T,R}} f^{\alpha_t}(s_0 + t, x_t)e^{-\varphi_t}\,dt + v(s_0 + \tau_{T,R}, x_{\tau_{T,R}})e^{-\varphi_{\tau_{T,R}}}\right]$$

$$+ \mathsf{M}_{s_0,x}^{\alpha[p]}\int_0^{\tau_{T,R}} h^p(s_0 + t, x_t)e^{-\varphi_t}\,dt, \tag{2}$$

where

$$\tau_{T,R} = \inf\{t \geq 0:(s_0 + t, x_t) \notin [0,T) \times S_R\}.$$

By Theorem 2.2.4, the absolute value of the last mathematical expectation does not exceed

$$N\|h^p\|_{d+1,C_{T,R}} \leq N\varepsilon,$$

where N does not depend on ε. Hence it tends to zero as $\varepsilon \downarrow 0$.

Further, by virtue of the equality $v(T,x) = g(x)$ the first expression in the right side of (2) equals

$$v^{\alpha[p]}(s_0,x) - \mathsf{M}_{s_0,x}^{\alpha[p]}\left[\int_{\tau_{T,R}}^{T-s_0} f^{\alpha_t}(s_0 + t, x_t)e^{-\varphi_t}\,dt + g(x_{T-s_0})e^{-\varphi_{T-s_0}}\chi_{\tau_{T,R}<T-s_0}\right]$$

$$+ \mathsf{M}_{s_0,x}^{\alpha[p]}v(s_0 + \tau_{T,R}, x_{\tau_{T,R}})e^{-\varphi_{\tau_{T,R}}}\chi_{\tau_{T,R}<T-s_0}.$$

For proving the theorem it suffices to show that the two last terms tend to zero as $R \to \infty$ uniformly with respect to ε. By virtue of the growth estimates, as $|x| \to \infty$, of the functions $f^\alpha(t,x)$, $g(x)$, $v(t,x)$, to do as indicated we need only to prove that

$$\lim_{R\to\infty}\sup_{\varepsilon>0}\mathsf{M}_{s_0,x}^{\alpha[p]}\left(1 + \sup_{t\leq T-s_0}|x_t|\right)^m (T - s_0 + 1)\chi_{\tau_{T,R}<T-s_0} = 0. \tag{3}$$

Note that on the set $\{\tau_{T,R}^{\alpha[p],s_0,x} < T - s_0\}$ the inequality

$$1 \leq \frac{1}{R+1}\left(1 + \sup_{t\leq T-s_0}|x_t^{\alpha[p],s_0,x}|\right)$$

is satisfied.

Hence

$$\mathsf{M}_{s_0,x}^{\alpha[p]}\left(1 + \sup_{t\leq T-s_0}|x_t|\right)^m \chi_{\tau_{T,R}<T-s_0} \leq \frac{1}{R+1}\sup_{\alpha\in\mathfrak{A}}\mathsf{M}_{s_0,x}^{\alpha}\left(1 + \sup_{t\leq T-s_0}|x_t|\right)^{m+1}$$

Furthermore, (3) follows from the estimates of moments of solutions of stochastic equations, thus completing the proof of the theorem. □

3. Theorem. *Suppose the assumption made in the preceding theorem is satisfied Then $w_{(M)} = w$ on $\bar{H}_T$. Moreover, we fix $s_0 \in [0,T]$, put $p = (C_{T,R},w,\varepsilon)$, define the Markov strategy $\alpha[p]$ using Eq. (1), and, finally, denote by $\tau_0^{\alpha[p],s_0,x}$ the time of first exit of the process $(s_0 + t, x_t^{\alpha[p],s_0,x})$ from $Q_0 = \{(t,y) \in \bar{H}_T: w(t,y) > g(t,y)\}$. Then for each $x \in E_d$*

$$\lim_{R\to\infty}\lim_{\varepsilon\downarrow 0} v^{\alpha[p],\tau_0}(s_0,x) = w(s_0,x).$$

PROOF. This proof follows closely the preceding one. By Theorem 4.7.7, $F[w] = 0$ (Q_0-a.e.). Therefore, by the definition of $\alpha[p](t,x)$

$$\|h^p\|_{d+1,Q_0\cap C_{T,R}} \leq \varepsilon,$$

where

$$h^p(t,x) = -L^{\alpha[p](t,x)}(t,x)w(t,x) - f^{\alpha[p](t,x)}(t,x).$$

According to Ito's formula

$$w(s_0,x) = \mathsf{M}^{\alpha[p]}_{s_0,x}\left[\int_0^{\bar\tau_{T,R}} f^{\alpha_t}(s_0 + t, x_t)e^{-\varphi_t}\,dt + w(s_0 + \bar\tau_{T,R}, x_{\bar\tau_{T,R}})e^{-\varphi_{\bar\tau_{T,R}}}\right]$$
$$+ \mathsf{M}^{\alpha[p]}_{s_0,x}\int_0^{\tau_{T,R}} h^p(s_0 + t, x_t)e^{-\varphi_t}\,dt, \tag{4}$$

where

$$\bar\tau_{T,R} = \inf\{t \geq 0 : (s_0 + t, x_t) \notin Q_0 \cap ([0,T) \times S_R)\}.^1$$

By Theorem 2.2.4, the last term does not exceed $N\|h^p\|_{d+1,Q_0\cap C_{T,R}}$ and tends to zero as $\varepsilon \downarrow 0$. The first term in (4) is equal to

$$v^{\alpha[p],\tau_0}(s_0,x) - \mathsf{M}^{\alpha[p]}_{s_0,x}\left[\int_{\bar\tau_{T,R}}^{\tau_0} f^{\alpha_t}(s_0 + t, x_t)e^{-\varphi_t}\,dt + g(s_0 + \tau_0, x_{\tau_0})e^{-\varphi_{\tau_0}}\chi_{\bar\tau_{T,R}<\tau_0}\right]$$
$$+ \mathsf{M}^{\alpha[p]}_{s_0,x}w(s_0 + \bar\tau_{T,R}, x_{\bar\tau_{T,R}})e^{-\varphi_{\bar\tau}}T,R\chi_{\bar\tau_{T,R}<\tau_0},$$

where the mathematical expectations tend to zero as $R \to \infty$ uniformly with respect to ε, which fact can be proved in the same way as the corresponding fact from the preceding proof, since $\bar\tau_{T,R} \leq \tau_0 \leq T - s_0$, and since, in addition, the inequality $\bar\tau_{T,R} < T - s_0$ is satisfied on the set $\{\bar\tau_{T,R} < \tau_0\}$.

The analysis carried on Eq. (4) implies the assertions of the theorem, thus completing the proof of the theorem. □

2. ε-Optimal Markov Strategies. The Bellman Equation in the Presence of Degeneracy

Theorems 1.2 and 1.3 provide the technique for finding ε-optimal Markov strategies if the strong nondegeneracy condition is satisfied: $(a(\alpha,t,x)\lambda,\lambda) \geq \delta_R|\lambda|^2$ for all $\alpha \in A$, $(t,x) \in \bar C_{T,R}$, $\lambda \in E_d$, $R > 0$, where $\delta_R > 0$. If we reject this condition, we shall not know how to construct ε-optimal "pure" Markov strategies. In some cases considered in this section it is possible, however, to construct ε-optimal "mixed" Markov strategies making no assumption about nondegeneracy. In the cases mentioned we prove the existence of (usual) ε-optimal Markov strategies.

[1] The superscripts $\alpha[p]$, s_0, x are omitted here and below in the proof.

In this section we assume that the assumptions made in the preceding section are satisfied. In particular, we assume that A is a convex set in a certain Euclidean space. We denote by (s_0,x_0) a fixed point of $\bar{H}_T$.

In addition to the basic d_1-dimensional Wiener process $(\mathbf{w}_t,\mathscr{F}_t)$ we shall need a d-dimensional Wiener process $(\hat{\mathbf{w}}_t,\hat{\mathscr{F}}_t)$ as well as a $(d + d_1)$-dimensional Wiener process $(\tilde{\mathbf{w}}_t,\tilde{\mathscr{F}}_t)$. We assume that the processes listed can be defined on the probability spaces $(\Omega,\mathscr{F},\mathrm{P})$, $(\hat{\Omega},\hat{\mathscr{F}},\hat{\mathrm{P}})$, $(\tilde{\Omega},\tilde{\mathscr{F}},\tilde{\mathrm{P}})$, respectively (we permit these spaces to be equivalent). The last d coordinates of the vector $\tilde{\mathbf{w}}_t$ form a d-dimensional Wiener process which we denote by $\tilde{\mathbf{w}}_t''$. We denote by $\tilde{\mathbf{w}}_t'$ a d_1-dimensional Wiener process composed of the first d_1 coordinates of the vector $\tilde{\mathbf{w}}_t$.

Recall that we agreed in Section 1 to denote by the letter p the triple Q, u and ε, where Q is the bounded subregion of H_T, $u \in W^{1,2}(Q)$, and $\varepsilon > 0$. As in Section 1, we denote here by $\alpha[p](t,x)$ some smooth function on E_{d+1} with values in A such that its first derivatives with respect to x are bounded, the functions

$$\sigma(\alpha[p](t,x),t,x), \qquad b(\alpha[p](t,x),t,x)$$

satisfy the Lipschitz condition with respect to x uniformly with respect to t, and

$$\|F[u] - [L^{\alpha[p]}u + f^{\alpha[p]}]\|_{d+1,Q} \leq \varepsilon.$$

The existence of the function $\alpha[p]$ having the properties listed was proved in Lemma 1.1.

If $p = (Q,u,\varepsilon)$, z_t is a (nonrandom) continuous function which is given on $[0, T - s_0]$ and assumes values in E_d, we can define the Markov strategy $\alpha[p,z)$ using the formulas

$$\begin{aligned} \alpha_t[p,z](x_{[0,t]}) &= \alpha[p](s_0 + t, x_t + \varepsilon z_t), \qquad & t \leq T - s_0, \\ \alpha_t[p,z](x_{[0,t]}) &= \alpha[p](s_0 + t, x_t + \varepsilon z_{T-s_0}), \qquad & t > T - s_0. \end{aligned} \tag{1}$$

We consider the equation

$$x_t = x_0 + \int_0^t \sigma_r(x_r)\,d\mathbf{w}_r + \int_0^t b_r(x_r)\,dr, \tag{2}$$

where

$$\begin{aligned} \sigma_r(x) &= \sigma(\alpha[p](s_0 + r, x + \varepsilon z_r), s_0 + r, x), \\ b_r(x) &= b(\alpha[p](s_0 + r, x + \varepsilon z_r), s_0 + r, x). \end{aligned}$$

In the same way as we did in proving Lemma 1.1, we show here that the coefficients of Eq. (2) satisfy the Lipschitz condition with respect to x with a constant $K(1 + N_1)$, where N_1 has been taken from the proof of Lemma 1.1. Therefore, Eq. (2) is solvable, and, furthermore, the Markov strategy $\alpha[p,z]$ is admissible at a point (s_0,x_0). Using the notation from Section 3.1, we write the solution of Eq. (2) as $x_t^{\alpha[p,z],s_0,x_0}$. Since s_0, x_0 are fixed, we write in short $x_t^{\alpha[p,z]}$.

It is seen that $v^{\alpha[p,\hat{\mathbf{w}}]}(s_0,x_0) \le v(s_0,x_0)$ everywhere on $\hat{\Omega}$. We shall prove below (see Corollary 2) that $v^{\alpha[p,\hat{\mathbf{w}}]}(s_0,x_0)$ is a random variable. Hence for any sets $p = (Q,u,\varepsilon)$

$$\hat{\mathsf{M}} v^{\alpha[p,\hat{\mathbf{w}}]}(s_0,x_0) \le v(s_0,x_0), \tag{3}$$

where $\hat{\mathsf{M}}$ denotes the mathematical expectation associated with the measure $\hat{\mathsf{P}}$.

We can interpret the mathematical expectation in (3) to be the payoff obtained by means of a mixed (weighted, randomized) Markov strategy. We explain this without defining concretely the mixed strategy. Assume that a probability measure is given on a set $\mathfrak{A}_M(s_0,x_0)$. Also, we assume that first in correspondence with this measure we have a game involving the Markov strategy α and, next, the process is controlled by means of this strategy α. The average payoff of the control is equal to $v^\alpha(s_0,x_0)$. The integral of $v^\alpha(s_0,x_0)$ with respect to a probability measure on $\mathfrak{A}_M(s_0,x_0)$ represents the total average payoff of the control of this type. In the case when the probability distribution on $\mathfrak{A}_M(s_0,x_0)$ is given by a random element $\alpha[p,\hat{\mathbf{w}}]$, the total average payoff equals the left-hand side of (3). From a practical viewpoint, the technique of controlling a process by means of a random Markov strategy is no less legitimate than that of controlling a process by means of a (nonrandom, pure) Markov strategy.

The fact that the left-hand side can be expressed in other terms is very important for the further discussion. On the probability space $(\tilde{\Omega},\tilde{\mathscr{F}},\tilde{\mathsf{P}})$ we consider the following equation:

$$x_t = x_0 + \int_0^t \tilde{\sigma}_r(x_r)\, d\tilde{\mathbf{w}}_r' + \int_0^t \tilde{b}_r(x_r)\, dr, \tag{4}$$

where

$$\tilde{\sigma}_r(x) = \sigma(\alpha[p](s_0 + r, x + \varepsilon\tilde{\mathbf{w}}_r''), s_0 + r, x),$$
$$\tilde{b}_r(x) = b(\alpha[p](s_0 + r, x + \varepsilon\tilde{\mathbf{w}}_r''), s_0 + r, x).$$

Eq. (4) has a unique solution. In fact, $\tilde{\sigma}_r(0)$ and $\tilde{b}_r(0)$ are bounded, since $\sigma(\alpha,t,0)$ and $b(\alpha,t,0)$ are bounded uniformly with respect to (α,t). Furthermore, the functions $\tilde{\sigma}_r(x)$, $\tilde{b}_r(x)$ satisfy the Lipschitz condition with respect to x, having the constant $K(1 + N_1)$, where N_1 has been taken from the proof of Lemma 1.1. Finally, the processes $\tilde{\sigma}_r(x)$ and $\tilde{b}_r(x)$ are progressively measurable with respect to $\{\tilde{\mathscr{F}}_t\}$.

We introduce a convenient notation for a solution of Eq. (4). If x_t is a solution of Eq. (4), we put

$$\beta_t[p] = \alpha[p](s_0 + t, x_t + \varepsilon\tilde{\mathbf{w}}_t'').$$

The process $\beta[p]$ is a strategy with respect to a system of the σ-algebras of $\{\tilde{\mathscr{F}}_t\}$ in the sense of Definition 3.1.1. It is seen that x_t satisfies the equation

$$x_t = x_0 + \int_0^t \sigma(\beta_r[p], s_0 + r, x_r)\, d\tilde{\mathbf{w}}_r' + \int_0^t b(\beta_r[p], s_0 + r, x_r)\, dr.$$

Using the standard notations, we can write the following: $x_t = x_t^{\beta[p],s_0,x_0}$. This also enables us to apply usual short notations, write the indices $\beta[p]$, s_0, x_0 only for the sign of the mathematical expectation, and introduce $v^{\beta[p]}(s_0,x_0)$, $v^{\beta[p],\tau}(s_0,x_0)$ in writing mathematical expectations of functionals of a solution of (4). In addition, since s_0, x_0 are fixed, we shall write $x_t^{\beta[p]}$ instead of $x_t^{\beta[p],s_0,x_0}$.

The following lemma gives an obvious formula with which we can modify the left-hand side of (3).

1. Lemma. *Let $F(z,x_{[0,T-s_0]})$ be a measurable function given on $C^2([0, T - s_0], E_d)$ and such that*

$$|F(z,x_{[0,T-s_0]})| \leq N\left(1 + \sup_{t \leq T - s_0} |x_t|\right)^n \tag{5}$$

for some constants N, n and also for all z, $x_{[0,T-s_0]} \in C([0, T - s_0], E_d)$. Let a bounded region $Q \subset H_T$, let a function $u \in W^{1,2}(Q)$ and let a number $\varepsilon > 0$. Using Lemma 1.1., we construct the function $\alpha[p](t,x)$ on the basis of the set $p = (Q,u,\varepsilon)$. Using formula (1) we introduce Markov strategies $\alpha[p,z]$ for $z \in C([0, T - s_0], E_d)$. Furthermore, we define the strategy $\beta[p]$ using the formula $\beta_t[p] = \alpha[p](s_0 + t, x_t + \varepsilon\tilde{\mathbf{w}}_t'')$, where x_t is a solution of Eq. (4). Then the function

$$\Phi(z) \equiv \mathsf{M}_{s_0,x_0}^{\alpha[p,z]} F(z,x_{[0,T-s_0]})$$

is bounded and measurable with respect to z for $z \in C([0, T - s_0], E_d)$,

$$\hat{\mathsf{M}}\Phi(\hat{\mathbf{w}}) = \tilde{\mathsf{M}}_{s_0,x_0}^{\beta[p]} F(\tilde{\mathbf{w}}'',x_{[0,T-s_0]}), \tag{6}$$

where $\hat{\mathsf{M}}$ ($\tilde{\mathsf{M}}$) denotes the mathematical expectation associated with the measure $\hat{\mathsf{P}}$ ($\tilde{\mathsf{P}}$).

PROOF. The boundedness of $\Phi(z)$ follows from (5) and the fact that due to familiar estimates of moments of solutions of stochastic equations

$$\sup_{\alpha \in \mathfrak{A}} \mathsf{M}_{s_0,x_0}^{\alpha} \sup_{t \leq T - s_0} |x_t|^n < \infty.$$

In proving (6) and the measurability of $\Phi(z)$, we use the results obtained in Section 2.9. We note that it is possible to solve Eq. (4) in a different way.

We denote by $\tilde{\mathscr{F}}_t'$ the completion (with respect to measure $\tilde{\mathsf{P}}$) of the smallest σ-algebra containing $\tilde{\mathscr{F}}_t$ as well as all sets of the form $\{\tilde{\mathbf{w}}_r'' \in \Gamma\}$, where $r \leq T - s_0$ and Γ denotes a Borel subset of E_d. We draw the reader's attention to the fact that r runs through the entire interval $[0, T - s_0]$.

It is easily seen that the processes $\tilde{\sigma}_t(x)$, $\tilde{b}_t(x)$ from Eq. (4) are progressively measurable with respect to new σ-algebras, and also that $(\tilde{\mathbf{w}}_t',\tilde{\mathscr{F}}_t')$ is a d_1-dimensional Wiener process. We note as well that the solution of Eq. (4) does not change on $[0, T - s_0]$ when we pass from $(\tilde{\mathbf{w}}_t',\tilde{\mathscr{F}}_t)$ to $(\tilde{\mathbf{w}}_t',\tilde{\mathscr{F}}_t')$. This obvious fact follows, for instance, from the uniqueness of a solution of Eq. (4) and the fact that the former solution x_t is progressively measurable with respect to

$\{\tilde{\mathscr{F}}_t\}$ and, in addition, is progressively measurable with respect to $\{\tilde{\mathscr{F}}'_t\}$ by virtue of the inclusion $\tilde{\mathscr{F}}_t \subset \tilde{\mathscr{F}}'_t (t \in [0, T - s_0])$.

Further, we use Theorem 2.9.4 in the case where

$$Z = C([0, T - s_0], E_d), \qquad \zeta = \tilde{\mathbf{w}}''_{[0, T-s_0]}, \xi = x_0$$

and for $z \in Z$

$$\sigma^z_t(x) = \sigma(\alpha[p](s_0 + t, x + \varepsilon z_t), s_0 + t, x),$$
$$b^z_t(x) = b(\alpha[p](s_0 + t, x + \varepsilon z_t), s_0 + t, x).$$

Then the solution of Eq. (4) is a solution of the equation

$$x_t = \xi + \int_0^t \sigma^\zeta_r(x_r)\, d\tilde{\mathbf{w}}'_r + \int_0^t b^\zeta_r(x_r)\, dr.$$

By Theorem 2.9.4 (also, see Remark 2.9.9)

$$\tilde{\mathsf{M}}^{\beta[p]}_{s_0,x_0}\{F(\tilde{\mathbf{w}}'',x_{[0, T-s_0]}) | \tilde{\mathscr{F}}'_0\} = \tilde{\Phi}(\tilde{\mathbf{w}}''),$$

where

$$\tilde{\Phi}(z) = \tilde{\mathsf{M}}F(z,\tilde{x}^{\alpha[p,z]}_{[0, T-s_0]}),$$

$\tilde{x}^{\alpha[p,z]}_t$ is a solution of Eq. (2), in which $\mathbf{w}_r$ is replaced by $\tilde{\mathbf{w}}'_r$. By Corollary 2.9.3, $\tilde{\Phi}(z) = \Phi(z)$. Hence

$$\tilde{\mathsf{M}}^{\beta[p]}_{s_0,x_0}F(\tilde{\mathbf{w}}'',x_{[0, T-s_0]}) = \tilde{\mathsf{M}}\Phi(\tilde{\mathbf{w}}'') = \hat{\mathsf{M}}\Phi(\hat{\mathbf{w}}).$$

We have proved Eq. (6). We can easily derive the measurability of $\tilde{\Phi}$, therefore, of $\tilde{\Phi}$, from Remark 2.9.5, if we write F as $F_+ - F_-$. This proves the lemma. □

2. Corollary. *For $\alpha = \alpha[p,z]$ or $\alpha = \beta[p]$ let $\tau = \tau^\alpha$ be the time of first exit of the process $(s_0 + t, x^\alpha_t)$ from a region $Q_1 \subset (-1,T) \times E_d$.*

Then, the function $v^{\alpha[p,z],\tau}(s_0,x_0)$ is measurable, bounded with respect to z on $C([0, T - s_0], E_d)$, and, finally,

$$\hat{\mathsf{M}}v^{\alpha[p,\hat{\mathbf{w}}],\tau}(s_0,x_0) = v^{\beta[p],\tau}(s_0,x_0).$$

Furthermore, the function $v^{\alpha[p,z]}(s_0,x_0)$ is measurable, bounded with respect to z on $C([0, T - s_0], E_d)$, and also

$$\hat{\mathsf{M}}v^{\alpha[p,\hat{\mathbf{w}}]}(s_0,x_0) = v^{\beta[p]}(s_0,x_0). \tag{7}$$

Indeed, the second assertion is a particular case of the first assertion ($Q_1 = (-1,T) \times E_d$, $g(t,x) \equiv g(x)$). In order to prove the first assertion, we introduce the function $\tau(z)$ for $z \in C([0, T - s_0], E_d)$ to be the time of first exit of the curve $(s_0 + t, z_t)$ from Q_1. Since $Q_1 \subset (-1,T) \times E_d$, $\tau(z) \le T - s_0$. It is easy to prove that $\underline{\lim}_{z_n \to z} \tau(z_n) \ge \tau(z)$. Therefore, the function $\tau(z)$ is lower semicontinuous and, in addition, is measurable with respect to z.

Further, we consider the function

$$F(z,x_{[0, T-s_0]}) \equiv F_1(\tau(x_{[0, T-s_0]}),z,x_{[0, T-s_0]}),$$

where

$$F_1(\tau,z,x_{[0,T-s_0]}) \equiv g(s_0+\tau,x_\tau)\exp\left[-\int_0^\tau c^{\alpha[p](s_0+r,x_r+\varepsilon z_r)}(s_0+r,x_r)\,dr\right]$$
$$+\int_0^\tau f^{\alpha[p](s_0+t,x_t+\varepsilon z_t)}(s_0+t,x_t)$$
$$\times\exp\left[-\int_0^t c^{\alpha[p](s_0+r,x_r+\varepsilon z_r)}(s_0+r,x_r)\,dr\right]dt.$$

The last function is obviously continuous on $[0,T-s_0]\times C^2([0,T-s_0],E_d)$. Hence $F(z,x_{[0,T-s_0]})$ is measurable, being the composition of measurable functions. The application of the lemma to the function $F(z,x_{[0,T-s_0]})$ leads us immediately to the first assertion of the corollary.

3. Remark. We have discussed above the technique for controlling a process by means of an initial randomized Markov strategy. This technique provides an average payoff equal to the left-hand side of (7). From Eq. (7), we get another possibility to obtain the same payoff. Suppose that we have realized the d-dimensional Wiener process $\hat{\mathbf{w}}_t$ so that it is observable and independent of $\mathbf{w}_t$. The pair $(\mathbf{w}_t,\hat{\mathbf{w}}_t)$ forms a (d_1+d)-dimensional Wiener process. Furthermore, the pair $(x_t^{\alpha,s_0,x_0},z_t)$, where $z_t\equiv\hat{\mathbf{w}}_t$, satisfies the equation

$$x_t=x_0+\int_0^t\sigma(\alpha_r,s_0+r,x_r)\,d\mathbf{w}_r+\int_0^t b(\alpha_r,s_0+r,x_r)\,dr,$$
$$z_t=0+\int_0^t 1\,d\hat{\mathbf{w}}_t.$$

We have thus obtained a $2d$-dimensional controlled process, for which the function $\tilde{\alpha}\equiv\alpha[p](s_0+t,x_t+\varepsilon z_t)$ is a Markov strategy (we mentioned the observability property of $\hat{\mathbf{w}}_t$ because trajectories of a controlled process are taken as observable). If we assume the process $(\mathbf{w}_t,\hat{\mathbf{w}}_t)$ to be $\tilde{\mathbf{w}}_t$, we can easily see that

$$\mathsf{M}_{s_0,x_0}^{\tilde{\alpha}}\left[\int_0^{T-s_0} f^{\tilde{\alpha}_t}(s_0+t,x_t)e^{-\varphi_t}\,dt+g(x_{T-s_0})e^{-\varphi_{T-s_0}}\right]=v^{\beta[p]}(s_0,x_0).$$

Therefore, we can obtain the left-hand side of (7) by mixing Markov strategies as well as by applying a strategy which is Markov with respect to a complete controlled process. In the next section we call strategies of this kind adjoint Markov strategies.

4. Lemma. *We take the functions v_ε from Section 4.6. Further, for $\varepsilon\neq 0$ let $p=(C_{T,R},v_\varepsilon,\varepsilon^1)$. Using Lemma 1.1, we construct the function $\alpha[p](t,x)$ on the basis of the set p. Using Eq. (1), we introduce Markov strategies $\alpha[p,z]$ for $z\in C([0,T-s_0],E_d)$. On a probability space $(\tilde{\Omega},\tilde{\mathscr{F}},\tilde{\mathsf{P}})$ we define the strategy $\beta[p]$ using the formula $\beta_t[p]=\alpha[p](s_0+t,x_t+\varepsilon\tilde{\mathbf{w}}_t'')$, where x_t is a solution* *of Eq.* (4). *Finally, we assume that*

$$\lim_{\varepsilon\to 0}\lim_{R\to\infty}\lim_{\varepsilon^1\downarrow 0}\tilde{\mathsf{M}}^{\beta[p]}_{s_0,x_0}\Bigg\{g(x_{T-s_0}+\varepsilon\tilde{\mathbf{w}}''_{T-s_0})$$
$$\times\exp\left[-\int_0^{T-s_0}c^{\beta_r[p]}(s_0+r,x_r+\varepsilon\tilde{\mathbf{w}}''_r)\,dr\right]$$
$$+\int_0^{T-s_0}f^{\beta_t[p]}(s_0+t,x_t+\varepsilon\tilde{\mathbf{w}}''_t)$$
$$\times\exp\left[-\int_0^t c^{\beta_r[p]}(s_0+r,x_r+\varepsilon\tilde{\mathbf{w}}''_r)\,dr\right]dt\Bigg\}\geq v(s_0,x_0). \tag{8}$$

Then $v_{(M)}(s_0,x_0)=v(s_0,x_0)$. *Moreover,*

$$\lim_{\varepsilon\to 0}\lim_{R\to\infty}\lim_{\varepsilon^1\downarrow 0}\hat{\mathsf{M}}v^{\alpha[p,\hat{\mathbf{w}}]}(s_0,x_0)=v(s_0,x_0), \tag{9}$$

and also for each $\delta>0$

$$\lim_{\varepsilon\to 0}\lim_{R\to\infty}\lim_{\varepsilon^1\downarrow 0}\hat{\mathsf{P}}\{v^{\alpha[p,\hat{\mathbf{w}}]}(s_0,x_0)<v(s_0,x_0)-\delta\}=0. \tag{10}$$

PROOF. It follows from (10) that $v_{(M)}(s_0,x_0)=v(s_0,x_0)$. In turn, (10) follows from (9) since, according to Chebyshev's inequality, the probability given in (10) does not exceed

$$\frac{1}{\delta}\hat{\mathsf{M}}[v(s_0,x_0)-v^{\alpha(p,\hat{\mathbf{w}})}(s_0,x_0)].$$

Therefore, we need to prove only (9). First we note that for $\varepsilon\neq 0$ the nondegeneracy of the processes $x_t^{\tilde{\alpha},s,x}(\varepsilon)$ (see Section 4.6, Inequality (4.6.2)) ensures the existence of generalized derivatives $v_{\varepsilon x^i}$, $v_{\varepsilon x^i x^j}$, $(\partial/\partial t)v_\varepsilon$ as well as the boundedness of these derivatives in each cylinder $C_{T,R}$. This implies, in particular, that the function $\alpha[p](t,x)$ is definite.

We take an arbitrary strategy $\beta=\beta_t$ which is progressively measurable with respect to $\{\tilde{\mathscr{F}}_t\}$. Further, we consider the expression

$$u(\varepsilon)\equiv\tilde{\mathsf{M}}\Bigg\{\int_0^{T-s_0}f^{\beta_t}(s_0+t,x_t+\varepsilon\tilde{\mathbf{w}}''_t)$$
$$\times\exp\left[-\int_0^t c^{\beta_r}(s_0+r,x_r+\varepsilon\tilde{\mathbf{w}}''_r)\,dr\right]dt$$
$$+g(x_{T-s_0}+\varepsilon\tilde{\mathbf{w}}''_{T-s_0})\exp\left[-\int_0^{T-s_0}c^{\beta_r}(s_0+r,x_r+\varepsilon\tilde{\mathbf{w}}''_r)\,dr\right]\Bigg\},$$

where x_t is a solution of the following equation (having coefficients not depending on ε):

$$x_t=x_0+\int_0^t\sigma(\beta_r,s_0+r,x_r)\,d\tilde{\mathbf{w}}'_r+\int_0^t b(\beta_r,s_0+r,x_r)\,dr.$$

Differentiating $u(\varepsilon)$ over ε, bringing the notation of the derivative under the sign of the mathematical expectation and the integral, using the fact that the derivatives f, c, and g increase with respect to x not faster than a certain power, and, finally, applying the familiar estimates of moments of solutions of stochastic equations, we conclude that there exists a constant $N(x_0,K,T,m)$ for which $|u'(\varepsilon)| \leq N(x_0,K,T,m)$ for $|\varepsilon| \leq 1$. Hence $|u(0) - u(\varepsilon)| \leq N(x_0,K,T,m)|\varepsilon|$ for $|\varepsilon| \leq 1$. It is crucial that the constant N does not depend on the strategy β.

Due to the result thus obtained, we can replace the expression $x_t + \varepsilon\tilde{\mathbf{w}}_t''$ by x_t everywhere in (8). However, by Corollary 2 the left-hand side of (8) thus modified coincides with the left-hand side of (9). Therefore, the left-hand side of (9) is not smaller than $v(s_0,x_0)$. Since $v^\alpha(s_0,x_0) \leq v(s_0,x_0)$ for each strategy $\alpha \in \mathfrak{A}$, the left-hand side of (9) is, on the other hand, not greater than $v(s_0,x_0)$. We have thus proved the lemma. □

Now we can prove the main result of this section.

5. Theorem. *Suppose that at least one of the following conditions is satisfied*:

a. $\sigma(\alpha,t,x)$ *and* $b(\alpha,t,x)$ *do not depend on* x;
b. $a(\alpha,t,x)$ *does not depend on* x;
c. *for all* $t \in [0,T]$, $x \in E_d$, $\lambda \neq 0$

$$\sup_{\alpha \in A} (a(\alpha,t,x)\lambda,\lambda) > 0.$$

Then, using the notations of the preceding lemma, we have that inequality (8) *as well as the assertions of this lemma hold true.*

PROOF. Presumably, if condition (a) is satisfied, condition (b) will be satisfied. We included condition (a) in the statement of the theorem for completeness. If (a) is satisfied, the proof of the theorem becomes very simple.

In fact, if (a) is satisfied, then, (see Eq. (4)) the process $x_t^{\beta[p]} + \varepsilon\tilde{\mathbf{w}}_t''$ is a solution of the equation

$$x_t = x_0 + \int_0^t \sigma(\alpha[p](s_0 + r, x_r), s_0 + r, x_r)\, d\tilde{\mathbf{w}}_r' + \varepsilon\tilde{\mathbf{w}}_t''$$
$$+ \int_0^t b(\alpha[p](s_0 + r, x_r), s_0 + r, x_r)\, dr.$$

If we introduce the matrix δ_ε in the same way as we did in Section 4.6, we can easily turn the last equation into Eq. (4.6.1). Therefore, $x_t^{\beta[p]} + \varepsilon\tilde{\mathbf{w}}_t'' = x_t^{\alpha[p],s_0,x_0}(\varepsilon)$ for all t almost surely. Further, from formulas (4.6.3) and the definition of $\alpha[p] = \alpha[C_{T,R},v_\varepsilon,\varepsilon^1]$, we have

$$F_\varepsilon[u] - [L_\varepsilon^\alpha u + f^\alpha] = F[u] - [L^\alpha u + f^\alpha],$$
$$\|F_\varepsilon[v_\varepsilon] - [L_\varepsilon^{\alpha[p]} v_\varepsilon + f^{\alpha[p]}]\|_{d+1,C_{T,R}} \leq \varepsilon^1. \tag{11}$$

From this, fixing $\varepsilon \neq 0$ and applying Theorem 1.2 to the controlled process $x_t^{\tilde{\alpha},s,x}(\varepsilon)$, we have that the expression in (8) under the sign of the lower limit with respect to ε is equal to $v_\varepsilon(s_0,x_0)$. By Theorem 4.6.1, $v_\varepsilon \to v$; therefore, we have proved inequality (8) and, by this token, the assertions of Lemma 4 for (a).

If (a) is not satisfied, the equality $x_t^{\beta[p]} + \varepsilon\tilde{\mathbf{w}}_t'' = x_t^{\alpha[p]s_0,x_0}(\varepsilon)$ does not hold, in general. Thus, we cannot apply Theorem 1.2 for proving (8). In cases (b) and (c), to be considered at the same time, we can prove Eq. (8) almost in the same way as Theorem 1.2.

We assume everywhere below that $\varepsilon \neq 0$, $|\varepsilon| \leq 1$. It can easily be seen that the process $y_t^p \equiv x_t^{\beta[p]} + \varepsilon\tilde{\mathbf{w}}_t''$ satisfies the equation

$$y_t^p = x_0 + \int_0^t \sigma_\varepsilon(\alpha[p](s_0 + r, y_r^p), s_0 + r, y_r^p - \varepsilon\tilde{\mathbf{w}}_r'')\, d\tilde{\mathbf{w}}_r$$

$$+ \int_0^t b(\alpha[p](s_0 + r, y_r^p), s_0 + r, y_r^p - \varepsilon\tilde{\mathbf{w}}_r'')\, dr.$$

By Theorem 4.7.7, $F_\varepsilon[v_\varepsilon] = 0$ (H_T-a.e.). From this as well as from (11) it follows that

$$\|h^p\|_{d+1,C_{T,R}} \leq \varepsilon^1,$$

where $h^p \equiv -L_\varepsilon^{\alpha[p]}v_\varepsilon - f^{\alpha[p]}$. Since the matrix $\sigma_\varepsilon\sigma_\varepsilon^*$ is uniformly nonsingular, we can apply Ito's formula to the expression

$$v_\varepsilon(s_0 + t, y_t^p)\exp\left[-\int_0^t c^{\beta_r[p]}(s_0 + r, y_r^p)\, dr\right].$$

Using Ito's formula, for each $R^1 \geq 0$ we obtain

$$v_\varepsilon(s_0,x_0) = I_1^p(R^1) + I_2^p(R^1) + I_3^p(R^1) + I_4^p(R^1), \tag{12}$$

where, letting

$$\varphi_t^p = \int_0^t c^{\beta_r[p]}(s_0 + r, y_r^p)\, dr,$$

$\tau_R^{p^1}$ denotes the first exit time of the process $(s_0 + t, y_t^p)$ from $[0,T) \times S_{R^1}$, we write the variables $I_i^p(R^1)$ as follows:

$$I_1^p(R^1) = \tilde{\mathsf{M}}\left\{\int_0^{\tau_{R^1}^p} f^{\beta_t[p]}(s_0 + t, y_t^p)e^{-\varphi_t^p}\, dt + v_\varepsilon(s_0 + \tau_{R^1}^p, y_{\tau_{R^1}^p}^p)\exp(-\varphi_{\tau_{R^1}^p}^p)\right\},$$

$$I_2^p(R^1) = \tilde{\mathsf{M}}\int_0^{\tau_{R^1}^p} h^p(s_0 + t, y_t^p)e^{-\varphi_t^p}\, dt,$$

$$I_3^p(R^1) = \tilde{\mathsf{M}}\int_0^{\tau_{R^1}^p} [b(\alpha[p](s_0 + t, y_t^p), s_0 + t, y_t^p)$$
$$- b(\alpha[p](s_0 + t, y_t^p), s_0 + t, y_t^p - \varepsilon\tilde{\mathbf{w}}_t'')]$$
$$\times \operatorname{grad}_x v_\varepsilon(s_0 + t, y_t^p)e^{-\varphi_t^p}\, dt,$$

$$I_4^p(R^1) = \tilde{\mathsf{M}}\int_0^{\tau_{R^1}^p} \sum_{i,j=1}^d [a^{ij}(\alpha[p](s_0 + t, y_t^p), s_0 + t, y_t^p)$$
$$- a^{ij}(\alpha[p](s_0 + t, y_t^p), s_0 + t, y_t^p - \varepsilon\tilde{\mathbf{w}}_t'')]$$
$$\times v_{\varepsilon x^i x^j}(s_0 + t, y_t^p)e^{-\varphi_t^p}\, dt.$$

As in proving Theorem 1.2, we show that $\lim_{\varepsilon^1 \downarrow 0} I_\varepsilon^p(R^1) = 0$ if $R > R^1$;

$$\lim_{R^1 \to \infty} \sup_{|\varepsilon| \le 1} \sup_{R > 0} \sup_{\varepsilon^1 > 0} \left| I_1^p(R^1) - \tilde{\mathsf{M}} \left\{ g(y_{T-s_0}^p) e^{-\varphi_{T-s_0}^p} + \int_0^{T-s_0} f^{\beta_t[p]}(s_0 + t, y_t^p) e^{-\varphi_t^p} \, dt \right\} \right| = 0. \tag{13}$$

Next, we turn to the variables $I_2^p(R^1)$. By Theorem 4.1.1, it is easy to obtain for $|\varepsilon| \le 1$ that $|\operatorname{grad}_x v_\varepsilon(t,x)| \le N(K,T,m)(1 + |x|)^{2m}$ (H_T-a.e.). Suppose that the last inequality is satisfied on a set Γ_ε such that means $(H_T \backslash \Gamma_\varepsilon) = 0$. We put the sum

$$\chi_{\Gamma_\varepsilon}(s_0 + t, y_t^p) + \chi_{H_T \backslash \Gamma_\varepsilon}(s_0 + t, y_t^p)$$

before dt in the formula for $I_3^p(R^1)$. Furthermore, we split $I_3^p(R^1)$ into two terms in an appropriate manner. Applying Theorem 2.2.4 to the second term, we can see that it equals to zero. The first term as well as, therefore, $I_3^p(R^1)$ does not exceed

$$N(K,T,m)\tilde{\mathsf{M}} \int_0^{T-s_0} |\varepsilon \tilde{\mathbf{w}}_t''|(1 + |y_t^p|)^{2m} \, dt.$$

Therefore,

$$\lim_{\varepsilon \to 0} \sup_{R > 0} \sup_{\varepsilon^1 > 0} |I_3^p(R^1)| = 0.$$

If (b) is satisfied, then $I_4^p(R^1) = 0$. If (c) is satisfied, according to Remark 4.7.6 the derivatives $v_{\varepsilon x^i x^j}$ are bounded in C_{T,R^1} (a.e.) by a constant not depending on ε. In addition, for each $l \in E_d$

$$\|a_{(l)}(\alpha,t,x)\| = \tfrac{1}{2}\|\sigma_{(l)}(\alpha,t,x)\sigma^*(\alpha,t,x) + \sigma(\alpha,t,x)\sigma_{(l)}^*(\alpha,t,x)\| \le N(d,d_1)K^2(1 + |x|).$$

Hence

$$\|a(\alpha,t,x) - a(\alpha,t,y)\| \le N(d,d_1)K^2(1 + |x| + |y|)|x - y|,$$

$$|I_4^p(R^1)| \le N\tilde{\mathsf{M}} \int_0^{T-s_0} |\varepsilon \tilde{\mathbf{w}}_t''|(1 + |y^p| + |\tilde{\mathbf{w}}_t''|) \, dt,$$

where N does not depend on ε, R, ε^1 (although N depends, for example, on R^1). This indicates that

$$\lim_{\varepsilon \to 0} \sup_{R > 0} \sup_{\varepsilon^1 > 0} |I_4^p(R^1)| = 0$$

in both cases (b) and (c).

Finally, from (12) and the properties investigated of $I_i^p(R^1)$ we conclude that for each $R^1 \ge 0$

$$\varlimsup_{\varepsilon \to 0} v_\varepsilon(s_0,x_0) \le \varliminf_{\varepsilon \to 0} \lim_{R \to \infty} \varliminf_{\varepsilon^1 \downarrow 0} \tilde{\mathsf{M}} \left\{ g(y_{T-s_0}^p) e^{-\varphi_{T-s_0}^p} + \int_0^{T-s_0} f^{\beta_t[p]}(s_0 + t, y_t^p) e^{-\varphi_t^p} \, dt \right\} + \gamma(R^1),$$

where $\gamma(R^1)$ is the expression in (13) under the sign of the limit. Letting $R^1 \to \infty$ and noting that, by Theorem 4.6.1, the left side of the last inequality is $v(s_0,x_0)$, we arrive at inequality (8) (the notations have slightly been changed). The theorem is proved. □

We suggest the reader should prove a similar theorem for the optimal stopping problem as an exercise.

6. Exercise

Let one of conditions (a), (b), or (c) of Theorem 5 be satisfied. Also, let $p = (C_{T,R},w_\varepsilon,\varepsilon^1)$, $\tau_\delta = \tau_\delta^{\alpha[p,z],s_0,x_0}$ be the time of first exit of the process $(s_0 + t, x_t^{\alpha[p,z]})$ from $Q_\delta = \{(t,x) \in \bar{H}_T : w(t,x) > g(t,x) + \delta\}$.

As an analog of Lemma 4, prove that

$$\lim_{\delta\downarrow 0}\lim_{\varepsilon\to 0}\lim_{R\to\infty}\lim_{\varepsilon^1\downarrow 0} \hat{\mathsf{M}} v^{\alpha[p,\hat{w}],\tau_\delta}(s_0,x_0) = w(s_0,x_0) \tag{14}$$

and that for each $\delta^1 > 0$

$$\overline{\lim_{\delta\downarrow 0}}\,\overline{\lim_{\varepsilon\to 0}}\lim_{R\to\infty}\overline{\lim_{\varepsilon^1\downarrow 0}}\, \hat{\mathsf{P}}\{v^{\alpha[p,\hat{w}],\tau_\delta}(s_0,x_0) < w(s_0,x_0) - \delta^1\} = 0.$$

Draw a conclusion that $w_M = w$ in $\bar{H}_T$.

7. Exercise

Consider a 1-dimensional process: $d = d_1 = 1$, $T = 1$, $A = [-1,1]$, $\sigma(\alpha,s,x) = \sigma(x + \alpha)$, where

$$\sigma(x) = \begin{cases} 1 & \text{for } x \geq 1, \\ x & \text{for } x \in [-1,1], \\ -1 & \text{for } x \leq -1, \end{cases}$$

$$b(\alpha,s,x) = c^\alpha(s,x) = f^\alpha(s,x) = 0, \qquad g(x) = x^2.$$

Show that $v(s,x) = x^2 + 1 - s$. Let $\alpha_n(x) = n\zeta(nx) * \operatorname{sgn} x$.

Prove that for a point (0,0) ε-optimal strategies can be found among Markov strategies of the form $\alpha_t(x_{[0,t]}) = \alpha_n(x_t + z_t)$ for an appropriate choice of n and a continuous function z_t. Is this assertion true if one takes $\alpha_n(x_t)$ instead of $\alpha_n(x_t + z_t)$?

3. The Payoff Function and Solution of the Bellman Equation: The Uniqueness of the Solution of the Bellman Equation

The problem of finding a payoff function is one of the main problems in optimal control theory. As we have seen in Sections 1 and 2, the knowledge of a payoff function enables us, for example, to construct ε-optimal strategies. According to the results obtained in Section 4.7 it is natural to seek for a payoff function treated as a solution of the Bellman equation. Assume that we have found a solution of the Bellman equation. A question immediately arises whether the solution found is equal to the payoff function. If it is known a priori that the payoff function satisfies the Bellman equation, the question posed above is equivalent to a question about uniqueness of a solution of the Bellman equation. In the general case, the positive answer to the latter question contains an assertion of uniqueness of a solution of the Bellman equation.

In this section we show that a "smooth" solution of the Bellman equation which does not increase too rapidly as $|x| \to \infty$ is equal to a payoff function.

We assume that the assumptions made in Section 3.1 are satisfied. Note that, as Exercise 4.3.1 shows, the assumptions mentioned do not ensure either the existence of derivatives of a payoff function or that the payoff function satisfies the Bellman equation.

Furthermore, we assume that on an initial probability space we are given a d-dimensional Wiener process (with respect to σ-algebras of $\{\mathscr{F}_t\}$) $\hat{\mathbf{w}}_t$ not depending on $\mathbf{w}_t$. It is always easy to have this assumption satisfied if we consider a direct product of an initial space by a space on which a d-dimensional Wiener process is defined. It is crucial to emphasize the fact that a payoff function does not depend on the expansion of a probability space (see Remarks 3.3.10 and 3.4.10).

The fact that the solution of the Bellman equation is a payoff function will be proved in two stages. First, we prove that the solution is not smaller than the payoff function. Second, we shall prove the converse. At the same time, we shall elucidate the general question as to how a function has to be related to the Bellman equation in order that we may assert that it is greater (smaller) than the payoff function.

Our argument includes a function $u(t,x)$ given on $\bar{H}_T$, about which we sometimes assume that there exist constants N and $p \geq 0$ such that

$$|u(t,x)| \leq N(1 + |x|)^p. \tag{1}$$

1. Definition. Let Q be a subregion of H_T. We write $u \in W^{1,2}_{\text{loc}}(Q)$ if $u \in W^{1,2}(Q')$ for any bounded region Q' which together with its closure $\bar{Q}'$ lies in Q.

The definitions which follows include the operator F defined in the introduction to Chapter 4.

2. Definition. A function u is said to be *excessive* (with respect to the operator F) *in a certain region* $Q \subset H_T$ if $u \geq 0$ in Q, $u \in W^{1,2}_{\text{loc}}(Q) \cap C(\bar{Q})$, $F[u] \leq 0$ (a.e. on Q). A function excessive in H_T is said to be an *excessive* function.

3. Definition. A function u is said to be *superharmonic* (with respect to the operator F) *in a certain region* $Q \subset H_T$ if u satisfies inequality (1) in a region Q, $u \in W^{1,2}_{\text{loc}}(Q) \cap C(\bar{Q})$, $F[u] \leq 0$ (a.e. on Q). A function superharmonic in H_T is said to be a *superharmonic* function.

The property of excessive and superharmonic functions, which is of our main concern, is to be found in the following lemma.

4. Lemma. *Let u be a superharmonic (or excessive) function in a region $Q \subset H_T$. Then, for any $(s,x) \in \bar{Q}$, $\alpha \in \mathfrak{A}$, $\tau \in \mathfrak{M}(T - s)$*

$$u(s,x) \geq \mathsf{M}_{s,x}^{\alpha}\left[\int_0^{\tau \wedge \tau_Q} f^{\alpha_t}(s + t, x_t)e^{-\varphi_t}\,dt + u(s + \tau \wedge \tau_Q, x_{\tau \wedge \tau_Q})e^{-\varphi_{\tau \wedge \tau_Q}}\right], \tag{2}$$

where $\tau_Q = \tau_Q^{\alpha,s,x}$ is the time of first exit of the process $(s + t, x_t^{\alpha,s,x})$ from Q.

PROOF. If (s,x) lies on the boundary of Q, then $\tau_Q = 0$, and therefore the assertion is obvious. Let $(s,x) \in Q$. Also, let

$$\Gamma = \{(t,x) \in Q : F[u](t,x) \leq 0\}.$$

Then meas $(Q \backslash \Gamma) = 0$ and, in addition, for all $(t,x) \in \Gamma$ for any ω

$$\chi_\Gamma(t,x)[L^{\alpha_t}u(t,x) + f^{\alpha_t}(t,x)] \leq \chi_\Gamma(t,x)F[u](t,x) \leq 0.$$

Further, we make use of Theorem 2.10.2, taking into account that by that theorem the region Q is assumed to be bounded. We have that inequality (2) is satisfied if we replace τ_Q by the time of first exit of the process $(s + t, x_t^{\alpha,s,x})$ from $Q \cap C_{T,R}$. We write this time as $\tau_{R,Q}$. Replacing τ_Q by $\tau_{R,Q}$, we write inequality (2). Let $R \to \infty$. If u is a superharmonic function, then $|u| \leq N(1 + |x|)^p$ and, furthermore, the expressions under the sign of mathematical expectation in inequality (2) thus modified can be estimated in terms of a summable quantity

$$\sup_{t \leq T-s} (1 + |x_t^{\alpha,s,x}|)^{m+p}.$$

Since, as can be seen, $\tau_{R,Q} \to \tau_Q$ and $x_{\tau \wedge \tau_{R,Q}} \to x_{\tau \wedge \tau_Q}$, we can complete the proof of our lemma applying the Lebesgue theorem. If u is an excessive function, $u \geq 0$, and, in addition, one should apply Fatou's lemma instead of the Lebesgue theorem. The lemma is proved. □

5. Theorem. *Let a function u be given and let it be continuous in $\bar{H}_T$. Let region $Q \subset H_T$. Finally, let u be a superharmonic (excessive) function in the region Q. Then*

a. *if $u(s,x) \geq g(s,x)$ in Q, $u \geq w$ in $H_T \backslash Q$, then $u \geq w$ in $\bar{H}_T$;*
b. *if $u \geq v$ in $H_T \backslash Q$, $u(T,x) \geq g(x)$ in E_d, then $u \geq v$ in $\bar{H}_T$.*

PROOF. (a) Due to the continuity of the functions u and w it suffices to prove the inequality $u(s,x) \geq w(s,x)$ for $(s,x) \in H_T$. Since this inequality is satisfied outside Q by assumption, we can admit that $(s,x) \in Q$. In addition, let us take $\alpha \in \mathfrak{A}$, $\tau \in \mathfrak{M}(T - s)$. Applying Lemma 4, we consider the relation

$$\begin{aligned} &u(s + \tau \wedge \tau_Q, x_{\tau \wedge \tau_Q})e^{-\varphi_{\tau \wedge \tau_Q}} \\ &\quad = u(s + \tau, x_\tau)e^{-\varphi_\tau}\chi_{\tau \leq \tau_Q} + u(s + \tau_Q, x_{\tau_Q})e^{-\varphi_{\tau_Q}}\chi_{\tau_Q < \tau}, \end{aligned} \tag{3}$$

where the indices α, s, x are omitted for the sake of brevity. For $\tau \le \tau_Q$ we have here: $u(s+\tau, x_\tau) \ge g(s+\tau, x_\tau)$ since $(s+\tau, x_\tau) \in \bar{Q}$ and the inequality $u \ge g$ which is satisfied on Q by assumption will be satisfied on $\bar{Q}$ as well due to the continuity of u and g.

Next, if $\tau_Q < T - s$, then

$$(s+\tau_Q, x_{\tau_Q}) \in H_T \backslash Q,$$
$$u(s+\tau_Q, x_{\tau_Q}) \ge w(s+\tau_Q, x_{\tau_Q}).$$

If $\tau_Q = T - s$, then due to the continuity of u, g and, in addition, by virtue of the inequalities $u \ge g$ in Q, $w \ge g$ in H_T and $u \ge w$ in $H_T \backslash Q$ we have: $u \ge g$ in H_T, $u \ge g$ in $\bar{H}_T$, $u(T,x) \ge g(T,x)$,

$$u(s+\tau_Q, x_{\tau_Q}) \ge g(T, x_{\tau_Q}) = w(T, x_{\tau_Q}) = w(s+\tau_Q, x_{\tau_Q}).$$

From Lemma 4 and Eq. (3) analyzed above, we conclude that

$$u(s,x) \ge \mathsf{M}_{s,x}^{\alpha}\Bigg[\int_0^{\tau \wedge \tau_Q} f^{\alpha_t}(s+t, x_t)e^{-\varphi_t}\,dt$$
$$+ g(s+\tau, x_\tau)e^{-\varphi_\tau}\chi_{\tau \le \tau_Q} + w(s+\tau_Q, x_{\tau_Q})e^{-\varphi_{\tau_Q}}\chi_{\tau_Q < \tau}\Bigg].$$

Computing in the last inequality the upper bound with respect to $\tau \in \mathfrak{M}(T-s)$ and $\alpha \in \mathfrak{A}$, we obtain in accord with Theorem 3.1.9 that $u(s,x) \ge w(s,x)$. We have proved assertion (a). Applying Theorem 3.1.6 for $\tau = \tau_Q$, $r_t = 0$ we can prove assertion (b) in a similar way. This completes the proof of the theorem. □

6. Corollary. *In the case where w and v are superharmonic functions (see Section 4.7), the function w is the smallest superharmonic majorant of $g(s,x)$ and the function v is the smallest superharmonic function which majorizes $g(x)$ for $s = T$.*

7. Corollary. *Let $u \in W_{\text{loc}}^{1,2}(H_T) \cap C(\bar{H}_T)$. Also, let (1) be satisfied. Then, if*

$$(F[u](s,x) + u(s,x) - g(s,x))_+ + g(s,x) - u(s,x) \le 0, \qquad (H_T\text{-a.e.}),$$

$u(s,x) \ge w(s,x)$ in $\bar{H}_T$.

In fact, since the positive part of a number is greater than zero, by hypotheses, $g \le u$ (H_T-a.e.). Further, since g and u are continuous functions, $g \le u$ everywhere in H_T. Next, if the inequality $F[u](s,x) > 0$ were satisfied at a point $(s,x) \in H_T$, it is seen that $F[u] + u - g > 0$ at this point, hence $(F[u] + u - g)_+ + g - u = F[u] > 0$. Therefore, $F[u] \le 0$ (H_T-a.e.), and also u is a superharmonic majorant of $g(s,x)$.

Theorem 5 and Corollary 7 enable us to find upper estimates for a payoff function. In order to prove the theorem on lower estimates, we need three auxiliary results.

8. Lemma. *Let* $(s_0,x_0) \in H_T$, *let* $\alpha(s,x)$ *be a Borel function on* H_T *with values in* A, *and let* $\delta > 0$. *Further, let*

$$\sigma_n(s,x,z) = n^d\zeta(nz) * \sigma(\alpha(s,z),s,x),$$

$$\sigma_n(s,z) = \frac{n}{n+|z|^2}\,\sigma_n(s,z,z),$$

$$b_n(s,x,z) = n^d\zeta(nz) * b(\alpha(s,z),s,x),$$

$$b_n(s,z) = \frac{n}{n+|z|^2}\,b_n(s,z,z).$$

Let us find a strategy $\alpha^{n,\delta}$ *using the formula*

$$\alpha_t^{n,\delta}(\omega) = \alpha(s_0 + t, z_t^{n,\delta}(\omega)),$$

where $z_t^{n,\delta}(\omega)$ *is a solution of the equation*

$$z_t = x_0 + \int_0^t \sigma_n(s_0 + r, z_r)\,d\mathbf{w}_r + \delta\hat{\mathbf{w}}_t + \int_0^t b_n(s_0 + r, z_r)\,dr. \tag{4}$$

Then for all $q \geq 1$

$$\sup_{\delta\in[0,1]}\sup_{n\geq 1} \mathsf{M} \sup_{t\leq T-s_0} |z_t^{n,\delta}|^{2q} < \infty, \tag{5}$$

$$\overline{\lim_{\delta\downarrow 0}}\,\overline{\lim_{n\to\infty}}\, \mathsf{M} \sup_{t\leq T-s_0} |z_t^{n,\delta} - x_t^{\alpha^{n,\delta},s_0,x_0}|^{2q} = 0. \tag{6}$$

PROOF. As can easily be seen, the functions

$$\frac{n}{n+|x|^2}\,\sigma_n(s,x,z), \qquad \frac{n}{n+|x|^2}\,b_n(s,x,z)$$

are differentiable over z and, furthermore, the derivatives of these functions do not exceed Nn, with N not depending on n, s, x, z. Moreover, they satisfy the Lipschitz condition with respect to x since, for example, using the simple inequality

$$\|\sigma_n(s,x,z)\| \leq \sup_\alpha \|\sigma(\alpha,s,x)\| \leq K(1 + |x|),$$

we find

$$\left\|\frac{n}{n+|x_1|^2}\,\sigma_n(s,x_1,z) - \frac{n}{n+|x_2|^2}\,\sigma_n(s,x_2,z)\right\|$$

$$\leq \left|\frac{n}{n+|x_1|^2} - \frac{n}{n+|x_2|^2}\right| K(1 + |x_1|) + \|\sigma_n(s,x_1,z) - \sigma_n(s,x_2,z)\|$$

$$\leq 2K|x_1 - x_2| + K|x_1 - x_2| = 3K|x_1 - x_2|.$$

This implies that the coefficients of Eq. (4) satisfy the Lipschitz condition and Eq. (4) is solvable. Due to the familiar estimates of moments of solutions of stochastic equations (5) follows from the inequality

$$\|\sigma_n(s,z)\| \le \sup_y \|\sigma_n(s,z,y)\| \le \sup_\alpha \|\sigma(\alpha,s,z)\| \le K(1+|z|)$$

and a similar inequality for $|b_n(t,z)|$.

Further, the process $x_t^{\alpha^{n,\delta},s_0,x_0}$ is a solution of the following equation:

$$x_t = x_0 + \int_0^t \sigma(\alpha_r^{n,\delta}, s_0 + r, x_r)\, d\mathbf{w}_r + \int_0^t b(\alpha_r^{n,\delta}, s_0 + r, x_r)\, dr.$$

Comparing the last equation with Eq. (4), we have in accord with Theorem 2.5.9 that

$$\begin{aligned}
\mathsf{M} \sup_{t \le T - s_0} \left|z_t^{n,\delta} - x_t^{\alpha^{n,\delta},s_0,x_0}\right|^{2q} \le N\mathsf{M} &\sup_{t \le T - s_0} |\delta \hat{\mathbf{w}}_t|^{2q} \\
&+ N\mathsf{M} \int_0^{T-s_0} \|\sigma_n(s_0 + t, z_t^{n,\delta}) \\
&- \sigma(\alpha_t^{n,\delta}, s_0 + t, z_t^{n,\delta})\|^{2q}\, dt \\
&+ N\mathsf{M} \int_0^{T-s_0} |b_n(s_0 + t, z_t^{n,\delta}) \\
&- b(\alpha_t^{n,\delta}, s_0 + t, z_t^{n,\delta})|^{2q}\, dt,
\end{aligned}$$

where N depends only on q, K, $T - s_0$. It remains to show that the last two terms tend to zero as $n \to \infty$. We fix $\delta > 0$. Also, we consider the latter term only. Let τ_R^n be the time of first exit of the process $z_t^{n,\delta}$ from S_R. We have

$$\begin{aligned}
\mathsf{M} &\int_{\tau_R^n \wedge (T-s_0)}^{T-s_0} |b_n(s_0 + t, z_t^{n,\delta}) - b(\alpha_t^{n,\delta}, s_0 + t, z_t^{n,\delta})|^{2q}\, dt \\
&\le T(2K)^{2q} \mathsf{M} \chi_{\tau_R^n < T - s_0} \left(1 + \sup_{t \le T - s_0} |z_t^{n,\delta}|\right)^{2q} \\
&\le T(2K)^{2q} \frac{1}{1+R} \mathsf{M} \left(1 + \sup_{t \le T - s_0} |z_t^{n,\delta}|\right)^{2q+1} \to 0
\end{aligned}$$

as $R \to \infty$ uniformly with respect to n. Therefore, we can complete proving the lemma if we prove that for each $R > 0$

$$I_{n,R} \equiv \mathsf{M} \int_0^{\tau_R^n \wedge (T - s_0)} |b_n(s_0 + t, z_t^{n,\delta}) - b(\alpha(s_0 + t, z_t^{n,\delta}), s_0 + t, z_t^{n,\delta})|^{2q}\, dt \to 0$$

as $n \to \infty$. We note that the process $z_t^{n,\delta}$ is a solution of the equation

$$z_t = x_0 + \int_0^t \tilde{\sigma}_n(s_0 + r, z_r)\, d\tilde{\mathbf{w}}_r + \int_0^t b_n(s_0 + r, z_r)\, dr,$$

where the matrix $\tilde{\sigma}_n$ is obtained as a result of writing to σ_n from the right of the unit matrix of dimension $d \times d$, which is multiplied by δ, $\tilde{\mathbf{w}}_r = (\mathbf{w}_r, \hat{\mathbf{w}}_r)$. It is not hard to see that for each λ, $|\tilde{\sigma}_n^* \lambda|^2 = (\tilde{\sigma}_n \tilde{\sigma}_n^* \lambda, \lambda) = (\sigma_n \sigma_n^* \lambda, \lambda) + \delta \lambda^2 = |\sigma_n^* \lambda|^2 + \delta \lambda^2 \ge \delta \lambda^2$.

Therefore, by Theorem 2.2.4,

$$I_{n,R} \le N \|b_n(t,z) - b(\alpha(t,z),t,z)\|_{2q(d+1), C_{T,R}}^{2q}, \tag{7}$$

where N does not depend on n. Finally, the functions bounded in $C_{T,R}$ uniformly with respect to n

$$
\begin{aligned}
&|b_n(t,z) - b(\alpha(t,z),t,z)| \\
&\quad \le \left| \int_{|y|\le 1} b\left(\alpha\left(t, z - \frac{1}{n} y\right), t, z\right)\zeta(y)\,dy \right. \\
&\qquad \left. - b(\alpha(t,z),t,z) \right| + \left|1 - \frac{n}{n + |z|^2}\right| |b_n(t,z)| \\
&\quad \le \left| \int_{|y|\le 1} b\left(\alpha\left(t, z - \frac{1}{n} y\right), t, z - \frac{1}{n} y\right)\zeta(y)\,dy - b(\alpha(t,z),t,z) \right| + N\frac{1}{n}.
\end{aligned}
$$

Therefore, according to the well-known properties of convolutions (see Section 2.1) the foregoing functions tend to zero as $n \to \infty$ for each t for almost all z. Thus, by the Lebesgue theorem, the right side of (7) tends to zero as $n \to \infty$. The lemma is proved. □

Repeating the arguments given in the proof of Theorem 3.1.12 after Eq. (3.1.14) and, in addition, using the current lemma, we arrive at the following (also see Corollary 3.1.13).

9. Lemma. *Let us carry out the construction as was described in the formulation of Lemma 8. We take arbitrary Markov times $\tau^{n,\delta} \in \mathfrak{M}(T - s_0)$. Then*

$$
\begin{aligned}
&\varlimsup_{\delta\downarrow 0}\varlimsup_{n\to\infty} \left| \mathsf{M}\left\{ g(s_0 + \tau^{n,\delta}, z^{n,\delta}_{\tau^{n,\delta}}) \right.\right. \\
&\qquad \times \exp\left[-\int_0^{\tau^{n,\delta}} c^{\alpha^{n,\delta}_r}(s_0 + r, z^{n,\delta}_r)\,dr \right] + \int_0^{\tau^{n,\delta}} f^{\alpha^{n,\delta}_t}(s_0 + t, z^{n,\delta}_t) \\
&\qquad \left.\left. \times \exp\left[-\int_0^t c^{\alpha^{n,\delta}_r}(s_0 + r, z^{n,\delta}_r)\,dr \right] dt \right\} - v^{\alpha^{n,\delta},\tau^{n,\delta}}(s_0,x_0) \right| = 0,
\end{aligned}
$$

$$
\begin{aligned}
&\varlimsup_{\delta\downarrow 0}\varlimsup_{n\to\infty} \left| \mathsf{M}\left\{ g(z^{n,\delta}_{T-s_0}) \exp\left[-\int_0^{T-s_0} c^{\alpha^{n,\delta}_r}(s_0 + r, z^{n,\delta}_{\tau^{n,\delta}})dr \right] \right.\right. \\
&\qquad + \int_0^{T-s_0} f^{\alpha^{n,\delta}_t}(s_0 + t, z^{n,\delta}_t) \\
&\qquad \left.\left. \times \exp\left[-\int_0^t c^{\alpha^{n,\delta}_r}(s_0 + r, z^{n,\delta}_r)\,dr \right] dt \right\} - v^{\alpha^{n,\delta}}(s_0,x_0) \right| = 0.
\end{aligned}
$$

10. Lemma. *Let a bounded region $Q \subset H_T$. Let a function u satisfy inequality* (1) *in H_T, $u \in W^{1,2}(Q) \cap C(\bar{H}_T)$, let $F[u] \ge 0$* (a.e. *on Q*), *and let $\varepsilon > 0$. We take a Borel function $\alpha(s,x)$ given on H_T which assumes values from A and such that*

$$
L^{\alpha(s,x)}u(s,x) + f^{\alpha(s,x)}(s,x) \ge -\varepsilon \qquad \text{(a.e. on } Q\text{)}. \tag{8}
$$

We fix $(s_0,x_0) \in Q$. *Using Lemma 8 let us define the strategies* $\alpha^{n,\delta}$. *Then*

$$u(s_0,x_0) \le \varliminf_{\delta \downarrow 0} \varliminf_{n \to \infty} \mathsf{M}^{\alpha^{n,\delta}}_{s_0,x_0}\left[u(s_0 + \tau^{n,\delta}, x_{\tau^{n,\delta}})e^{-\varphi_{\tau^{n,\delta}}} + \int_0^{\tau^{n,\delta}} f^{\alpha_t^{n,\delta}}(s_0 + t, x_t)e^{-\varphi_t}\,dt \right] + \varepsilon(T - s_0),$$

where $\tau^{n,\delta}$ *is an arbitrary Markov time not exceeding the time of first exit of the process* $(s_0 + t, z_t^{n,\delta})$ *from the region* Q.

PROOF. First we note that, in the same way as we did in proving Lemmas 1.4.9 and 4.5.5, we can establish here the existence of a function $\alpha(s,x)$ such that for all $(s,x) \in Q$ the left side of (8) is greater than $F[u](s,x) - \varepsilon$. Since $F[u] \ge 0$ (a.e. on Q), the function α satisfies inequality (8).

Applying Ito's formula, we easily find

$$\begin{aligned} u(s_0,x_0) = \mathsf{M}\Big\{ & u(s_0 + \tau^{n,\delta}, z^{n,\delta}_{\tau^{n,\delta}}) \exp\left[-\int_0^{\tau^{n,\delta}} c^{\alpha_r^{n,\delta}}(s_0 + r, z_r^{n,\delta})\,dr \right] \\ & - \int_0^{\tau^{n,\delta}} \exp\left[-\int_0^t c^{\alpha_r^{n,\delta}}(s_0 + r, z_r^{n,\delta})\,dr \right] \\ & \times L^{\alpha_t^{n,\delta}}(s_0 + t, z_t^{n,\delta}) u(s_0 + t, z_t^{n,\delta})\,dt \Big\} + I^{n,\delta}, \end{aligned}$$

where

$$\begin{aligned} I^{n,\delta} = \mathsf{M} \int_0^{\tau^{n,\delta}} & \exp\left[-\int_0^t c^{\alpha_r^{n,\delta}}(s_0 + r, z_r^{n,\delta})\,dr \right] \Big\{ \sum_{i,j=1}^d \Big[a^{ij}(\alpha_t^{n,\delta}, s_0 + t, z_t^{n,\delta}) \\ & - a_n^{ij}(s_0 + t, z_t^{n,\delta}) - \frac{1}{2}\delta^2 \delta^{ij} \Big] u_{x^i x^j}(s_0 + t, z_t^{n,\delta}) \\ & + \sum_{i=1}^d [b^i(\alpha_t^{n,\delta}, s_0 + t, z_t^{n,\delta}) - b_n^i(s_0 + t, z_t^{n,\delta})] u_{x^i}(s_0 + t, z_t^{n,\delta}) \Big\}\,dt, \end{aligned}$$

$$a_n^{ij}(t,z) = \frac{1}{2}\sigma_n(t,z)\sigma_n^*(t,z).$$

By virtue of the inequality $\tau^{n,\delta} \le T - s_0$ we obtain from the above as well as (8) that

$$\begin{aligned} u(s_0,x_0) \le \mathsf{M}\Big\{ & u(s_0 + \tau^{n,\delta}, z^{n,\delta}_{\tau^{n,\delta}}) \\ & \times \exp\left[-\int_0^{\tau^{n,\delta}} c^{\alpha_r^{n,\delta}}(s_0 + r, z_r^{n,\delta})\,dr \right] + \int_0^{\tau^{n,\delta}} f^{\alpha_t^{n,\delta}}(s_0 + t, z_t^{n,\delta}) \\ & \times \exp\left[-\int_0^t c^{\alpha_r^{n,\delta}}(s_0 + r, z_r^{n,\delta})\,dr \right] dt \Big\} + \varepsilon(T - s_0) + I^{n,\delta}. \end{aligned}$$

We take the limit here as $n \to \infty$, $\delta \downarrow 0$. Assuming $g(s,x) \equiv u(s,x)$ in Lemma 9, we conclude that for proving this lemma we need only to show that

$$\overline{\lim_{\delta \downarrow 0}}\ \overline{\lim_{n \to \infty}}\ I^{n,\delta} = 0.$$

Since the process $z_t^{n,\delta}$ is nondegenerate and, in addition,

$$\det\left[a_n(t,z) + \frac{1}{2}\delta^2(\delta^{ij})\right] \geq \frac{1}{2^d}\delta^{2d}, \qquad |b_n(t,z)| \leq N \cdot 1$$

for some constant N and all $(t,z) \in Q$, $n \geq 1$, by Theorem 2.2.2 (compare with the proof of Theorem 2.10.2)

$$\frac{1}{2}\delta^2\left|\mathsf{M}\int_0^{\tau^{n,\delta}} \Delta u(s_0 + t, z_t^{n,\delta}) \exp\left[-\int_0^t c^{\alpha_r^{n,\delta}}(s_0 + r, z_r^{n,\delta})\,dr\right]dt\right|$$

$$\leq \frac{1}{2}\delta^2 e^T \mathsf{M}\int_0^{\tau^{n,\delta}} |\Delta u(s_0 + t, z_t^{n,\delta})| e^{-t}\,dt$$

$$\leq \frac{1}{2}\delta^2 e^T\left(\frac{2^d}{\delta^{2d}}\right)^{1/(d+1)} N\|u\|_{W^{1,2}(Q)} \leq \delta^{2/(d+1)}N,$$

where N does not depend on n, δ. Therefore, the estimated term belonging to $I^{n,\delta}$ tends to zero as $n \to \infty$, $\delta \downarrow 0$. Further, by Theorem 2.2.2,

$$\left|\mathsf{M}\int_0^{\tau^{n,\delta}} \exp\left[-\int_0^t c^{\alpha_r^{n,\delta}}(s_0 + r, z_r^{n,\delta})\,dr\right]\right.$$

$$\times \sum_{i,j=1}^{d} [a^{ij}(\alpha(s_0 + t, z_t^{n,\delta}), s_0 + t, z_t^{n,\delta})$$

$$\left. - a_n^{ij}(s_0 + t, z_t^{n,\delta})]u_{x^i x^j}(s_0 + t, z_t^{n,\delta})\,dt\right|$$

$$\leq \delta^{-2d/(d+1)}N \sum_{i,j=1}^{d} \|[a^{ij}(\alpha(t,z),t,z) - a_n^{ij}(t,z)]u_{x^i x^j}(t,x)\|_{d+1,Q}, \tag{9}$$

where N does not depend on n, δ. In proving Lemma 8 (see the reasoning carried out after Eq. (7)) we showed that $b_n(t,z) \to b(\alpha(t,z),t,z)$ (H_T-a.e.). In completely similar fashion, $\sigma_n(t,z) \to \sigma(\alpha(t,z),t,z)$; therefore, $a_n(t,z) \to a(\alpha(t,z),t,z)$ (H_T-a.e.). Since the totality of functions $a_n(t,z)$ is bounded on Q and, in addition, the derivatives $u_{x^i x^j} \in \mathscr{L}_{d+1}(Q)$, by the Lebesgue theorem the right side of (9) tends to zero as $n \to \infty$ for each δ. In the same way we can estimate the term containing $b^i - b_n^i$, which belongs to $I^{n,\delta}$. We have proved the lemma. □

The following theorem enables us to find lower estimates for payoff functions.

11. Theorem. *Let* $u \in W^{1,2}_{\text{loc}}(H_T) \cap C(\bar{H}_T)$ *and satisfy inequality* (1) *in* H_T. *Then*:

a. *if* $(F[u] + u - w)_+ + w - u \geq 0$ (H_T-a.e.), $u(T,x) \leq w(T,x)$ *for all* $x \in E_d$, *then* $u \leq w$ *in* $\bar{H}_T$;
b. *if* $(F[u] + u - v)_+ + v - u \geq 0$ (H_T-a.e.), $u(T,x) \leq v(Tx)$ *for all* $x \in E_d$, *then* $u \leq v$ *in* $\bar{H}_T$.

PROOF. Assertion (b) follows from (a). In fact, temporarily let us take $g(s,x) \equiv v(s,x)$. By Theorem 3.1.6, $w \equiv v$ and also the inequality $u \leq w$ from (a) implies that $u \leq v$.

Proof of (*a*). For each $\varepsilon > 0$

$$L^\alpha(u - \varepsilon) + f^\alpha = L^\alpha u + f^\alpha + \varepsilon c^\alpha \geq L^\alpha u + f^\alpha.$$

Therefore, $F[u - \varepsilon] \geq F[u]$. Furthermore, we note that for any real a the function $(a + t)_+ - t$ decreases with respect to t, thus implying that

$$\begin{aligned}(F[u - \varepsilon] + (u - \varepsilon) - w)_+ + w - (u - \varepsilon)\\ \geq (F[u] + u - w)_+ + w - u \geq 0 \qquad (H_T\text{-a.e.}).\end{aligned}$$

It is seen that $u(T,x) - \varepsilon < w(T,x)$. Therefore the function $u - \varepsilon$ satisfies (a), for $s = T$ this function being strictly smaller than w. If assertion (a) has been proved for such functions, then $u - \varepsilon \leq w$, from which, in turn, we conclude as $\varepsilon \downarrow 0$ that $u \leq w$. Therefore, we may assume that $u(T,x) < w(T,x)$ for all $x \in E_d$. We note also that due to the continuity of u and w it suffices to prove the inequality $u \leq w$ in H_T.

Let

$$Q' = \{(s,x) \in H_T : u(s,x) > w(s,x)\}.$$

We wish to prove that the region Q' is an empty set. Assume the converse. We take $(s_0,x_0) \in Q'$; in addition, let $R > |x_0|$ and $Q = Q' \cap C_{T,R} \cap \{(s,x) \in H_T : s > s_0/2\}$.

By virtue of the inequality $u(T,x) < w(T,x)$ we have: $\bar{Q} \subset H_T$, hence $u \in W^{1,2}(Q)$. The expression $w - u$ is negative on the region Q. Therefore, it follows from the inequality $(F[u] + u - w)_+ + w - u \geq 0$ that $(F[u] + u - w)_+ > 0$. Then, almost everywhere on Q

$$0 \leq (F[u] + u - w)_+ + w - u = F[u] + u - w + w - u = F[u].$$

Now we can apply the preceding lemma, thus obtaining for fixed $\varepsilon > 0$

$$\begin{aligned}u(s_0,x_0) \leq \varliminf_{\delta\downarrow 0}\ \varliminf_{n\to\infty} \mathsf{M}^{\alpha^{n,\delta}}_{s_0,x_0}\Bigg[w(s_0 + \tau^{n,\delta}, x_{\tau^{n,\delta}})e^{-\varphi_{\tau^{n,\delta}}}\\ + \int_0^{\tau^{n,\delta}} f^{\alpha_t^{n,\delta}}(s_0 + t, x_t)e^{-\varphi_t}\,dt\Bigg] + \varepsilon(T - s_0)\\ + \varlimsup_{\delta\downarrow 0}\ \varlimsup_{n\to\infty} \mathsf{M}^{\alpha^{n,\delta}}_{s_0,x_0}\big|u(s_0 + \tau^{n,\delta}, x_{\tau^{n,\delta}}) - w(s_0 + \tau^{n,\delta}, x_{\tau^{n,\delta}})\big|, \quad (10)\end{aligned}$$

where $\tau^{n,\delta}$ is the time of first exit of $(s_0 + t, z_t^{n,\delta})$ from the region Q.

If we take in Lemma 9 $|u(s,x) - w(s,x)|$ instead of $g(s,x)$ and if, in addition, we put $c^\alpha(s,x) \equiv 0$, we shall see that the last term in (10) is equal to

$$\overline{\lim_{\delta \downarrow 0}}\ \overline{\lim_{n \to \infty}}\ \mathsf{M}\big|u(s_0 + \tau^{n,\delta}, z^{n,\delta}_{\tau^{n,\delta}}) - w(s_0 + \tau^{n,\delta}, z^{n,\delta}_{\tau^{n,\delta}})\big|.$$

In order to estimate the last expression, we note that if $|z^{n,\delta}_{\tau^{n,\delta}}| < R$, a point $(s_0 + \tau^{n,\delta}, z^{n,\delta}_{\tau^{n,\delta}})$ lies on that part of the boundary of Q, where $u = w$. In short, if $|z^{n,\delta}_{\tau^{n,\delta}}| < R$, then

$$u(s_0 + \tau^{n,\delta}, z^{n,\delta}_{\tau^{n,\delta}}) = w(s_0 + \tau^{n,\delta}, z^{n,\delta}_{\tau^{n,\delta}}).$$

Therefore, the last term in (10) does not exceed

$$N \sup_{\delta,n} \mathsf{M}(1 + |z^{n,\delta}_{\tau^{n,\delta}}|)^{m+p}\chi_{|z^{n,\delta}_{\tau^{n,\delta}}| \geq R}$$

$$\leq \frac{1}{1+R} N \sup_{\delta,n} \mathsf{M}\left(1 + \sup_{t \leq T - s_0} |z^{n,\delta}_t|\right)^{m+p+1} \leq \frac{N}{1+R},$$

where, by virtue of (5) the constant N does not depend on R.

Finally, the first term in the right-hand side of (10) is smaller than $w(s_0,x_0)$ according to Theorem 3.1.11. Thus we obtain from (10)

$$u(s_0,x_0) \leq w(s_0,x_0) + \varepsilon(T - s_0) + \frac{N}{1+R}.$$

Here the numbers $R > |x_0|$, $\varepsilon > 0$ are arbitrary, N not depending on R. Letting $R \to \infty$, $\varepsilon \downarrow 0$, we conclude that $u(s_0,x_0) \leq w(s_0,x_0)$. However, this is impossible for a point $(s_0,x_0) \in Q'$. Hence the set Q' is empty. We have thus proved the theorem. □

Using the inequality $w(s,x) \geq g(s,x)$ as well as the fact that for any real a the function $(a - t)_+ + t$ increases with respect to t, we have

12. Corollary. *Let $u \in W^{1,2}_{\text{loc}}(H_T) \cap C(\bar{H}_T)$, let u satisfy inequality* (1) *in H_T, $u(T,x) \geq g(T,x)$ for all $x \in E_d$,*

$$(F[u](s,x) + u(s,x) - g(s,x))_+ + g(s,x) - u(s,x) \geq 0 \qquad (H_T\text{-a.e.}).$$

Then $u \leq w$ in $\bar{H}_T$.

Noting that for $a \geq 0$ the function $(a - t)_+ + t \geq 0$ for all t, we have

13. Corollary. *Let $u \in W^{1,2}_{\text{loc}}(H_T) \cap C(\bar{H}_T)$ and let it satisfy inequality* (1) *in H_T, $u(T,x) \geq g(x)$ for all $x \in E_d$, $F[u] \leq 0$ (H_T-a.e.). Then $u \leq v$ in $\bar{H}_T$.*

Combining Theorem 5 and Corollary 7 with the last two corollaries we can immediately assert that any solution of the Bellman equation is equal to a payoff function.

14. Theorem. *Let* $u \in W^{1,2}_{\text{loc}}(H_T) \cap C(\bar{H}_T)$ *and let* u *satisfy inequality* (1) *in* H_T. *Then*:

a. *if* $(F[u](s,x) + u(s,x) - g(s,x))_+ + g(s,x) - u(s,x) = 0$ (H_T-a.e.), $u(T,x) = g(T,x)$ *on* E_d, *then* $u = w$ *in* $\bar{H}_T$;

b. *if* $F[u] = 0$ (H_T-a.e.), $u(T,x) = g(x)$ *on* E_d, *then* $u = v$ *in* $\bar{H}_T$.

Lemma 4, Lemma 10, and Corrollary 13 enable us to prove Theorems 3.4.13 and 3.4.14.

15. Proof of Theorem 3.4.13. Since $na_+ - na \geq 0$, by the hypotheses of the theorem

$$\sup_{\alpha \in A}[L^\alpha g - ng + \tilde{f}^\alpha] \geq 0 \qquad (H_T\text{-a.e.}).$$

where $\tilde{f}^\alpha = n\tilde{v}_n + n(g - \tilde{v}_n)_+ + f^\alpha + K(1 + |x|)^m$, which implies, by Corollary 13, that

$$\begin{aligned} g(s,x) &\leq \sup_{\alpha \in \mathfrak{A}} \mathsf{M}^\alpha_{s,x}\left\{\int_0^{T-s} \tilde{f}^{\alpha_t}(s + t, x_t)e^{-\varphi_t - nt}\,dt \right. \\ &\qquad \left. + g(T,x_{T-s})e^{-\varphi_{T-s} - n(T-s)}\right\} \\ &\leq \sup_{\alpha \in \mathfrak{A}} \mathsf{M}^\alpha_{s,x}\left\{\tilde{v}_n(T,x_{T-s})e^{-\varphi_{T-s} - n(T-s)}\right. \\ &\qquad \left. + \int_0^{T-s} [f^{\alpha_t} + n(g - \tilde{v}_n)_+ + n\tilde{v}_n](s + t, x_t)e^{-\varphi_t - nt}\,dt\right\} \\ &\qquad + \sup_{\alpha \in \mathfrak{A}} \mathsf{M}^\alpha_{s,x} K \sup_{t \leq T-s} (1 + |x_t|)^m \int_0^{T-s} e^{-nt}\,dt. \end{aligned}$$

According to Lemma 3.4.3, the first term of the last expression is equal to $\tilde{v}_n(s,x)$. According to Corollary 2.5.12, the second term of the last expression does not exceed

$$N(K,m,T)(1 + |x|)^m \int_0^{T-s} e^{-nt}\,dt \leq \frac{1}{n} N(K,m,T)(1 + |x|)^m.$$

Therefore, $n(g - \tilde{v}_n) \leq N(1 + |x|)^m$. Further, we compare the definition of w with Lemma 3.4.3b. Using the fact that $|g - g_n| = (g - \tilde{v}_n)_+ \leq (1/n)N(1 + |x|)^m$, we have

$$\begin{aligned} |w(s,x) - \tilde{v}_n(s,x)| &\leq \sup_{\alpha \in \mathfrak{A}} \sup_{\tau \in \mathfrak{M}(T-s)} \mathsf{M}^\alpha_{s,x}|g - g_n|(s + \tau, x_\tau) \\ &\leq \frac{1}{n} N \sup_{\alpha \in \mathfrak{A}} \mathsf{M}^\alpha_{s,x} \sup_{t \leq T-s} (1 + |x_t|)^m. \end{aligned}$$

By invoking Corollary 2.5.12, we complete the proof of Theorem 3.4.13. □

16. Proof of Theorem 3.4.14. First we show that $Q \subset Q'_0$. Let the opposite be true: a point $(s_0,x_0) \in Q \backslash Q'_0$. It is seen that $w(s_0,x_0) = g(s_0,x_0)$. We consider without loss of generality that $x_0 = 0$. Furthermore, let $\varepsilon_1 = \frac{1}{2}h(s_0,0)$. Since $\varepsilon_1 > 0$ and $h(t,x)$ is a continuous function, for all sufficiently small R,ρ on the cylinder

$$\tilde{C}_{\rho,R} \equiv (s_0 - \rho, s_0 + \rho) \times S_R$$

the function $h(t,x)$ is greater than ε_1. We choose appropriate values of $R > 0$ and $\rho > 0$. Next, we diminish ρ so that the inequality

$$\mathsf{M} \sup_{t \le \rho} |z_t| \le \frac{R}{2} \tag{11}$$

is satisfied for all processes z_t of the form

$$z_t = \int_0^t \sigma_r \, d\mathbf{w}_r + \int_0^t b_r \, dr \tag{12}$$

for which $\|\sigma_r\| + |b_r| \le 2K(1 + R)$ for all r, ω. We can arrange for (11) to be satisfied by choosing $\rho > 0$ due to Corollary 2.5.12. We note immediately the implication of (11):

$$\mathsf{M}\tau \ge \tfrac{1}{2}\rho, \tag{13}$$

where τ is the time of first exit of $(s_0 + t, z_t)$ from $\tilde{C}_{\rho,R}$. For deriving (13) from (11) we note that

$$\left\{\sup_{t \le \rho} |z_t| < R\right\} \subset \{\tau = \rho\}.$$

Hence

$$\mathsf{M}\tau \ge \rho \mathsf{P}\left\{\sup_{t \le \rho} |z_t| < R\right\} = \rho\left(1 - \mathsf{P}\left\{\sup_{t \le \rho} |z_t| \ge R\right\}\right)$$
$$\ge \rho\left(1 - \frac{1}{R}\mathsf{M} \sup_{t \le \rho} |z_t|\right) \ge \frac{1}{2}\rho.$$

Further, we denote by $\tilde{h}(t,x)$ a continuous function which has compact support in H_T and is equal to $h(t,x)$ on $\tilde{C}_{\rho,R}$. By hypothesis of the theorem, $\sup_{\alpha \in A}[L^\alpha g + \tilde{f}^\alpha] \ge 0$ ($\tilde{C}_{\rho,R}$-a.e.), where $\tilde{f}^\alpha = f^\alpha - \tilde{h}$. Next, we fix $\varepsilon > 0$ and apply Lemma 10 replacing in this lemma f^α by $\tilde{f}^\alpha$. Then we can find strategies $\alpha^{n,\delta}$ such that

$$g(s_0,0) \le \varliminf_{\delta \downarrow 0} \varliminf_{n \to \infty} \mathsf{M}_{s_0,0}^{\alpha^{n,\delta}}\Bigg[g(s_0 + \tau^{n,\delta}, x_{\tau^{n,\delta}})e^{-\varphi_{\tau^{n,\delta}}}$$
$$+ \int_0^{\tau^{n,\delta}} \tilde{f}^{\alpha_t^{n,\delta}}(s_0 + t, x_t)e^{-\varphi_t}\,dt\Bigg] + \varepsilon(T - s_0),$$

where $\tau^{n,\delta}$ is the time of first exit of the process $(s_0 + t, z_t^{n,\delta})$ from $\tilde{C}_{\rho,R}$. Recalling the definition of w as well as the equality $g(s_0,0) = w(s_0,0)$, we derive from the inequality given above

$$\varliminf_{\delta\downarrow 0}\ \varliminf_{n\to\infty}\ \mathsf{M}_{s_0,0}^{\alpha^{n,\delta}}\int_0^{\tau^{n,\delta}}\tilde{h}(s_0+t,x_t)e^{-\varphi_t}\,dt\le\varepsilon(T-s_0).$$

Using Lemma 9 (in which we assume $g=0$, $f^\alpha=\tilde{h}$), we find

$$\varliminf_{\delta\downarrow 0}\ \varliminf_{n\to\infty}\ \mathsf{M}\int_0^{\tau^{n,\delta}}\tilde{h}(s_0+t,z_t^{n,\delta})\times\exp\left[-\int_0^t c^{\alpha_r^{n,\delta}}(s_0+r,z_r^{n,\delta})\,dr\right]dt\le\varepsilon(T-s_0).\tag{14}$$

Note that $|z_r^{n,\delta}|\le R$ for $r\le t\le\tau^{n,\delta}$. Hence $c^\alpha(s_0+r,z_r^{n,\delta})\le K(1+R)^m$. Furthermore, $(s_0+t,z_t^{n,\delta})\in\tilde{C}_{\rho,R}$ for $t\le\tau^{n,\delta}$. Therefore, in (14) we can replace $\tilde{h}$ by h. Finally, using the inequality $h>\varepsilon_1$ on $\tilde{C}_{\rho,R}$, we obtain from (14)

$$\varepsilon_1 e^{-K\rho(1+R)^m}\varliminf_{\delta\downarrow 0}\ \varliminf_{n\to\infty}\ \mathsf{M}\tau^{n,\delta}\le\varepsilon(T-s_0).\tag{15}$$

The time $\tau^{n,\delta}$ is the time of first exit of the process $(s_0+t,z_{t\wedge\tau^{n,\delta}}^{n,\delta})$ from $\tilde{C}_{\rho,R}$. In this case

$$z_{t\wedge\tau^{n,\delta}}^{n,\delta}=\int_0^t\chi_{r<\tau^{n,\delta}}\sigma_n(s_0+r,z_r^{n,\delta})\,d\mathbf{w}_r+\int_0^t\chi_{r<\tau^{n,\delta}}b_n(s_0+r,z_r^{n,\delta})\,dr,$$

$$\|\chi_{r<\tau^{n,\delta}}\sigma_n(s_0+r,z_r^{n,\delta})\|\le\chi_{r<\tau^{n,\delta}}K(1+|z_r^{n,\delta}|)\le K(1+R),$$

Also, the expression containing b_n can be estimated in a similar way. Therefore, by virtue of (13), $\mathsf{M}\tau^{n,\delta}\ge\frac{1}{2}\rho$ and, moreover, from (15) we have

$$\frac{1}{2}\rho\varepsilon_1 e^{-K\rho(1+R)^m}\le\varepsilon(T-s_0).\tag{16}$$

In (16), ρ, R, ε_1 do not depend on ε, ε being arbitrarily small. Therefore, the left-hand side of (16) is equal to zero. However, this contradicts the inequalities $\rho>0$ and $\varepsilon_1>0$, thus proving that $Q\subset Q_0'$.

We take now a connected component $\tilde{Q}_0'$ of the region Q_0' and prove that this component contains at least one connected component of the set Q. Assume the converse. Then it follows from the inclusion $Q\subset Q_0'$ that $\tilde{Q}_0'\cap Q=\varnothing$. Therefore, $F[g]\le 0$ almost everywhere on $\tilde{Q}_0'$. Thus, by Lemma 4 on $\tilde{Q}_0'$

$$g(s,x)\ge\sup_{\alpha\in\mathfrak{A}}\ \sup_{\gamma\in\mathfrak{M}(T-s)}\mathsf{M}_{s,x}^\alpha\left\{\int_0^{\tau\wedge\gamma}f^{\alpha_t}(s+t,x_t)e^{-\varphi_t}\,dt+g(s+\tau\wedge\gamma,x_{\tau\wedge\gamma})e^{-\varphi_{\tau\wedge\gamma}}\right\},\tag{17}$$

where $\tau=\tau^{\alpha,s,x}$ is the time of first exit of $(s+t,x_t^{\alpha,s,x})$ from $\tilde{Q}_0'$. It is seen that for $(s,x)\in\tilde{Q}_0'$ $g(s+\tau,x_\tau)=w(s+\tau,x_\tau)$. Hence

$$g(s+\tau\wedge\gamma,x_{\tau\wedge\gamma})e^{-\varphi_{\tau\wedge\gamma}}=g(s+\gamma,x_\gamma)e^{-\varphi_\gamma}\chi_{\gamma\le\tau}+w(s+\tau,x_\tau)e^{-\varphi_\tau}\chi_{\tau<\gamma}.$$

Using Theorem 3.1.9, we conclude from (17) $g = w$ on $\tilde{Q}'_0$, which fact contradicts the inequality $g < w$ on Q, thus proving the theorem. □

Many constructions carried out in this and preceding sections involve strategies of a special form.

17. Definition. Let $(s_0,x_0) \in H_T$. A strategy $\alpha_t(\omega)$ is said to be an (s_0,x_0)-*adjoint Markov* strategy if on $[0,T] \times E_d \times E_d$ we can define a matrix $\sigma(t,x,z)$ of dimension $d \times d_1$, a d-dimensional vector $b(t,x,z)$, a function $\alpha(t,x,z)$ with values in A, and, finally, a number δ such that there exists (with respect to $\{\mathscr{F}_t\}$) a progressively measurable solution $(x_t; z_t)$ of the system of equations

$$\begin{aligned} x_t &= x_0 + \int_0^t \sigma(\alpha(s_0 + r, x_r, z_r), s_0 + r, x_r)\, d\mathbf{w}_r \\ &\quad + \int_0^t b(\alpha(s_0 + r, x_r, z_r), s_0 + r, x_r)\, dr, \\ z_t &= x_0 + \int_0^t \sigma(s_0 + r, x_r, z_r)\, d\mathbf{w}_r + \delta \hat{\mathbf{w}}_t \\ &\quad + \int_0^t b(s_0 + r, x_r, z_r)\, dr, \end{aligned} \tag{18}$$

and, furthermore, for this solution $\alpha_t(\omega) = \alpha(s_0 + t, x_t(\omega), z_t(\omega))$ for all $t \in [0, T - s_0]$, ω. We denote by $\mathfrak{A}_{\Pi M}(s_0,x_0)$ the set of all (s_0,x_0)-adjoint Markov strategies.

Note that if a strategy α from $\mathfrak{A}_{\Pi M}(s_0,x_0)$ was constructed with the aid of Eq. (18), then $x_t = x_t^{\alpha,s_0,x_0}$ for $t \le T - s_0$. In fact, the first relation in (18) can be written as follows:

$$x_t = x_0 + \int_0^t \sigma(\alpha_r, s_0 + r, x_r)\, d\mathbf{w}_r + \int_0^t b(\alpha_r, s_0 + r, x_r)\, dr.$$

It is not hard to see that the strategies $\alpha^{n,\delta}$ which were constructed in Lemma 8 are adjoint Markov. If we take in Section 2 $(\mathbf{w}_t; \hat{\mathbf{w}}_t)$ for $\tilde{\mathbf{w}}_t$, and if, in addition, we assume in Eq. (18) that $\alpha(t,x,z) = \alpha[p](t, x + \varepsilon(z - x_0))$, $\sigma = 0$, $b = 0$, and $\delta = 1$, it turns out that the strategy $\beta[p]$ which was introduced prior to Lemma 2.1 is adjoint Markov.

It is clear that from a practical point of view we can interpret the employment of adjoint Markov strategies to be the supplement to the controlled system considered of "adjoint" coordinates z_t, using as a result Markov strategies with respect to an expanded controlled system $(x_t; z_t)$. We note also that in the deterministic case the process z_t is said to be a guiding process (see Krassovsky and Subbotin [27]).

18. Exercise

Assume that $v \in W^{1,2}_{\mathrm{loc}}(H_T)$, $F[v] = 0$ (H_T-a.e.). We fix $\varepsilon > 0$ and take a Borel function $\alpha(s,x)$ on H_T with values in A, such that

$$L^{\alpha(s,x)} v(s,x) + f^{\alpha(s,x)}(s,x) \ge -\varepsilon \qquad (H_T\text{-a.e.}).$$

For $(s_0,x_0) \in H_T$ we define the (s_0,x_0)-adjoint Markov strategy in the same way as we did in Lemma 8. Prove that

$$v(s_0,x_0) \leq \varliminf_{\delta \downarrow 0} \varliminf_{n \to \infty} v^{\alpha^{n,\delta}}(s_0,x_0) + \varepsilon(T - s_0).$$

19. Exercise

(Compare with Exercise 2.7.) Consider a one-dimensional controlled process; $d = d_1 = 1$, $T = 1$, $A = \{-1\} \bigcup \{+1\}$, $\sigma(\alpha,s,x) = \sigma(x + \alpha)$, where $\sigma(x) = \operatorname{sgn} x$ for $|x| \geq 1$, $\sigma(x) = x$ for $|x| \leq 1$. Let $b \equiv c \equiv f \equiv 0$, $g(x) = x^2$.

Show that $v(s) = x^2 + 1 - s$. Also show that one can find ε-optimal strategies for a point (0,0) among strategies of the form $\alpha_t = \operatorname{sgn} z_t$, where z_t is a solution of the equation

$$z_t = \int_0^t \alpha_n(z_t)\, d\mathbf{w}_t + \delta \hat{\mathbf{w}}_t,$$

$\alpha_n(x) = n\zeta(nx) * \operatorname{sgn} x$, n is a sufficiently large number, δ is a sufficiently small positive number.

20. Exercise

Show that in the situations described in Exercises 2.7 and 19 there exists no optimal strategy if $\mathscr{F}_t$ is the completion of the σ-algebra generated by $\mathbf{w}_s$ for $s \in [0,t]$. The question as to whether the equality $v_{(M)}(0,0) = v(0,0)$ is true in Exercise 19 still remains open.

Notes

Sections 1,2. Markov strategies are frequently understood as Borel functions $\alpha(t,x)$ for which the appropriate stochastic equation has at least a weak solution. The existence of ε-optimal strategies in the class of such strategies is proved in Krylov [36], Nisio [53, 59], and Portenko and Skorokhod [61]. Fleming in [15] constructed Markov (in the sense of the definition adopted in our book) optimal strategies for one class of problems. Some problems of finding optimal strategies have been solved in Beneš [6]. Our discussion closely follows [35].

Section 3. If (each) solution of the Bellman equation is equal to a payoff function, the Bellman equation can have only one solution. The theorems on uniqueness of solutions of nonlinear equations, unrelated to optimal control theory, can be found in Bakel'man [3], Ladyzhenskaja and Uraltseva [46], Ladyzhenskaja, Solonnikov, and Uraltseva [47], and [33, 43].

Controlled Processes with Unbounded Coefficients: The Normed Bellman Equation 6

In Chapters 3, 4, and 5 we studied controlled processes on a finite time interval under the assumption that the initial variables $\sigma(\alpha,t,x)$, $b(\alpha,t,x)$, $c^\alpha(t,x)$, and $f^\alpha(t,x)$ are bounded functions of α for each (t,x). The objective of Chapter 6 is to carry the results obtained in Chapters 3–5 over to controlled processes with coefficients unbounded with respect to α, and also to consider controlled processes on an infinite time interval.

1. Generalizations of the Results Obtained in Section 3.1

Let E_d be a Euclidean space of dimension d, let $T \in (0,\infty)$, let d_1 be an integer, and, finally, let $(\mathbf{w}_t,\mathscr{F}_t)$ be a d_1-dimensional Wiener process. We denote by A a separable metric space. We fix a representation of A as the union of nonempty increasing sets A_n: $A = \bigcup_{n=1}^{\infty} A_n$, $A_{n+1} \supset A_n$ (possibly, $A_1 = A_2 = \cdots = A$).

For $t \geq 0$, $x \in E_d$, and $\alpha \in A$ we assume that the functions $\sigma(\alpha,t,x)$, $b(\alpha,t,x)$, $c^\alpha(t,x) \geq 0$, $f^\alpha(t,x)$, $g(x)$, and $g(t,x)$, to be given, have the same meaning as the functions given in Section 3.1 have.

We assume that the functions σ, b, c and f are continuous with respect to (α,x). We also assume that for each t and n the foregoing functions are continuous with respect to x uniformly with respect to $\alpha \in A_n$ and that they are Borel with respect to (α,t,x). Furthermore, for each n let there exist constant $m_n \geq 0$ and $K_n \geq 0$ such that for all x and $y \in E_d$, $t \geq 0$ and $\alpha \in A_n$

$$\|\sigma(\alpha,t,x) - \sigma(\alpha,t,y)\| + |b(\alpha,t,x) - b(\alpha,t,y)| \le K_n|x - y|, \tag{1}$$

$$\|\sigma(\alpha,t,x)\| + |b(\alpha,t,x)| \le K_n(1 + |x|), \tag{2}$$

$$|c^\alpha(t,x)| + |f^\alpha(t,x)| \le K_n(1 + |x|)^{m_n}. \tag{3}$$

We assume that $g(x)$ and $g(t,x)$ are continuous and, in addition, for some constant $m \ge 0$ and $K \ge 0$ for all t, x

$$|g(x)| + |g(t,x)| \le K(1 + |x|)^m. \tag{4}$$

1. Definition. Let $n \ge 1$. We write $\alpha \in \mathfrak{A}_n$ if the process $\alpha = \alpha_t(\omega)(t \ge 0)$ is progressively measurable with respect to $\{\mathscr{F}_t\}$ and assumes values from A_n for $t \in [0,T]$. Let $\mathfrak{A} = \bigcup_n \mathfrak{A}_n$. The elements of a set $\mathfrak{A}$ are said to be strategies.

Fixing n and considering the strategies α from $\mathfrak{A}_n$ only, we have the scheme which we considered in Sections 3.1–3.4. Using here the usual notations given in Chapter 3, we put

$$v_n(t,x) = \sup_{\alpha \in \mathfrak{A}_n} v^\alpha(t,x),$$

$$w_n(t,x) = \sup_{\alpha \in \mathfrak{A}_n} \sup_{\tau \in \mathfrak{M}(T-t)} v^{\alpha,\tau}(t,x).$$

For each n we define in an obvious manner the set of natural strategies admissible at a point (t,x). We denote this set by $\mathfrak{A}_{n,E}(t,x)$. Let $\mathfrak{A}_E(t,x) = \bigcup_n \mathfrak{A}_{n,E}(t,x)$.

Further, let

$$v(t,x) = \sup_{\alpha \in \mathfrak{A}} v^\alpha(t,x),$$

$$w(t,x) = \sup_{\alpha \in \mathfrak{A}} \sup_{\tau \in \mathfrak{M}(T-t)} v^{\alpha,\tau}(t,x).$$

It is seen that $v_n \to v$ and $w_n \to w$ as $n \to \infty$. In addition, $v_n \le v$ and $w_n \le w$ for each n. From this and from Theorems 3.1.5 and 3.1.8 which enable us to estimate $|v_1|$ and $|w_1|$ we find

$$v(t,x) \ge -N(1 + |x|)^{m_1}, \qquad w(t,x) \ge -N(1 + |x|)^{m_1} \tag{5}$$

for $(t,x) \in \bar{H}_T$, where $N = N(K_1,m_1,T)$. It is clear that the functions v and w can in general take on values equal to $+\infty$. In order to give a sufficient condition for the functions v and w to be finite, we make use of Definition 5.3.2 of excessive functions and Definition 5.3.3 of superharmonic functions. Note that the operator F appearing in the above definitions can be represented by the same formula as that given in the introduction to Chapter 4.

2. Lemma. *Let a function u be given and let this function be continuous on $\bar{H}_T$. Let a region $Q \subset H_T$. Assume that u is a superharmonic (or excessive) function in the region Q. Then, if:*

a. $u(t,x) \geq g(t,x)$ *in* Q, $u \geq w$ *in* $H_T \, Q$, *then* $u \geq w$ *in* $\bar{H}_T$;
b. $u \geq v$ *in* $H_T \backslash Q$, $u(T,x) \geq g(x)$ *in* E_d, *then* $u \geq v$ *in* $\bar{H}_T$.

This lemma follows immediately from the inequalities $w_n \leq u$, $v_n \leq u$ (Theorem 5.3.5) and the fact that $w_n \to w$ and $v_n \to v$ as $n \to \infty$.

In some cases the condition of the lemma can easily be verified. We show how this can be done if, for example, the assumptions made in Section 3.1 are satisfied. Our arguments will not lead us to a new result. However, they will be useful from the viewpoint of methodology. Thus, we take the numbers K and m given in Section 3.1 and, furthermore, we put

$$u(t,x) = 2^m K e^{N_1(T-t)}(1 + |x|^2)^{m/2},$$

choosing the constant N_1 later. We have

$$\begin{aligned} L^\alpha u(t,x) + 2^m K(1 + |x|^2)^{m/2} = \Big\{ &\frac{m}{2}(1 + |x|^2)^{-1}\|\sigma\|^2 \\ &+ m\left(\frac{m}{2} - 1\right)(1 + |x|^2)^{-2}|\sigma^* x|^2 + m(1 + |x|^2)^{-1} bx \\ &- (N_1 + c)\Big\} u(t,x) + 2^m K(1 + |x|^2)^{m/2}, \end{aligned}$$

where $\sigma = \sigma(\alpha,t,x)$, etc. Since σ and b satisfy condition (3.1.2) (that is, condition (2) for $K_n = K$),

$$\begin{aligned} L^\alpha u(t,x) + 2^m K(1 + |x|^2)^{m/2} &\leq (N_2 - N_1)u(t,x) + 2^m K(1 + |x|^2)^{m/2} \\ &\leq (N_2 + 1 - N_1)u(t,x). \end{aligned}$$

From this we see that the left-hand side of the last inequality is negative for $N_1 = 1 + N_2$. In addition, in H_T

$$g(x) \vee g(t,x) \leq K(1 + |x|)^m \leq 2^m K(1 + |x|^2)^{m/2} \leq u(t,x),$$
$$L^\alpha u(t,x) + f^\alpha(t,x) \leq L^\alpha u(t,x) + 2^m K(1 + |x|^2)^{m/2} \leq 0.$$

Therefore, the function $u(t,x)$ satisfies at once all the conditions of Lemma 2, in this case, however, it is possible to take $Q = H_T$. From Lemma 2 we obtain the result which is familiar to us from Chapter 3: if the conditions of Section 3.1 are satisfied, then $v \leq u$, $w \leq u$.

In the particular case considered we applied Lemma 2 for $Q = H_T$. Generally speaking, in applying Lemma 2 in the specific cases, it would be natural to make an attempt to find a superharmonic (or excessive) function for which the conditions of Lemma 2 would be satisfied for $Q = H_T$. In such situations $H_T \backslash Q$ is empty and, in addition, the formulation of the lemma contains no conditions to be imposed on the values (in general, unknown) of v and w in H_T. However, it is not always possible to construct such a function u.

3. Exercise

We consider a one-dimensional case: $d = d_1 = T = 1$, $A = [0,\infty)$, $A_n = [0,n]$, $\sigma = \alpha x$, $b = c = f = 0$, and $g(x)$ is an arbitrary continuous function satisfying the inequality $|x| \le g(x) \le 2|x| + 1$.

Prove that there exist no superharmonic functions (in the sense of Definition 5.3.3) in H_T for which $u(1,x) \ge g(x)$ for all x. (At the same time, the function $2|x| + 1$ is superharmonic in $Q = H_1 \setminus \{(t,x): x = 0\}$. For $t = 1$ this function is not smaller than $g(x)$, and for $x = 0$ it is equal to $1 \ge g(0) = v(t,0)$. Hence the function $2|x| + 1$ together with the region Q satisfies the conditions of Lemma 2.)

In practice, it is natural to consider strategies only from $\mathfrak{A}$, that is, the strategies $\alpha_t(\omega)$ which assume values from some A_n that is the same for all (t,ω). In this connection, we give the following

4. Definition. Let $(s,x) \in \bar{H}_T$. The process $\alpha_t(\omega)$ which is progressively measurable with respect to $\{\mathscr{F}_t\}$ and has values from A is said to be a strategy admissible at a point (s,x), if

a. for $t \in [0, T - s]$ there exists (at least one) solution of the equation

$$x_t = x + \int_0^t \sigma(\alpha_r, s + r, x_r)\, d\mathbf{w}_r + \int_0^t b(\alpha_r, s + r, x_r)\, dr, \tag{6}$$

for which

b. $\mathsf{M} \int_0^{T-s} f_-^{\alpha_t}(s + t, x_t) \exp(-\int_0^t c^{\alpha_r}(s + r, x_r)\, dr)\, dt < \infty$,

c. $\mathsf{M} \sup_{t \le T-s} |x_t|^{m \vee m_1} < \infty$,

d. there is an integer-valued function $n(R) = n^\alpha(R)$ such that $\alpha_t \in A_n n(\sup_{r \le t} |x_r|)$ for all $t \in [0, T - s]$, ω.

We denote by $\mathfrak{A}(s,x)$ the set of all strategies admissible at a point (s,x). If $\alpha \in \mathfrak{A}(s,x)$, we denote by $x_t^{\alpha,s,x}$ a certain (fixed forever) solution of Eq. 6 which satisfies conditions (b), (c), (d).[1] For the strategies $\alpha \in \mathfrak{A}(s,x)$ we shall use the abbreviated notation $\varphi_t^{\alpha,s,x}$, $\mathsf{M}_{s,x}^\alpha$, etc., giving them the same meaning as we did before. We note that the expressions $v^\alpha(s,x)$ and $v^{\alpha,\tau}(s,x)$ for $\alpha \in \mathfrak{A}(s,x)$ and $\tau \in \mathfrak{M}(T - s)$ have been defined in accord with Definition 4b,c as well as condition (4), although these expressions are possibly equal to $+\infty$.

Under the assumptions made in Section 3.1 the sets $\mathfrak{A}$ and $\mathfrak{A}(s,x)$ coincide if we put $A_1 = A_2 = \cdots = A$. In fact, comparing Definition 4 and Definition 3.1.1, we have that $\mathfrak{A}(s,x) \subset \mathfrak{A}$. On the other hand, for $\alpha \in \mathfrak{A}$ condition (d) is a consequence of the equality of A_n. Conditions (b) and (c) follow from the estimates of moments of solutions of stochastic equations (see Section 2.5), which fact shows as well that in the general case $\mathfrak{A}(s,x) \supset \mathfrak{A}_n$, $\mathfrak{A}(s,x) \supset \mathfrak{A}$. Hence

$$v(s,x) \le \sup_{\alpha \in \mathfrak{A}(s,x)} v^\alpha(s,x),$$

[1] Indeed, it is not hard to show that there is only one solution to within an equivalence.

$$w(s,x) \le \sup_{\alpha \in \mathfrak{A}(s,x)} \sup_{\tau \in \mathfrak{M}(T-s)} v^{\alpha,\tau}(s,x).$$

Before formulating the next theorem we note that due to (5) and Definition 4b,c the expressions in (7) (see below) standing under the sign of the upper bound have been defined, and furthermore, they are greater than $-\infty$.

5. Theorem. (a) $v(s,x) = \sup_{\alpha \in \mathfrak{A}(s,x)} v^{\alpha}(s,x)$ *for all* $(s,x) \in \bar{H}_T$.

(b) *The function* $v(s,x)$ *is lower semicontinuous on* $\bar{H}_T$, $v(T,x) = g(x)$.

(c) *If* $(s,x) \in \bar{H}_T$ *and if for each* $\alpha \in \mathfrak{A}(s,x)$ *the time* $\tau = \tau^{\alpha} \in \mathfrak{M}(T-s)$ *and a bounded (with respect to* (t,ω)*) nonnegative progressively measurable (with respect to* $\{\mathscr{F}_t\}$*) process* $r_t = r_t^{\alpha}$ *are defined, then*

$$v(s,x) = \sup_{\alpha \in \mathfrak{A}(s,x)} \mathsf{M}_{s,x}^{\alpha}\left\{\int_0^{\tau}[f^{\alpha_t}(s+t,x_t) + r_t v(s+t,x_t)] \times \exp\left(-\varphi_t - \int_0^t r_u\,du\right)dt + v(s+\tau,x_\tau)\exp\left(-\varphi_\tau - \int_0^{\tau} r_u\,du\right)\right\}. \tag{7}$$

(d) *In* (a) *and in* (7) *we can replace the set* $\mathfrak{A}(s,x)$ *by* $\mathfrak{A}_E(s,x)$ *as well as by* $\mathfrak{A}$.

PROOF. Since the equality $v(T,x) = g(x)$ is obvious, (a) follows from (c) for $r_t^{\alpha} \equiv 0$, $\tau^{\alpha} = T - s$. From the continuity of $v_n(s,x)$ (see Theorem 3.1.5) we deduce

$$\varliminf_{(t,y)\to(s,x)} v(t,y) = \varliminf_{(t,y)\to(s,x)} \sup_n v_n(t,y) \ge \sup_n \varliminf_{(t,y)\to(s,x)} v_n(t,y) = \sup_n v_n(s,x) = v(s,x).$$

Therefore, $v(s,x)$ is lower semicontinuous. We have proved (b). Let

$$\Psi_{(\pm)}^{\alpha,s,x}(u,\tau) \equiv \int_0^{\tau}[f_{\pm}^{\alpha_t}(s+t,x_t^{\alpha,s,x}) + r_t^{\alpha}u_{\pm}(s+t,x_t^{\alpha,s,x})] \times \exp\left(-\varphi_t^{\alpha,s,x} - \int_0^t r_p^{\alpha}\,dp\right)dt + u_{\pm}(s+\tau,x_\tau^{\alpha,s,x}) \times \exp\left(-\varphi_\tau^{\alpha,s,x} - \int_0^{\tau} r_p^{\alpha}\,dp\right).$$

Also, we introduce $\Psi^{\alpha,s,x}(u,\tau)$, omitting the signs $\pm$ in the last formula.

By Theorem 3.1.7, for each n

$$v_n(s,x) = \sup_{\alpha \in \mathfrak{A}_{n,E}(s,x)} \mathsf{M}_{s,x}^{\alpha}\Psi(v_n,\tau) \le \sup_{\alpha \in \mathfrak{A}_E(s,x)} \mathsf{M}_{s,x}^{\alpha}\Psi(v,\tau).$$

Therefore,

$$v(s,x) \le \sup_{\alpha \in \mathfrak{A}_E(s,x)} \mathsf{M}_{s,x}^{\alpha}\Psi(v,\tau) \le \sup_{\alpha \in \mathfrak{A}} \mathsf{M}_{s,x}^{\alpha}\Psi(v,\tau) \le \sup_{\alpha \in \mathfrak{A}(s,x)} \mathsf{M}_{s,x}^{\alpha}\Psi(v,\tau),$$

which immediately implies that in order to prove the theorem, it suffices to prove the inequality

$$v(s,x) \ge \mathsf{M}_{s,x}^{\alpha}\Psi(v,\tau) \tag{8}$$

for all $\alpha \in \mathfrak{A}(s,x)$, $\tau \in \mathfrak{M}(T-s)$. We fix $\alpha \in \mathfrak{A}(s,x)$. Further, for $R > 0$ we define the strategy β using the formula $\beta_t = \alpha_{t \wedge \tau_R}$, where τ_R is the time of first exit of $x_t^{\alpha,s,x}$ from S_R. It is seen that $\beta \in \mathfrak{A}_{n(R)}$, where $n(R) = n^\alpha(R)$ has been taken from Definition 4. Furthermore, the processes $x_{t\wedge\tau_R}^{\beta,s,x}$ and $x_{t\wedge\tau_R}^{\alpha,s,x}$ satisfy the (same) equation:

$$x_t = x + \int_0^t \chi_{r<\tau_R}\sigma(\beta_r, s+r, x_r)\,d\mathbf{w}_r + \int_0^t \chi_{r<\tau_R} b(\beta_r, s+r, x_r)\,dr.$$

Hence the foregoing processes coincide for all $t \in [0, T-s]$. In particular, $x_t^{\beta,s,x} = x_t^{\alpha,s,x}$ for all $t \le \tau_R$ almost surely. Since $\beta \in \mathfrak{A}_{n(R)+j}$ for $j = 1, 2, \ldots$, we have by Theorem 3.1.6 that

$$\begin{aligned} v(s,x) \ge v_{n(R)+j}(s,x) &\ge \mathsf{M}_{s,x}^{\beta}\Psi(v_{n(R)+j}, \tau\wedge\tau_R) \\ &= \mathsf{M}_{s,x}^{\alpha}\Psi(v_{n(R)+j}, \tau\wedge\tau_R). \end{aligned} \tag{9}$$

Let $j \to \infty$, $R \to \infty$. We note that when $R \to \infty$, $\tau\wedge\tau_R \to \tau$, $\tau\wedge\tau_R = \tau$ for each ω for all sufficiently large R. By Fatou's lemma

$$\varliminf_{R\to\infty}\ \varliminf_{j\to\infty} \mathsf{M}_{s,x}^{\alpha}\Psi_{(+)}(v_{n(R)+j}, \tau\wedge\tau_R) \ge \mathsf{M}_{s,x}^{\alpha}\Psi_{(+)}(v,\tau).$$

Furthermore, by virtue of inequalities (5) and Definition 4b,c the totality of variables $\Psi_{(-)}^{\alpha,s,x}(v_{n(R)+j}, \tau\wedge\tau_R)$ is bounded by a summable quantity. From this, using the Lebesgue theorem, we find

$$\lim_{R\to\infty}\ \lim_{j\to\infty} \mathsf{M}_{s,x}^{\alpha}\Psi_{(-)}(v_{n(R)+j}, \tau\wedge\tau_R) = \mathsf{M}_{s,x}^{\alpha}\Psi_{(-)}(v,\tau).$$

Since $\Psi = \Psi_{(+)} - \Psi_{(-)}$, the above relations enable us to obtain (8) from (9), thus completing the proof of the theorem. □

6. Exercise

In Exercise 3 we take $g(s,x) = [x^2/(|x|+1)](1-s)$.

Prove that the assumptions of Lemma 2a can be satisfied for $u(s,x) = |x|$, $Q = H_1 \setminus \{(s,0)\}$. Prove also that $w(s,x) = |x|(1-s)$, and, furthermore, that if $\varepsilon \ge 0$, $\tau_\varepsilon^{\alpha,s,x}$ is the time of first exit of $(s+t, x_t^{\alpha,s,x})$ from the region

$$Q_\varepsilon = \{(s,x) \in H_1 : w(s,x) > g(s,x) + \varepsilon\},$$

then for $(s,x) \in Q_\varepsilon$

$$\sup_{\alpha\in\mathfrak{A}} \mathsf{M}_{s,x}^{\alpha} g(s+\tau_\varepsilon, x_{\tau_\varepsilon}) + \varepsilon = \sup_{\alpha\in\mathfrak{A}} \mathsf{M}_{s,x}^{\alpha} w(s+\tau_\varepsilon, x_{\tau_\varepsilon}) = \varepsilon(|x|+1) < w(s,x).$$

This exercise demonstrates that not all the assertions of Theorems 3.1.10 and 3.1.11 are true in general. In this connection, we do not prove theorems on ε-optimal stopping times in the general case. We note nevertheless that it is convenient to search for ε-optimal stopping times using Theorem 3.1.10 and approximating $w(s,x)$ with the aid of $w_n(s,x)$.

In the next theorem, it is useful to keep in mind the remark made before Theorem 5.

7. Theorem. (a) *For all* $(s,x) \in \bar{H}_T$

$$w(s,x) = \sup_{\alpha \in \mathfrak{A}(s,x)} \sup_{\tau \in \mathfrak{M}(T-s)} v^{\alpha,\tau}(s,x).$$

(b) *The function* $w(s,x)$ *is lower semicontinuous on* $\bar{H}_T$, $w(s,x) \geq g(s,x)$ *on* $\bar{H}_T$.

(c) *For any* $\alpha \in \mathfrak{A}(s,x)$, $\tau \in \mathfrak{M}(T-s)$ *and the nonnegative bounded progressively measurable processes (with respect to* $\{\mathscr{F}_t\}$*)* r_t *the inequality*

$$w(s,x) \geq \mathsf{M}^\alpha_{s,x}\left\{w(s+\tau, x_\tau)\exp\left(-\varphi_\tau - \int_0^\tau r_p\,dp\right)dt\right.$$
$$\left.+ \int_0^\tau [f^{\alpha_t}(s+t,x_t) + r_t w(s+t,x_t)]\exp\left(-\varphi_t - \int_0^t r_p\,dp\right)dt\right\},$$

is satisfied.

(d) *We have assertion (3.1.9) of Theorem 3.1.9, in which we can replace the upper bound with respect to* $\alpha \in \mathfrak{A}$ *by the upper bound with respect to* $\alpha \in \mathfrak{A}(s,x)$ *as well as by the upper bound with respect to* $\alpha \in \mathfrak{A}_E(s,x)$.

PROOF. We prove (b) exactly in the same way as we proved (b) in the preceding theorem. Using the notation from the proof of Theorem 5, we can write (c) as follows:

$$w(s,x) \geq \mathsf{M}^\alpha_{s,x}\Psi(w,\tau).$$

We can prove this in the same way as we prove an analogous inequality for v, making use, however, of Theorem 3.1.11 instead of Theorem 3.1.6, by which for all n, $\beta \in \mathfrak{A}_n$

$$w_n(s,x) \geq \mathsf{M}^\beta_{s,x}\Psi(w_n,\tau).$$

Proof of (d). For each $\alpha \in \mathfrak{A}(s,x)$ let a time $\tau = \tau^\alpha \in \mathfrak{M}(T-s)$ be defined. Putting in (c) $r_t \equiv 0$ and also, noting that $w(s,x) \geq g(s,x)$, we obtain

$$w(s,x) \geq \sup_{\alpha \in \mathfrak{A}(s,x)} \sup_{\gamma \in \mathfrak{M}(T-s)} \mathsf{M}^\alpha_{s,x}\Psi(w, \tau \wedge \gamma)$$
$$\geq \sup_{\alpha \in \mathfrak{A}(s,x)} \sup_{\gamma \in \mathfrak{M}(T-s)} \mathsf{M}^\alpha_{s,x}\left\{\int_0^{\tau \wedge \gamma} f^{\alpha_t}(s+t,x_t)e^{-\varphi_t}\,dt\right.$$
$$\left.+ g(s+\gamma,x_\gamma)e^{-\varphi_\gamma}\chi_{\gamma \leq \tau} + w(s+\tau,x_\tau)e^{-\varphi_\tau}\chi_{\tau<\gamma}\right\}.$$

Let us extend the above inequalities, replacing first $\mathfrak{A}(s,x)$ by $\mathfrak{A} \subset \mathfrak{A}(s,x)$, and next by $\mathfrak{A}_E(s,x) \subset \mathfrak{A}$. Finally, let us take into account that the last resulting expression will not be smaller than the following expression:

$$\sup_{\alpha \in \mathfrak{A}_{n,E}(s,x)} \sup_{\gamma \in \mathfrak{M}(T-s)} \mathsf{M}^\alpha_{s,x}\left\{\int_0^{\tau \wedge \gamma} f^{\alpha_t}(s+t,x_t)e^{-\varphi_t}\,dt\right.$$
$$\left.+ g(s+\gamma,x_\gamma)e^{-\varphi_\gamma}\chi_{\gamma \leq \tau} + w_n(s+\tau,x_\tau)e^{-\varphi_\tau}\chi_{\tau<\gamma}\right\} = w_n(s,x)$$

for all n. Here we have equality by Theorem 3.1.9. To complete proving (d) we need to let $n \to \infty$ in the sequence of inequalities obtained if we proceed as indicated above.

Assertion (a) follows from (d) if we assume $\tau^\alpha = T - s$. We have proved the theorem. □

In Theorems 5 and 7 we asserted that $v(T,x) = g(x)$ and $w(T,x) = g(T,x)$. In other words, we know the boundary values of v and w for $s = T$. Regretfully, it may turn out that these boundary values of the functions v and w are weakly related to the values of these functions for $s < T$ (see Exercise 11). Thus, the question arises as to when the boundary values are assumed, i.e., when

$$\lim_{s\uparrow T} v(s,x) = g(x), \qquad \lim_{s\uparrow T} w(s,x) = g(T,x).$$

The following lemma shows that always

$$\varliminf_{s\uparrow T} v(s,x) \geq g(x), \qquad \lim_{s\uparrow T} w(s,x) \geq g(T,x).$$

8. Lemma. *For all $R > 0$ the limits*

$$\lim_{s\uparrow T} \sup_{|x|\leq R} [v(s,x) - g(x)]_-, \qquad \lim_{s\uparrow T} \sup_{|x|\leq R} [w(s,x) - g(T,x)]_-$$

are equal to zero.

PROOF. By Theorem 3.1.5,

$$\lim_{s\uparrow T} \sup_{|x|\leq R} |v_1(s,x) - g(x)| = 0.$$

Hence the assertion of the lemma follows from

$$v_1 \leq v, \qquad [v(s,x) - g(x)]_- \leq [v_1(s,x) - g(x)]_- \leq |v_1(s,x) - g(x)|.$$

We can consider the function w in a similar way, thus completing the proof of the lemma. □

9. Theorem. (a) *For each $\varepsilon > 0$ and $R > 0$ let there be $\delta > 0$ and a superharmonic (or excessive) function $u(s,x)$ in a region $(T - \delta, T) \times E_d$ such that in this region $u(s,x) \geq g(s,x)$ and $u(T,x) \leq g(T,x) + \varepsilon$ for $|x| \leq R$. Then for all $R > 0$*

$$\lim_{s\uparrow T} \sup_{|x|\leq R} |w(s,x) - g(T,x)| = 0.$$

(b) *For each $\varepsilon > 0$ and $R > 0$ let there be $\delta > 0$ and a superharmonic (or excessive) function $u(s,x)$ in a region $(T - \delta, T) \times E_d$ such that $g(x) \leq u(T,x)$ for all x, $u(T,x) \leq g(x) + \varepsilon$ for $|x| \leq R$. Then for all $R > 0$*

$$\lim_{s\uparrow T} \sup_{|x|\leq R} |v(s,x) - g(x)| = 0.$$

PROOF OF (a). We take $\varepsilon > 0$ and $R > 0$. Also, we find $\delta > 0$ and the corresponding function $u(s,x)$. Applying Lemma 2a, we consider instead of the strip H_T the strip $(T - \delta, T) \times E_d$ and, furthermore, we take the latter strip for Q. We thus have $u(s,x) \geq w(s,x)$ in $[T - \delta, T] \times E_d$, which yields

$$[u(s,x) - g(s,x)]_+ \geq [w(s,x) - g(s,x)]_+.$$

Further, due to the continuity of $u(s,x)$ and $g(s,x)$ we have

$$\overline{\lim_{s \uparrow T}} \sup_{|x| < R} [w(s,x) - g(s,x)]_+ \leq \sup_{|x| \leq R} [u(T,x) - g(T,x)]_+.$$

By hypothesis, the last expression does not exceed ε. Since ε is arbitrary,

$$\lim_{s \uparrow T} \sup_{|x| \leq R} [w(s,x) - g(s,x)]_+ = 0.$$

Comparing this result with Lemma 8 and noting that $|a| = a_+ + a_-$, we have proved assertion (a). Assertion (b) can be proved in a similar way, thus proving the theorem. □

10. Remark. We have applied the version of Lemma 2 in which $Q = H_T$. We could have used Lemma 2 in its general form but the statement of the appropriate theorem would then become too cumbersome.

Using the scheme investigated in Section 3.1 as an example, we show how to apply Theorem 9. We shall see that the assumptions of Theorem 9 can always be satisfied if we take the controlled processes given in Section 3.1. We consider only the second assertion of Theorem 9. Assume that $\varepsilon > 0$ and $R > 0$. First we find an infinitely differentiable function $\bar{g}(x)$ such that $g(x) \leq \bar{g}(x)$ for all x, $\bar{g}(x) \leq g(x) + \varepsilon$ for $|x| \leq R$. It is seen that since $|g(x)| \leq 2^m K(1 + |x|^2)^{m/2}$, one can take $\bar{g}(x) = 2^m K(1 + |x|^2)^{m/2}$ for $|x| > 2R$. Let

$$\rho(s,x) = (1 + |x|^2)^{m/2} \sqrt{T - s}, \qquad u(s,x) = \rho(s,x) + \bar{g}(x).$$

The computations, similar to those carried out after Lemma 2, yield

$$\begin{aligned} &L^\alpha u(s,x) + 2^m K(1 + |x|^2)^{m/2} \\ &\quad = \left\{ \frac{m}{2}(1 + |x|^2)^{-1} \|\sigma\|^2 + m\left(\frac{m}{2} - 1\right)(1 + |x|^2)^{-2} |\sigma^* x|^2 \right. \\ &\qquad \left. + m(1 + |x|^2)^{-1} bx - c - \frac{1}{2(T - s)} \right\} \rho(s,x) \\ &\qquad + L^\alpha \bar{g}(s,x) + 2^m K(1 + |x|^2)^{m/2}. \end{aligned}$$

Using the assumptions made in Section 3.1 about the growth order of σ and b, we easily find that in H_T

$$L^\alpha u(s,x) + 2^m K(1 + |x|^2)^{m/2} \leq \left[\bar{N} - \frac{1}{2(T - s)} \right] \rho(s,x) + \bar{N}(1 + |x|^2)^{m/2}.$$

The last expression is smaller than zero for $s \in (T - \delta, T)$ if δ is such that

$$\bar{N} + \bar{N}\frac{1}{\sqrt{\delta}} - \frac{1}{2\delta} \leq 0.$$

Hence the function u satisfies the hypotheses of Theorem 9b.

11. Exercise

We consider a 1-dimensional case: $d = d_1 = T = 1$, $A[0,\infty)$, $A_n = [0,n]$, $\sigma(\alpha,s,x) = \alpha$, $b = c = f = 0$. Let $g(x)$ be a bounded continuous function.

Prove that for $s < 1$

$$v(s,x) = \sup_y g(y), \qquad \lim_{s \uparrow 1} v(s,x) = \sup_y g(y).$$

12. Exercise

Prove Theorems 3.1.10 and 3.1.11 for $\varepsilon > 0$ in the general case if it is known that each connected component of the region Q_ε is bounded. (*Hint*: It is necessary to prove first that Q_ε is in fact a region, and second, that for each connected component of Q_ε there is a number n_0 after which it is contained in the sets $\{(s,x): w_n(s,x) > g(s,x) + (\varepsilon/2)\}$.)

13. Exercise

Prove Theorems 3.1.10 and 3.1.11 for $\varepsilon > 0$ in the general case if it is known that $|w(s,x)| + |f^\alpha(s,x)| \leq N(1 + |x|)^m$ for all α, s, x and that, in addition, for some $r > 0$ for all s, x, $\varepsilon > 0$

$$\sup_{\alpha \in \mathfrak{A}} \mathsf{M}_{s,x}^\alpha \sup_{t \leq \tau_\varepsilon} |x_t|^{m+r} < \infty.$$

2. General Methods for Estimating Derivatives of Payoff Functions

In Chapter 4 estimates of derivatives of payoff functions played an essential role. They were used, in particular, in deriving the Bellman equations. The proof of the estimates given in Chapter 4 was based on the estimates of moments of derivatives of solutions of stochastic equations with respect to the basis of the initial data in Section 2.8. In this section we give more precise estimates of derivatives of solutions of stochastic equations as well as more precise estimates of derivatives of payoff functions, omitting the proof, which will appear in a forthcoming publication.

We introduce a vector $\gamma^\alpha(t,x)$ of dimension $d \times (d_1 + d + 4)$, whose coordinates are given by the variables $\sigma^{ij}(\alpha,t,x)$ $(i = 1,\ldots,d, j = 1,\ldots,d_1)$,

$b^i(\alpha,t,x)$ $(i = 1, \ldots, d)$, $c^\alpha(t,x)$, $f^\alpha(t,x)$, $g(x)$, $g(t,x)$. We assume repeatedly in this section that the assumptions made in the previous section have been satisfied. We also assume that for each $\alpha \in A$, $t \in [0,T]$ the vector $\gamma^\alpha(t,x)$ is twice continuously differentiable in x and, in addition, for all n, $\alpha \in A_n$, $l \in E_d$, $(t,x) \in \bar{H}_T$

$$|\gamma^\alpha_{(l)}(t,x)| + |\gamma^\alpha_{(l)(l)}(t,x)| \leq K_n(1 + |x|)^{m_n}. \tag{1}$$

Besides these (tacit in the assertions of this section), we shall need additional assumptions. For convenience of reference we enumerate them.

We fix $\delta \in (0,\frac{1}{2}]$. Also, let

$$L_1^\alpha(t,x) = \frac{\partial}{\partial t} + \sum_{i,j=1}^d a^{ij}(\alpha,t,x)\frac{\partial^2}{\partial x^i\,\partial x^j} + \sum_{i=1}^d b^i(\alpha,t,x)\frac{\partial}{\partial x^i} - 2(1-\delta)c^\alpha(t,x),$$

$$L_2^\alpha(t,x) = \frac{\partial}{\partial t} + \sum_{i,j=1}^d a^{ij}(\alpha,t,x)\frac{\partial^2}{\partial x^i\,\partial x^j} + \sum_{i=1}^d b^i(\alpha,t,x)\frac{\partial}{\partial x^i} - 2\delta c^\alpha(t,x).$$

We fix five nonnegative functions $u_{-1}(t,x)$, $u_1(t,x)$, $u_{-2}(t,x)$, $u_2(t,x)$, $u_0(t,x)$, which together with two derivatives with respect to x and one derivative with respect to t are given and, in addition, are continuous in $\bar{H}_T$. On $\bar{H}_T$ let

$$L_1^\alpha u_{\pm 1} \leq 0, \qquad L_2^\alpha u_{\pm 2} \leq 0, \qquad L^\alpha u_0 \leq 0. \tag{2}$$

We note that such functions always exist; for example, $u_i \equiv 0$. However, we shall require more than inequality (2). A list of the conditions follows. Not all are required to be satisfied simultaneously. The assertions indicate specifically the various sublists of conditions.

1. Assumption (on first derivatives of σ and b). For all $(t,x) \in \bar{H}_T$, $\alpha \in A$, $l \in E_d$

$$u_1\left\{2\frac{2+\delta}{2-\delta}|\sigma^*_{(l)}l|^2 + (1+\delta)|l|^2\|\sigma_{(l)}\|^2 + 2|l|b_{(l)}l\right\}$$
$$+ 2|l|l\sigma_{(l)}\sigma^*\operatorname{grad}_x u_1 \leq -\frac{1-\delta}{2}|l|^2 L_1^\alpha u_1, \tag{3}$$

where we have omitted for the sake of brevity the arguments (t,x) in the function $u_1(t,x)$ and the operator $L_1^\alpha(t,x)$, and also the arguments (α,t,x) in the functions $\sigma(\alpha,t,x)$, $b(\alpha,t,x)$.

We shall omit the arguments in the other assumptions in a similar way.

2. Assumption (on second derivatives of σ and b). For all $(t,x) \in \bar{H}_T$, $\alpha \in A$, $\alpha \in A$, $l \in E_d$

$$u_1\|\sigma_{(l)(l)}\|^2 \leq (-L_1^\alpha u_1)^{1-(\delta/2)}(-L_1^\alpha u_{-1})^{\delta/2}, \tag{4}$$

$$|u_1 b_{(l)(l)} + \sigma_{(l)(l)}\sigma^*\operatorname{grad}_x u_1| \leq (-L_1^\alpha u_1)^{1-(\delta/4)}(-L_1^\alpha u_{-1})^{\delta/4}. \tag{5}$$

3. Assumption (on first derivatives of c, f, g). (a) For all $(t,x) \in \bar{H}_T$, $\alpha \in A_n$, $l \in E_d$, $u = v_n$, w_n $(n = 1,2, \ldots)$

$$|f^{\alpha}_{(l)} - c^{\alpha}_{(l)}u| \leq (-L^{\alpha}_1 u_1)^{1/2}(-L^{\alpha}_2 u_2)^{1/2}, \tag{6}$$

$$\sqrt{u_1 u_2}|c^{\alpha}_{(l)}| \leq (-L^{\alpha}_1 u_1)^{(1/2)-(\delta/4)}(-L^{\alpha}_2 u_{-2})^{(1/2+(\delta/4)}, \tag{7}$$

$$|g_{(l)}(t,x)| \leq \sqrt{u_1 u_2(t,x)}, \qquad |g_{(l)}(x)| \leq \sqrt{u_1 u_2(T,x)} \tag{8}$$

or

(b) $c^{\alpha}(t,x)$ does not depend on x. There exists a linear subspace $E^{\mathscr{L}}_d$ of the space E_d such that for each $l' \in E^{\mathscr{L}}_d$, $l'' \perp E^{\mathscr{L}}_d$, $\alpha \in A$, $(t,x) \in \bar{H}_T$

$$\sigma^{*}_{(l'')}(\alpha,t,x)l' = 0, \qquad l'b_{(l'')}(\alpha,t,x) = 0. \tag{9}$$

Furthermore, for each $l' \in E^{\mathscr{L}}_d$, $l \in E_d$, $\alpha \in A$, $(t,x) \in \bar{H}_T$

$$\sigma^{*}_{(l)(l)}(\alpha,t,x)l' = 0, \qquad l'b_{(l)(l)}(\alpha,t,x) = 0, \tag{10}$$

Finally, for all $(t,x) \in \bar{H}_T$, $\alpha \in A$, $l'' \perp E^{\mathscr{L}}_d$

$$|f^{\alpha}_{(l'')}| \leq (-L^{\alpha}_1 u_1)^{1/2}(-L^{\alpha}_2 u_2)^{1/2}, \tag{11}$$

$$|g_{(l'')}(t,x)| \leq \sqrt{u_1 u_2(t,x)}, \qquad |g_{(l'')}(x)| \leq \sqrt{u_1 u_2(T,x)}. \tag{12}$$

4. Assumption (on second derivatives of c, f, g). For all $(t,x) \in \bar{H}_T$, $\alpha \in A_n$, $l \in E_d$, $u = v_n$, w_n $(n = 1,2, \ldots)$

$$[f^{\alpha}_{(l)(l)} - c^{\alpha}_{(l)(l)}u]_- \leq (-L^{\alpha}_1 u_1)^{(1/2-(\delta/4)}(-L^{\alpha}_2 u_2)^{(1/2+(\delta/4)}, \tag{13}$$

$$\begin{aligned} [g_{(l)(l)}(t,x)]_- &\leq u_1^{(1/2)-(\delta/4)}u_2^{(1/2)+(\delta/4)}(t,x), \\ [g_{(l)(l)}(x)]_- &\leq u_1^{(1/2-(\delta/4)}u_2^{(1/2)+(\delta/4)}(T,x). \end{aligned} \tag{14}$$

We shall formulate later two more assumptions. However, right now we discuss some techniques for verifying the assumptions listed in various places.

First we note that if in a specific case it is possible to choose functions u_2 so that each function satisfies only one of the inequalities (6), (8), or (11)–(14), the sum of these functions can be used as u_2 in all the inequalities. The functions u_{-1}, u_{-2} appear only on the right-hand sides of the inequalities (4), (5), and (7). The left-hand sides of the inequalities contain u_1 only, with the exception of inequality (7) in the left-hand side of which u_2 appears also. The greatest number of conditions is imposed on the function u_1. Assumption 1 being the most stringent. Inequality (3) being satisfied depends only on the appropriate choice of the function u_1. In the remaining inequalities, the right-hand sides contain functions which do not appear in the left-hand sides so that these inequalities can be satisfied by choosing these functions appropriately. In this connection, it is useful to bear in mind that since one can multiply or divide these inequalities by $|l|^2$, (3) can be satisfied for all $l \in E_d$ if and only if (3) is satisfied for all unit vectors l.

Further, we can easily verify Assumption 1 if first derivatives of σ and b with respect to x are bounded by the same constant for all α, t, x. In fact, in this situation we assume that $u_1(t,x) = e^{-Nt}$. Then, the left-hand side of (3) does not, obviously, exceed $N_1 e^{-Nt}|l|^2$, where the constant N_1 does not depend on α, t, x. The right-hand side of (3) is equal to

$$\frac{1-\delta}{2}|l|^2 N e^{-Nt} + (1-\delta)^2 |l|^2 c^\alpha(t,x) e^{-Nt}.$$

Since $c^\alpha \geq 0$, one can, for example, assume that $\delta = \frac{1}{2}$, $N = 4N_1$. In this case the function u_1 constructed satisfies (3).

The constant T does not appear in an explicit manner in the assumptions made, which fact enables us to employ our assumptions for control over an infinite interval. Note that for $T = \infty$ it is sometimes inconvenient to take a function of the form e^{-Nt} as the function u_1. In such cases one can keep in mind that Assumption 1 will be satisfied if, for example, $c^\alpha(t,x)$ is "sufficiently" large compared to first derivatives of σ and b with respect to x, more precisely, if

$$2\frac{2+\delta}{2-\delta}|\sigma^*_{(l)} l|^2 + (1+\delta)\|\sigma_{(l)}\|^2 + 2b_{(l)} l \leq (1-\delta)^2 c^\alpha \tag{15}$$

for all $(t,x) \in \bar{H}_T$, $\alpha \in A$ and for unit $l \in E_d$. Indeed, (15) coincides with (3) for $u_1 \equiv 1$, $|l| = 1$. (Inequality (15) is discussed in more detail in the notes to this chapter.)

We have given two cases in which Assumption 1 is satisfied. We note that the first case holds in the scheme investigated in Chapters 3–5. Indeed, if the constant K_n in (1.1) does not depend on n and, in addition, is equal to K, the assumption on differentiability of σ and b enables us to rewrite (1.1) as $\|\sigma_{(l)}\| + |b_{(l)}| \leq K$. In this connection, we show that if the conditions under which Theorems 4.7.4, 4.7.5, and 4.7.7 on smoothness of a payoff function and, also, on the Bellman equation were proved, Assumptions 1–4 will be satisfied as well. In other words, we wish to show that if $A_1 = A_2 = \cdots = A$, $K_1 = K_2 = \cdots = K$, $m_1 = m_2 = \cdots = m$, there always exist functions u_i and a number δ which satisfy Assumptions 1–4. Here as well as in a similar situation in Section 1, our objective is purely methodological.

In order to have (3) satisfied, we take $\delta = \frac{1}{2}$, $u_1 = e^{-N_1 t}$, and furthermore, we choose an appropriate N_1. Since

$$u_1\|\sigma_{(l)(l)}\|^2 \leq K^2(1+|x|)^{2m}, \qquad -L_1^\alpha u_1 \geq N_1 e^{-N_1 T},$$
$$|u_1 b_{(l)(l)} + \sigma_{(l)(l)}\sigma^* \operatorname{grad}_x u_1| = u_1|b_{(l)(l)}| \leq K(1+|x|)^m,$$

for verifying Assumption 2 we need only to choose u_{-1} from the condition

$$N_2(1+|x|)^{8m} \leq -L_1^\alpha u_{-1}, \qquad L_1^\alpha u_{-1} + N_2(1+|x|)^{8m} \leq 0,$$

where N_2 is a constant. Repeating the arguments given after Lemma 1.2[2] in Section 1, we easily convince ourselves that for u_{-1} we can take a function of the form $N_3 e^{-N_4 t}(1+|x|^2)^{4m}$.

[2] In these arguments one needs to take $8m$ instead of m.

We see from Chapter 3 that $|v|, |w| \le N(1 + |x|)^m$. Since for $u = v, w$

$$|f^\alpha_{(l)} - c^\alpha_{(l)}u| \le |f^\alpha_{(l)}| + |c^\alpha_{(l)}| \cdot |u| \le N(1 + |x|)^{2m},$$

the function u_2 for inequality (6) can be sought from the condition

$$L^\alpha_2 u_2 + N(1 + |x|)^{4m} \le 0,$$

where N is an appropriate constant. It is seen that the function u_2, equal to $N_5 e^{-N_6 t}(1 + |x|^2)^{2m}$, is applicable for a certain choice of constants N_5, N_6. Similarly, to have (13) satisfied, it suffices to solve an inequality of the form

$$L^\alpha_2 u_2 + N(1 + |x|)^{8m/(2+\delta)} \le 0.$$

As we saw in Section 1, this can easily be done. Finally, choosing functions of the form $N_7 e^{-N_8 t}(1 + |x|^2)^n$ which satisfy inequalities (8), (14), and next adding up all the expressions thus found for u_2, we obtain the function to be taken as u_2 in inequalities (6), (8), (13), and (14) in the large. Since we had dealt with an expression of the form $N_9 e^{-N_{10} t}(1 + |x|^2)^n$, each time, the final version of the function u_2 thus obtained satisfies the inequality $|u_2| \le N(1 + |x|)^n$, using which we can easily find the function u_{-2} satisfying (7). Therefore, Assumptions 1–4 have been satisfied.

5. Exercise

Summarizing the above arguments, show that in the case considered one can take

$$\delta = \tfrac{1}{2},$$
$$u_1 = e^{N(T-t)}, \qquad u_{-1} = Ne^{N(T-t)}(1 + |x|^2)^{4m},$$
$$u_2 = Ne^{N(T-t)}(1 + |x|^2)^{2m},$$
$$u_{-2} = Ne^{N(T-t)}(1 + |x|^2)^{(12/5)m}.$$

Also, we write down what Assumptions 1–4 become if we take constants for the functions u_i and if, in addition, we consider only Assumption 3a. Here (3) becomes (15). Inequality (4) can obviously be satisfied if there exists a constant N such that

$$\|\sigma_{(l)(l)}\| \le (c^\alpha)^{1-(\delta/2)}(Nc^\alpha)^{\delta/2} = N^{\delta/2} c^\alpha.$$

In other words, it suffices to take a constant N for which

$$\|\sigma_{(l)(l)}\| \le Nc^\alpha. \tag{16}$$

Considering the remaining inequalities in a similar way, we can see that Assumptions 1–4 can be satisfied if inequalities (15) and (16) are satisfied and, furthermore, if there exists a constant N such that

$$\begin{gathered} |b_{(l)(l)}| \le Nc^\alpha, \qquad |c^\alpha_{(l)}| \le Nc^\alpha, \qquad |g_{(l)}(t,x)| \le N, \\ |g_{(l)}(x)| \le N, \qquad [g_{(l)(l)}(t,x)]_- \le N, \qquad [g_{(l)(l)}(x)]_- \le N, \\ |f^\alpha_{(l)} - c^\alpha_{(l)}u| \le Nc^\alpha, \qquad [f^\alpha_{(l)(l)} - c^\alpha_{(l)(l)}u]_- \le Nc^\alpha \end{gathered} \tag{17}$$

for all $(t,x) \in \bar{H}_T$, $\alpha \in A_n$, $l \in E_d$, $u = v_n$, w_n $(n = 1,2,\ldots)$. In this case we can say that c^α is sufficiently large and that the derivatives of the functions $g(t,x)$, $g(x)$ are bounded.

On the face of it, the last inequalities in (17) as well as inequalities (6) and (13) seem to be rather strange. These inequalities contain the functions v_n and w_n, which are in general unknown. In this connection, we note that

$$|f^\alpha_{(l)} - c^\alpha_{(l)}u| \le |f^\alpha_{(l)}| + |c^\alpha_{(l)}| \cdot |u|,$$
$$[f^\alpha_{(l)(l)} - c^\alpha_{(l)(l)}u]_- \le [f^\alpha_{(l)(l)}]_- + [c^\alpha_{(l)(l)}u]_+ \le [f^\alpha_{(l)(l)}]_- + |c^\alpha_{(l)(l)}| \cdot |u|.$$

Hence, if, for example, we could, using (1.5) and Lemma 1.2, estimate the functions v_n and w_n and also, we could prove that $|v_n| \le \tilde{u}$ and $|w_n| \le \tilde{u}$ for a function $\tilde{u}$, (6) and (13) will be satisfied if

$$|f^\alpha_{(l)}| \le (-L^\alpha_1 u_1)^{1/2}(-L^\alpha_2 u_2)^{1/2}, \qquad |c^\alpha_{(l)}|\tilde{u} \le (-L^\alpha_1 u_1)^{1/2}(-L^\alpha_2 u_2)^{1/2},$$
$$|f^\alpha_{(l)(l)}| \le (-L^\alpha_1 u_1)^{(1/2)-(\delta/4)}(-L^\alpha_2 u_2)^{(1/2)+(\delta/4)},$$
$$|c^\alpha_{(l)(l)}|\tilde{u} \le (-L^\alpha_1 u_1)^{(1/2)-(\delta/4)}(-L^\alpha_2 u_2)^{(1/2)+(\delta/4)}.$$

The last remarks concerning (6), (13), and (17) are unnecessary if $c^\alpha(t,x)$ does not depend on x. In such a case $c^\alpha_{(l)} = 0$ and $c^\alpha_{(l)(l)} = 0$, the functions v_n and w_n do not belong to (6), (13), and (17), and, finally, these functions need not be estimated. This fact shows how convenient it is to write (6) and (13) as we did above.

In the case where $c^\alpha(t,x)$ does not depend on x, Assumption 3 is regarded as satisfied if we succeed in satisfying conditions (b). Let us discuss this case. Inequalities (11) and (12) are particular cases of inequalities (6) and (8) since $c_{(l)} \equiv 0$ and also since in (11) and (12) we consider not all $l'' \in E_d$ but only those which are orthogonal to a subspace $E^{\mathscr{L}}_d$. We can posit two extreme possibilities: $E^{\mathscr{L}}_d$ contains only a zero vector; $E^{\mathscr{L}}_d$ coincides with the entire space E_d.

If the first possibility is realized, equalities (9) and (10) hold for arbitrary σ and b, any vector $l'' \perp E^{\mathscr{L}}_d$, and (11) and (12) must be satisfied for all $(t,x) \in \bar{H}_T$, $\alpha \in A^d$, $l'' \in E_d$. Also, in this case conditions (a) are satisfied. The advantage of considering conditions (b) along with conditions (a) is obvious when $E^{\mathscr{L}}_d = E_d$. Here only the zero vector is orthogonal to $E^{\mathscr{L}}_d$; furthermore, inequalities (11) and (12) are automatically satisfied ($u_{(l)} \equiv 0$ for $l = 0$), as well as Eq. (9).

Therefore, if c^α does not depend on x, Assumption 3 is satisfied if, for example, for all l', $l \in E_d$ (10) is satisfied; in this case we can assume $E^{\mathscr{L}}_d = E_d$. We note that, as can easily be seen, (10) holds for all l', $l \in E_d$, $\alpha \in A$, $(t,x) \in \bar{H}_T$ if and only if the functions $\sigma(\alpha,t,x)$ and $b(\alpha,t,x)$ are linear with respect to x for all $\alpha \in A$, $t \in [0,T]$. It is seen that Assumption 2 is satisfied as well.

In order to consider the intermediate possibility $0 \ne E^{\mathscr{L}}_d \ne E_d$, let us imagine that the space $E^{\mathscr{L}}_d$ is generated by the first coordinate vectors $e_1, e_2, \ldots, e_{d_0}$, where $1 \le d_0 < d$. Equalities (9) and (10) have then to be satisfied for the vectors l' with arbitrary first d_0 coordinates as well as coordinates numbered $d_0 + 1, \ldots, d$, which are equal to zero. From this condition it readily follows that for all $l'' \perp E_d^{\mathscr{L}}$, $l \in E_d$, $i = 1, \ldots, d_0$; $j = 1, \ldots, d_1$

$$\begin{aligned} \sigma^{ij}_{(l'')}(\alpha,t,x) &= 0, \qquad & b^i_{(l'')}(\alpha,t,x) &= 0, \\ \sigma^{ij}_{(l)(l)}(\alpha,t,x) &= 0, \qquad & b^i_{(l)(l)}(\alpha,t,x) &= 0. \end{aligned} \tag{18}$$

The second relations in (18) imply that the first d_0 rows of the matrix σ and the first d_0 coordinates of the vector b depend on the coordinates of x in a linear way. The first relations in (18) show that the elements of the matrix σ and the elements of the vector b do not depend on $x^{d_0+1}, \ldots, x^d$. Therefore, the system

$$dx_t = \sigma(\alpha_t, s + t, x_t)\, d\mathbf{w}_t + b(\alpha_t, s + t, x_t)\, dt$$

splits into two parts: the "upper" part for the coordinates x_t^i ($i = 1, \ldots, d_0$) and the "lower" part for the coordinates x_t^i ($i = d_0 + 1, \ldots, d$), the "upper" system being linear with respect to $x_t^1, \ldots, x_t^{d_0}$ and, in addition, containing no other unknowns.

We shall complete this discussion of Assumptions 1–4 by suggesting the reader should do

6. Exercise

Show that Assumptions 1–4 can be satisfied if σ, b, c do not depend on x and, also, for all $(t,x) \in \bar{H}_T$, $\alpha \in A$, $l \in E_d$

$$\begin{aligned} [f^\alpha_{(l)(l)}]_- &\le (-L_1^\alpha u_1)^{(1/2)-(\delta/4)}(-L_2^\alpha u_2)^{(1/2)+(\delta/4)}, \\ [g_{(l)(l)}(t,x)]_- &\le u_1^{(1/2)-(\delta/4)} u_2^{(1/2)+(\delta/4)}(t,x), \\ [g_{(l)(l)}(x)]_- &\le u_1^{(1/2)-(\delta/4)} u_2^{(1/2)+(\delta/4)}(T,x). \end{aligned}$$

Note that the last inequalities are automatically satisfied if f^α, $g(t,x)$, $g(x)$ are downward convex with respect to x.

7. Assumption (on derivatives with respect to t). (a) The functions f, g, σ, b, c are continuously (with respect to (t,x)) differentiable with respect to (t,x) for all $(t,x) \in \bar{H}_T$, n, $\alpha \in A_n$,

$$\left\| \frac{\partial}{\partial t}\sigma \right\| + \left| \frac{\partial}{\partial t} b \right| + \left| \frac{\partial}{\partial t} c \right| + \left| \frac{\partial}{\partial t} f \right| + \left| \frac{\partial}{\partial t} g(t,x) \right| \le K_n(1 + |x|)^{m_n} \tag{19}$$

for all $(t,x) \in \bar{H}_T$, $\alpha \in A$

$$u_1 \left\| \frac{\partial}{\partial t}\sigma \right\|^2 \le -L_1^\alpha u_{-1}, \tag{20}$$

$$\left| u_1 \frac{\partial}{\partial t} b + \left[\frac{\partial}{\partial t}\sigma \right] \sigma^* \operatorname{grad}_x u_1 \right| \le (-L_1^\alpha u_1)^{1/2}(-L_1^\alpha u_{-1})^{1/2}. \tag{21}$$

(b) Furthermore, for all $(t,x) \in \bar{H}_T$, $\alpha \in A_n$, $u = v_n$, w_n $(n = 1,2, \ldots)$

$$\left[\frac{\partial}{\partial t} f^\alpha - u \frac{\partial}{\partial t} c^\alpha\right]_- \leq -L^\alpha u_0, \tag{22}$$

$$\left[\frac{\partial}{\partial t} g(t,x)\right]_- \leq u_0(t,x). \tag{23}$$

(c) Finally, the derivatives $g_{x^i x^j}(t,x)$, $g_{x^i x^j}(x)$ are continuous in $\bar{H}_T$ and, also, for all $x \in E_d$

$$\begin{gathered} -\frac{\partial}{\partial t} g(T,x) + L^\alpha g(T,x) + f^\alpha(T,x) \leq u_0(T,x), \\ L^\alpha(T,x)g(x) + f^\alpha(T,x) \leq u_0(T,x). \end{gathered} \tag{24}$$

Much of what has been said above could easily be repeated for Assumption 7. We note only that in verifying that Assumption 7 is satisfied, one need not attempt to find immediately the function u_{-1} which simultaneously satisfies inequalities (20) and (21) (as well as inequalities (4) and (5)). One can find one function for (20) and another function for (21), and, next, one can take as u_{-1} the sum of the functions thus obtained. The same holds for u_0 from (22)–(24). Moreover, one should keep in mind that $L^\alpha g(T,x)$ in (24) is the quantity at a point (T,x), obtained as a result of application of the operator $L^\alpha(t,x)$ to the function $g(t,x)$, and, in addition, that $L^\alpha(T,x)g(x)$ is to be regarded as the value of the operator $L^\alpha(T,x)$ evaluated on the function $g(x)$.

8. Assumption (summarizing). Assumptions 1, 2, 3b, 4, and 7 are satisfied, also, for $l' \in E_d$, $\alpha \in A$, $(t,x) \in \bar{H}_T$,

$$\frac{\partial}{\partial t} \sigma^*(\alpha,t,x)l' = 0, \qquad l' \frac{\partial}{\partial t} b(\alpha,t,x) = 0, \tag{25}$$

or Assumptions 1, 2, 3a, 4, and 7 are satisfied.

We see from the results obtained in Chapter 4 that derivatives of the functions v and w can be estimated only on the sets on which these derivatives belong to the operator $F[u]$. In this connection, here as well as in Section 4.7, let $Q^* = \{(t,x) \in H_T : \sup_{\alpha \in A} (a(\alpha,t,x)\lambda,\lambda) > 0 \text{ for all } \lambda \neq 0\}$,

$$n^\alpha(t,x) = (1 + \operatorname{tr} a(\alpha,t,x) + |b(\alpha,t,x)| + c^\alpha(t,x) + |f^\alpha(t,x)|)^{-1},$$

$$\mu = \mu(t,x) = \inf_{|\lambda|=1} \sup_{\alpha \in A} n^\alpha(t,x)(a(\alpha,t,x)\lambda,\lambda).$$

We also introduce Q_n^* and μ_n $(n = 1,2, \ldots)$ using similar relations in which we put A_n instead of A.

9. Lemma. $\mu_n \leq \mu_{n+1}$, $\mu = \lim_{n\to\infty} \mu_n$, *the function* $\mu(t,x)$ *is lower semicontinuous in* $\bar{H}_T$, *the equalities*

$$Q^* = \{(t,x) \in H_T : \mu(t,x) > 0\} = \bigcup_{n=1}^{\infty} Q_n^*$$

are satisfied, the set Q^* is open, and, finally, the function $\mu^{-1}(t,x)$ is locally bounded on Q^*.

PROOF. The inequality $\mu_n \le \mu_{n+1}$ is obvious. Further, due to the boundedness of $n^\alpha(t,x)a(\alpha,t,x)$, which is uniform with respect to (α,t,x), the quadratic forms $n^\alpha(t,x)(a(\alpha,t,x)\lambda,\lambda)$ and the functions

$$\sup_{\alpha\in A_n} n^\alpha(t,x)(a(\alpha,t,x)\lambda,\lambda), \qquad \sup_{\alpha\in A} n^\alpha(t,x)(a(\alpha,t,x)\lambda,\lambda)$$

are continuous with respect to $\lambda \in S_1$ uniformly with respect to α, t, x, n. In accord with Dini's theorem it follows from the obvious relation

$$\sup_{\alpha\in A_n} n^\alpha(t,x)(a(\alpha,t,x)\lambda,\lambda) \uparrow \sup_{\alpha\in A} n^\alpha(t,x)(a(\alpha,t,x)\lambda,\lambda)$$

that in this case the convergence is uniform with respect to $\lambda \in \partial S_1$ for each t, x. In particular, the lower bounds of the foregoing expressions converge, that is, $\mu_n(t,x) \to \mu(t,x)$ as $n \to 0$. Since the limit of an increasing sequence of continuous functions is lower semicontinuous and, furthermore, by Lemma 4.7.3 the functions $\mu_n(t,x)$ are continuous, $\mu(t,x)$ is lower semicontinuous. The equality $Q^* = \bigcup Q_n^*$ follows from the fact that $\mu_n(t,x) \uparrow \mu(t,x)$. We can prove the other equality fc.: Q^* by repeating the appropriate arguments from the proof of Lemma 4.7.3. Finally, the last assertions follow from the continuity of $\mu_n(t,x)$ and also from the monotone convergence of $\mu_n(t,x)$ to $\mu(t,x)$. The lemma is proved. □

We formulate now the main results obtained in estimating derivatives of the functions v and w. It should be borne in mind that we shall give some assertions of Theorems 10–12 in a short form, in other words, we assert, for example, in Theorem 11 that for each $x \in E_d$ the function $v(t,x)$ is left continuous with respect to t on $[0,T]$, and that the function

$$\psi(t,x) \equiv v(t,x) + \int_0^t \bar{u}_2(r,x)\,dx$$

increases (does not decrease) with respect to t on $[0,T]$.

We fix $x \in E_d$. The function $\psi(t,x)$, as any increasing function, can have not more than a countable set of discontinuity points; all the discontinuities are of the first kind and, in addition, $\psi(t-,x) \le \psi(t,x) \le \psi(t+,x)$ for each t. Since an integral with respect to r is a continuous function of t, we conclude that for each $x \in E_d$ the function $v(t,x)$, being the function of t, can have not more than a countable set of discontinuity points, all of whose discontinuities are of first kind, and, in addition, that $v(t,x) \le v(t+,x)$ for each t, that is, the graph of $v(t,x)$ can have upward jumps only.

The properties of $v(t,x)$ listed above do not appear in Theorem 11. A similar situation is observed for the corresponding assertions of Theorems 10 and 12. In Theorem 10, the appropriate auxiliary function decreases, and hence $v(t,x)$ has downward jumps only.

Finally, we note that from the inequality $\psi(t-,x) \le \psi(t,x)$ follows the inequality $v(t-,x) \le v(t,x)$. Furthermore, by Theorem 1.5, the function $v(t,x)$

is lower semicontinuous. Therefore, $v(t,x) \le v(t-,x)$. Comparing the last inequality with the previous one, we obtain $v(t,x) = v(t-,x)$. Thus, the assertion about the left continuity of $v(t,x)$ with respect to t in Theorem 11 is a consequence of Theorem 1.5 and of the fact that $\psi(t,x)$ increases with respect to t.

10. Theorem. *Suppose that a function v is bounded above in each cylinder $\bar{C}_{T,R}$ and also that Assumptions 1–4 are satisfied. Then:*

a. *for each l_1, $l_2 \in E_d$ inside H_T there exist generalized derivatives $v_{(l_1)(l_2)}(t,x)(dt\,dx)$ (see Definition 2.1.2), and also for each $l \in E_d$ inside H_T*

$$v_{(l)(l)}(t,x)(dt\,dx) \ge \bar{u}_1(t,x)\,dt\,dx,$$

where

$$\bar{u}_1 = 4\delta^{-3/2} u_1^{(1/2)-(\delta/4)}\left[u_2^{(1/2)+(\delta/4)} + u_{-1}^{\delta/4} u_2^{1/2} + u_{-2}^{(1/2)+(\delta/4)}\right];$$

b. *for each $t \in [0,T]$ the function $v(t,x)$ is continuous with respect to x, has a generalized derivative with respect to x (in the sense of Definition 2.1.1), and, in addition, for each $R > 0$, $\gamma \in (0,1]$, $t \in [0,T]$ almost everywhere in S_R*

$$|\operatorname{grad}_x v(t,x)| \le \frac{2}{\gamma}\|v(t,\cdot)\|_{B(S_{R+1})} + \frac{\gamma}{2}\|\bar{u}_1(t,\cdot)\|_{B(S_{R+1})};$$

c. *inside H_T there exists a derivative $(\partial/\partial t)v(t,x)(dt\,dx)$,*

$$\frac{\partial}{\partial t} v(dt\,dx) \le \inf_{\beta \in A} |L^\beta|(\bar{u}_1 + |\operatorname{grad}_x v| + v_+ + 1)\,dt\,dx,$$

$$L^\beta v(dt\,dx) + f^\beta\,dt\,dx \le 0^3$$

for each $\beta \in A$; for each $l \in E_d$

$$\mu v_{(l)(l)}(dt\,dx) \le (\bar{u}_1 + |\operatorname{grad}_x v| + v_+ + 1)\,dt\,dx + \left(\frac{\partial}{\partial t} v\right)_-(dt\,dx);$$

d. *for each $x \in E_d$ the function $v(t,x)$ is right continuous with respect to t on $[0,T]$, and, furthermore, for each $R > 0$ there is a constant $N \ge 0$ such that for all $x \in S_R$ the function $v(t,x) - Nt$ decreases with respect to t on $[0,T]$.*

Finally, if we replace v by w in the above formulations, assertions (a)–(d) *will still hold.*

11. Theorem. *Let Assumptions* 1, 3b, *and* 7 *be satisfied; in this case for all $l' \in E_d^{\mathscr{L}}$, $\alpha \in A$, $(t,x) \in \bar{H}_T$ let equalities* (25) *be satisfied. Or let Assumptions* 1, 3a, *and* 7 *be satisfied. Then:*

a. *the function $v(t,x)$ is bounded in each cylinder $\bar{C}_{T,R}$, for each $x \in E_d$ the function $v(t,x)$ is left continuous with respect to t on $[0,T]$, and, finally,*

[3] The meaning of $|L^\beta|$ is explained in the statement of Theorem 4.3.3; $L^\beta u(dt\,dx)$ is defined prior to Theorem 4.2.7.

for each $x \in E_d$ *the function*

$$v(t,x) + \int_0^t \bar{u}_2(r,x)\,dr$$

increases with respect to t on $[0,T]$, *where* $\bar{u}_2 = (7\delta^{-1}u_{-1}u_2)^{1/2} + u_0$;

b. *inside* H_T *there exists a derivative* $(\partial/\partial t)v(dt\,dx)$ *and*

$$\frac{\partial}{\partial t} v(t,x)(dt\,dx) \geq -\bar{u}_2(t,x)\,dt\,dx;$$

c. *for each* $R \geq 0$

$$\lim_{t \uparrow T} \sup_{|x| \leq R} |v(t,x) - g(x)| = 0.$$

Assertions (a)–(c) *will remain true if we replace in them v by w and* $g(x)$ *by* $g(T,x)$.

The next theorem holds if the assumptions given in the preceding section only are satisfied. We could have formulated it in Section 1; however, it is more closely related to the discussion in this section.

12. Theorem. *Let there exist a continuous function* $\varphi(t,x)$ *given on* $\bar{H}_T$, *and also a number* $\delta_1 > 0$ *such that for each* $t \in (0,T]$ *the function* $u_t(s,x) \equiv v(t,x) + \int_s^t \varphi(r,x)\,dr$, *being a function of the variables* (s,x), *is superharmonic (or excessive) in the region* $H_T \cap (t - \delta_1, t) \times E_d$ *(see Definitions* 5.3.2, 5.3.3). *Then, we can prove Theorem* 11a,b,c *if we replace* $\bar{u}_2$ *by* φ.

13. Theorem. *Let the summarizing Assumption* 8 *be satisfied. Then the functions* v_n, v, w_n *and* w *are uniformly bounded and also equicontinuous in each cylinder* $\bar{C}_{T,R}$. *Generalized first derivatives of these functions with respect to* (t,x) *are uniformly bounded in each cylinder* $\bar{C}_{T,R}$. *Moreover, each of the foregoing functions satisfies the inequalities*

$$-\bar{u}_2 \leq \frac{\partial}{\partial t} u \leq \inf_{\beta \in A_1} |L^\beta|(\bar{u}_1 + |\mathrm{grad}_x u| + u_+ + 1) \qquad (H_T\text{-a.e.}),$$

$$|\mathrm{grad}_x u| \leq \frac{2}{\gamma} \|u\|_{B(C_{T,R+1})} + \frac{\gamma}{2} \|\bar{u}_1\|_{B(C_{T,R+1})} \qquad (C_{T,R}\text{-a.e.})$$

for any $R > 0$, $\gamma \in (0,1]$. *Finally,* $v_n \uparrow v$ *and* $w_n \uparrow w$ *as* $n \to \infty$ *uniformly in* $\bar{C}_{T,R}$ *for each* $R > 0$.

14. Theorem. *Let the summarizing Assumption* 8 *be satisfied. Then in the region* Q^* *the functions* $v(t,x)$ *and* $w(t,x)$ *have all generalized second derivatives with respect to x, which are locally bounded in* Q^*. *Moreover, for each* $l \in E_d$

$$-\bar{u}_1 \le v_{(l)(l)} \le \frac{1}{\mu}\left[\bar{u}_1 + \left(\frac{\partial}{\partial t} v\right)_- + |\operatorname{grad}_x v| + v_+ + 1\right],$$

$$-\bar{u}_1 \le w_{(l)(l)} \le \frac{1}{\mu}\left[\bar{u}_1 + \left(\frac{\partial}{\partial t} w\right)_- + |\operatorname{grad}_x w| + w_+ + 1\right]$$

almost everywhere in Q^*. Finally, for each n_0 second derivatives of the functions v_n and w_n with respect to x for $n \ge n_0$ are uniformly bounded with respect to x and, also, $n \ge n_0$ in each bounded region which together with its closure lies in $Q^*_{n_0}$.

Theorems 10–12 are of an auxiliary nature in relation to Theorems 13 and 14. The former theorems can in turn be derived from the following estimates of derivatives of solutions of stochastic equations.

For $\alpha \in \mathfrak{A}$, $|l| = 1$, $(s,x) \in \bar{H}_T$ let

$$y_t^{\alpha,s,x} = \mathscr{L}B\text{-}\frac{\partial}{\partial l} x_t^{\alpha,s,x}, \qquad z_t^{\alpha,s,x} = \mathscr{L}B\text{-}\frac{\partial^2}{\partial l^2} x_t^{\alpha,s,x}.$$

Since by the definition of the set $\mathfrak{A}$ it follows from the inclusion $\alpha \in \mathfrak{A}$ that $\alpha \in \mathfrak{A}_n$ for a certain n, due to the results obtained in Section 2.8 the processes $y_t^{\alpha,s,x}$, $z_t^{\alpha,s,x}$ exist and are continuous on $[0,T]$. Let

$$q_t^{\alpha,s,x} = \mathscr{L}B\text{-}\frac{\partial}{\partial s} x_t^{\alpha,s,x}.$$

This process is defined for $\alpha \in \mathfrak{A}$, $(s,x) \in H_T$ on a time interval $[0, T\text{-}s]$ if assumption 7a is satisfied.

15. Theorem. *Let Assumptions 1 and 2 be satisfied. Then for each $(s,x) \in H_T$, $\tau \in \mathfrak{M}(T-s)$, $\alpha \in \mathfrak{A}$*

$$10\delta^{-3} u_1^{1-(\delta/2)} u_{-1}^{\delta/2}(s,x) \ge \mathsf{M}_{s,x}^{\alpha}\left\{u_1(s+\tau, x_\tau)|z_\tau|^2 e^{-2(1-\delta)\varphi_\tau} - \int_0^\tau |z_t|^2 e^{-2(1-\delta)\varphi_t} L_1^{\alpha_t} u_1(s+t, x_t)\,dt\right\}.$$

Furthermore, for each $t_1 \in [0, T-s]$, $\gamma \in [0, 8/(2-\delta)]$ on a set $\{\tau \ge t_1\}$ almost surely

$$2\delta^{-1} u_1(s+t_1, x_{t_1}^{\alpha,s,x})|y_{t_1}^{\alpha,s,x}|^\gamma e^{-2(1-\delta)\varphi_{t_1}^{\alpha,s,x}} \ge \mathsf{M}_{s,x}^{\alpha}\left\{u_1(s+\tau, x_\tau)|y_\tau|^\gamma e^{-2(1-\delta)\varphi_\tau} - \int_{t_1}^\tau |y_t|^\gamma e^{-2(1-\delta)\varphi_t} L_1^{\alpha_t} u_1(s+t, x_t)\,dt \,\Big|\, \mathscr{F}_{t_1}\right\}.$$

16. Theorem. *Let Assumptions* 1 *and* 7a *be satisfied. Then for each* $(s,x) \in H_T$, $\tau \in \mathfrak{M}(T-s)$, $\alpha \in \mathfrak{A}$

$$7\delta^{-1}u_{-1}(s,x) \geq \mathsf{M}^{\alpha}_{s,x}\left\{u_1(s+\tau, x_\tau)|q_\tau|^2 e^{-2(1-\delta)\varphi_\tau} - \int_0^\tau |q_t|^2 e^{-2(1-\delta)\varphi_t} L_1^{\alpha_t} u_1(s+t, x_t)\,dt\right\}.$$

17. Exercise

Making use of Exercise 5 and Theorems 13 and 14, prove Theorem 4.7.4 and, also, that in that theorem

$$|\text{grad}_x u| \leq N(1+|x|)^{2m} e^{N(T-s)} \qquad (H_T\text{-a.e.}).$$

3. The Normed Bellman Equation

As is seen from Section 1.2 (also, see below Exercise 15), a payoff function need not satisfy the Bellman equation. In this case the initial functions σ, b, c, f, g may be as smooth as desired. The objective of the present section consists in deriving a correct normed Bellman equation for a payoff function. We shall see that such an equation holds in a very broad class of cases.

In addition to the assumptions given in Section 6.1, let us assume the following. We denote by $\gamma^\alpha(t,x)$ the vector of dimension $d \times (d_1 + d + 4)$, having the coordinates $\sigma^{ij}(\alpha,t,x)$ $(i = 1, \ldots, d, j = 1, \ldots, d_1)$, $b^i(\alpha,t,x)$ $(i = 1, \ldots, d)$, $c^\alpha(t,x)$, $f^\alpha(t,x)$, $g(x)$, $g(t,x)$. We assume that for each $\alpha \in A$, $l \in E_d$ the derivatives $\gamma^\alpha_{(l)}(t,x)$, $\gamma^\alpha_{(l)(l)}(t,x)$ as well as the derivative $(\partial/\partial t)\gamma^\alpha(t,x)$ exist and are continuous on $\bar{H}_T$. Assume that for all n, $\alpha \in A_n$, $l \in E_d$, $(t,x) \in \bar{H}_T$

$$\left|\frac{\partial}{\partial t}\gamma^\alpha(t,x)\right| + |\gamma^\alpha_{(l)}(t,x)| + |\gamma^\alpha_{(l)(l)}(t,x)| \leq K_n(1+|x|)^{m_n}. \tag{1}$$

Let

$$Q_n^* = \left\{(t,x) \in H_T : \sup_{\alpha \in A_n}(a(\alpha,t,x)\lambda,\lambda) > 0 \text{ for all } \lambda \neq 0\right\},$$

$$Q^* = \left\{(t,x) \in H_T : \sup_{\alpha \in A}(a(\alpha,t,x)\lambda,\lambda) > 0 \text{ for all } \lambda \neq 0\right\}.$$

By Lemma 2.9 the sets Q^* and Q_n^* are open, $Q_{n+1}^* \supset Q_n^*$, $Q^* = \bigcup_n Q_n^*$. To state our last assumption we assume that for each n_0 and for each bounded region Q' which together with its closure lies in $Q^*_{n_0}$ there exists a constant N such that for all $n \geq n_0, j = 1, \ldots, d$

$$|v_{nx^i}| + |v_{nx^ix^j}| + \left|\frac{\partial}{\partial t} v_n\right| \leq N \qquad \text{(a.e. on } Q'), \tag{2}$$

$$|w_{nx^i}| + |w_{nx^ix^j}| + \left|\frac{\partial}{\partial t} w_n\right| \leq N \qquad \text{(a.e. on } Q'), \tag{3}$$

$$|v_n(t,x)| \leq N, \ |w_n(t,x)| \leq N, \qquad (t,x) \in Q'. \tag{4}$$

As was shown in Section 4.7, the functions v_n and w_n in the region Q_n^* have generalized first and second derivatives with respect to x as well as a generalized first derivative with respect to t, which appear in the inequalities given above. We also know from Section 4.7 that these derivatives are locally bounded in Q_n^*. We require that the local boundedness of the derivatives in Q_n^* be uniform with respect to $n \geq n_0$. Finally, we note that due to Theorems 2.13 and 2.14 the assumption about the validity of (2)–(4) can be satisfied if the summarizing Assumption 8 in Section 2 is satisfied.

It follows from (2)–(4) that the functions v_n and w_n are equicontinuous and uniformly bounded in $\bar{Q}'$. This together with the obvious relations $v_n \to v$, $w_n \to w$ enables us to conclude that v and w are continuous in Q^*. Furthermore, from (2), (3), and the convergence of $v_n \to v$, $w_n \to w$ it follows that the functions v and w have two generalized derivatives with respect to x, one derivative with respect to t, and, in addition, these derivatives are locally bounded in Q^*.

If $m_\alpha(t,x)$ is a nonnegative function given for $\alpha \in A$, $t \in [0,T)$, $x \in E_d$, let

$$G^{m_\alpha}(u_0,u_{ij},u_i,u,t,x) = \sup_{\alpha \in A} m_\alpha(t,x)\left[u_0 + \sum_{i,j=1}^{d} a^{ij}(\alpha,t,x)u_{ij} + \sum_{i=1}^{d} b^i(\alpha,t,x)u_i - c^\alpha(t,x)u + f^\alpha(t,x)\right]. \tag{5}$$

1. Definition. A nonnegative function $m_\alpha(t,x)$ ($\alpha \in A$, $t \in [0,T)$, $x \in E_d$) is said to be a normalizing multiplier if for all u_0, u_{ij}, u_i, u, $t \in [0,T)$, $x \in E_d$

$$G^{m_\alpha}(u_0,u_{ij},u_i,u,t,x) < \infty.$$

The normalizing multiplier $m_\alpha(t,x)$ is called regular if there exists a function $N(t,x) < \infty$ such that for all α, t, x the inequality $m_{\alpha 0}(t,x) \leq N(t,x)m_\alpha(t,x)$ can be satisfied, where

$$m_{\alpha 0}(t,x) \equiv \left[1 + \sum_{i,j=1}^{d} |a^{ij}(\alpha,t,x)|^2 + \sum_{i=1}^{d} |b^i(\alpha,t,x)|^2 + |c^\alpha(t,x)|^2 + |f^\alpha(t,x)|^2\right]^{-1/2}.$$

2. Exercise

Prove that $m_{\alpha 0}(t,x)$ is a normalizing multiplier and, furthermore, that for each nonnegative function $N(t,x)$ the function $N(t,x)m_{\alpha 0}(t,x)$ is a normalizing multiplier.

The following theorem is a theorem on the normed Bellman equation.

3. Theorem. *Let $m_\alpha(t,x)$ be a normalizing multiplier. Then*:
a. $G^{m_\alpha}[v] = 0 \qquad$ (a.e. *on* Q^*). (6)
b. $G^{m_\alpha}[w] \leq 0$ (a.e. on Q^*), $w(t,x) \geq g(t,x)$ *in the region* Q^*, $G^{m_\alpha}[w] = 0$ *almost everywhere in the region* $Q^0 = \{(t,x) \in Q^*\colon w(t,x) > g(t,x)\}$. *In short,*

$$(G^{m_\alpha}[w] + w - g)_+ + g - w = 0 \qquad \text{(a.e. } on\ Q^*); \tag{7}$$

4. Definition. We call Eq. (6) the normed Bellman equation. We call Eq. (7) the normed Bellman equation for the optimal stopping problem.

In order to prove Theorem 3, we need four lemmas.

5. Lemma. *Assertions* (a) *and* (b) *in Theorem* 3 *hold true if* $m_\alpha(t,x) = m_{\alpha 0}(t,x)$.

Proof. First, we prove Theorem 3b. Taking advantage of the continuity of $w(t,x)$ and $g(t,x)$, the reader can easily prove that the first assertion in (b) is equivalent to the second assertion. Hence we shall prove the first assertion only.

We note first that $w_n \geq g$ (see, for example, Theorem 4.7.5) and $w \geq w_n$. Therefore $w \geq g$. Further, we denote by $G_n^{m_\alpha}(u_0,u_{ij},u,t,x)$ the right side in (5) if we replace A by A_n. Due to the nonnegativeness and boundedness of $m_{\alpha 0}$ ($m_{\alpha 0} \leq 1$) it can be seen that if for some $n \geq 1$, u_0, u_{ij}, u_i, u, t, x

$$\sup_{\alpha \in A_n}\left[u_0 + \sum_{i,j=1}^{d} a^{ij}(\alpha,t,x)u_{ij} + \sum_{i=1}^{d} b^i(\alpha,t,x)u_i - c^\alpha(t,x)u + f^\alpha(t,x)\right] \leq 0, \tag{8}$$

then $G_n^{m_{\alpha 0}}(u_0,u_{ij},u_i,u,t,x) \leq 0$. If we have equality in (8), then

$$G_n^{m_{\alpha 0}}(u_0,u_{ij},u_i,t,x) = 0.$$

Then, by Theorem 4.7.5, we have that $G_n^{m_\alpha 0}[w_n] \leq 0$ (Q_n^*-a.e.), $G_n^{m_\alpha 0}[w_n] = 0$ ($Q_n^0 = \{(t,x) \in Q_n^*\colon w_n(t,x) > g(t,x)\}$-a.e.). Next, we take a bounded region $Q' \subset \bar{Q}' \subset Q^*$ and, furthermore, we choose a number n_0 so that $\bar{Q}' \subset Q_{n_0}^*$. The existence of such a number follows from the theorem on separation of the finite covering of the compactum $\bar{Q}'$ from the covering of $\bar{Q}'$ by expanding regions Q_n^*.

For $n \geq n_0$ we have: $\bar{Q}' \subset Q_n^*$, $G_{n_0}^{m_{\alpha 0}}[w_n] \leq G_n^{m_{\alpha 0}}[w_n]$, $w_{n_0} \leq w_n$, $Q_{n_0}^0 \subset Q_n^0$. Moreover, $G^{m_{\alpha 0}}[w_n] \geq G_n^{m_{\alpha 0}}[w_n]$, which together with what has been proved above enables us to conclude that for $n \geq n_0$

$$G_{n_0}^{m_{\alpha 0}}[w_n] \leq 0 \qquad (Q'\text{-a.s.}) \qquad G^{m_{\alpha 0}}[w_n] \geq 0 \qquad (Q' \cap Q_{n_0}^0\text{-a.e.}). \tag{9}$$

We take now the limit in (9) as $n \to \infty$. We note that the functions $m_{\alpha 0}(t,x)$, $m_{\alpha 0}(t,x)a(\alpha,t,x)$, $m_{\alpha 0}(t,x)$, $b(\alpha,t,x)$, $m_{\alpha 0}(t,x)c^{\alpha}(t,x)$, $m_{\alpha 0}(t,x)f^{\alpha}(t,x)$ are bounded. Further, we assume that inequalities (2)–(4) are satisfied. Applying Theorem 4.5.1, we see that the passage to the limit is possible in (9) and that

$$G_{n_0}^{m_{\alpha 0}}[w] \leq 0 \qquad (Q'\text{-a.e.}), \qquad G^{m_{\alpha 0}}[w] \geq 0 \qquad (Q' \cap Q_{n_0}^0\text{-a.e.}).$$

Letting $n_0 \to \infty$ and using the fact that $G_{n_0}^{m_{\alpha 0}} \uparrow G^{m_{\alpha 0}}$, $w_{n_0} \uparrow w$, $Q^0 = \bigcup_n Q_n^0$, we find

$$G^{m_{\alpha 0}}[w] \leq 0 \qquad (Q'\text{-a.e.}), \qquad G^{m_{\alpha 0}}[w] \geq 0 \qquad (Q' \cap Q^0\text{-a.e.}).$$

Comparing the last inequalities, we have $G^{m_{\alpha 0}}[w] = 0$ $(Q' \cap Q^0\text{-a.e.})$. Taking advantage of the arbitrariness of Q', we complete the proof of Theorem 3b. We can prove Theorem (a) in a similar way. The lemma is thus proved. □

The above arguments included many properties of controlled processes. It turns out that Theorem 3 can be deduced from the lemma on the basis of the fact that if $m_{\alpha}(t,x)$ is a normalizing multiplier, any solution of the equation (inequality) $G^{m_{\alpha 0}}(u_0,u_{ij},u_i,u,t,x) = 0$ (≤ 0) is also a solution of the equation (inequality) $G^{m_{\alpha}}(u_0,u_{ij},u_i,u,t,x) = 0$ (≤ 0). For proving this, we need the following.

6. Lemma. *Let d_2 be an integer and on A let there be given two functions: l^{α} with values in E_{d_2} and a numerical function h^{α}. We assume that the equality $|l^{\alpha}|^2 + |h^{\alpha}|^2 = 0$ can be satisfied for no $\alpha \in A$. For $u \in E_{d_2}$ let*

$$F(u) = \sup_{\alpha \in A}(l^{\alpha}u + h^{\alpha}),^4 \qquad G(u) = \sup_{\alpha \in A} n_{\alpha 0}(l^{\alpha}u + h^{\alpha}),$$

where $n_{\alpha 0} = (|l^{\alpha}|^2 + |h^{\alpha}|^2)^{-1/2}$. Then the set $\Gamma = \{u: F(u) \leq 0\}$ is convex, closed (possibly, empty) and, furthermore, the set $\Gamma_0 = \{u: G(u) = 0\}$ is the boundary of the former set.

PROOF. It is seen that the inequality $F(u) \leq 0$ is equivalent to the inequality $G(u) \leq 0$. Hence

$$\Gamma = \{u: G(u) \leq 0\}. \tag{10}$$

Further, the upper bound of a set of linear functions is downward convex. Therefore, the function $G(u)$ $(F(u))$ is downward convex. The inequality $h_{\alpha 0}l^{\alpha}u + n_{\alpha 0}h^{\alpha} \leq |u| + 1$ implies the finiteness of $G(u)$; the properties of finiteness and convexity in turn imply the continuity of $G(u)$. This together with (10) enables us to assert that Γ is a closed convex set and that $\Gamma_0 \supset \partial\Gamma$. It remains only to prove that $\Gamma_0 \subset \partial\Gamma$.

Let the converse be true. Then there is a point $u_0 \in \Gamma_0$ such that $u_0 \notin \partial\Gamma$. Note that, obviously, $\Gamma_0 \subset \Gamma$. Therefore, $u_0 \in \Gamma$. Since u_0 does not lie on the boundary of the set Γ, u_0 is an interior point of Γ. We consider without loss

[4] It is possible that $F(u)$ assumes the value $+\infty$ at some points or everywhere.

of generality that $u_0 = 0$. Also, let the number $r > 0$ be so small that $U = \{u : |u| < r\} \subset \Gamma$. Then $G(u) \leq 0$ in U, $G(0) = G(u_0) = 0$.

We choose a sequence α_i such that $n_{\alpha_i 0} h^{\alpha_i} \to 0$. This can be done since $G(0) = 0$. The vectors $n_{\alpha_i 0} l^{\alpha_i}$ lie in a unit sphere. Hence, one can choose from the sequence of the vectors $n_{\alpha_i 0} h^{\alpha_i}$ a convergent subsequence. We denote the limit of this sequence by e. Since $|n_{\alpha 0} l^{\alpha}|^2 + |n_{\alpha 0} h^{\alpha}|^2 = 1$ for all $\alpha \in A$ and since $n_{\alpha_i 0} h^{\alpha_i} \to 0$, $||^2 = 1$. Finally, it follows from the inequalities $n_{\alpha_i 0} l^{\alpha_i} u + n_{\alpha_i 0} h^{\alpha_i} \leq G(u) \leq 0$ which hold for all $u \in U$ that $eu \leq 0$ for all $u \in U$. However, if we take $u = (1/r)e$, we obtain $e^2 \leq 0$, which is impossible because of the equality $|e|^2 = 1$. The contradiction thus obtained proves the lemma. □

We fix $(t,x) \in H_T$, some function $m_\alpha(t,x) > 0$, and also we use this lemma in order to study the inequality $G^{m_\alpha}(u_0, u_{ij}, u_i, u, t, x) \leq 0$. It is natural to regard the set $\mathbf{u} = (u_0, u_{ij}, u_i, u)$ $(i, j = 1, \ldots, d)$ as a point of the Euclidean space E_{d_2}, where $d_2 = 2 + d + d^2$. Let

$$h^\alpha = m_\alpha(t,x) f^\alpha(t,x),$$

$$l^\alpha = (m_\alpha(t,x),\ m_\alpha(t,x) a^{ij}(\alpha,t,x),\ m_\alpha(t,x) b^i(\alpha,t,x),\ -m_\alpha(t,x) c^\alpha(t,x)).$$

Then, as can easily be seen,

$$G^{m_\alpha}(\mathbf{u},t,x) = \sup_{\alpha \in A} (l^\alpha \mathbf{u} + h^\alpha).$$

The function $G^{m_\alpha}(\mathbf{u},t,x)$ is therefore good enough to replace $F(u)$. It is seen that if $m_\alpha > 0$, the function $n_{\alpha 0}$ from Lemma 6 is equal to $[m_\alpha(t,x)]^{-1} m_{\alpha 0}(t,x)$, which fact indicates that in the given case the function $G(u)$ from Lemma 6 is equal to $G^{m_{\alpha 0}}(\mathbf{u},t,x)$. Using Lemma 6, we immediately arrive at the following.

7. Lemma. *Let $(t,x) \in H_T$ and let some function $m_\alpha(t,x) > 0$ for all $\alpha \in A$. Then the set*

$$\Gamma^{m_\alpha}(t,x) \equiv \{(u_0, u_{ij}, u_i, u) : G^{m_\alpha}(u_0, u_{ij}, u_i, u, t, x) \leq 0\}$$

being a set in E_{2+d+d^2}, is convex, closed, and, in addition, the set

$$\Gamma_0(t,x) \equiv \{(u_0, u_{ij}, u_i, u) : G^{m_{\alpha 0}}(u_0, u_{ij}, u_i, u, t, x) = 0\}$$

is the boundary of the former set.

8. Lemma. *Let $m_\alpha(t,x)$ be a normalizing multiplier. Then all solutions of the equation (inequality) $G^{m_{\alpha 0}}(u_0, u_{ij}, u_i, u, t, x) = 0$ (≤ 0) are also solutions of the equation (inequality) $G^{m_\alpha}(u_0, u_{ij}, u_i, u, t, x) = 0$ (≤ 0). If $m_\alpha(t,x)$ is a regular normalizing multiplier, the converse also holds, that is, the foregoing equations (inequalities) are equivalent.*

PROOF. The assertions about the above inequalities easily follow from the fact that $m_{\alpha 0}(t,x) > 0$, and, further, if $m_\alpha(t,x)$ is a regular multiplier, then $m_\alpha(t,x) > 0$.

Let us prove the first assertion of the lemma about the equations. We fix some t, x; but we do not write t, x in the arguments of the functions. By Lemma 7, in any neighborhood of a point (u'_0,u'_{ij},u'_i,u') such that $G^{m_{\alpha 0}}(u'_0,u'_{ij},u'_i,u') = 0$ there are points (u_0,u_{ij},u_i,u) at which $F > 0$. It is seen that $G^{m_\alpha} \geq 0$ at the same points. By Definition 1, the function G^{m_α} is finite. Furthermore, since the function G^{m_α} is convex, it is also continuous. Comparing the last two assertions, we conclude that $G^{m_\alpha}(u'_0,u'_{ij},u'_i,u') \geq 0$. The converse inequality is obvious, therefore $G^{m_\alpha}(u'_0,u'_{ij},u'_i,u') = 0$. We have thus proved the first assertion.

In order to prove the second assertion of the lemma, we note that if $G^{m_\alpha}(u'_0,u'_{ij},u'_i,u') = 0$, the expression appearing in (5) under the sign of the upper bound is negative. Therefore, having taken $N = N(t,x)$ from Definition 1, we have

$$0 = NG^{m_\alpha}(u'_0,u'_{ij},u'_i,u') \leq G^{m_{\alpha 0}}(u'_0,u'_{ij},u'_i,u').$$

On the other hand, it follows from the equality $G^{m_\alpha} = 0$ that $G^{m_\alpha} \leq 0$ and (since $m_\alpha > 0$) $G^{m_{\alpha 0}} \leq 0$. The lemma is proved.

The comparison of Lemma 5 with Lemma 8 immediately proves Theorem 3. □

Lemma 8 shows that all the information on payoff functions, which can be obtained from the normed Bellman equations having different $m_\alpha(t,x)$, is contained in an equation corresponding to the normalizing multiplier $m_{\alpha 0}(t,x)$. Despite the fact that $m_{\alpha 0}(t,x)$ plays therefore a particularly essential role, it is frequently convenient to consider in practice other normalizing multipliers. The second assertion of Lemma 8 shows that the application of regular normalizing multipliers induces no loss of the information on payoff functions.

9. Example. Let $d = 1$, $A = (-\infty,\infty)$, $a(\alpha,t,x) = 1$, $b(\alpha,t,x) = 2\alpha$, $c^\alpha = 0$, $f^\alpha(t,x) = -\alpha^2 f(t,x)$, where $f(t,x) > 0$.

In this case

$$G^{m_\alpha}(u_0,u_{11},u_1,u,t,x) = \sup_{\alpha \in (-\infty,\infty)} m_\alpha(t,x)[u_0 + u_{11} + 2\alpha u_1 - \alpha^2 f(t,x)],$$

$$m_{\alpha 0}(t,x) = [2 + 4\alpha^2 + \alpha^4 f^2(t,x)]^{-1/2}.$$

The equation $G^{m_{\alpha 0}} = 0$ seems to be, at least at first sight, inconvenient. Let $m_\alpha(s,x) \equiv 1$. Since $f > 0$, $2\alpha u_1 - \alpha^2 f$ is the bounded function of α and, furthermore, 1 is a normalizing multiplier. Obviously, this multiplier is even regular. The calculations of G^1 reduce to finding the vertex of the parabola $2\alpha u_1 - \alpha^2 f$. We have

$$G^1(u_0,u_{11},u_1,u,t,x) = u_0 + u_{11} + u_1^2 f^{-1}(t,x).$$

Therefore, in this case the equation $G^1 = 0$ (equivalent to the equation $G^{m_{\alpha 0}} = 0$) becomes

$$u_0 + u_{11} + u_1^2 f^{-1/2}(t,x) = 0.$$

It seems much more complicated to transform the equation $G^{m_{\alpha 0}} = 0$ to the equation given above when we do not use Lemma 8.

10. Exercise

Prove that the function $m_\alpha(t,x) \equiv 1$ is a normalizing multiplier if and only if for any $r \geq 0$, $t \in [0,T)$, $x \in E_d$

$$\sup_{\alpha \in A} \{r[\operatorname{tr} a(\alpha,t,x) + |b(\alpha,t,x)| + c^\alpha(t,x)] + f^\alpha(t,x)\} < \infty.$$

Turning again to Lemma 7, we see how the function G^{m_α} behaves on a boundary of the set Γ. If the point $(u'_0,u'_{ij},u'_i,u') \in \Gamma_0$ and if, in addition, the function G^{m_α} is finite in some neighborhood of the foregoing point, here as well as in the proof of Lemma 8 the function G^{m_α} is continuous at the point (u'_0,u'_{ij},u'_i,u'), and, also, in any neighborhood of this point there are points at which $G^{m_\alpha} > 0$ and $G^{m_\alpha}(u'_0,u'_{ij},u'_i,u',t,x) = 0$. It is also possible that $G^{m_\alpha}(u'_0,u'_{ij},u'_i,u',t,x) \leq 0$, and that in any neighborhood of the point (u'_0,u'_{ij},u'_i,u') there are points at which $G^{m_\alpha} = +\infty$.

11. Definition. We say that at the point (u'_0,u'_{ij},u'_i,u') the function $G^{m_\alpha}(u_0,u_{ij},u_i,u,t,x)$ (t, x are fixed) has a zero crossing if:

a. $G^{m_\alpha}(u'_0,u'_{ij},u'_i,u',t,x) = 0$, the function G^{m_α} is continuous at the point (u'_0,u'_{ij},u'_i,u'), and, furthermore, in any neighborhood of the point (u'_0,u'_{ij},u'_i,u') there are points at which $G^{m_\alpha} > 0$; or
b. $G^{m_\alpha}(u'_0,u'_{ij},u'_i,u',t,x) \leq 0$, and in any neighborhood of the point (u'_0,u'_{ij},u'_i,u') there are points at which $G^{m_\alpha} = +\infty$.

Recall that

$$F(u_0,u_{ij},u_i,u,t,x) \equiv \sup_{\alpha \in A}\left[u_0 + \sum_{i,j=1}^{d} a^{ij}(\alpha,t,x)u_{ij} + \sum_{i=1}^{d} b^i(\alpha,t,x)u_i - c^\alpha(t,x)u + f^\alpha(t,x)\right],$$

$$F[u](t,x) = F\left(\frac{\partial}{\partial t}u(t,x), u_{x^ix^j}(t,x), u_{x^i}(t,x), u(t,x), t, x\right).$$

It is seen that $F(u_0,u_{ij},u_i,u,t,x) = G^1(u_0,u_{ij},u_i,u,t,x)$. Combining what has been said prior to Definition 11 with the assertion of Lemma 7, we have that the function $F(u_0,u_{ij},u_i,u,t,x)$ has a zero crossing at each point of the set $\Gamma_0(t,x)$. Then, due to Lemma 5 the following theorem holds.

12. Theorem. (a) *For almost all $(t,x) \in Q^*$ the function $F(u_0,u_{ij},u_i,u,t,x)$, being the function of (u_0,u_{ij},u_i,u), has a zero-crossing at the point*

$$\left(\frac{\partial}{\partial t}v(t,x), v_{x^ix^j}(t,x), v_{x^i}(t,x), v(t,x)\right).$$

(b) $F[w] \leq 0$ (Q^*-a.e.), $w(t,x) \geq g(t,x)$ *everywhere on* Q^*, *and for almost all* $(t,x) \in Q^0$ *the function* $F(u_0,u_{ij},u_i,u,t,x)$, *being the function of* (u_0,u_{ij},u_i,u), *has a zero crossing at the point* $((\partial/\partial t)w(t,x),w_{x^ix^j}(t,x),w_{x^i}(t,x),w(t,x))$.

This theorem enables us to make the Bellman equation meaningful even if the equation cannot be satisfied. Indeed, we write

$$F(u_0,u_{ij},u_i,u,t,x) \asymp 0$$

if F crosses zero at the point (u_0,u_{ij},u_i,u). Then assertion (a) in Theorem 12 implies that

$$F[v] \asymp 0 \qquad (Q^*\text{-a.e.})$$

From Lemmas 5 and 7 we have the following.

13. Theorem. (a) *For almost all* $(t,x) \in Q^*$ *the point*

$$\left(\frac{\partial}{\partial t} v(t,x), v_{x^ix^j}(t,x), v_{x^i}(t,x), v(t,x)\right)$$

lies on the boundary $\Gamma_0(t,x)$ *of the set* $\Gamma^1(t,x)$.

(b) *For almost all* $(t,x) \in Q^0$ *the point*

$$\left(\frac{\partial}{\partial t} w(t,x), w_{x^ix^j}(t,x), w_{x^i}(t,x), w(t,x)\right)$$

lies on the boundary $\Gamma_0(t,x)$ *of the set* $\Gamma^1(t,x)$.

Let us indicate here an example of the application of Theorem 12.

14. Example. $d = 1$, $A = [0,\infty)$, $f^\alpha(t,x) = \alpha g(t,x)$, $L^\alpha = (\partial/\partial t) + (\partial^2/\partial x^2) - \alpha$.

The Bellman equation is here the following:

$$\sup_{\alpha\in[0,\infty)} \left[\frac{\partial}{\partial t} v + v_{xx} + \alpha(g(t,x) - v)\right] = 0. \tag{11}$$

This equation may be not satisfied. Theorem 12 provides the correct interpretation of Eq. (11). According to Theorem 12 the function

$$F(u_0,u_{11},u_1,u,t,x) \equiv \sup_{\alpha\in[0,\infty)} [u_0 + u_{11} + \alpha(g(t,x) - u)]$$

has a zero-crossing at the point $((\partial/\partial t)v,v_{xx},v_x,v)$. According to Definition 11 there exist two ways for a function to cross zero. The first way implies that

$$F[v] \equiv \sup_{\alpha\in[0,\infty)} \left[\frac{\partial}{\partial t} v + v_{xx} + \alpha(g(t,x) - v)\right] = 0 \tag{12}$$

and, furthermore, for any (u_0,u_{11},u_1,u) close to $((\partial/\partial t)v,v_{xx},v_x,v)$ the function $F(u_0,u_{11},u_1,u,t,x)$ is finite. It immediately follows from (12) that $g(t,x) \leq v$,

$(\partial/\partial t)v + v_{xx} = 0$. Since any small variation of v in (12) need not make the left side of (12) go to infinity, $g(t,x) < v$. Therefore, the first way for a function to cross zero implies the relations

$$g(t,x) < v, \qquad \frac{\partial}{\partial t} v + v_{xx} = 0 \tag{13}$$

to be satisfied.

In the other case, i.e., on the set (t,x) where (13) is not satisfied, the left side of (12) does not exceed zero, and, furthermore, some arbitrarily small variations of $(\partial/\partial t)v$, v_{xx}, v make the left side of (12) go to infinity, which is possible only if $g(t,x) = v$. The inequality $F[v] \leq 0$ yields $(\partial/\partial t)v + v_{xx} \leq 0$. Therefore, either (13) is satisfied or

$$g(t,x) = v, \qquad \frac{\partial}{\partial t} v + v_{xx} \leq 0. \tag{14}$$

It is clear that (13) or (14) hold almost everywhere in H_T rather than for all $(t,x) \in H_T$. We also note that by Theorem 12 the function $w(t,x)$ constructed on the basis of $g(t,x)$, $f^\alpha(t,x) \equiv 0$, $L^\alpha = (\partial/\partial t) + (\partial^2/\partial x^2)$ satisfies either (13) or (14) almost at each point (t,x).

We can obtain (13) and (14) in a somewhat different way if we take the normalizing multiplier $m_\alpha(s,x) = 1/(1 + \alpha)$ and, in addition, if we let $\beta = \alpha/(1 + \alpha)$. Then the equation $G^{m_\alpha}[v] = 0$ can be written as

$$\sup_{\beta \in [0,1)} \left[(1 - \beta)\left(\frac{\partial}{\partial t} v + v_{xx} \right) + \beta(g(t,x) - v) \right] = 0.$$

The expression under the sign of the upper bound is a linear function of β. Using the fact that a linear function on an interval attains the upper bound at either end point of the interval, we immediately have

$$\frac{\partial}{\partial t} v + v_{xx} \leq 0 \quad \text{if } g(t,x) - v \leq 0,$$

$$\frac{\partial}{\partial t} v + v_{xx} = 0 \quad \text{if } g(t,x) - v < 0,$$

which precisely means that either (13) or (14) is satisfied.

15. Exercise

Show that in Example 14, Eq. (11) cannot be satisfied almost everywhere in H_T if $g(T,x) = 0$ and, also, if $g(t,x)$ is a function which is bounded and continuous in $\bar{H}_T$ and satisfies the inequality $g(t,x) > 0$ for $(t,x) \in H_T$.

16. Exercise

Let $f^\alpha(t,x) \geq 0$ for all α, t, and x, and let $m_\alpha(t,x)$ be a normalizing multiplier. Prove that there exists a function $N(t,x)$ such that $m_\alpha(t,x) \leq N(t,x) m_{\alpha 0}(t,x)$ for all α, t, x.

17. Exercise

Prove all the theorems given in Section 6.3 in the case where, in the assumption associated with inequalities (2)–(4), inequalities (2) and (3) are replaced respectively by

$$\left|\frac{\partial}{\partial t} v_n + \Delta v_n\right| \leq N \quad \text{(a.s. on } Q'), \qquad \left|\frac{\partial}{\partial t} w_n + \Delta w_n\right| \leq N \quad \text{(a.e. on } Q').$$

4. The Optimal Stopping of a Controlled Process on an Infinite Interval of Time

In this section we investigate the limiting behavior of a payoff function as $T \to \infty$ in the problem of optimal stopping of a controlled process. We assume that the basic inequalities (1.1)–(1.4) are satisfied for all $\alpha \in A_n$ in each trip $\bar{H}_T$ for constant K_n, m_n, K, m, which depend in general on T. We vary T as well as the sets $\mathfrak{A}_n$ and $\mathfrak{A}$ and also the functions $w_n(t,x)$ and $w(t,x)$ (see Section 1) associated with the controlled process in H_T. Hence, it is natural to attribute the index T to $\mathfrak{A}_n$, $\mathfrak{A}$, w_n, w, defined in Section 1. Thus, we consider the sets of strategies $\mathfrak{A}_n^T$, $\mathfrak{A}^T$ as well as the functions $w_n(T,t,x)$, $w(T,t,x)$.

In addition to the assumptions that $K_n = K_n(T)$, $m_n = m_n(T)$, $K = K(T)$, $m = m(T)$, we state and use some other assumptions which will be given after Exercise 11 in this section.

We denote by $\mathfrak{M}$ the set of all Markov times with respect to $\{\mathscr{F}_t\}$, $\mathfrak{A} = \bigcap_T \mathfrak{A}^T$. It is seen that $\mathfrak{A}$ is the set of all progressively measurable (with respect to $\{\mathscr{F}_t\}$) functions α_t having values in A, for each of which for $T > 0$ there is a number n such that $\alpha_t(\omega) \in A_n$ for all $t \leq T$, $\omega \in \Omega$.

Obviously, $\mathfrak{M}(T - s) \supset \mathfrak{M}(T' - s)$ for $T \geq T'$. Therefore,

$$\begin{aligned} w(T,s,x) &= \sup_{\alpha \in \mathfrak{A}^T} \sup_{\tau \in \mathfrak{M}(T-s)} v^{\alpha,\tau}(s,x) \\ &\geq \sup_{\alpha \in \mathfrak{A}^T} \sup_{\tau \in \mathfrak{M}(T'-s)} v^{\alpha,\tau}(s,x). \end{aligned} \tag{1}$$

If $\tau \in \mathfrak{M}(T' - s)$, the values of α_t for $t > T'$ (even for $t > T' - s$) are of no consequence for the computation of $v^{\alpha,\tau}(s,x)$. This implies that the right side of (1) is equal to

$$\sup_{\alpha \in \mathfrak{A}^{T'}} \sup_{\tau \in \mathfrak{M}(T'-s)} v^{\alpha,\tau}(s,x) = w(T',s,x).$$

Therefore, $w(T,s,x)$ increases with respect to T and has the (possibly, infinite) limit

$$\lim_{T \to \infty} w(T,s,x) \equiv w(s,x).$$

It is seen that for $s \leq S$

$$w(s,x) = \sup_{T \geq S} w(T,s,x) = \sup_{T \geq S} \sup_n w_n(T,s,x).$$

It is clear that the function $w(s,x)$ is lower semicontinuous with respect to (s,x) as the upper bound of continuous functions $w_n(T,s,x)$. In particular, $w(s,x)$ is a Borel function.

It is seen that for arbitrary $\alpha \in \mathfrak{A}$, $\tau \in \mathfrak{M}$

$$\mathsf{M}_{s,x}^{\alpha}\left\{\int_0^{\tau} e^{-\varphi_t} f^{\alpha_t}(s+t,x_t)\,dt + e^{-\varphi_\tau} g(s+\tau,x_\tau)\right\} \tag{2}$$

might be infinite. We denote by $|\mathfrak{A} \times \mathfrak{M}|(s,x)$ the set of pairs $(\alpha,\tau) \in \mathfrak{A} \times \mathfrak{M}$, for which at least one of the following expressions is finite:

$$v_{(+)}^{\alpha,\tau}(s,x) = \mathsf{M}_{s,x}^{\alpha}\left\{\int_0^{\tau} e^{-\varphi_t} f_+^{\alpha_t}(s+t,x_t)\,dt + e^{-\varphi_\tau} g_+(s+\tau,x_\tau)\right\},$$

$$v_{(-)}^{\alpha,\tau}(s,x) = \mathsf{M}_{s,x}^{\alpha}\left\{\int_0^{\tau} e^{-\varphi_t} f_-^{\alpha_t}(s+t,x_t)\,dt + e^{-\varphi_\tau} g_-(s+\tau,x_\tau)\right\}.$$

It is natural to put $v^{\alpha,\tau}(s,x) = v_{(+)}^{\alpha,\tau}(s,x) - v_{(-)}^{\alpha,\tau}(s,x)$ for $(\alpha,\tau) \in |\mathfrak{A} \times \mathfrak{M}|(s,x)$. Such a definition of $v^{\alpha,\tau}$ agrees with the former definition since for $\alpha \in \mathfrak{A}^T$, $\tau \in \mathfrak{M}(T-s)$

$$\mathsf{M}_{s,x}^{\alpha}\left\{\int_0^{\tau} e^{-\varphi_t} |f(s+t,x_t)|\,dt + e^{-\varphi_\tau} |g(s+\tau,x_\tau)|\right\} < \infty.$$

This also implies that $\mathfrak{A} \times \mathfrak{M}(T-s) \subset |\mathfrak{A} \times \mathfrak{M}|(s,x)$ for any $s \le T$, $x \in E_d$, and that the set $|\mathfrak{A} \times \mathfrak{M}|(s,x)$ is nonempty. It is seen that $|\mathfrak{A} \times \mathfrak{M}|(s,x) = \mathfrak{A} \times \mathfrak{M}$ for all s and x if the functions $f^{\alpha}(s,x)$ and $g(s,x)$ for all values of the arguments are greater than zero or, conversely, always smaller than zero.

We investigate next how the function $w(s,x)$ is related to the function

$$\sup_{(\alpha,\tau) \in |\mathfrak{A} \times \mathfrak{M}|(s,x)} v^{\alpha,\tau}(s,x),$$

which it is natural to call a payoff function in the problem of optimal stopping of a controlled process at an infinite time interval.

1. Lemma. *Let a function $u(s,x)$ be nonnegative, continuous in $[0,\infty) \times E_d$, and, in addition, let this function belong to $W_{\text{loc}}^{1,2}((0,\infty) \times E_d)$. Let $h^{\alpha}(s,x)$ be measurable with respect to s, continuous with respect to (α,x), and nonnegative. Finally, let us assume that $L^{\alpha}u + h^{\alpha} \le 0$ $((0,\infty) \times E_d$-a.e.) for each $\alpha \in A$. Then for each $\alpha \in \mathfrak{A}$, $(s,x') \in [0,\infty) \times E_d$, $\tau \in \mathfrak{M}$*

$$\mathsf{M}_{s,x}^{\alpha}\left[\int_0^{\tau} e^{-\varphi_t} h^{\alpha_t}(s+t,x_t)\,dt + e^{-\varphi_\tau} u(s+\tau,x_\tau)\right] \le u(s,x). \tag{3}$$

PROOF. The assertion of the lemma resembles that of Lemma 5.3.4; therefore, we prove it in a similar way. Let $\Gamma = \{(t,x): L^{\alpha}u(t,x) + h^{\alpha}(t,x) \le 0$ for all $\alpha \in A'\}$, where A' is some everywhere countable dense set in the set A. It is seen that $\text{meas}((0,\infty) \times E_d \backslash \Gamma) = 0$. Due to the continuity of h^{α} and the coefficients L^{α} with respect to α for all $(t,x) \in \Gamma$, $\alpha \in A$ we have $L^{\alpha}u(t,x) +$

$h^\alpha(t,x) \leq 0$. Then, for any $\alpha \in \mathfrak{A}$, ω, $(t,x) \in \Gamma$

$$\chi_\Gamma(t,x)[L^{\alpha_t}u(t,x) + h^{\alpha_t}(t,x)] \leq 0.$$

This implies that according to Remark 2.10.6 and Theorem 2.10.2 inequality (3) is satisfied for each $R > 0$, $T > 0$, $(s,x) \in \bar{C}_{T,R}$ if we replace τ by $\tau \wedge \tau_{T,R}$, where $\tau_{T,R}$ is the time of first exit of $(s + t, x_t)$ from $[0,T) \times S_R$. Applying next Fatou's lemma, letting $R \to \infty$, $T \to \infty$, and, finally, making use of the nonnegativeness of h^α and u, we complete the proof of (3). The lemma is proved. □

2. Lemma. *Let $u_1(s,x)$ and $u_2(s,x)$ be nonnegative continuous functions on $[0,\infty) \times E_d$. We assume that $u_2 \in W^{1,2}_{\text{lo}}((0,\infty) \times E_d)$. We also assume that either of the following conditions is satisfied:*

a. $L^\alpha u_2 \leq 0$ ($(0,\infty) \times E_d$-a.e.) *for all* $\alpha \in A$,

$$\lim_{T\to\infty} \sup_{x \in E_d} \frac{u_1(T,x)}{u_2(T,x)} = 0,$$

where an expression of the form 0/0 is assumed to be equal to zero;

b. $L^\alpha u_2 + u_1 \leq 0$ ($(0,\infty) \times E_d$-a.e.) *for all* $\alpha \in A$.

Then for any $s \geq 0$, $x \in E_d$

$$\lim_{T\to\infty} \sup_{\alpha \in \mathfrak{A}} \sup_{\tau \in \mathfrak{M}} \mathsf{M}^\alpha_{s,x} \chi_{\tau \geq T-s} u_1(s + \tau, x_\tau) e^{-\varphi_\tau} = 0 \tag{4}$$

if (a) *is satisfied, and, further, for any* $\alpha \in \mathfrak{A}$

$$\lim_{T\to\infty} \mathsf{M}^\alpha_{s,x} u_1(T, x_{T-s}) e^{-\varphi_{T-s}} = 0 \tag{5}$$

if (b) *is satisfied.*

Proof. Let

$$\varepsilon(T) = \sup_{T' > T} \sup_{x \in E_d} \frac{u_1(T',x)}{u_2(T',x)}.$$

Under assumption (a), $\varepsilon(T) \downarrow 0$ as $T \to \infty$. Using Lemma 1, we find

$$\begin{aligned}\mathsf{M}^\alpha_{s,x} \chi_{\tau \geq T-s} u_1(s + \tau, x_\tau) e^{-\varphi_\tau} &\leq e(T) \mathsf{M}^\alpha_{s,x} u_2(s + \tau, x_\tau) e^{-\varphi_\tau} \\ &\leq \varepsilon(T) u_2(s,x).\end{aligned}$$

We have thereby proved equality (4). By Lemma 1

$$\int_0^\infty \mathsf{M}^\alpha_{s,x} u_1(s + t, x_t) e^{-\varphi_t}\, dt \leq u_2(s,x) < \infty$$

if (b) is satisfied. Since the integral with respect to t converges in the last equation, the integrand tends to zero. Therefore, (5) is satisfied. This completes the proof of the lemma. □

3. Remark. Lemma 2b will be satisfied if, for example, $u_1 \leq N_1 c^\alpha$ for all α for some constant N_1. In this case we may take $u_2 \equiv N_1$. Lemma 2a will be satisfied if $u_1 \in W^{1,2}_{\text{loc}}((0,\infty) \times E_d)$ and $L^\alpha u_1 + \varepsilon u_1 \leq 0$ ($(0,\infty) \times E_d$-a.e.) for all α for some $\varepsilon > 0$, in which case the function $u_1(s,x)e^{s\varepsilon}$ can be taken for u_2.

It is convenient to use Lemma 2 in verifying the next theorem.

4. Theorem. *Let $s \geq 0$, $x \in E_d$ and, also, for each $\alpha \in \mathfrak{A}$ let*

$$\varliminf_{T\to\infty} \mathsf{M}^\alpha_{s,x} e^{-\varphi_{T-s}} g_-(T,x_{T-s}) = 0.$$

Then

$$w(s,x) = \sup_{(\alpha,\tau)\in|\mathfrak{A}\times\mathfrak{M}|(s,x)} v^{\alpha,\tau}(s,x).$$

PROOF. It is seen that

$$\begin{aligned} w(s,x) &= \lim_{T\to\infty} \sup_{\alpha\in\mathfrak{A}^T} \sup_{\tau\in\mathfrak{M}(T-s)} v^{\alpha,\tau}(s,x) \\ &= \lim_{T\to\infty} \sup_{\alpha\in\mathfrak{A}} \sup_{\tau\in\mathfrak{M}(T-s)} v^{\alpha,\tau}(s,x) \\ &\leq \sup_{T>0} \sup_{(\alpha,\tau)\in|\mathfrak{A}\times\mathfrak{M}|(s,x)} v^{\alpha,\tau}(s,x) = \sup_{(\alpha,\tau)\in|\mathfrak{A}\times\mathfrak{M}|(s,x)} v^{\alpha,\tau}(s,x). \end{aligned}$$

Let us prove the converse. To this end, as can easily be seen, it suffices to show that for all $(\alpha,\tau) \in |\mathfrak{A} \times \mathfrak{M}|(s,x)$

$$v^{\alpha,\tau}(s,x) \leq \varlimsup_{T\to\infty} v^{\alpha,\tau\wedge(T-s)}(s,x). \tag{6}$$

By definition, $g(s+\tau, x_\tau) = 0$ if $\tau = \infty$. Then, by the theorem on monotone convergence

$$\begin{aligned} v^{\alpha,\tau}_{(+)}(s,x) &= \varliminf_{T\to\infty} \mathsf{M}^\alpha_{s,x}\left\{\int_0^{\tau\wedge(T-s)} e^{-\varphi_t} f^{\alpha_t}_+(s+t,x_t)\,dt \right. \\ &\qquad \left. + \chi_{\tau\leq T-s} e^{-\varphi_\tau} g_+(s+\tau, x_\tau)\right\} \\ &\leq \varliminf_{T\to\infty} \mathsf{M}^\alpha_{s,x}\left\{\int_0^{\tau\wedge(T-s)} e^{-\varphi_t} f^{\alpha_t}_+(s+t, x_t)\,dt \right. \\ &\qquad \left. + \chi_{\tau\leq T-s} e^{-\varphi_\tau} g_+(s+\tau, x_\tau) + \chi_{\tau> T-s} e^{-\varphi_{T-s}} g_+(T,x_{T-s})\right\} \\ &= \varliminf_{T\to\infty} v^{\alpha,\tau\wedge(T-s)}_{(+)}(s,x). \end{aligned}$$

Similarly,

$$\begin{aligned} v^{\alpha,\tau}_{(-)}(s,x) &\geq \varlimsup_{T\to\infty} v^{\alpha,\tau\wedge(T-s)}_{(-)}(s,x) - \varliminf_{T\to\infty} \mathsf{M}^\alpha_{s,x}\chi_{\tau>T-s} e^{-\varphi_{T-s}} g_-(T,x_{T-s}) \\ &= \varlimsup_{T\to\infty} v^{\alpha,\tau\wedge(T-s)}_{(-)}(s,x). \end{aligned}$$

Subtracting the last inequalities, we find (6). The theorem is proved. □

5. Exercise

Consider the process $x_t = -e^{\mathbf{w}_t - (1/2)t}$ where $\mathbf{w}_t$ is a one-dimensional Wiener process. The process x_t is a solution of the stochastic equation

$$x_t = -1 + \int_0^t x_t \, d\mathbf{w}_t.$$

Prove that

$$\sup_{\tau \in \mathfrak{M}} \mathsf{M} x_\tau = 0, \qquad \sup_{\tau \in \mathfrak{M}(T)} \mathsf{M} x_\tau = -1$$

for each $T > 0$.

This exercise shows that the assertion of Theorem 4 is false, if we impose no restrictions on the controlled process.

Theorem 4 answers the question as to when a payoff function in the problem of optimal stopping of a controlled process on an infinite time interval can be obtained as the limit of payoff functions associated with finite time intervals. The presence of the set $|\mathfrak{A} \times \mathfrak{M}|(s,x)$ in the statement of this theorem is rather inconvenient. In this connection we consider the case where $|\mathfrak{A} \times \mathfrak{M}|(s,x) = \mathfrak{A} \times \mathfrak{M}$.

6. Theorem. *Let there be nonnegative functions $\bar{w}_1, \bar{w}_2 \in W^{1,2}_{\text{loc}}((0,\infty) \times E_d)$ continuous in $[0,\infty) \times E_d$ and such that $\bar{w}_1 \geq g$ in $[0,\infty) \times E_d$ and, in addition, for all $\alpha \in A$ let*

$$L^\alpha \bar{w}_1 \leq 0, \qquad L^\alpha \bar{w}_2 + f^\alpha_+ \leq 0 \qquad ((0,\infty) \times E_d)\text{-a.e.}$$

Then $|\mathfrak{A} \times \mathfrak{M}|(s,x) = \mathfrak{A} \times \mathfrak{M}$ for each s, x. Furthermore, $v^{\alpha,\tau}_{(+)}(s,x) \leq \bar{w}_1(s,x) + \bar{w}_2(s,x)$ for all $\alpha \in \mathfrak{A}$, $\tau \in \mathfrak{M}$, $s \geq 0$, $x \in E_d$; $w \leq \bar{w}_1 + \bar{w}_2$, and also the function w is finite.

The assertions of this theorem follow immediately from the inequalities

$$v^{\alpha,\tau}_{(+)}(s,x) \leq \mathsf{M}^\alpha_{s,x}\left\{\int_0^\tau e^{-\varphi_t} f^{\alpha_t}_+(s + t, x_t)\, dt + e^{-\varphi_\tau}[\bar{w}_1 + \bar{w}_2](s + \tau, x_\tau)\right\}$$

$$\leq \bar{w}_1(s,x) + \bar{w}_2(s,x) < \infty.$$

Here the second inequality has been obtained by Lemma 1.

7. Exercise

Let $p, q > 0$, and let $\mathbf{w}_t$ be a d-dimensional Wiener process

$$w(T,s,x) = \sup_{\tau \in \mathfrak{M}(T-s)} \mathsf{M} \frac{|x + \mathbf{w}_\tau|^p}{(1 + s + \tau)^q}.$$

In this case σ denotes a unit matrix, $b = 0$, $c = f = 0$, $g(s,x) = |x|^p/(1 + s)^q$. Using Theorem 4, show that

$$w(s,x) = \sup_{\tau \in \mathfrak{M}} \mathsf{M} \frac{|x + \mathbf{w}_\tau|^p}{(1 + s + \tau)^q}.$$

Prove that $w(s,x) = \infty$ for $p \geq 2q$ and $w(s,x) < \infty$ for $p < 2q$ for all s, x. (In the first case use the law of the iterated logarithm; in the second case use Theorem 6, assuming $\bar{w}_2 \equiv 0$,

$$\bar{w}_1(s,x) = N \frac{1}{(1+s)^{d/2}} e^{x^2/2(1+s)} + 1$$

and choosing an appropriate N.)

Theorem 4 together with the equality

$$w(s,x) = \lim_{T\to\infty} \lim_{n\to\infty} w_n(T,s,x)$$

as well as the results obtained in Chapter 5 enables us to find strategies and Markov times for which $v^{\alpha,\tau}(s,x)$ differs arbitrarily slightly from a payoff function in the problem of optimal stopping of a controlled process on an infinite time interval. In some cases, one can indicate an optimal stopping time.

8. Theorem. *Let a set A consist of a single point.*

Assume that there exists a nonnegative function $\bar{w} \in W^{1,2}_{\mathrm{loc}}((0,\infty) \times E_d)$ *continuous in* $[0,\infty) \times E_d$ *and such that*

$$\lim_{T\to\infty} \sup_{x\in E_d} \frac{|g(T,x)|}{\bar{w}(T,x)} = 0, \qquad L\bar{w} \leq 0^5 \quad ((0,\infty) \times E_d)\text{-a.e.}$$

Also, we assume that for each $s \geq 0$, $x \in E_d$ *either of the following inequalities is satisfied*:

$$\begin{aligned} &\mathsf{M}_{s,x} \int_0^\infty e^{-\varphi_t} f_+(s+t, x_t)\, dt < \infty, \\ &\mathsf{M}_{s,x} \int_0^\infty e^{-\varphi_t} f_-(s+t, x_t)\, dt < \infty. \end{aligned} \tag{7}$$

We denote by $\tau_0^{s,x}$ *the time of first exit of the process* $(s+t, x_t^{s,x})$ *from* $Q_0 = \{(s,x) \in [0,\infty) \times E_d : w(s,x) > g(s,x)\}$.

Then, for each $\tau \in \mathfrak{M}$ *the variable* $v^\tau(s,x)$ *is defined*, $v^\tau(s,x) < \infty (v^\tau(s,x) > -\infty)$, *if the first (second) inequality* (7) *is satisfied*,

$$w(s,x) = \sup_{\tau\in\mathfrak{M}} v^\tau(s,x), \qquad w(s,x) = v^{\tau_0}(s,x),$$

for all (s,x).

Proof. First we note that, since the function $w(s,x) - g(s,x)$ is lower semicontinuous, the set Q_0 is in fact a region. Further, by Lemma 2

$$\sup_{\tau\in\mathfrak{M}} \mathsf{M}_{s,x} e^{-\varphi_\tau} \chi_{\tau \geq T-s} |g(s+\tau, x_\tau)| \to 0 \tag{8}$$

as $T \to \infty$. From this it follows, in particular, that for each $\tau \in \mathfrak{M}$ for large T

$$\mathsf{M}_{s,x} e^{-\varphi_\tau} \chi_{\tau > T-s} |g(s+\tau, x_\tau)| < \infty.$$

[5] We omit here and henceforth the index α, which assumes only one value; we assume that the expression $0^\circ 0$ is equal to zero.

Furthermore, since $|g(s,x)| \le K(T)(1 + |x|)^{m(T)}$ for $s \le T$, due to the estimates of moments of stochastic equations for any T we have

$$\mathsf{M}_{s,x} e^{-\varphi_\tau} \chi_{\tau \le T - s} |g(s + \tau, x_\tau)| < \infty.$$

Therefore, for any τ

$$\mathsf{M}_{s,x} e^{-\varphi_\tau} |g(s + \tau, x_\tau)| < \infty. \tag{9}$$

It is also seen that

$$\mathsf{M}_{s,x} \int_0^\tau e^{-\varphi_t} f_\pm(s + t, x_t)\, dt \le \mathsf{M}_{s,x} \int_0^\infty e^{-\varphi_t} f_\pm(s + t, x_t)\, dt.$$

The last two inequalities immediately imply our assertions concerning $v^\tau(s,x)$. Now we can write $|\mathfrak{A} \times \mathfrak{M}|(s,x) = \mathfrak{A} \times \mathfrak{M}$. Further, by Theorem 4,

$$w(s,x) = \sup_{\tau \in \mathfrak{M}} v^\tau(s,x).$$

We prove that the time $\tau_0^{s,x}$ is optimal. We denote by $\tau_0^{s,x}(T)$ the time of first exit of $(s + t, x_t)$ from $Q_0(T) = \{(s,x) \in \bar{H}_T : w(T,s,x) > g(s,x)\}$. By Theorem 3.1.10 for each T on $\bar{H}_T$

$$w^T(s,x) = \mathsf{M}_{s,x} \left\{ \int_0^{\tau_0(T)} e^{-\varphi_t} f(s + t, x_t)\, dt + e^{-\varphi_{\tau_0(T)}} g(s + \tau_0(T), x_{\tau_0(T)}) \right\}.$$

Let $T \to \infty$. By virtue of the relation $w(T,s,x) \uparrow w(s,x)$ we have $Q_0(T) \subset Q_0(T')$ for $T' > T$, $Q_0 = \bigcup_T Q_0(T)$, $\tau_0^{s,x}(T) \uparrow \tau_0^{s,x}$. The continuity of $x_t^{s,x}$ with respect to t also implies that on a set $\{\tau_0^{s,x} < \infty\}$

$$\lim_{T \to \infty} x^{s,x}_{\tau_0^{s,x}(T)} = x^{s,x}_{\tau_0^{s,x}}.$$

By the monotone convergence theorem,

$$\begin{aligned} \mathsf{M}_{s,x} \int_0^{\tau_0(T)} e^{-\varphi_t} f(s + t, x_t)\, dt &= \mathsf{M}_{s,x} \int_0^{\tau_0(T)} e^{-\varphi_t} f_+(s + t, x_t)\, dt \\ &\quad - \mathsf{M}_{s,x} \int_0^{\tau_0(T)} e^{-\varphi_t} f_-(s + t, x_t)\, dt \\ &\to \mathsf{M}_{s,x} \int_0^{\tau_0} e^{-\varphi_t} f_+(s + t, x_t)\, dt \\ &\quad - \mathsf{M}_{s,x} \int_0^{\tau_0} e^{-\varphi_t} f_-(s + t, x_t)\, dt \\ &= \mathsf{M}_{s,x} \int_0^{\tau_0} e^{-\varphi_t} f(s + t, x_t)\, dt. \end{aligned}$$

Applying here the estimates of solutions of stochastic equations as well as the dominated convergence theorem, we conclude that for each T'

$$\begin{aligned} w(s,x) &= \mathsf{M}_{s,x} \int_0^{\tau_0} e^{-\varphi_t} f(s + t, x_t)\, dt \\ &\quad + \mathsf{M}_{s,x} e^{-\varphi_{\tau_0}} \chi_{\tau_0 \le T' - s}\, g(s + \tau_0, x_{\tau_0}) \\ &\quad + \lim_{T \to \infty} \mathsf{M}_{s,x} e^{-\varphi_{\tau_0(T)}} \chi_{\tau_0(T) > T' - s}\, g(s + \tau_0(T), x_{\tau_0(T)}). \end{aligned} \tag{10}$$

As $T' \to \infty$ the last expression tends to zero by virtue of (8). By the dominated convergence theorem, it follows from (9) that

$$\lim_{T' \to \infty} \mathsf{M}_{s,x} e^{-\varphi_{\tau_0}} \chi_{\tau_0 \le T' - s} g(s + \tau_0, x_{\tau_0}) = \mathsf{M}_{s,x} e^{-\varphi_{\tau_0}} g(s_0 + \tau_0, x_{\tau_0}).$$

Thus, letting T' in (10) to infinity, we obtain: $w(s,x) = v^{\tau_0}(s,x)$, which proves the theorem. □

However, the condition associated with (7) cannot be removed.

9. Example. Let $c = 0$, $f(s,x) = (d/ds)(s \sin s)$, $g(s,x) = 0$. Then, as can easily be seen, $w(s,x) = \infty$, $\tau_0 = \infty$, and, in addition,

$$\mathsf{M}_{s,x} \int_0^\infty f(s + t, x_t) e^{-\varphi_t}\, dt$$

is not defined.

10. Example. Let $c = 0$, $f = 0$, $g(s,x) = -1/(1 + s)^2$. In this case we can take $1/(1 + s)$ for $\bar{w}(s,x)$. Obviously, $w(s,x) = 0$, $\tau_0 = \infty$, and, in addition, by Theorem 8,

$$\mathsf{M}_{s,x} g(s + \tau_0, x_{\tau_0}) = \mathsf{M}_{s,x} \chi_{\tau_0 < \infty} g(s + \tau_0, x_{\tau_0}) = 0.^{6}$$

11. Exercise

Let us take $p < 2q$ in Exercise 7. Show that in such case one can take $\bar{w}_1$ given in Hint to Exercise 7 for $\bar{w}$ in Theorem 8. Using the fact that the process $\sqrt{c}\mathbf{w}_{(t/c)}$ is a Wiener process for any constant $c > 0$ (self-scaling property of a Wiener process), prove that

$$w(s,x) = (1 + s)^{(p/2) - q} w\left(0, \frac{x}{\sqrt{1 + s}}\right).$$

Deduce from this, noting that

$$g(s,x) = (1 + s)^{(p/2) - q} g\left(\frac{x}{\sqrt{1 + s}}\right).$$

and also applying the fact that the problem is spherically symmetric, that Q_0 together with each point (s_0,x_0) contains a part of the paraboloid

$$\left\{(s,x): \frac{|x|}{\sqrt{1 + s}} = \frac{|x_0|}{\sqrt{1 + s_0}}\right\}$$

which lies in $[0,\infty) \times E_d$. Using the almost obvious inequality

$$w(s,x) \ge (1 + s_0)^q (1 + s)^{-q} w(s_0,x) \qquad \text{for} \quad s \ge s_0$$

[6] By definition.

prove that if $(s_0,x_0) \in Q_0$ (i.e., $w(s_0,x_0) > g(s_0,x_0)$), then $(s,x_0) \in Q_0$ for all $s \geq s_0$. Combining this result with the preceding one, we arrive at the assertion that for some constant $c_0 \geq 0$

$$Q_0 = \{(s,x): s \geq 0, x \in E_d, |x| < c_0 \sqrt{1+s}\}.$$

Prove that $w(s,0) > 0$, and then $Q_0 \neq \varnothing, c_0 > 0$.

We proceed now to derive the normed Bellman equation. In addition to the assumptions formulated at the start of this section, we shall impose the following conditions.

Let $\gamma^\alpha(s,x)$ be a vector of dimension $d \times (d_1 + d + 3)$ having coordinates $\sigma^{ij}(\alpha,s,x)$ $(i = 1, \ldots, d,\ j = 1, \ldots, d_1)$, $b^i(\alpha,s,x)$ $(i = 1, \ldots, d)$, $c^\alpha(s,x)$, $f^\alpha(s,x)$, $g(s,x)$. Assume that the vector $\gamma^\alpha(s,x)$ is one time continuously differentiable with respect to s, twice continuously differentiable with respect to x on $[0,\infty) \times E_d$ for each $\alpha \in A$, and for all $n = 1, 2, \ldots, \alpha \in A_n$, $l \in E_d$, $T > 0$, $(s,x) \in \bar{H}_T$

$$\left|\frac{\partial}{\partial s}\gamma^\alpha(s,x)\right| + |\gamma^\alpha_{(l)}(s,x)| + |\gamma^\alpha_{(l)(l)}(s,x)| \leq K_n(T)(1+|x|)^{m_n(T)}. \tag{11}$$

Let

$$Q_n^*(T) = \{(s,x) \in H_T : \sup_{\alpha \in A_n} (a(\alpha,s,x)\lambda, \lambda) > 0 \text{ for all } \lambda \neq 0\},$$

$$Q^* = \{(s,x) \in (0,\infty) \times E_d : \sup_{\alpha \in A} (a(\alpha,s,x)\lambda, \lambda) > 0 \text{ for all } \lambda \neq 0\}.$$

We assume that for each n_0, $T_0 > 0$ and for each bounded region Q' which together with its closure lies in $Q^*_{n_0}(T_0)$, there exists a constant N such that for all $n \geq n_0$, $T > T_0$, $i, j = 1, \ldots, d$ almost everywhere in Q'

$$|w_n(T,s,x)| + |w_{nx^i}(T,s,x)| + |w_{nx^ix^j}(T,s,x)| + \left|\frac{\partial}{\partial s} w_n(T,s,x)\right| \leq N. \tag{12}$$

Thus, we assume that the assumptions made in Section 3 are satisfied in each strip H_T. In this connection we note that we can obtain the required estimates of derivatives in various cases if we apply the results obtained in Section 2.

Finally, we introduce the concept of a normalizing multiplier in the same way as we did in Section 3, letting $T = \infty$ in Definition 3.1.

12. Theorem. *The function $w(s,x)$ in the region Q^* is continuous, has a generalized first derivative with respect to s, and two generalized derivatives with respect x; all these derivatives are locally bounded in Q^*. For any normalizing multiplier $m_\alpha(s,x)$ we have $G^{m_\alpha}[w] \leq 0$ (Q^*-a.e.), $w(s,x) \geq g(s,x)$ in the region Q^*, $G^{m_\alpha}[w] = 0$ ($Q^0 = \{(s,x) \in Q^* : w(s,x) > g(s,x)\}$-a.e.).*

The proof of this theorem follows the proof of Theorem 3.3b. Hence we restrict ourselves to some hints only. One can easily derive the assertion of the theorem for an arbitrary normalizing multiplier from the assertion for $m_\alpha(s,x) = m_{\alpha 0}(s,x)$ with the aid of Lemma 3.8. According to Theorem 3.3 this theorem holds if we replace w by $w(T) \equiv w(T,s,x)$, and, furthermore, if we replace Q^* by $Q^*(T) = \bigcup_n Q_n^*(T)$. By Theorem 4.5.1, it is possible to take the limit in equalities of the type $G^{m_{\alpha 0}}[w(T)] = 0$ and in inequalities of the type $G^{m_{\alpha 0}}[w(T)] \le 0$ as $T \to \infty$.

13. Remark. As in the previous section assertions of the type of Theorems 3.12b and 3.13b hold here. The translation of the preceding arguments requires no changes.

We consider an important particular case of the problem of optimal stopping of a controlled process on an infinite time interval where the functions $\sigma(\alpha,s,x)$, $b(\alpha,s,x)$, $c^\alpha(s,x)$, $f^\alpha(s,x)$, $g(s,x)$ are time homogeneous, i.e., they do not depend on s. In this case for any $\alpha \in \mathfrak{A}$ and for bounded $\tau \in \mathfrak{M}$ the function $v^{\alpha,\tau}(s,x)$ does not depend on s. In order to convince oneself that this is the case, it suffices to write (2) in an explicit form and to note that $x_t^{\alpha,s,x}$ does not depend on s. It follows from the equality

$$w(T,s,x) = \sup_{\alpha \in \mathfrak{A}} \sup_{\tau \in \mathfrak{M}(T-s)} v^{\alpha,\tau}(s,x)$$

that $w(T,s,x)$ depends only on x and $T - s$. This, in turn, implies that $w(s,x) = \lim_{T\to\infty} w(T,s,x)$ does not depend on s.

It is clear that if $m_\alpha(s,x) = m_\alpha(x)$, then $G^{m_\alpha}(u_0,u_{ij},u_i,u,s,x)$ does not depend on s. Here we put

$$\bar{G}^{m_\alpha}(u_{ij},u_i,u,x) = \bar{G}^{m_\alpha}(0,u_{ij},u_i,u,s,x),$$
$$\bar{G}^{m_\alpha}[u](x) = \bar{G}^{m_\alpha}(u_{x^ix^j}(x),u_{x^i}(x),u(x),x).$$

It is natural that we omit now the argument s in the functions which do not depend on s in the case considered.

14. Theorem. *The function $w(x)$ in the region*

$$D^* = \left\{x \in E_d : \sup_{\alpha \in A}(a(\alpha,x)\lambda, \lambda) > 0 \textit{ for all } \lambda \neq 0\right\}$$

is continuous and, furthermore, has two generalized derivatives with respect to x, which are locally bounded in D^*. In addition, if the nonnegative function $m_\alpha(x)$ is such that for all x, u_{ij}, u_i, u

$$\sup_\alpha m_\alpha(x) < \infty, \qquad \bar{G}^{m_\alpha}(u_{ij},u_i,u,x) < \infty, \tag{13}$$

then $\bar{G}^{m_\alpha}[w] \le 0$ (D^*-a.e.), $w(x) \ge g(x)$ in D^*, $\bar{G}^{m_\alpha}[w] = 0$ ($D^0 = \{x \in D^*: w(x) > g(x)\}$-a.e.).

This theorem follows from Theorem 12 since, obviously,

$$Q^* = (0,\infty) \times D^*, \qquad Q^0 = (0,\infty) \times D^0, \qquad \frac{\partial}{\partial s} w(x) = 0$$

and, by virtue of the inequality

$$G^{m_\alpha}(u_0,u_{ij},u_i,u,x) \le |u_0| \sup_\alpha m_\alpha(x) + \bar{G}^{m_\alpha}(u_{ij},u_i,u,x) < \infty,$$

the function $m_\alpha(s,x) \equiv m_\alpha(x)$ is a normalizing multiplier.

15. Remark. In the stationary case one can write K_n and m_n instead of $K_n(T)$ and $m_n(T)$ in (11), since the vector γ does not depend on s. In verifying (12) using the results obtained in Section 2, it is natural to seek the functions u_i depending only on x. The same is applicable to the functions $\bar{w}_1$ and $\bar{w}_2$ from Theorem 6.

16. Exercise

Returning to Exercises 7 and 11, we assume that $p < 2q$.

Show that we can change the function $g(s,x)$ inside the region $\{(s,x): s \ge 0, |x| < (c_0/2)\sqrt{1+s}\}$ in such a way (mainly, smooth out) that the payoff function does not change and, furthermore, Assumption 2.8 can be satisfied for $u_0 = u_1 = u_2 = \bar{w}_1$ ($\bar{w}_1$ given in Hint to Exercise 7), $u_{-1} = u_{-2} = 0$ for each T. Derive from the above that $w(s,x)$ has one generalized derivative with respect to s and two generalized derivatives with respect to x, these derivatives being locally bounded in $(0,\infty) \times E_d$,

$$\frac{\partial w}{\partial s} + \frac{1}{2}\Delta w + h = 0 \qquad ((0,\infty) \times E_d\text{-a.e.}), \tag{14}$$

where $h = 0$ in Q^0, $h = -(\partial g/\partial s) - \frac{1}{2}\Delta g$ outside Q^0. Regarding (14) as an equation with respect to w and noting that $w \le (1+s)^{-\varepsilon}\bar{w}_1$ for some $\varepsilon > 0$, prove that

$$w(0,x) = \mathsf{M} \int_0^\infty h(t, x + \mathbf{w}_t)\, dt.$$

Putting here $x = x_0 = (c_0, 0, \ldots, 0)$,

$$h(s,x) = \left[-\frac{\partial g(s,x)}{\partial s} - \frac{1}{2}\Delta g(s,x) \right] \chi_{|x| > c_0\sqrt{1+s}},$$

and using the fact that $w(0,x_0) = g(0,x_0)$ and also, that the distribution of $\mathbf{w}_t$ is known, write an equation for c_0. Prove that such an equation has a unique solution with respect to c_0.

5. Control on an Infinite Interval of Time

Let the functions $\sigma(\alpha,t,x)$, $b(\alpha,t,x)$, $c^\alpha(t,x)$, $f^\alpha(t,x)$ be given for all $\alpha \in A$, $t \ge 0$, $x \in E_d$. Also, in each strip of H_T let the functions given satisfy Assumptions (1.1)–(1.3), with constant K_n and m_n, which depend, generally speaking, on T. As in the previous section, we introduce here the sets of strategies $\mathfrak{A}_n^T$, $\mathfrak{A}^T = \bigcup_n \mathfrak{A}_n^T$, $\mathfrak{A} = \bigcap_T \mathfrak{A}^T$. Let

$$v_n(T,s,x) = \sup_{\alpha \in \mathfrak{A}_n^T} \mathsf{M}_{s,x}^\alpha \int_0^{T-s} e^{-\varphi_t} f^{\alpha_t}(s+t, x_t)\, dt,$$

$$v(T,s,x) = \sup_n v_n(T,s,x).$$

The objective of this section consists in investigating the limit behavior of $v(T,s,x)$ as $T \to \infty$. We denote by $|\mathfrak{A}|(s,x)$ the set of all $\alpha \in \mathfrak{A}$ for which at least one of the following expressions is finite:

$$v^{\alpha}_{(+)}(s,x) = \mathsf{M}^{\alpha}_{s,x} \int_0^{\infty} e^{-\varphi_t} f^{\alpha_t}_{+}(s+t, x_t)\, dt,$$

$$v^{\alpha}_{(-)}(s,x) = \mathsf{M}^{\alpha}_{s,x} \int_0^{\infty} e^{-\varphi_t} f^{\alpha_t}_{-}(s+t, x_t)\, dt.$$

For $\alpha \in |\mathfrak{A}|(s,x)$ let $v^{\alpha}(s,x) = v^{\alpha}_{(+)}(s,x) - v^{\alpha}_{(-)}(s,x)$.

Throughout this section we assume that there exists $\underline{\alpha} \in A$ and also, that $\underline{v}(s,x)$ and $\underline{v}_1(s,x)$ are nonnegative continuous functions given on $[0,\infty) \times E_d$, which belong to $W^{1,2}_{\text{loc}}((0,\infty) \times E_d)$ and such that

$$L^{\underline{\alpha}} \underline{v} + f^{\underline{\alpha}}_{-} \le 0 \qquad ((0,\infty) \times E_d\text{-a.e.}),$$

$$\lim_{T\to\infty} \sup_{x \in E_d} \frac{\underline{v}(T,x)}{\underline{v}^1(T,x)} = 0,$$

where a relation of the form 0/0 is assumed to be equal to zero, and, in addition, for all $\alpha \in A$ $L^{\alpha} \underline{v}^1 \le 0$ $((0,\infty) \times E_d$-a.e.).

This assumption was discussed in Remark 4.3. Furthermore, after Theorem 4 we add some conditions to those listed above.

1. Theorem. *For all $s \ge 0$, $x \in E_d$ we have*

$$v(s,x) \equiv \lim_{T\to\infty} v(T,s,x), \qquad v(s,x) \ge -\underline{v}(s,x), \qquad v(s,x) = \sup_{\alpha \in |\mathfrak{A}|(s,x)} v^{\alpha}(s,x). \quad (1)$$

Proof. By Lemma 4.2, for any $s \ge 0$, $x \in E_d$

$$\lim_{T\to\infty} \sup_{\alpha\in\mathfrak{A}} \mathsf{M}^{\alpha}_{s,x} \underline{v}(T, x_{T-s}) e^{-\varphi_{T-s}} = 0. \quad (2)$$

Let

$$\tilde{v}(T,s,x) = \sup_{\alpha \in \mathfrak{A}^T} \mathsf{M}^{\alpha}_{s,x} \left\{ \int_0^{T-s} e^{-\varphi_t} f^{\alpha_t}(s+t, x_t)\, dt - e^{-\varphi_{T-s}} \underline{v}(T, x_{T-s}) \right\}. \quad (3)$$

Due to (2),

$$\lim_{T\to\infty} |v(T,s,x) - \tilde{v}(T,s,x)| = 0. \quad (4)$$

Further, let $T' > T$, and let $\alpha \in \mathfrak{A}^{T'}$ and $\alpha_t = \underline{\alpha}$ for $t \in [T-s, T'-s]$. By Theorem 2.9.7, almost surely

$$\mathsf{M}^{\alpha}_{s,x} \left\{ \int_{T-s}^{T'-s} e^{-\varphi_t} f^{\alpha_t}(s+t, x_t)\, dt - e^{-\varphi_{T'-s}} \underline{v}(T', x_{T'-s}) \Big| \mathscr{F}_{T-s} \right\}$$

$$= \mathsf{M}^{\alpha}_{T, x^{\alpha,s,x}_{T-s}} \left\{ \int_0^{T'-T} e^{-\varphi_t} f^{\alpha_t}(T+t, x_t)\, dt - e^{-\varphi_{T'-T}} \underline{v}(T', x_{T'-T}) \right\}.$$

By Lemma 4.1, the last expression is greater than $(-1)\underline{v}(T,x_{T-s}^{\alpha,s,x})$. Therefore, for strategies α of the given type

$$\mathsf{M}_{s,x}^{\alpha}\left\{\int_0^{T'-s} e^{-\varphi_t} f^{\alpha_t}(s+t,x_t)\,dt - e^{-\varphi_{T'-s}}\underline{v}(T',x_{T'-s})\right\}$$

$$\geq \mathsf{M}_{s,x}^{\alpha}\left\{\int_0^{T-s} e^{-\varphi_t} f^{\alpha_t}(s+t,x_t)\,dt - e^{-\varphi_{T-s}}\underline{v}(T,x_{T-s})\right\}. \tag{5}$$

Computing here the upper bounds, we obtain, obviously, $\tilde{v}(T,s,x) \leq \tilde{v}(T',s,x)$. Therefore, the function $\tilde{v}(T,s,x)$ increases with respect to T. From this and (4) it follows that the limits

$$\lim_{T\to\infty} \tilde{v}(T,s,x), \qquad \lim_{T\to\infty} v(T,s,x)$$

exist and are equal. Furthermore, these limits are greater than $\tilde{v}(s,s,x) = -v(s,x)$.

Let us prove (1). On one hand, by the monotone convergence theorem for $\alpha \in |\mathfrak{A}|(s,x)$

$$\begin{aligned} v^{\alpha}(s,x) &= v_{(+)}^{\alpha}(s,x) - v_{(-)}^{\alpha}(s,x) \\ &= \lim_{T\to\infty} \mathsf{M}_{s,x}^{\alpha}\int_0^{T-s} e^{-\varphi_t} f^{\alpha_t}(s+t,x_t)\,dt \\ &\leq \lim_{T\to\infty} v(T,s,x) = v(s,x). \end{aligned}$$

Thus, the right side of (1) does not exceed $v(s,x)$. On the other hand, let $s \geq 0$, $x \in E_d$, $T > s$, $\alpha \in \mathfrak{A}^T$. Also, we define $\alpha' \in \mathfrak{A}$ using the formula $\alpha_t' = \alpha_t$ for $t < T - s$, $\alpha_t' = \underline{\alpha}$ for $t \geq T - s$. For $T' > T$, as above,

$$\begin{aligned} \mathsf{M}_{s,x}^{\alpha'}\int_0^{T'-s} e^{-\varphi_t} f_-^{\alpha_t}(s+t,x_t)\,dt \leq{}& \mathsf{M}_{s,x}^{\alpha'}\int_0^{T-s} e^{-\varphi_t} f_-^{\alpha_t}(s+t,x_t)\,dt \\ &+ \mathsf{M}_{s,x}^{\alpha'}\mathsf{M}\left\{\int_{T-s}^{T'-s} e^{-\varphi_t} f_-^{\alpha_t}(s+t,x_t)\,dt \right. \\ &\left. + e^{-\varphi_{T-s}}\underline{v}(T',x_{T'-s}) \middle| \mathscr{F}_{T-s}\right\} \\ \leq{}& \mathsf{M}_{s,x}^{\alpha}\int_0^{T-s} e^{-\varphi_t} f_-^{\alpha_t}(s+t,x_t)\,dt \\ &+ \mathsf{M}_{s,x}^{\alpha} e^{-\varphi_{T-s}}\underline{v}(T,x_{T-s}), \end{aligned}$$

where, by virtue of (2) for a sufficiently large T, the right side is finite and, therefore, the primary expression is bounded uniformly with respect to T'. Hence, if T is sufficiently large, $\alpha' \in |\mathfrak{A}|(s,x)$.

From (2) and (5) we have

$$v^{\alpha'}(s,x) = \lim_{T'\to\infty} \mathsf{M}_{s,x}^{\alpha'}\left\{\int_0^{T'-s} e^{-\varphi_t} f^{\alpha_t}(s+t,x_t)\,dt - e^{-\varphi_{T'-s}}v(T',x_{T'-s})\right\}$$

$$\geq \mathsf{M}_{s,x}^{\alpha}\left\{\int_0^{T-s} e^{-\varphi_t} f^{\alpha_t}(s+t,x_t)\,dt - e^{-\varphi_{T-s}}v(T,x_{T-s})\right\}. \tag{6}$$

Therefore, for sufficiently large T

$$\sup_{\alpha \in |\mathfrak{A}|(s,x)} v^{\alpha}(s,x) \geq \tilde{v}(T,s,x).$$

As $T \to \infty$ we have that $v(s,x)$ does not exceed the right side of (1), thus proving the theorem. □

2. Remark. The above proof can be used in finding ε-optimal strategies on an infinite time interval if $v(s,x) < \infty$. Namely, we take first a large T such that the limiting expression in (2) becomes smaller than $\varepsilon/3$ and that $|v(T,s,x) - v(s,x)| < \varepsilon/3$. Next, we choose $\alpha \in \mathfrak{A}$ such that

$$\mathsf{M}_{s,x}^{\alpha} \int_0^{T-s} e^{-\varphi_t} f^{\alpha_t}(s+t, x_t)\, dt \geq v(T,s,x) - \frac{\varepsilon}{3}.$$

Further, as we did in the proof given, we costruct α' on the basis of α. Then, by (6)

$$v^{\alpha'}(s,x) \geq v(T,s,x) - \frac{2\varepsilon}{3} \geq v(s,x) - \varepsilon.$$

3. Remark. We can write (1) as follows:

$$v(s,x) \equiv \lim_{T\to\infty} \lim_{n\to\infty} v_n(T,s,x) = \sup_{\alpha \in |\mathfrak{A}|(s,x)} v^{\alpha}(s,x).$$

It turns out that we can interchange the limit with respect to n and the limit with respect to T. In fact, let us denote by $\tilde{v}_n(T,s,x)$ the right side of (3), in which $\mathfrak{A}^T$ is replaced by $\mathfrak{A}_n^T$. Here as well as in our theorem we prove that $\tilde{v}_n(T,s,x)$ increases with respect to T, if $A_n \ni \alpha$. Let this condition be satisfied for $n \geq n^*$. Then from (2) we have

$$\begin{aligned} v(s,x) &= \lim_{T\to\infty} \lim_{n\to\infty} \tilde{v}_n(T,s,x) = \sup_{T\geq s} \sup_{n\geq n*} \tilde{v}_n(T,s,x) \\ &= \sup_{n\geq n*} \sup_{T\geq s} \tilde{v}_n(T,s,x) = \sup_{n\geq n*} \lim_{T\to\infty} \tilde{v}_n(T,s,x) \\ &= \sup_{n\geq n*} \lim_{T\to\infty} v_n(T,s,x) = \lim_{n\to\infty} \lim_{T\to\infty} v_n(T,s,x). \end{aligned}$$

Remark 2 contains the inequality $v(s,x) < \infty$. The sufficient condition for $v(s,x)$ to be finite can easily be found from Lemma 4.1.

4. Theorem. *Let there exists a nonnegative function* $\bar{v} \in W_{\mathrm{loc}}^{1,2}((0,\infty) \times E_d)$ *which is continuous in* $[0,\infty) \times E_d$ *and such that for all* $\alpha \in A$ $L^{\alpha}\bar{v} + f_+^{\alpha} \leq 0$ (a.e.). *Then*

$$|\mathfrak{A}|(s,x) = \mathfrak{A}, \qquad -v_-(s,x) \leq v(s,x) \leq \bar{v}(s,x)$$

for all $s \geq 0$, $x \in E_d$.

We derive now the normed Bellman equation. We shall assume in future that the assumptions made in the preceding section after Exercise 4.11 are satisfied if we replace in those assumptions $g(s,x)$ by 0, and, in addition, if we replace in (4.12) the function $w_n(T,s,x)$ by $v_n(T,s,x)$.

Here as well as in Section 4, the following holds true.

5. Theorem. *The function $v(s,x)$ in the region Q^* is continuous, has a generalized first derivative with respect to s and two generalized derivatives with respect to x, and, in addition, all these derivatives are locally bounded in Q^*. For each normalizing multiplier $m_\alpha(s,x)$*

$$G^{m_\alpha}[v] = 0 \qquad \text{(a.e. on } Q^*).$$

We consider now in more detail the case where the functions σ, b, c, f do not depend on s. It is seen that here the function $v(s,x)$ does not depend on s: $v(s,x) = v(x)$. Here as well as in the preceding section, we introduce $\bar{G}^{m_\alpha}(u_{ij},u_i,u,x)$ $\bar{G}^{m_\alpha}[u](x)$ and a region D^*. From Theorem 5 we immediately have the following.

6. Theorem. *The function $v(x)$ in the region D^* is continuous and has two generalized derivatives with respect to x. These derivatives are locally bounded in D^*. If the nonnegative function $m_\alpha(x)$ is such that for all x, u_{ij}, u_i, u*

$$\sup_\alpha m_\alpha(x) < \infty, \qquad \bar{G}^{m_\alpha}(u_{ij},u_i,u,x) < \infty, \tag{7}$$

then $\bar{G}^{m_\alpha}[v] = 0$ (D^*-a.e.).

The first condition in (7) (also, (4.13)) is superfluous in some cases.

7. Theorem. *For all $n \geq 1$, $x \in D^*$, let*

$$\inf_{\alpha \in A_n} [\operatorname{tr} a(\alpha,x) + |b(\alpha,x)| + c^\alpha(x) + |f^\alpha(x)|] > 0. \tag{8}$$

Then $\bar{G}^{m_\alpha}[v] = 0$ (D^*-a.e.) *for any nonnegative function $m_\alpha(x)$ such that the inequality $\bar{G}^{m_\alpha}(u_{ij},u_i,u,x) < \infty$ is satisfied for all x, u_{ij}, u_i, u.*

PROOF. Let

$$v_n(x) = \lim_{T\to\infty} v_n(T,s,x).$$

As follows from Theorem 1, for all n for which $\underline{\alpha} \in A_n$, this limit exists. Let

$$\bar{m}_{\alpha 0}(x) = \left\{ \sum_{i,j=1}^d [a^{ij}(\alpha,x)]^2 + \sum_{i=1}^d [b^i(\alpha,x)]^2 + [c^\alpha(x)]^2 + [f^\alpha(x)]^2 \right\}^{-1/2},$$

$$\bar{G}_n^{m_\alpha}(u_{ij},u_i,u,x) = \sup_{\alpha \in A_n} m_\alpha(x)\left[\sum_{i,j=1}^d a^{ij}(\alpha,x)u_{ij} + \sum_{i=1}^d b^i(\alpha,x)u_i - c^\alpha(x)u + f^\alpha(x) \right],$$

$$\bar{G}_n^{m_\alpha}[u](x) = G_n^{m_\alpha}(u_{x^ix^j}(x),u_i(x),u(x),x).$$

We note that by (8)

$$\sup_{\alpha \in A_n} \bar{m}_{\alpha 0}(x) < \infty.$$

Let n be such that $\underline{\alpha} \in A_n$ and let temporarily $A = A_n$. Then, by Theorem 6 we have that $\bar{G}_n^{\bar{m}_{\alpha 0}}[v_n] = 0$ (D^*-a.e.), where

$$D_n^* = \{x \in E_d : \sup_{\alpha \in A_n} (a(\alpha,x)\lambda, \lambda) > 0 \text{ for all } \lambda \neq 0\}.$$

We take a bounded region $D' \subset D' \subset D^*$. Also, we choose a large number n_0 such that $\bar{D}' \subset D_{n_0}^*$, $\underline{\alpha} \in A_{n_0}$. For $n \geq n_0$

$$\bar{G}^{\bar{m}_{\alpha 0}}[u] \geq \bar{G}_n^{\bar{m}_{\alpha 0}}[u] \geq \bar{G}_{n_0}^{\bar{m}_{\alpha 0}}[u];$$

hence

$$\bar{G}_{n_0}^{\bar{m}_{\alpha 0}}[v_n] \leq 0, \qquad \bar{G}^{\bar{m}_{\alpha 0}}[v_n] \geq 0 \qquad (D'\text{-a.e.}). \tag{9}$$

Let here $n \to \infty$. By hypothesis, the ordinary derivatives of the functions $v_n(T,s,x)$ as well as the functions themselves are bounded in $(0,1) \times D'$ uniformly for $n \geq n_0$, $T \geq 2$. This implies that for $n \geq n_0$ the function $v_n(x)$ has two generalized derivatives which are uniformly bounded in D'. According to Remark 3 and Theorem 4.5.1 we have from (9) as $n \to \infty$

$$\bar{G}_{n_0}^{\bar{m}_{\alpha 0}}[v] \leq 0, \qquad \bar{G}^{\bar{m}_{\alpha 0}}[v] \geq 0 \qquad (D'\text{-a.e.}).$$

Since D' is an arbitrary subregion of D^* and furthermore, $\bar{G}_{n_0}^{\bar{m}_{\alpha 0}} \uparrow \bar{G}^{\bar{m}_{\alpha 0}}$ as $n_0 \to \infty$,

$$\bar{G}^{\bar{m}_{\alpha 0}}[v] = 0 \qquad (D^*\text{-a.e.}).$$

Next, repeating almost word-for-word the proof of Lemma 3.8, it is easy to see that all solutions of the equation $\bar{G}^{\bar{m}_{\alpha 0}}(u_{ij},u_i,u,x) = 0$ are solutions of the equation $\bar{G}^{m_\alpha}(u_{ij},u_i,u,x) = 0$ if the function $\bar{G}^{m_\alpha}$ is finite. Therefore, in this case $\bar{G}^{m_\alpha}[v] = 0$ (D^*-a.e.), thus proving the theorem. □

Sometimes the application of Theorem 7 yields more extensive information about a payoff function than the application of Theorem 6 does.

8. Example. Let $A = (0,1]$, $A_n = [1/n,1]$, and let the function $f(x) \geq 0$. It is not hard to visualize a situation in which

$$\bar{G}^{m_\alpha}[u](x) = \sup_{\alpha \in A} m_\alpha(x)[\alpha \Delta u(x) - \alpha u(x) + \alpha^2 f(x)].$$

If the function $m_\alpha(x)$ is bounded with respect to α for each x and if, in addition, $m_\alpha(x) > 0$, the relation $\bar{G}^{m_\alpha}[v](x) = 0$ holds if and only if $\Delta v(x) - v(x) + f(x) \leq 0$. In particular, this inequality is equivalent to the Bellman equation ($m_\alpha = 1$). At the same time, if we take $m_\alpha(x) = 1/\alpha$, it is easily seen that $\bar{G}^{m_\alpha}$ is a finite function and also (by Theorem 7 if the appropriate conditions are satisfied)

$$0 = \sup_{\alpha \in A}(\Delta v - v + \alpha f) = \Delta v - v + f.$$

9. Exercise

Let $A = (0,1]$, $A_n = [1/n,1]$. Furthermore, let $f(x)$ be a smooth function with compact support, and let

$$\bar{G}^{m_\alpha}[u](x) = \sup_{\alpha \in A} m_\alpha(x)[\alpha \Delta u(x) - \alpha u(x) + \alpha f(x)].$$

Prove that $\Delta v - v + f \leq 0$ (a.e.), $v \geq 0$ and $\Delta v - v + f = 0$ almost everywhere in the set $\{x : v(x) > 0\}$. Prove also that if $f \leq 0$, then $v = 0$ and, in addition, if we take the normalizing multiplier α^{-1}, we obtain the false relation $\Delta v - v + f = 0$. Explain why Theorem 7 is inapplicable in this case. (*Hint*: See the assumptions made at the start of this section.)

10. Remark. If condition (8) is satisfied, it is possible to prove the theorem on zero crossing or the theorem that $(v_{x^i x^j}, v_{x^i}, v)$ belongs to a boundary of some set, which are similar to Theorems 3.12 and 3.13.

Notes

Section 1. The idea of Exercise 6 is similar to that of an example due to Dynkin.

Section 2. In [36] and [37], it is required that c be sufficiently large compared to the first and second derivatives of σ, b, c, f. It is thus assumed that (15) and (17) are satisfied. In [22], an example illustrates that if $T = \infty$ and, in addition, inequality (15) is violated, the second derivative of the payoff function can be unbounded. It is known about a diffusion process (A consists of a single point; see Freidlin [19]) that if "killing" is insignificant, the "payoff function" for $T = \infty$ need not possess smoothness. It is also known that as c as well as the smoothness of initial functions increases, the smoothness of the "payoff function" increases. (See Freidlin [19]). It is interesting to note that the increase of smoothness need not occur in controlled processes. For instance, let $(\mathbf{w}_t, \mathscr{F}_t)$ be a one-dimensional Wiener process, $A = [0,\nu]$, $T = \infty$,

$$v(x) = \sup_{\alpha \in 0} \mathsf{M} \int_0^\infty e^{-\lambda t} \cos\left(x + \int_0^t \sqrt{2\alpha_s}\, d\mathbf{w}_s\right) dt,$$

where $\lambda > 0$. It is not hard to show that

$$v(x) = \frac{1}{\lambda} \cos x, \qquad x \in [0,z],$$

$$v(x) = \frac{1}{\lambda + \nu} \cos x + \frac{\nu}{\lambda(\lambda + \nu)} \frac{\cos z}{\operatorname{ch} \sqrt{\frac{\lambda}{\nu}}(z - \pi)} \operatorname{ch} \sqrt{\frac{\lambda}{\nu}}(x - \pi). \qquad x \in |z,\pi|,$$

where z is the solution on $(0,\pi)$ of the equation

$$\tan z = \sqrt{\frac{\lambda}{\nu}} \operatorname{th} \sqrt{\frac{\lambda}{\nu}} (\pi - z).$$

Hence $v''(z-) = -(1/\lambda)\cos z \neq 0 = v''(z+)$ and, further, for each λ the second derivative v is discontinuous. Concerning the increase of smoothness of a payoff function, see also the remarks made prior to Theorem 1.4.15.

Sections 3, 4, 5. In these sections the results obtained in [37] have been developed further. Theorem 4.8 can be obtained by the methods used by Shiryayev in [69]. There are other well-known methods for finding a stopping boundary, different from the method described in Exercise 4.16, see [44, 56, 68].

Appendix 1

Some Properties of Stochastic Integrals

We shall mention some facts from the stochastic integral theory, omitting the proof. The latter can however be found in Doob [9], Dynkin [11], and Shiryayev [51], and Gikhman and Skorokhod [23, 24].

Let $(\Omega,\mathscr{F})$ be a measurable space, and let $\{\mathscr{F}_t, t \geq 0\}$ be an increasing family of σ-algebras (a flow of σ-algebras) which satisfies the condition $\mathscr{F}_t \subset \mathscr{F}$ for all $t \geq 0$. A process $\xi_t(\omega)$ given for $t \geq 0$, $\omega \in \Omega$ with values in E_d, is said to be progressively measurable (with respect to $\{\mathscr{F}_t\}$) if for each $s > 0$ the function $\xi_t(\omega)$ considered for $t \in [0,s]$, $\omega \in \Omega$ is measurable with respect to the direct product of a σ-algebra of Borel subsets of an interval $[0,s]$ and $\mathscr{F}_s$. It is a known fact that a continuous process ξ_t, which is measurable with respect to $\mathscr{F}_t$ for each t, is progressively measurable. A nonnegative function τ given on Ω is said to be a Markov time (with respect to $\{\mathscr{F}_t\}$), if for any $s \geq 0$ the set $\{\omega:\tau(\omega) > s\} \in \mathscr{F}_s$. It is also a known fact that the times of first exit of continuous progressively measurable processes ξ_t from open sets are Markov times.

Let a probability measure P be given on $(\Omega,\mathscr{F})$. A continuous process $\mathbf{w}_t = (\mathbf{w}_t^1, \ldots, \mathbf{w}_t^{d_1})$ defined for $t \geq 0$, $\omega \in \Omega$ is referred to as a d_1-dimensional Wiener process if $\mathbf{w}_0 = 0$, the increments $\mathbf{w}_t$ on nonoverlapping intervals are independent, and, finally, if the distribution $\mathbf{w}_t - \mathbf{w}_s (t > s)$ is normal with parameters 0, $(t - s)I$, where I denotes the unit matrix of dimension $d_1 \times d_1$. If, furthermore, for any $t \geq 0$ the variable $\mathbf{w}_t$ is $\mathscr{F}_t$-measurable, the increment $\mathbf{w}_{t+h} - \mathbf{w}_t$ for $h \geq 0$ does not depend on $\mathscr{F}_t$ and if the σ-algebras of $\mathscr{F}_t$ are complete, we say that the pair $(\mathbf{w}_t,\mathscr{F}_t)$ is a d_1-dimensional Wiener process or that $\mathbf{w}_t$ is a d_1-dimensional Wiener process with respect to $\{\mathscr{F}_t\}$. It should be mentioned here that any Wiener process $\mathbf{w}_t$ is Wiener with respect to the completion of its own σ-algebras of $\mathscr{F}'_t = \sigma\{\mathbf{w}_s : s \leq t\}$.

Let $(\mathbf{w}_t, \mathscr{F}_t)$ be a d_1-dimensional Wiener process, and let σ_t be a random matrix of dimension $d \times d_1$ which is progressively measurable with respect to $\{\mathscr{F}_t\}$ and such that for each $t \geq 0$

$$\mathsf{M} \int_0^t \|\sigma_s\|^2 \, ds < \infty. \tag{1}$$

Then the stochastic integral $\int_0^t \sigma_s \, d\mathbf{w}_s$ is defined, and is a continuous progressively measurable process (with respect to $\{\mathscr{F}_t\}$) satisfying the condition

$$\mathsf{M} \left| \int_0^t \sigma_s \mathbf{w}_s \right|^2 = \mathsf{M} \int_0^t \|\sigma_s\|^2 \, ds. \tag{2}$$

The stochastic integral $\int_0^t \sigma_s \, d\mathbf{w}_s$ can be constructed in the following way. Let a process σ_t be a step process, i.e., there exist numbers $0 = t_0 < t_1 < t_2 < \cdots < t_n = \infty$ such that $\sigma_t = \sigma_{t_i}$ for $t \in [t_i, t_{i+1})$, $i = 0, \ldots, n-1$. Then $\int_0^t \sigma_s \, d\mathbf{w}_s$ can be defined as the corresponding sum: if $t \in [t_i, t_{i+1})$, then

$$\int_0^t \sigma_s \, d\mathbf{w}_s = \sum_{j=0}^{i-1} \sigma_{t_j}(\mathbf{w}_{t_{j+1}} - \mathbf{w}_{t_j}) + \sigma_{t_i}(\mathbf{w}_t - \mathbf{w}_{t_i}). \tag{3}$$

In this case equality (2) can immediately be verified.

In the general case we prove that there exists a sequence of progressively measurable step processes $\sigma_t(n)$ such that for each $t > 0$

$$\lim_{n \to \infty} \mathsf{M} \int_0^t \|\sigma_s - \sigma_s(n)\|^2 \, ds = 0.$$

Hence, by the Cauchy criterion and equality (2), the sequence of stochastic integrals $\int_0^t \sigma_s(n) \, d\mathbf{w}_s$ is Cauchy in L_2 (mean square sense), and, hence, convergent in L_2. Let us denote the limit $\int_0^t \sigma_s \, d\mathbf{w}_s$, where the limit for each t is determined of course only to within an equivalence. We also prove that for each t, ω one can choose values of $\int_0^t \sigma_s \, d\mathbf{w}_s$ such that the process thus obtained becomes continuous with respect to t. Hence by the integral $\int_0^t \sigma_s \, d\mathbf{w}_s$ we usually mean a continuous process.

It turns out that for any Markov time τ for all $t \geq 0$ almost surely

$$\int_0^{t \wedge \tau} \sigma_s \, d\mathbf{w}_s = \int_0^t \chi_{s \leq \tau} \sigma_s \, d\mathbf{w}_s.$$

A stochastic integral is in general not the limit of integral sums, similar to the Riemann–Stieltjes sums. It is, however, known that for each process σ_t there exists a sequence of integers $i(n)$ which tends to infinity as $n \to \infty$ and is such that for all $T > 0$ and almost all $s \in [0,1]$

$$\mathsf{P}\left\{ \lim_{n \to \infty} \sup_{t \leq T} \left| \int_0^t \sigma_r \, d\mathbf{w}_r - \int_0^t \sigma_{\kappa_{i(n)}(r+s)-s} \, d\mathbf{w}_r \right| = 0 \right\} = 1, \tag{4}$$

where $\kappa_i(t) = 2^{-i}[t2^i]$, $[a]$ denoting the largest integer $\leq a$. We note that the second integral in (4) is an integral of a step function and, therefore, is of the form (3).

A stochastic integral can be defined not only for the functions σ_t satisfying condition (1). The characteristic feature of a stochastic integral under condition (1) is that

$$\mathsf{M}\left\{\int_t^\tau \sigma_s\,d\mathbf{w}_s \middle| \mathscr{F}_t\right\} = 0 \tag{5}$$

almost surely on the set $\{\omega:\tau(\omega) \geq t\}$ for any bounded Markov time τ and for any t.

Let τ be a Markov time, let σ_s be a process with values in a set of matrices of dimension $d \times d_1$, and let b_s be a d-dimensional process. We assume that $\chi_{s\leq\tau}\sigma_s$, $\chi_{s\leq\tau}b_s$ are progressively measurable. Furthermore, we assume that the integrals $\int_0^t \chi_{s\leq\tau}\sigma_s\,d\mathbf{w}_s$, $\int_0^t \chi_{s\leq\tau}b_s\,ds$ are defined. If the process ξ_t satisfies the relation

$$\mathsf{P}\left\{\sup_{t\leq\tau}\left|\xi_t - \left(x + \int_0^t \chi_{s\leq\tau}\sigma_s\,d\mathbf{w}_s + \int_0^t \chi_{s\leq\tau}b_s\,ds\right)\right| = 0\right\} = 1, \tag{6}$$

it is convenient to write $d\xi_t = \sigma_t\,d\mathbf{w}_t + b_t\,dt$, $t \leq \tau$, $\xi_0 = x$. The formal expression $\sigma_t\,d\mathbf{w}_t + b_t\,dt$ is called the stochastic differential of ξ_t. The notation $d\xi_t = \sigma_t\,d\mathbf{w}_t + b_t\,dt$, $t \leq \tau$, $\xi_0 = x$ is only the short-hand representation of Eq. (6), and is more convenient than (6) when we encounter the need to write $u(\xi_t)$, where u is some function, using the stochastic integrals. It is sufficient to find the stochastic differential $du(\xi_t)$.

It turns out that for any twice continuously differentiable function u given on E_d, we have the following Ito's formula:

$$du(\xi_t) = \sum_{i=1}^d x_{x^i}(\xi_t)\,d\xi_t^i + \frac{1}{2}\sum_{i,j=1}^d u_{x^ix^j}(\xi_t)\,d\xi_t^i\,d\xi_t^i,$$
$$t \leq \tau, \qquad u(\xi_0) = u(x), \tag{7}$$

where the first term is understood as

$$\sum_{i=1}^d u_{x^i}(\xi_t)\left(\sum_{j=1}^{d_1}\sigma_t^{ij}\,d\mathbf{w}_t^j + b_t^i\,dt\right) = \sum_{j=1}^{d_1}\left(\sum_{i=1}^d u_{x^i}(\xi_t)\sigma_t^{ij}\right)d\mathbf{w}_t^j + \left(\sum_{i=1}^d u_{x^i}(\xi_t)b_t^i\right)dt.$$

We can write the first term in (7) in a short form: grad $u(\xi_t)\,d\xi_t$ or grad $u(\xi_t)$ $\sigma_t\,d\mathbf{w}_t$ + grad $u(\xi_t)b_t\,dt$. In order to compute $d\xi_t^i\,d\xi_t^j$ in the second term in (7), one is to apply the usual rules for removing parentheses as well as the following rules for the product of stochastic differentials: $(d\mathbf{w}_t^i)^2 = dt$, $d\mathbf{w}_t^i\,d\mathbf{w}_t^j = 0$ for $i \neq j$, $d\mathbf{w}_t^i\,dt = 0$, $(dt)^2 = 0$. In short,

$$d\xi_t^i\,d\xi_t^j = \sum_{k=1}^{d_1}\sigma_t^{ik}\sigma_t^{jk}\,dt = (\sigma_t\sigma_t^*)^{ij}\,dt.$$

Putting together in (7) the terms with dt and the terms with $d\mathbf{w}_t$, respectively, we can rewrite Ito's formula as follows:

$$du(\xi_t) = \operatorname{grad} u(\xi_t)\sigma_t\,d\mathbf{w}_t + L^{\sigma_t,b_t}u(\xi_t)\,dt. \tag{8}$$

Equation (7) is an analog of Taylor's formula with two terms, and can be as well called a formula for the stochastic differential of a composite function, the integral version of which (compare with (8)),

$$u(\xi_t) = u(x) + \int_0^t \chi_{s\le\tau} \operatorname{grad} u(\xi_s)\sigma_s\, d\mathbf{w}_s + \int_0^t \chi_{s\le\tau} L^{\sigma_s,b_s} u(\xi_s)\, ds, \qquad t \le \tau,$$

is known as the change of variables formula.

Sometimes the need arises to find the stochastic differential of $u(t,\xi_t)$. In such a case we can add to the process ξ_t another coordinate, assuming $\xi_t^{d+1} = \int_0^t ds$. We thus reduce the problem to the case when u depends explicitly only on a process which is $(d+1)$-dimensional. Then,

$$du(t,\xi_t) = \operatorname{grad}_x u(t,\xi_t)\sigma_t\, d\mathbf{w}_t + \left(\frac{\partial}{\partial t} + L^{\sigma_t,b_t}\right)u(t,\xi_t)\, dt. \tag{9}$$

Finally, if c_t is a nonnegative progressively measurable process such that $\int_0^\tau c_t\, dt < \infty$ and $\varphi_t = \int_0^t c_s\, ds$, then

$$\begin{aligned} d[e^{-\varphi_t}u(t,\xi_t)] &= u(t,\xi_t)\, de^{-\varphi_t} + e^{-\varphi_t}\, du(t,\xi_t) + (de^{-\varphi_t})\, du(t,\xi_t) \\ &= e^{-\varphi_t} \operatorname{grad}_x u(t,\xi_t)\sigma_t\, d\mathbf{w}_t + e^{-\varphi_t}\left(\frac{\partial}{\partial t} + L^{\sigma_t,b_t} - c_t\right)u(t,\xi_t)\, dt. \end{aligned} \tag{10}$$

Equations (9) and (10) hold for $t \le \tau$ with probability one for every function $u(t,x)$ which together with its two derivatives with respect to x and a first derivative with respect to t is continuous with respect to (t,x) in the closure of some region in the space of the variables (t,x) containing with probability one the trajectories (t,ξ_t) before time τ.

It is often necessary to use a corollary obtained upon integrating (10), taking the mathematical expectation and applying property (5):

$$e^{-\varphi_t}u(t,\xi_t) = \mathsf{M}\left\{e^{-\varphi_\tau}u(\tau,\xi_\tau) + \int_t^\tau e^{-\varphi_s}\left(\frac{\partial}{\partial s} + L^{\sigma_s,b_s} - c_s\right)u(s,\xi_s)\, ds \middle| \mathscr{F}_t\right\}$$

$$\text{(a.s. on } \{\tau \ge t]),$$

if τ is bounded, the mathematical expectation exists and, finally,

$$\mathsf{M}\int_0^\tau |\sigma_t^* \operatorname{grad}_x u(t,\xi_t)|^2 e^{-2\varphi_t}\, dt < \infty.$$

Furthermore, let σ_t, b_t depend on t, ω as well as the point $x \in E_d$: $\sigma_t = \sigma_t(x)$, $b_t = b_t(x)$. For each $x \in E_d$ let the processes $\sigma_t(x)$, $b_t(x)$ be defined for all $t \ge 0$, $\omega \in \Omega$ and, furthermore, let these processes be progressively measurable. We assume that there exist two constant K_1 and K_2 such that for all possible values of the arguments

$$\|\sigma_t(x) - \sigma_t(y)\| + |b_t(x) - b_t(y)| \le K_1|x - y|,$$
$$\|\sigma_t(0)\| + |b_t(0)| \le K_2.$$

We fix $x \in E_d$. The assertion of Ito's theorem implies that the stochastic integral equation

$$x_t = x + \int_0^t \sigma_s(x_s)\, d\mathbf{w}_s + \int_0^t b_s(x_s)\, ds$$

for the function $x_t = x_t(\omega)$ $(t \geq 0, \omega \in \Omega)$ has a unique (to within an equivalence) continuous (with respect to t) progressively measurable (with respect to $\{\mathscr{F}_t\}$) solution.

Appendix 2

Some Properties of Submartingales

Let $(\Omega,\mathscr{F},\mathsf{P})$ be a probability space, and let $\{\mathscr{F}_t, t \geq 0\}$ be an increasing family of σ-algebras satisfying condition $\mathscr{F}_t \subset \mathscr{F}$. The real-valued process $\xi_t(\omega)$ given for $t \in [0,T]$ is said to be a submartingale on the time interval $t \in [0,T]$ (with respect to the family of $\{\mathscr{F}_t\}$) if the random variables ξ_t are $\mathscr{F}_t$-measurable,

$$\mathsf{M}\xi_t^- < \infty, \qquad \mathsf{M}\{\xi_t|\mathscr{F}_s\} \geq \xi_s \qquad \text{(a.s.)} \tag{1}$$

for all t, $s \in [0,T]$, $s \leq t$. The process ξ_t is referred to as a martingale if the second inequality in (1) is an equality. The process ξ_t is referred to as a supermartingale if the process $(-\xi_t)$ is a submartingale. The properties of submartingales, martingales, and supermartingales are well known (see, for instance, [9, 51, 54]). We give some properties without proving them.

1. If ξ_t is a submartingale and, in addition, $\varphi(x)$ is an increasing convex downward function of the real x, then $\varphi(\xi_t)$ is a submartingale.

2. If ξ_t is a martingale and, in addition $\varphi(x)$ is a convex downward function, then $\varphi(\xi_t)$ is a submartingale.

3. If ξ_t is a separable submartingale, $p > 1$, then

$$\mathsf{M}\sup_{t \leq T}(\xi_t^+)^p \leq \left(\frac{p}{p-1}\right)^p \mathsf{M}(\xi_T^+)^p.$$

It is useful to bear in mind that $(p/(p-1))^p \leq 4$ for $p \geq 2$.

4. Let ξ_t be a right continuous supermartingale and let $\mathsf{M}|\xi_t| < \infty$ for $t \leq T$. If τ is a Markov time, $\xi_{t \wedge \tau}$ is a supermartingale. Further, if τ_1 and τ_2 are Markov times such that $\tau_1(\omega) \leq \tau_2(\omega) \leq T$ for all ω, then $\mathsf{M}\xi_{\tau_1} \geq \mathsf{M}\xi_{\tau_2}$.

We do not think that the following fact is well known, hence we shall prove it here.

Lemma. *Let κ_t be a supermartingale with continuous trajectories, and let* $\mathsf{M}\ \sup_{t \le T} \kappa_t| < \infty$. *Further, let $\Phi_t(\omega)$ be a nonnegative continuous progressively measurable process which increases in t for all ω or decreases in t for all ω and is bounded on $[0,T] \times \Omega$. Then:*

a. *the process $\rho_t \equiv \kappa_t \Phi_t - \int_0^t \kappa_s \, d\Phi_s$ is a supermartingale. Also, for any Markov time τ not exceeding T,*

$$\mathsf{M}\rho_\tau \ge \sup_{t,\omega} \Phi_t \mathsf{M}(\kappa_T - \kappa_0) + \mathsf{M}\kappa_0 \Phi_0;$$

b. *$\rho_t - \kappa_t \Phi_0$ is a supermartingale if Φ_t increases in t; $\rho_t - \kappa_t \Phi_0$ is a submartingale if Φ_t decreases in t.*

(Note that in the applications frequently

$$\Phi_t = \exp\left(- \int_0^t r_s \, ds \right), \qquad \kappa_t = \int_0^t f_s \, ds + v_t.$$

Here, as follows from Fubini's theorem,

$$\rho_t = \int_0^t (f_s + r_s v_s) \exp\left(- \int_0^s r_{s_1} \, ds_1 \right) ds + v_t \exp\left(- \int_0^t r_s \, ds \right).$$

PROOF. Writing κ_t as $\kappa_0 + \tilde{\kappa}_t$, it is easily seen that it suffices to consider the case where $\kappa_0 = 0$. We assume that $\kappa_0 = 0$ and κ_t, Φ_t are defined for $t \ge T$ according to the formulas $\kappa_t = \kappa_T$, $\Phi_t = \Phi_T$.

First we prove that

$$\rho_t = \lim_{r \downarrow 0} \frac{1}{r} \int_0^t (\kappa_{s+r} - \kappa_s) \Phi_s \, ds, \tag{2}$$

$$\mathsf{M} \sup_{t \le T, r \in (0,1)} \left| \frac{1}{r} \int_0^t (\kappa_{s+r} - \kappa_s) \Phi_s \, ds \right| < \infty. \tag{3}$$

We note that

$$\frac{1}{r} \int_0^t (\rho_{s+r} - \rho_s) \, ds = \frac{1}{r} \int_t^{t+r} \rho_s \, ds - \frac{1}{r} \int_0^r \rho_s \, ds,$$

$$\mathsf{M} \sup_{t \le T, r \in (0,1)} \left| \frac{1}{r} \int_0^t (\rho_{s+r} - \rho_s) \, ds \right| \le 2\mathsf{M} \sup_{t \le T} |\rho_t|$$

$$\le 4 \sup_{t,\omega} \Phi_t \mathsf{M} \sup_{t \le T} |\kappa_t| < \infty. \tag{4}$$

Furthermore, due to the equality $\rho_0 = 0$ as well as the continuity of ρ_t

$$\rho_t = \lim_{r \downarrow 0} \frac{1}{r} \int_0^t (\rho_{s+r} - \rho_s) \, ds. \tag{5}$$

Next, the absolute value of the difference between the limit expressions in (2) and (5) is equal to

$$\left|\frac{1}{r}\int_0^t ds \int_s^{s+r} (\kappa_{s+r} - \kappa_{s_1})\, d\Phi_{s_1}\right| \leq w(r)\left|\frac{1}{r}\int_0^t (\Phi_{s+r} - \Phi_s)\, ds\right| \leq w(r) 2 \sup_{t,\omega} \Phi_t,$$

where $w(r)$ denotes the modulus of continuity of the function κ. This estimate, the fact that $w(r)$ tends to zero as $r \to 0$, the inequality

$$w(r) \leq 2 \sup_{t \leq T} |\kappa_t|,$$

and relations (4) and (5) together yield (2) and (3).

Let $t_1 < t$. Using (2), let us find $\mathsf{M}\{\rho_t | \mathscr{F}_{t_1}\}$. Inequality (3) enables us here to take the limit after taking the mathematical expectation. The last inequality in (4) enables us to interchange the mathematical expectation and integration operations. In addition, we make use of the fact that $\Phi \geq 0$ and also, for $s \geq t_1$,

$$\mathsf{M}\{(\kappa_{s+r} - \kappa_s)\Phi_s | \mathscr{F}_{t_1}\} = \mathsf{M}\{\Phi_s \mathsf{M}\{\kappa_{s+r} - \kappa_s | \mathscr{F}_s\} | \mathscr{F}_{t_1}\} \leq 0.$$

Then we have

$$\mathsf{M}\{\rho_t | \mathscr{F}_{t_1}\} \leq \mathsf{M}\left\{\lim_{r \downarrow 0} \frac{1}{r}\int_0^{t_1} (\kappa_{s+r} - \kappa_s)\Phi_s\, ds \,\Big|\, \mathscr{F}_{t_1}\right\} = \mathsf{M}\{\rho_{t_1} | \mathscr{F}_{t_1}\} = \rho_{t_1}.$$

Therefore, ρ_t is a supermartingale.

Further,

$$\mathsf{M}\rho_\tau \geq \mathsf{M}\rho_T = \lim_{r \downarrow 0} \frac{1}{r}\int_0^T \mathsf{M}\Phi_s \mathsf{M}\{\kappa_{s+r} - \kappa_s | \mathscr{F}_s\}\, ds.$$

Here the conditional expectation is negative. Hence

$$\mathsf{M}\rho_\tau \geq \sup_{t,\omega} \Phi_t \mathsf{M} \lim_{r \downarrow 0} \frac{1}{r}\int_0^T (\kappa_{s+r} - \kappa_s)\, ds = \sup_{t,\omega} \Phi \mathsf{M}\kappa_T.$$

Assertion (a) is thus proved. We can prove (b) in a similar way, using the formula

$$\rho_t - \kappa_t \Phi_0 = \lim_{r \downarrow 0} \frac{1}{r}\int_0^t (\kappa_{s+r} - \kappa_s)(\Phi_s - \Phi_0)\, ds. \qquad \square$$

Bibliography

V. I. Arkin, V. A. Kolemayev, and A. N. Shiryayev, On finding optimal controls, *Trudy MAIN* **71**, (1964), 21–25. (in Russian).

K. G. Äström, *Introduction to Stochastic Control Theory* (Academic Press, New York, 1970).

I. Ya. Bakel'man, *Geometric Methods for Solving Elliptic Equations* (Nauka, Moscow, 1965) (in Russian).

R. S. Bellman, *Dynamic Programming* (Princeton University Press, Princeton, N.J., 1957).

R. S. Bellman and R. E. Kalaba, *Quasilinearization and Nonlinear Boundary-Value Problems* (Elsevier, New York, 1965).

V. E. Beneš, Full bang to reduce predicted miss is optimal, *SIAM J. Control and Optimization* **14**, (1976), 62–84.

R. S. Bucy and P. D. Joseph, *Filtering for Stochastic Processes with Applications to Guidance* (Wiley, New York, 1968).

C. Derman, *Finite State Markovian Decision Processes* (Academic Press, New York, 1970).

J. L. Doob, *Stochastic Processes* (Wiley, New York, 1953).

N. Dunford and J. T. Schwartz, *Linear Operators: General Theory* (Wiley–Interscience, New York, 1958).

E. B. Dynkin, *Markov Processes* (Academic Press, New York, 1965) (English translation).

E. B. Dynkin, and A. A. Yushkevich, *Controlled Markov Processes and Applications* (Nauka, Moscow, 1975) (in Russian). [English translation: Springer–Verlag, New York, 1979.]

W. H. Fleming, Some Markovian optimization problems, *J. Math. and Mech.* **12**(1) (1963), 131–140.

W. H. Fleming, The Cauchy problem for degenerate parabolic equations, *J. Math. and Mech.* **13** (1964), 987–1008.

W. H. Fleming, Duality and a priori estimates in Markovian optimization problems, *J. Math. Anal.* **16** (1966), 254–279.

W. H. Fleming, Stochastic control for small noise intensities, *SIAM J. Control* **9**(3) (1971), 437–515.

W. H. Fleming and R. W. Rishel, *Deterministic and Stochastic Optimal Control* (Springer–Verlag, Berlin, New York, 1975).

M. I. Freidlin, A note on the generalized solution of the Dirichlet problem, *Theory Prob. Appl.* **10**(1) (1965), 161–164 (English translation).

M. I. Freidlin, On the smoothness of solutions of degenerate elliptic equations, *Math. USSR Izvestija* **2**(6) (1968), 1337–1357 (English translation).

E. B. Frid, On the semiregularity of boundary points for nonlinear equations, *Math. USSR Sbornik* **23**(4) (1974), 483–507 (English translation).

A. Friedman, *Stochastic Differential Equations*, I, II (Academic Press, New York, 1975).

I. L. Genis and N. V. Krylov, An example of a one-dimensional controlled process, *Theory Prob. Appl.* **21**(1) (1976), 148–152 (English translation).

I. I. Gikhman and A. V. Skorokhod, *Stochastic Differential Equations* (Naukova Dumka, Kiev, 1968)(in Russian). [English translation: Springer-Verlag, Berlin, Heidelberg, New York, 1972.]

I. I. Gikhman and A. V. Skorokhod, *The Theory of Random Processes* Vol. 3 (Nauka, Moscow, 1975) (in Russian). [English translation: Springer-Verlag, Berlin, Heidelberg, New York, 1979.]

I. V. Girsanov, Minimax problems in the theory of diffusion processes, *Soviet Mathematics, Doklady Akademii Nauk SSSR* **136**(4)(1961), 118–121 (English translation).

R. A. Howard, *Dynamic Programming and Markov Processes* (Wiley, New York, 1960).

N. N. Krassovsky and A. I. Subbotin, *Closed-Loop Differential Games* (Nauka, Moscow, 1974) (in Russian), [English translation: Springer-Verlag, Berlin, Heidelberg, New York, forthcoming.]

N. V. Krylov, On Ito's stochastic integral equations, *Theory Prob. Appl.* **14**(2) (1969); 330–336 (English translation). [*See also:* Letter to the Editor: Correction to On Ito's stochastic integral equations, *Theory Prob. Appl.* **17**(1) (1972), 20–33 (English translation).]

N. V. Krylov, On a problem with two free boundaries for an elliptic equation and optimal stopping of a Markov process, *Soviet Mathematics, Doklady Akademii Nauk SSSR* **194**(6) (1970), 1370–1372 (English translation).

N. V. Krylov, Bounded inhomogeneous nonlinear elliptic and parabolic equations in the plane, *Math. USSR Sbornik* **11**(1) (1970), 89–99. (English translation).

N. V. Krylov, Control of Markov processes and *W*-spaces, *Math. USSR Izvestija*, **5**(1) (1971), 233–265 (English translation).

N. V. Krylov, An inequality in the theory of stochastic integrals, *Theory Prob. Appl.* **16**(3) (1971), 438–448 (English translation).

N. V. Krylov, On the theory of nonlinear degenerating elliptic equations, *Soviet Mathematics, Doklady Akademii Nauk SSSR* **201**(6) (1971), 1820–1823 (English translation).

N. V. Krylov, On uniqueness of the solution of Bellman's equation, *Math. USSR Izvestija* **5**(6) (1971); 1387–1398 (English translation).

N. V. Krylov, *Lectures on the Theory of Elliptic Differential Equations*, (Izd. Moskovsk. Gos. Univers., Moscow, 1972) (in Russian).

N. V. Krylov, Control of a solution of a stochastic integral equation, *Theory Prob. Appl.* **17**(1) (1972), 114–131 (English translation).

N. V. Krylov, On control of the solution of a stochastic integral equation with degeneration, *Math. USSR Izvestija* **6**(1) (1972), 249–262 (English translation).

N. V. Krylov, On the selection of a Markov process from a system of processes and the construction of quasi-diffusion processes, *Math. USSR Izestija* **7**(3)(1973), 691–708 (English translation).

N. V. Krylov, Some estimates in the theory of stochastic integrals, *Theory Prob. Appl.* **18**(1) (1973), 54–63 (English translation).

N. V. Krylov, Some estimates of the probability density of a stochastic integral, *Math. USSR Izvestija* **8**(1) (1974), 233–254 (English translation).

N. V. Krylov, On Bellman's equation, *Trudy Shkoly-Seminara po Teorii Sluchainykh Protsessov* (*Druskininkai, November 25–30, 1974*), *1* (Vilnius, 1975) (in Russian).

N. V. Krylov, Sequences of convex functions and estimates of the maximum of the solution of a parabolic equation, *Siberian Math. J.* **17**(2)(1976); 226–236 (English translation).

N. V. Krylov, The maximum principle for parabolic equations, *Uspekhi Matem. Nauk* **31**(4) (1976), 267–268 (In Russian).

R. Kudzhma, Optimal stopping of semistable Markov processes, *Litovsk. Matem. Sbornik* **13**(3) (1973), 113–117 (In Russian).

H. J. Kushner, *Stochastic Stability and Control* (Academic Press, New York, 1967).

O. A. Ladyzhenskaja and N. N. Uraltseva, *Linear and Quasi-Linear Equations of Elliptic Type* (Academic Press, New York, 1968) (English translation).

O. A. Ladyzhenskaja, V. A. Solonnikov and N. N. Uraltseva, *Linear and Quasi-Linear Equations of Parabolic Type*, (Transl. Math. Monographs, Vol. 23, Amer. Math. Soc., Providence, R.I., 1968).

H. Lewy and G. Stampacchia, On existence and smoothness of solutions of some noncoersive variational inequalities, *Arch. Rational Mech. Anal.* **41**(4) (1971), 242–253.

J. L. Lions, On inequalities in partial derivatives, *Uspekhi Matem. Nauk* **26**(2) (1971), 205–263 (in Russian).

J. L. Lions, *Optimal Control of Systems Governed by Partial Differential Equations* (Springer-Verlag, Berlin, Heidelberg, New York, 1971).

R. S. Liptser and A. N. Shiryayev, *Statistics of Random Processes*, 1, 2 (Springer-Verlag, Berlin, Heidelberg, New York, 1977–1978) (English translation).

P. Mandl, On optimal control of a nonstopped diffusion process, *Z. Warscheinlichkeitstheorie, und Verw. Gebiete* **4**(1) (1965), 1–9.

P. Mandl, On the control of a Wiener process for a limited number of switchings, *Theory Prob. Appl.* **12**(1) (1967), 68–76 (English translation).

P. A. Meyer, *Probability and Potentials* (Blaisdell, Waltham, Mass., 1966).

H. Mine and S. Osaki, *Markovian Decision processes* A (Elsevier, New York, 1970).

T. P. Miroshnichenko, Optimal stopping of the integral of a Wiener process, *Theory Prob. Appl.* **20**(2)(1975), 387–391 (English translation).

S. M. Nikolskii, *Approximation of Functions of Several Variables and Imbedding Theorems* (Springer-Verlag, Berlin, Heidelberg, New York, 1975) (English translation).

M. Nisio, Remarks on stochastic optimal controls, *Jap. J. Math.* **1**(1) (1975), 159–183.

M. Nisio, Some remarks on stochastic optimal controls, *Proc. Third USSR–Japan Sympos. Probab. Theory*, pp. 446–460 (Lecture Notes in Mathematics No. 550, Springer-Verlag, Berlin, Heidelberg, New York, 1976).

L. S. Pontryagin, V. G. Boltyansky, R. V. Gamkrelidze, and E. F. Mishchenko, *Mathematical Theory of Optimal Processes* (Nauka, Moscow, 1969) (in Russian).

I. I. Portenko and A. V. Skorokhod, *Existence of ε-Optimal Markov Strategies for Controlled Diffusion Processes. Problems of Statistics and Control of Random Processes* (Izd. Instituta Matematiki Akademii Nauk Ukrainskoi SSR, Kiev, 1973) (in Russian).

G. Pragarauskas, Some estimates of stochastic integrals, *Litovsk. Matem. Sbornik* **15**(3) (1975), 211–217 (in Russian).

G. Pragarauskas, On the control theory of discontinuous random processes, *Trudy Shkoly-Seminara po Teorii Sluchainykh Protsessov*, pp. 252–281 (*Druskininkai. November 25–30, 1974, 1*, (Vilnius, 1975) (in Russian).

Yu. V. Prokhorov, Control of a Wiener process when the number of switchings is bounded, *Trudy MIAN* **71** (1964), 82–87 (in Russian).

M. V. Safonov, Control of a Wiener process when the number of switches is bounded, *Theory Prob. Appl.* **21**(3) (1976), 593–599 (English translation).

M. V. Safonov, On a Dirichlet's problem for Bellman's equation in a plane domain, *Math. USSR Sbornik* **102**(2) (1977), 260–279 (In Russian).

L. Schwartz, *Théorie des Distributions*, 1 (Hermann, Paris, 1950).

L. A. Shepp, Explicit solutions to some problems of optimal stopping, *Ann. Math. Statist.* **40**(3) (1969), 993–1010.

A. N. Shiryayev, *Optimal Stopping Rules* (Springer-Verlag, Berlin, Heidelberg, New York, 1978) (English translation).

A. V. Skorokhod, *Studies in the Theory of Random Processes* (Scripta Technica, Washington 1965) (English translation).

V. I. Smirnov, *A Course of Higher Mathematics* (Pergamon, Oxford and New York, 1964) (English translation).

S. L. Sobolev, *Applications of Functional Analysis in Mathematical Physics* (Amer. Math. Soc., Providence, R.I., 1963) (English translation).

T. Tobias, The optimal stopping of diffusion processes and parabolic variational inequalities, *Differential Equations* **9**(4) (1973), 534–538 (English translation).

T. Tobias, On optimal stopping of diffusion processes with singular matrices of diffusion, *Izvestia of Akademii Nauk Estonskoi SSR* **23**(3) (1974), 199–202 (in Russian).

A. Yu. Veretennikov and N. V. Krylov, On explicit formulas for solutions of stochastic equations, *Math, USSR Sbornik* **29**(2)(1976), 239–256 (English translation).

W. M. Wonham, *Random Differential Equations in Control Theory: Probabilistic Methods in Applied Mathematics*, 2 (Academic Press, New York, 1970).

A. K. Zvonkin, On sequentially controlled Markov processes, *Math. USSR Sbornik* **15**(4)(1971) 607–617 (English translation).

A. K. Zvonkin and N. V. Krylov, On strong solutions of stochastic differential equations, *Trudy Shkoly-Seminara po Teorii Sluchainykh Protsessov* pp. 9–88 (*Druskininkai, November 25–30, 1974*), *2*, (Vilnius. 1975) (in Russian).

Index